Pharmaceutical Microbiological Quality Assurance and Control

Pharmaceutical Microbiological Quality Assurance and Control

Practical Guide for Non-Sterile Manufacturing

Edited by

David Roesti
Novartis Pharma Stein AG, Switzerland

Marcel Goverde
MGP Consulting GmbH, Switzerland

WILEY

This edition first published 2020

Registered Office
John Wiley & Sons, Inc., 111 River Street, Hoboken, NJ 07030, USA

Editorial Office
111 River Street, Hoboken, NJ 07030, USA

For details of our global editorial offices, customer services, and more information about Wiley products visit us at www.wiley.com.

Wiley also publishes its books in a variety of electronic formats and by print-on-demand. Some content that appears in standard print versions of this book may not be available in other formats.

Library of Congress Cataloging-in-Publication Data is applied for

Hardback ISBN: 9781119356073

Cover Design: Wiley
Cover Images: Courtesy of Christine Farrance, Senior Director of R&D Charles River Laboratories, Newark, Delaware; Courtesy of Sabrina Kuoni; Courtesy of Thomas Meindl, Labor LS

Set in 10/12pt Warnock by SPi Global, Pondicherry, India

FSC
www.fsc.org
MIX
Paper from responsible sources
FSC® C013604

Contents

Editors

David Roesti, PhD
Novartis Pharma Stein AG, Stein, Switzerland
david.roesti@novartis.com

Dr. David Roesti holds a PhD in microbial ecology from the University of Neuchâtel, Switzerland, and has more than 20 years of experience in the field of microbiology within various domains (drug product manufacturing, food microbiology, biogas production, and microbial interactions in the rhizosphere). Currently, he works at Novartis Pharma AG in Stein, Switzerland, in the Manufacturing Science & Technology department and is responsible to define the microbial control strategy at the site and is a global subject matter expert in microbiology for the Novartis group. Prior to this assignment, he led the Rapid Microbiological Methods team at Novartis Pharma AG and was the laboratory supervisor for the microbiological testing of non-sterile drug products at Novartis Pharma Stein AG. David Roesti is an elected member of the General Chapters Microbiology Expert Committee of the Unites States Pharmacopoeia 2015–2020 revision cycle and is a member of the advisory board of the European Compliance Academy Microbiology Group. Finally, David Roesti is main author or coauthor of many different publications in either peer-reviewed journals or book chapters and has regularly held presentations in scientific congresses or expert groups.

Marcel Goverde, PhD
MGP Consulting GmbH, Binningen, Switzerland
g@mgp.li

Dr. Marcel Goverde studied biology at the University of Basel, where he gained his PhD in 2001. He gathered first work experience as a scientific collaborator in the agro-biological section of Novartis. He is now associated with a development project in Costa Rica and teaching at different school levels. From 2002 to 2010 he led several QC labs for microbiology at F. Hoffmann-La Roche Ltd. From 2010 to 2011 he worked as a QC expert for microbiology at Novartis

Pharma Ltd. Since 2011 he has been running his own business for consulting, training, project, and deviation management named MGP Consulting GmbH. Marcel Goverde is a regular speaker at different institutions and has written several scientific papers. He has been the Swiss expert in the EDQM group for Modern Microbiological Methods since 2003, which was then integrated into Group 1 (Microbiological Methods and Statistical Analysis) in 2015. Marcel Goverde is deputy chair of the ECA Pharmaceutical Microbiology Working Group and holds a green belt in Lean Six Sigma.

List of Contributors

Ina Bach, PhD
Global Compliance Assistance, St. Gallen, Switzerland
ina.bach@dr-bach.ch

Dr. Ina Bach is an independent consultant specializing in the areas of quality assurance, compliance, and auditing, with particular application to the pharmaceutical and biotech industry. Ina Bach has worked over 15 years as a senior compliance auditor in a Global Auditing and Compliance group of one of the top 10 international pharmaceutical companies and as an inspector for the Swiss regulator authority. When she was working for the authorities, she also worked closely with international health organization as a nominated inspector in several of prequalification projects. Additionally, she can look back on many years of experience in the pharmaceutical industry in which she was in responsible positions involved in the development, manufacture of various dosage forms, in project management, as well as in quality assurance and quality control.

Tony Cundell, PhD
Microbiological Consulting, LLC, Scarsdale, NY, USA
tonycundell@gmail.com

Dr. Tony Cundell consults with a number of pharmaceutical, consumer health, and dietary supplement companies; microbiology instrument manufacturers; contract testing laboratories; and sterile compounding pharmacies in the areas of microbial risk assessment, regulatory affairs, and microbiological testing. Prior to November 2013, he worked for Merck Research Laboratories in Summit, New Jersey, as the Senior Principal Scientist in early phase drug development. Earlier in his career, Tony Cundell worked at a director level in Quality Control and Product Development organizations at the New York Blood Center, Lederle Laboratories, Wyeth Pharmaceuticals, and Schering-Plough. He is a member of the 2015–2020 USP Microbiology Committee of Experts where he takes a leadership role in the area of modern microbiology methods, cochairing the USP Expert Panel that published a stimuli article in the September–October 2017 Pharmacopeial Forum entitled *The Development of Compendial Rapid Sterility Tests.* Tony Cundell chaired the PDA task force responsible for the ground-breaking 2000 Technical Report No. 33 *The Development, Validation, and Implementation of New (Rapid) Microbiological Methods.* In June 2009, he coedited with Anthony Fontana a book entitled *Water Activity Applications in the Pharmaceutical Industry* and contributed two chapters to the book. He was cochair of the PDA task force responsible for 2014 Technical Report No. 67 *Exclusion of Objectionable Microorganisms from Non-Sterile Drug Products.* He received the 2016 PDA Martin Van Trieste Pharmaceutical Science Award for outstanding contributions to the advancement of pharmaceutical science. He has a PhD in microbiology from Lincoln University, Canterbury, New Zealand.

Christine E. Farrance, PhD
Senior Director of R&D and Scientific Affairs, Charles River Laboratories, Inc., Newark, DE, USA
Christine.Farrance@crl.com

Dr. Christine E. Farrance received a Ph.D. in Cellular and Molecular Biology from the University of Pennsylvania, a B.S. in Biochemistry from Cornell University, and has more than 20 years of experience conducting both applied and basic research using molecular, microbial, genetic, and biochemical techniques. Currently, Senior Director of R&D and Scientific Affairs at Charles River Laboratories, Inc., she manages the scientific development projects and priorities across a cross-functional team. She is also involved in coordinating research studies, and developing technical content for publications, presentations, and training. In addition, she provides her expertise to other business units within Charles River for technical trouble shooting and evaluation of new technologies. Before joining Accugenix/Charles River, she was a scientist at a nonprofit institute developing a plant-based production system for recombinant vaccines and protein therapeutics. Her work has also included research in the food safety industry, and in plant-based production systems for monoclonal antibodies as a Research Assistant Professor. Christine Farrance brings unique scientific and technical experiences to Charles River.

Oliver Gordon, PhD
The Francis Crick Institute, London, United Kingdom
Oliver.Gordon@crick.ac.uk

Dr. Oliver Gordon studied Molecular Biology (main focus in Microbiology and Infection Biology) at the Biocenter and the University Hospital in Basel, Switzerland. From 2010 to 2014, he was working at Novartis Pharma AG in Switzerland in the QA/QC-Microbiology department in the Launch Center for Rapid Microbiological Methods. As Rapid Microbiology Specialist, he played a leading role in evaluation and validation of Rapid Microbiological Methods and the subsequent replacement of Traditional Microbiological Methods, including method transfers to interested Novartis sites worldwide. In 2014, he started a PhD in the lab of Caetano Reis e Sousa at the Francis Crick Institute in London, where his research focuses on understanding how molecules released from dead cells modulate immune and tissue repair responses.

Ingo Grimm
Director Sales & Services, Labor LS, Bad Bocklet, Germany
ingo.grimm@labor-ls.de

Ingo Grimm is an experienced specialist in sales and business development activities in the field of microbiological and analytical testing of pharmaceutical products (small molecules and biotech) and medical devices. Within the last 15 years he became an expert in the needs of these industries, especially in the DACH-region. Currently, Ingo Grimm works as Head of Sales & Services at Labor LS SE & Co. KG which is one of the leading cGMP-laboratories in Europe. He graduated at the University of Applied Sciences Würzburg-Schweinfurt and holds a Master of Business Administration.

David Hussong, PhD
Chief Technical Officer, Eagle Analytical Services, Houston, TX, USA
David.Hussong@outlook.com

Dr. David Hussong is currently the Chief Technical Officer at Eagle Analytical Services (Houston, TX). He has previously served as a regulatory microbiology consultant (2015–2017). In 2014, David Hussong retired from the Commissioned Corps of the US Public Health Service after 30 years with the Food and Drug Administration (FDA), where he had served in many positions, culminating as the Associate Director for New Drug Microbiology in the Office of Pharmaceutical Science. In addition, he is currently the chair of the USP Microbiology Expert Committee for the 2015–2020 cycle. David Hussong earned his PhD in microbiology from the University of Maryland at College Park (UMCP), and he previously served as a research microbiologist at UMCP, the US Department of Agriculture, and the US Naval Medical Research Institute.

Laurent Leblanc
R&D Manager Health and Personal Care Business, bioMérieux
Laurent.LEBLANC@biomerieux.com

Laurent Leblanc is the R&D Manager of bioMérieux' Health and Personal Care business. He holds a Master's degree in Biotechnology from the University of Limoges, France. For the last 15 years, he worked in several biotechnology companies and before joining bioMérieux in 2008, he worked in microbiological control in the pharmaceutical industry. He is now involved in designing and bringing to the market the new innovative and efficient solutions dedicated to the pharmaceutical and cosmetic industries.

Marion Louis
Global Solution Manager for Pharmaceutical Culture Media of the Healthcare Business, bioMérieux
marion.louis@biomerieux.com

Marion Louis is a Global Solution Manager for Pharmaceutical Culture Media in the HealthCare Business of bioMérieux. Graduated with an Engineering degree in Biotechnologies from Polytech, graduate schools of engineering in France, Marion Louis holds as well a Master degree in Management from University of Clermont-Ferrand. With 10 years' experience in microbiology as a Senior Application Engineer in pharmaceutical industries, she is now responsible to provide performant solutions to pharmaceutical industries.

Thomas Meindl, PhD
Senior Consultant, Labor LS, Bad Bocklet, Germany
thomas.meindl@labor-ls.de

Dr. Thomas Meindl studied biology in Mainz and Tübingen, Germany, and received a PhD from the University of Basel. He started his career at Novartis, Basel, working on receptor–ligand interaction studies. Then, he moved to Sympore GmbH, a start-up company in Tübingen, where he helped to develop new drugs in the field of immune suppression and inflammation. From there he moved to SKM oncology, Fürth, Germany, a company active in clinical trials. Finally, he moved to Labor LS SE, formally known as Labor L + S AG, in Bad Bocklet, Germany. There he worked as head of department and division manager for several departments (assays, endotoxin testing, molecular biology, disinfectant testing, research and development, validation of computerized systems, and implementation of LIMS); until today he works as a senior consultant for release testing of pharmaceuticals.

Michael J. Miller, PhD
President, Microbiology Consultants, LLC, Lutz, USA
mjm@microbiologyconsultants.com

Dr. Michael J. Miller is an internationally recognized microbiologist and subject matter expert in pharmaceutical microbiology, contamination control, aseptic processing, sterilization, laboratory design, and the validation and implementation of rapid microbiological methods (RMMs). He is currently the President of Microbiology Consultants, LLC (microbiologyconsultants.com) and owner of rapidmicromethods.com, a website dedicated to the advancement of rapid methods within healthcare-related industries. For 30 years, he has held numerous R&D, manufacturing, quality, business development, and executive leadership roles at multinational firms such as Johnson & Johnson, Eli Lilly and Company, and Bausch & Lomb. In his current role, Michael Miller consults with multinational companies in providing technical, quality, regulatory, and training solutions in support of RMMs; sterile and non-sterile pharmaceutical manufacturing; contamination control; isolator technology; environmental monitoring; sterilization; and antimicrobial effectiveness. Michael Miller has authored more than 100 technical publications and presentations and is the editor of PDA's *Encyclopedia of Rapid Microbiological Methods*. He currently serves on the editorial and scientific review boards for American Pharmaceutical Review, European Pharmaceutical Review, and the PDA Journal of Science and Technology. Michael Miller also was the chairperson during the revision of PDA Technical Report #33: *Evaluation, Validation and Implementation of New Microbiological Testing Methods*. He currently serves as an advisor to the USP Microbiology Expert Committee in the area of rapid sterility testing. Michael Miller holds a PhD in Microbiology and Biochemistry from Georgia State University (GSU), a BA in Anthropology, and Sociology from Hobart College.

Félix A. Montero Julian, PhD
Scientific Director of the Healthcare Business, bioMérieux
felix.monterojulian@biomerieux.com

Dr. Félix Montero is a Scientific Director of the Healthcare Business of bioMérieux. He has over 25 years of experience in industrial and clinical diagnostics and previously served as the Chemunex R&D Director in bioMérieux. Félix Montero graduated from the Autonomous Metropolitan University in Mexico as Industrial Biochemistry Engineer and obtained a PhD in Immunology from the Aix Marseille II University in France. He is a member of different scientific organizations (PDA, ISAC) and served as an expert in a panel for the Development of Compendial Rapid Sterility Tests for the USP. Félix Montero has been and continues to be extensively involved in the implementation and acceptance of rapid and alternative microbiological methods. He is a prominent speaker at congresses and conferences and a regular contributor to bioMérieux whitepapers.

Tim Sandle, PhD
Head of Microbiology, Bio Products Laboratory, Hertfordshire, United Kingdom
Tim.Sandle@bpl.co.uk

Dr. Tim Sandle is a pharmaceutical microbiologist, with experience in microbiological testing, research and development, risk assessment, and investigation. Tim Sandle is Head of Microbiology at Bio Products Laboratory, United Kingdom and a visiting tutor at University College London and the School of Pharmacy and Pharmaceutical Sciences, University of Manchester. In addition, he serves on several national and international committees, including the microbiology society Pharmig. Tim Sandle has written over 600 book chapters, peer-reviewed papers, and technical articles, largely relating to microbiology and pharmaceutical sciences.

Robert Schwarz
University Lecturer, University of Applied Sciences, Wien, Austria
robert.schwarz@edu.fh-campuswien.ac.at

Robert Schwarz worked after his education as a medical-technical analyst and as a biomedical analyst at a medical laboratory in Vienna with his core responsibilities in clinical chemistry, hematology, and hemostaseology. From 2001 to 2005, he led the environmental monitoring team at Baxter Bioscience in Vienna with the core tasks microbiological and physical clean room monitoring and qualification. Meanwhile, he completed his studies at the University of Applied Sciences FH Campus Wien in parallel with his professional career and successfully graduated hereby with the double degree "Bioprocess Engineering" and "Biotechnological Quality Management." From 2005 to 2018, he was a validation specialist and was besides equipment qualification responsible for sterilization, decontamination, and cleaning validation as well. Since 2010, he has also been working as a university lecturer at FH Campus Wien in the field of biotechnology with a focus on validation/qualification, aseptic process methods, and clean room technology. He is now starting his business as a freelancing consultant and trainer for pharmaceutical industry.

Alexandra Stärk
MS&T Technical Steward Sterility Assurance, Novartis Pharma Stein AG, Stein, Switzerland
alexandra.staerk@novartis.com

After studying Hygiene Technology at the Technical University of Albstadt-Sigmaringen, Alexandra Stärk worked since 1995 at Novartis Pharma Stein AG in the microbiological QA/QC department, first as team leader and then 8 years as head of this department. Since 2016, she moved into a new role within Novartis Pharma Stein AG and is now responsible for a team of microbiological experts in the department of Manufacturing, Science & Technology which defined the microbiological control strategies for sterile, non-sterile, and cell-gene-therapy production on a global and local level.

Preface

The present book is targeted at microbiologists and those in charge of microbiological quality primarily working in pharmaceutical companies of every size and specialty. The functions held by the readers are microbiology laboratory heads, QA/QC departments, outsourcing departments, regulatory departments, CEOs of small to midsize companies, and health authorities.

The aim of the book is to deliver very special knowledge for microbiological control and its strategy for non-sterile products in a comprehensive way with practical examples. The focus of pharmaceutical microbiology is often on sterile products and aseptic processing. But especially in non-sterile manufacturing, microbiological issues are often present but neglected. Furthermore, many more companies are producing non-sterile than sterile products, and these companies often outsource their microbiological testing to third parties. Thus, they do not have the microbiological expertise in-house and therefore have difficulties interpreting the results they receive from the third party.

Whereas often neglected and considered less critical than for sterile pharmaceuticals, microbial contamination in products not required to be sterile may also cause a health hazard for the patient or may degrade the product thus impacting its therapeutic activity. Even if these products are not required to be free of microorganisms, only a low bioburden is generally accepted and no objectionable microorganisms should be present. In drug products not required to be sterile, regulators therefore expect cGMPs controlling microbial contamination to be followed and implemented.

This book provides the reader a thorough and modern approach to controlling and monitoring microbial contamination during the manufacturing of non-sterile pharmaceuticals. It covers state-of-the-art microbiology quality control (QC) tests as well as risk mitigation strategies so that readers can implement these methodologies in their own facility or laboratory to meet microbiology cGMPs. The latest developments in technology for microbiological testing are also discussed.

The chapter authors, who are international leaders in the topics they have written about, share their long experience in practicing microbiological QA/QC in different types of pharmaceutical companies or by health authorities.

The Book Is Outlined as Follows

Chapter 1 summarizes the different strategies, outlined in detail in the rest of the book, to control and monitor microbiological contamination during the manufacturing of non-sterile pharmaceuticals. The focus is on the six main factors *facility, procedures, product ingredients, utilities, equipment,* and *formulation,* which have an influence on the quality of the final drug product.

Chapter 2 presents the central importance of different approaches to microbial risk assessment and mitigation in non-sterile drug product manufacturing. Risk assessment should take place during product development as well as routine manufacturing, QC, and product release. Involving microbiological expertise at an early stage of development can help to provide a robust process to control microbiological contaminations. Furthermore, production processes that are already running can be optimized or critical control points can be elaborated with risk-based assessment tools. Different risk assessment tools such as impact matrix, failure mode and effects analysis (FMEA), and hazard analysis and critical control points (HACCP) are described, thereby considering the hierarchical risk of ingredients, dosage forms, and processing steps as well as the products' attributes. Finally, the potential of new emerging manufacturing technologies in terms of their microbial risks are addressed.

Chapter 3 introduces one of the most important aspects of receiving robust and reproducible results with microbiological product testing – the qualification of microbiological laboratory personnel and equipment. Since most testing is not yet automated, variability between analysts may affect the outcome of the test result. The chapter describes different approaches to laboratory personnel qualification and re-qualification. While for equipment, a classification of the equipment is needed to address its qualification. Descriptive practical examples are given that show the importance of correct and maintained qualification of laboratory personnel as well as equipment or methods.

Chapter 4 dives into the world of culture media that are the most relevant to any growth-based microbiological method. Therefore, the quality of these media is most important. In recent years more and more companies outsource the preparation of the growth media. In this chapter, challenges such as the development of a culture media, the quality of the raw materials, and the manufacturing process are described. There is a special focus on the QC and release of the manufactured or purchased culture media. Finally, several examples of issues and troubleshooting are given.

Chapter 5 outlines the microbiological test methods used to test non-sterile dosage forms, drug substances, and excipients. It provides the reader with a detailed understanding of procedures including practical tips as well as the rationale for setting acceptance criteria, internal out of expectation (OOE) levels, and testing frequencies. One of the most important aspects of testing is the verification of the suitability of the method used, which is addressed in detail. The whole chapter is supported with practical examples.

Chapter 6 addresses the microbiological testing of primary packaging. Primary packaging is in direct contact with the drug product or API. Therefore, its microbial status must be controlled or monitored. However, there are no regulations for microbial requirements in regard to primary packaging for non-sterile products, which must be developed internally by each company. This chapter gives guidance on the acceptance criteria and testing frequencies that may be used depending on the type of primary packaging. Furthermore, guidance to the testing of primary packaging material including the verification of suitability of the test method applied and handling of out of specification (OOS) and OOE results is provided with practical examples.

Chapter 7 looks at all different types of utilities and discusses how they are designed, qualified, and controlled. Instructions are given when utilities need to be upgraded, reconstructed, or renovated. Utilities such as compressed air, gas, water systems, clean steam, and cleanrooms are reviewed in terms of their performance from the microbiological contamination perspective. Aspects of regulatory requirements, monitoring, sampling, instruments, and methods used for sampling and testing are highlighted. Finally, cleaning of equipment, sanitization, and cleaning validation are also described.

Chapter 8 describes microbiological environmental monitoring in non-sterile manufacturing, which is executed to verify that the environment remains under acceptable microbiological control. Compared to sterile manufacturing, no clear regulatory requirements exist for monitoring levels and sampling frequency, thus user examples are given including the definition of sampling points. Different methods used for testing are presented with different approaches for proving their validation or suitability such as recovery rates, incubation conditions, culture media, or sample hold time. Furthermore, strategies for initial validation as well as revalidation for cleanrooms are elaborated and a clear strategy for deviations is described with some practical examples of an investigation.

Chapter 9 reviews microbiological identification systems used in the GMP environment and discusses their advantages and disadvantages depending on their usage. Precise identification of microorganisms has also gained high relevance for non-sterile product testing. Since non-sterile products must be shown to be free of objectionable microorganisms, the identification of each isolate down to the correct species level is needed. Finally, some examples of isolate identification and their challenges are provided.

Chapter 10 defines microbiological monitoring levels based on historical data and how to trend microbiological data. In general, microbiological counts are not regularly distributed which means that statistical methods assuming a regular distribution of data cannot be used to determine microbiological acceptance levels based on historical data. Alternative statistical methods using other distribution models should be used and are described in this chapter. In addition, microbiological data should be reviewed routinely and trended to assess the capability of the measures to control contamination and verify that no adverse trend is occurring. An adverse trend can be defined as repeating, higher-than-usual counts, or an increasing number of microorganisms or contamination occurrences over a certain time period. Different methods to trend microbiological data using either statistical or graphical approaches are described.

Chapter 11 gives guidance on the handling of objectionable microorganisms. In non-sterile manufacturing, low microbial counts are tolerated, and the final product does not necessarily need to be devoid of microorganisms. Nonetheless, some microorganisms are considered objectionable in the sense that they can adversely affect the appearance, physicochemical attributes, or therapeutic effects of a non-sterile product, or, due to their numbers and/or pathogenicity, may cause infection, allergic response, or toxemia in patients receiving the product. Findings with objectionable microorganisms represent the majority of microbiologically related FDA recalls of non-sterile products. This chapter focuses on microbiological risk assessments that evaluate whether a recovered microorganism is objectionable. Different strategies are given where the objectionability of the isolate found is regarded in relation to the criticality of the product (low-risk to high-risk products) and its patient population. Different sources of objectionable microorganisms are described. Finally, the chapter contains working examples of real cases with which readers would also be confronted.

Chapter 12 summarizes the complexity of the investigations of microbiological OOS cases or deviations that require high expertise. First, data integrity in microbiological laboratories is addressed including the implementation of the ALCOA+ principle for the laboratory. Second, definitions for OOS, OOE, out of trend (OOT), and exceedance of action or alert level are given. The two-level approach (investigation in the lab and investigation of product quality) is described with practical examples. For all general microbiological tests (environmental monitoring, water testing, growth promotion test, and product testing), a detailed procedure for handling deviations is given.

Chapter 13 provides a current overview of rapid microbiological methods (RMMs) that can be used in non-sterile product manufacturing. RMMs may significantly reduce the time-to-result of microbiological tests and therefore have the potential to shorten throughput time for drug product release. Other potential benefits of RMMs are, for instance, a reduction in inventory costs, faster stop or go decisions during manufacturing, decreased risk of stock-outs and supply bottlenecks, improved data integrity, automation, and introduction of a paperless laboratory. The chapter gives guidance on the validation approaches for RMMs with a focus on the three relevant guiding documents, USP chapter <1223>, Ph. Eur. chapter 5.1.6, and PDA TR No. 33. Finally, it shows how a business case for RMMs can be developed if you want to implement such a method in your facility.

Chapter 14 is the validation protocol of an RMM that was established for the microbiological examination of non-sterile and nonfilterable drug products, excipients, and APIs. The Celsis Advance system using ATP bioluminescence was used to do this and a thorough validation protocol covering robustness, ruggedness, repeatability, specificity, limit of detection, accuracy, and precision was developed. There was a special focus on the correct and sophisticated statistical evaluation approach for the validation data, which can be used as an example for the reader's own validation studies. Finally, equivalence in routine operation of the RMM with the compendial method is demonstrated as well as the suitability for product-specific testing.

Chapter 15 is an ex-regulator's view of the microbiological QA/QC functions in the pharmaceutical industry. The author starts with the beginning of pharmaceuticals in the fifteenth century and shows where and when the first legally authorized standards were published all the way up to modern GMP regulations. Microbes have the possibility to grow and survive in places that are often unique to a specific material or environment and therefore need special attention. Our testing of the product or environment should be seen as a snapshot of the contamination present and only continuous quality activities can give a reasonable control of microbiological product quality. Product quality is the responsibility of quality management made up of quality assurance (QA) and QC.

Chapter 16 provides an overview of the most important regulatory chapters and guidelines for non-sterile product manufacturing and testing in the EU. The author gives a general overview of audit assessment tools for a microbiological laboratory. Several drop-down lists for each subject (personnel, documentation, culture media, trends, methods, facility, equipment, and reference cultures) help the reader to prepare adequately for the next audit or inspection. The chapter closes with typical issues detected during the evaluation of microbiological laboratories.

Chapter 17 describes outsourcing strategies to contract laboratories. Outsourcing may result in other issues that could impact resources. The chapter provides some guidance on the most relevant points to verify if microbiological testing has been outsourced to a contractor. For example, which microbiological tests can better be performed in-house and what can be easily outsourced? It addresses the advantages and disadvantages of outsourcing and the business case for outsourcing. Most important of all, when outsourcing a process, is the quality agreement to ensure that both parties are talking about the same thing.

David Roesti
Novartis Pharma Stein AG, Switzerland

Marcel Goverde
MGP Consulting GmbH, Switzerland

Foreword

"*Another new textbook on Pharmaceutical Microbiology*," you, dear reader, may think. "*There are already so many textbooks on pharmaceutical, microbiological laboratories on the market – why do I need THIS textbook?*" could come to your mind in this context.

What makes this book so special that it is worth a second look? In my opinion, this book closes a gap that has existed until now: while there are some very good textbooks available for sterile products and their (aseptic) production, the focus of this textbook is on **non-sterile products and their production**.

As the expert reader knows, the microbiological laboratory is one of the pillars of quality control in the pharmaceutical industry, alongside to the analyticalchemical laboratory. In contrast to the analytical-chemical laboratory, the microbiological laboratory and the microbiological control concept for the production of non-sterile products may not have received the attention that would have been required in recent years. An indicator that supports this assumption is the increasing number of "major" and "critical" observations and "warning letters" issued by Health Authorities in recent years concerning microbiological control concepts and the pharmaceutical microbiological laboratory. This is certainly also due to the fact that the microbiological control concept and the microbiological laboratory are increasingly becoming the focus of Health Authority inspections.

The pharmaceutical microbiological laboratory is involved in all phases of the product life cycle: from research and development to the manufacture of clinical trial batches as well as batches for the commercial market. All steps of the manufacturing process are controlled by microbiological analyses. This includes the evaluation of the microbial quality of raw or primary packaging material, the detection of microbial contamination during the manufacturing process, the control of the production environment ("Environmental Monitoring"), and last, but not least, assessment of the final product to release the produced batch. In addition, microbiological testing is also required for equipment qualification, process validation, and cleaning validation.

For all these parameters and activities, limits or, if applicable, specifications must be defined. These limits or specifications may be based on historical data using adequate statistical tools. If these limits are exceeded or even if the specification is failed, a deviation management must be defined.

In my opinion, the textbook comes at the right time. Pharmaceutical microbiology is currently in a transition phase: some growth-based methods that have been used for several decades (but which still have their right to exist, are used in daily routine and are described in detail in this textbook) are currently being replaced by modern, automated methods that can generate the analytical result faster: the so-called alternative or rapid microbiological methods. With the help of these alternative microbiological methods not only analytical lead time for batch release can be significantly shortened (up to now the classical, growth-based microbiological tests were the time-limiting analyses of the release process) but also the manufacturing processes can be controlled more tightly. In the best case the analysis results are available in real time. This allows an immediate reaction if deviations occur during the manufacturing process.

In summary, this textbook, written by leading authors familiar with practice in GMP-processes, covers all of the above aspects of a modern microbiological control system for the manufacture of non-sterile products. All measures required to establish and assess the microbiological control concept, and which should be reflected in a modern microbiological quality control laboratory for the manufacture of non-sterile products, are described in detail in this textbook. Practical examples from "QC/QA everyday life" in large multinational pharmaceutical companies are included in this textbook. It is not only for readers with a microbiological background but also for "non-microbiologists," i.e. colleagues in other quality control units, the quality assurance departments, the regulatory departments, and other colleagues interested in the topic.

I wish you a few entertaining hours reading this comprehensive and informative textbook!

August 2019 *Dr. Sven Deutschmann*

Acknowledgments

The editors would like to express their sincere gratitude to all the chapter authors that have contributed to this endeavor on top of their numerous activities.

We are also grateful to Jonathan T. Rose, Senior Editor at Wiley for having approached us to write a book on pharmaceutical microbiology and his trust throughout this project as well as Aruna Pragasam, Project Editor, and Gayathree Sekar, Production Editor, at Wiley for their continuous support during the editing and production process.

We would also like to show our gratitude to Sabrina Kuoni for assisting in formatting all the chapters as well as Yvonne Böhler for the editing work on the images received.

Finally, our heartfelt thanks go to Anne Smiejan-Roesti and Maja Christina Goverde for their continuous encouragement and support during this time-consuming writing process.

The Editors

Messieurs, c'est les microbes qui auront le dernier mot.

(Gentlemen, it is the microbes who will have the last word.)

– *Louis Pasteur*

1

Microbiological Control Strategy

David Roesti[1] *and Marcel Goverde*[2]

[1] *Novartis Pharma Stein AG, Stein, Switzerland*
[2] *MGP Consulting GmbH, Binningen, Switzerland*

CONTENTS

1.1 Introduction

Microbiological controls in non-sterile pharmaceutical drug product manufacturing consist of preventing microorganisms from contaminating the final product and keeping their numbers low during the manufacturing process. By controlling the overall bioburden level, the probability of product contamination with an objectionable microorganism is also reduced. The effectiveness on the controls can be continuously evaluated with sound microbiological monitoring and trending of results.

Microbial controls can be defined with the support of risk management tools helping to find the right balance between excessive controls and contamination risk (Figure 1.1).

Pharmaceutical Microbiological Quality Assurance and Control: Practical Guide for Non-Sterile Manufacturing, First Edition. Edited by David Roesti and Marcel Goverde.

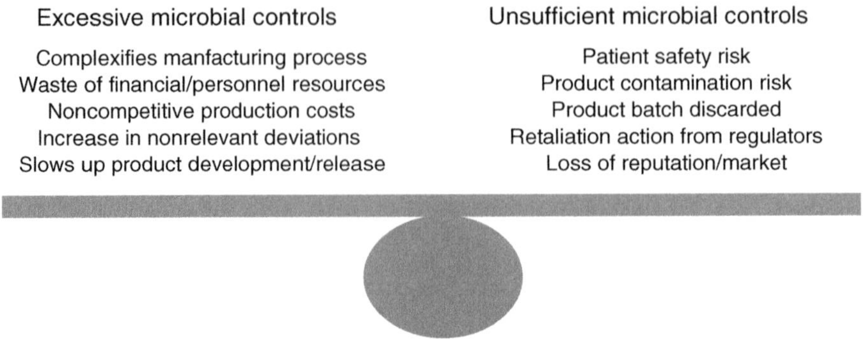

Figure 1.1 Balance between excessive and insufficient microbial controls.

The United States Pharmacopeia (USP) chapter <1115> provides the most recent and comprehensive guidance on factors affecting microbial controls and suggests a risk management approach to establish these controls. USP <1115> writes that *microbiological risk should be assessed on a case-by-case basis during the development of a new product and should be evaluated during the validation of the manufacturing process.*

The present chapter will likewise present high-level multiple microbial controls that can be introduced during production of non-sterile drug products. Details of the controls are further addressed in the corresponding chapters of this book.

A search of the FDA database for recalls (Enforcement Reports) for the category "drug products" using the keywords "microbiology," "microbiological," and "microbial" found 14 recalls for non-sterile drug products (since the 1st of January 2014). Of these recalls, 10 were due to out-of-specification (OOS) results for microbiological specifications or aerobic microbiological count, two were due to product released to market prior to microbiological testing, and two omitted testing for a specified microorganism (source FDA website, last visited on 14 December 2018). Further details on product recalls for non-sterile products between the years 1998 and 2006 Jimenez (2007), and 2004 and 2011 Sutton and Jimenez (2012).

1.2 Overview of a Microbial Control Strategy Program

A comprehensive microbial control program should allow to identify the risk of contamination to the product as well as the different mitigation steps to control this risk. It should also be based on the latest regulatory guidelines as well as industry best practices or current scientific knowledge and would inspire the design of facilities, equipment selection, choice of raw materials, cleaning and manufacturing procedures, etc. (Figure 1.2).

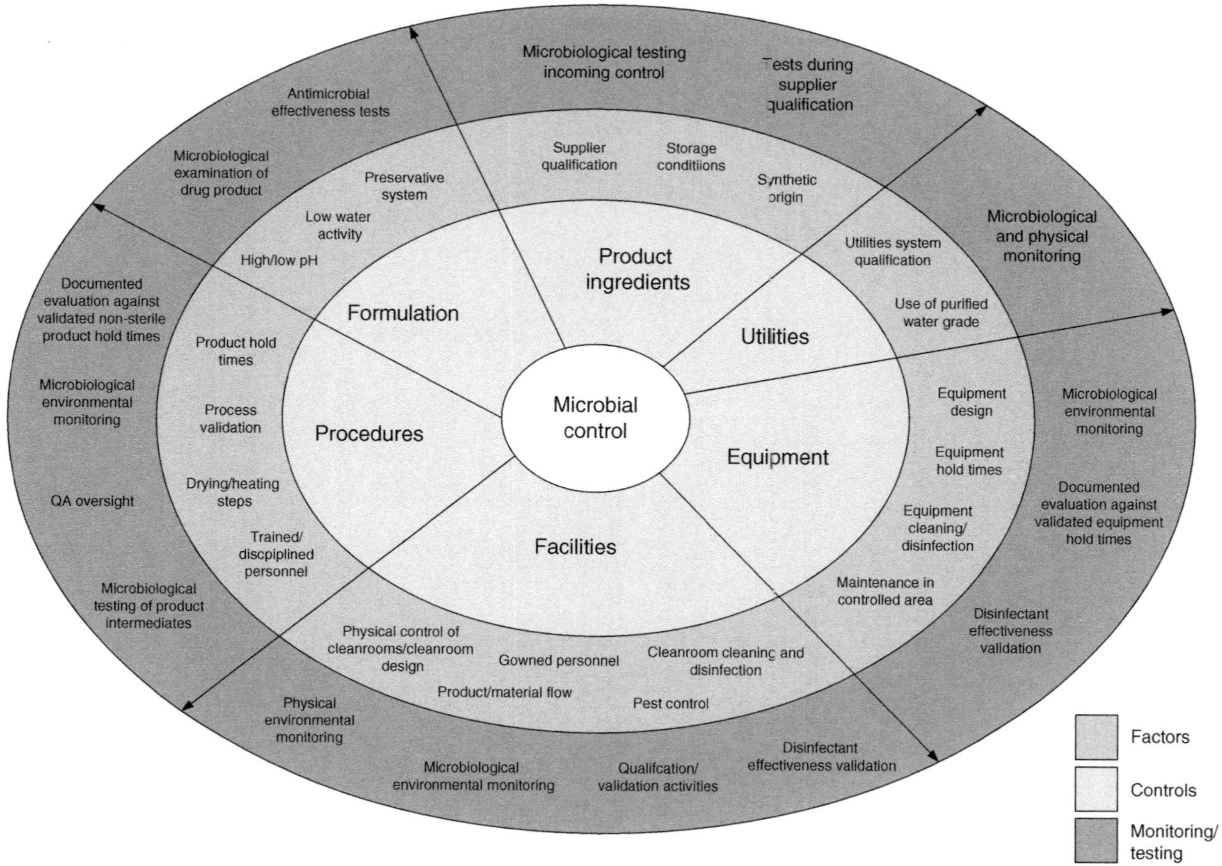

Figure 1.2 Examples of factors, controls, and monitoring implicated in a comprehensive microbial control program for non-sterile products.

A formalized risk assessment should initially define the contamination risks and control points. A risk-based approach is described in Chapter 2. The resulting control points and monitoring strategy serve then as basis for the initial microbial control program. The program covers the facility or product's life cycle and may be adapted following trend results and changes made in, for example, the facility design or process. It is recommended that on a regular basis (e.g. every two to three years or after relevant changes), the microbial control program is evaluated and if deemed necessary adapted. Trending of monitoring data strongly supports such evaluation (refer to Chapter 10 for a detailed review on trending of microbiological data). The following sections summarize the different controls to be included in a comprehensive microbial control program for non-sterile products.

1.3 Main Factors to Be Controlled

1.3.1 Controlled Facilities

Current good manufacturing practices require that facilities manufacturing medicinal products for human use should be effectively designed:

- to run operations in a controlled environment
- to prevent product cross-contamination
- to prevent microbiological contamination
- to provide sufficient space allowing operations running as intended
- to facilitate product, personnel, and material flow
- to permit effective cleaning and disinfection.

Facility design is part of all regulatory guidelines on current Good Manufacturing Practice (cGMP). For instance, the EU EudraLex volume 4 chapter 5 states that for medicinal products manufacturing areas the technical measures may include:

i) Dedicated manufacturing facility (premises and equipment);
ii) Self-contained production areas having separate processing equipment and separate heating, ventilation and air-conditioning (HVAC) systems. It may also be desirable to isolate certain utilities from those used in other areas;
iii) Design of manufacturing process, premises and equipment to minimize opportunities for cross-contamination during processing, maintenance and cleaning;
iv) Use of "closed systems" for processing and material/product transfer between equipment;
v) Use of physical barrier systems, including isolators, as containment measures;

vi) Controlled removal of dust close to source of the contaminant, e.g. through localized extraction;
vii) Appropriate use of air-locks and pressure cascade to confine potential air. borne contaminant within a specified area;
viii) Minimizing the risk of contamination caused by recirculation or re-entry of untreated or insufficiently treated air;
ix) Use of automatic clean in place systems of validated effectiveness;
x) For common general wash areas, separation of equipment washing, drying and storage areas.

For the United States, the guidelines on facilities can be found in the U.S. Code of Federal Regulations such as 211.42 Design and construction and 211.46 Ventilation, air filtration, air heating, and cooling. Other guidance on facility design can be found in PIC/S, WHO (e.g. WHO 2011) or engineering technical documents such as, for instance, the ISPE Baseline guideline on solid oral dosage forms (ISPE 2016).

For production areas of non-sterile products most regulatory guidelines do not enforce that the air cleanliness is classified in terms of particle concentration (e.g. as per ISO 14644-1). Also, the cleanroom classification of the EudraLex Annex 1 does not apply. However, there are a number of country-specific guiding documents that provide some requirements for microbiological air quality in non-sterile manufacturing facilities:

- The Chinese FDA GMP guideline (CFDA 2010) requires that *The exposed processing areas for oral liquid and solid preparations, drugs applied through tract (including recta), epidermal products, and other non-sterile products, as well as the exposed processing areas for handling immediate packaging materials should be designed as* ***Grade D*** *according to requirements in Annex 1 of Good Manufacturing Practice (GMP) for sterile products.* This would imply that the cleanrooms have to be classified with grade D maximum permitted number of particles equal to or greater than 0.5 μm of 3,520,000.
    - A similar requirement is given in the Mexicana NOM-059-SSA1-2013, where an **ISO Class 8** is required for preparation and primary packaging of non-sterile pharmaceutical formulations.
    - Annex 10 of the EudraLex also requires **at least a Grade D** environment for the manufacture of pressurized metered-dose aerosol preparations for inhalation.
    - In the introduction of the Brazilian Quality Guide (ANVISA 2013), it is stated that this guideline also applies to the manufacture of non-sterile medications, however, *The majority of production areas of non-sterile medications do not require this kind of classification, but they should always be designed and maintained as "controlled areas".* Thus, it will be the responsibility of the manufacturer to define the environmental requirements for such "controlled areas".

- The German Aide Mémoire (ZLG 2010) provides a table with requirements for the microbiological air quality for non-sterile manufacturing facilities.
- Furthermore, some guidance for microbiological environmental monitoring is given by different authors as in the ECA guideline (Goverde and Roesti 2018), Goverde (2018), Rieth (2017), Rieth and Krämer (2016), or Seyfarth (2002).

Nonetheless, in general, controlled not classified (CNC) areas are considered acceptable for areas in which non-sterile products are manufactured.

Controlled not classified (CNC) areas where HVAC systems are specifically designed to reduce airborne contaminants below the level of ambient environment and both temperatures and room humidity are controlled more tightly than the ambient environment (Goldschmidt and Farquharson (2017).

The tightness of controls would depend on the criticality of the product (e.g. route of administration and growth-promotion properties). Based on the tighter gowning procedures, cleaning steps, personnel and material flows, etc., it is expected that environmental bioburden is lower in more critical product manufacturing areas (e.g. products for inhalation) as compared to less critical ones (e.g. solid oral dosage forms).

In order to control the temperature, relative humidity, air pressure, and air particle level, HVAC systems are incorporated and incoming air is filtered with high efficiency particulate air (HEPA) filters. For non-sterile manufacturing areas, HEPA filters of H13 and H14 classes with a retention of less than 99.95% and less than 99.995%, respectively, of 0.3 μm airborne particles are used.

Rooms dedicated for the manufacturing of non-sterile products should be segregated from other operating areas and may only be entered via separate air locks for personnel and materials. Differential air pressure exists between cleanroom and unclassified areas. In the authors' experience both pathways (personnel as well as materials) represent a certain risk for the contamination of the cleanrooms. For personnel, hand washing and disinfection as well as the gowning procedure are in general well established, however, QA oversight activities can lead to improvements in both procedures – at the very least for individual staff members. Most people need assistance or retraining in both procedures since, for example, the correct hand disinfection may be neglected or the gowning procedure not correctly executed (e.g. overall touches the floor and hair net is placed as last piece of protective gowning). Contamination may also be brought in clean/controlled areas with the lock-in of materials. Here, typical contamination problems are absence of disinfection of the trolley used to carry materials or contamination originating from the material itself (e.g. introduction of cardboard boxes or wooden pallets). Thus, establishing a regular QA oversight program with a microbiological specialist can help to reduce contamination along these transfer areas.

The systems used to control the environment of the facility are qualified prior to running operations and are part of the facilities' life cycle. As per EU GMP Annex 15, *Qualification activities should consider all stages from initial development of the user requirements specification through to the end of use of the equipment, facility, utility or system.*

Modern nonporous construction materials (e.g. epoxide floor/wall surface) with smooth surfaces easy to clean should mitigate the buildup of microorganisms on walls, ceiling, and floors. The surfaces should be also resistant to the cleaning or disinfection solutions. Older buildings with decaying wall or floor surfaces are typically a source of environmental contamination.

Only trained personnel should enter the cleanroom areas where products are manufactured. They wear dedicated gowning which provides a physical barrier from the body to the working environment. Gowning consists of overalls (e.g. made of Tyvek or cotton material), shoe covers or dedicated shoes, hair/beard covers, and face masks in the most critical areas (in near vicinity to product exposed to the environment). Gloves regularly disinfected with 70% ethanol/isopropanol should also be worn in the most critical areas or at least close to the product (<2 m). Operators should be trained with focus on general hygiene practices and awareness. Some companies involve the quality assurance department to oversee operations in order to verify that good hygiene practices are followed (QA oversight).

In aseptic manufacturing, regular inspection by the quality assurance department (often also called QA oversight) is required by various regulations (e.g. FDA 2004; EudraLex 2017) and is established in many companies. There is no such requirement for non-sterile manufacturing, but the implementation of QA oversight can be a great advantage. Various concepts can be considered:

1) After deviations in environmental monitoring, oversight is carried out to find the root cause of the deviation or as CAPA to check if the root cause has been eliminated. This is then usually announced and planned.
2) Random oversight: At regular intervals the QA officer verifies the behavior, maintenance of hygiene rules, data integrity, etc.
3) Random monitoring: If the sampling for the environmental monitoring is carried out by production (delegated monitoring), monitoring by the QC without any warning at regular intervals (e.g. monthly or quarterly) is recommended in order to verify the plausibility of the environmental results when monitoring is delegated.

If this kind of concept is implemented, then it is recommended that the same persons carry out the oversight to maintain continuity and facilitate continuous improvement.

A cleanroom cleaning program should be in place to ensure that chemicals and microorganisms are removed from surfaces not only of operational but also of surrounding areas (e.g. transfer and storage areas). Such a program can be set up in the following way (see also Sandle 2012):

a) Definition of cleaning and disinfection agents (as well as the equipment such as vacuum cleaner, brushes, wipes, mops, bucket system, etc.) to be used. In case spore forming bacteria are in excessive numbers, use of sporicidal agents should be considered on a frequency basis.
b) Definition of agents' concentration as well as contact time.
c) Antimicrobial efficacy of disinfectants may be evaluated by executing disinfectant efficacy tests as per USP chapter <1072> or ISO norms.
d) Defined cleaning and disinfection procedure for each room or equipment (e.g. starting from the clean to the dirty side; starting from the top to the bottom; correct wetting of surface; contact time; protection of equipment or raw materials present in the room; handling of cleaning equipment before and after usage; disassembling of equipment).
e) Well-trained personnel, defined responsibilities.
f) Visual inspection after cleaning and disinfection.
g) Frequency of cleaning and disinfection (e.g. daily for critical area, weekly for surrounding areas, and after usage of equipment). The frequency should be established taking a risk-based approach and might be challenged by the results of the routine environmental microbiological monitoring (for further details see Chapter 8).
h) Inspection of equipment for cleanliness before use. Defined procedure for waste handling.
i) Definition of a cleanroom idle time after cleaning and disinfection based on a rationale and confirmed with experimental data. In case the idle time is exceeded, the cleanroom is cleaned/disinfected prior to restart of operations.
j) Agents used should regularly be checked for microbial contamination.
k) Definition of storage conditions and expiry dates of stored as well as opened agent containers.
l) Technical agreement with the supplier of the cleaning and disinfection agent.
m) Defined procedure after special activities (e.g. maintenance and after reconstruction).
n) GMP-compliant documentation.

Good housekeeping is important to avoid contamination sources. For instance, all materials from natural origin (e.g. wooden pallets and cardboard boxes) should be prohibited in cleanrooms since they can harbor a large number of microorganisms and cannot be sanitized properly. In addition, water is a hotspot of microbial contamination and special attention should be made that surfaces dry up completely after cleaning and that no stagnant water is present especially in washing rooms. Procedures for handling waste, contaminated water, or gowns as well as pest control procedures should also be in place. Use of disposable materials/utensils would also limit contamination risks.

Pest control is a clear requirement of the regulations, e.g. the EU GMP guideline states in its chapter 3, § 3.4 *Premises should be designed and equipped so as to afford maximum protection against the entry of insects or other animals* (EudraLex 2014a). The corresponding article for the US is 21 CFR 211.56. Any pest can affect raw materials, drug products, health, and hygiene. This can lead to economic damage, loss of image, and loss of trust for the company. Therefore, a pest management plan with its control is compulsory.

As precautionary measures, joints must be insect-proof, fly screens must be installed, and the HVAC must also have insect-proof front screens. Locks should be mutually lockable. Water drains must be closed or protected as inlet points. Poorly fitting joints or lamp sockets, etc., should be replaced. A regular optical control of protective measures as well as control of incoming materials is inevitable and in order to have an overview, bait traps must be set up and checked regularly. An expert should be consulted (e.g. a state-approved company). In the event of infestation, appropriate measures must be taken. Infested materials must be removed and cleaning is required as a minimum action.

Chemical, physical, and biological measures are available to combat the infestation (Rieth and Krämer 2016; Rieth 2017).

- **Chemical methods:** Use of biocides such as pyrethroids against insects or general biozides such as organophosphates or carbamates.
- **Physical methods:** UV lamps or bait traps with adhesive traps or electrical insect destruction, ultrasound.
- **Biological methods:** Use of natural enemies, insect viruses, or biological toxins (e.g. Bti toxin from the bacterium *Bacillus thuringiensis*).

Some further guidance on pest control can be found, for example, in the *Japan Affiliate: Pest Control Manual* of the ISPE (2018).

In order to evaluate the effectiveness of the different controls in the facility, its environment is microbiologically monitored. Performing air and surface environmental monitoring ensures that the overall microbial counts remain under a state of control and that hygienic and cleaning practices are respected. An excessive count or adverse trend observed in the routine environmental monitoring program might indicate a potential problem (e.g. HVAC system not functioning properly and cleaning procedure not followed) and would require immediate attention. Chapter 8 details an environmental monitoring program for non-sterile products.

1.3.2 Controlled Procedures

Whereas low-level ingress of microorganisms generally does not represent a major safety concern during the manufacturing process, their ability to multiply if the conditions are adequate represents a serious risk to the product. The main factors which favor microbial growth during a production process are ambient temperature, water availability or moisture level, nutrient traces, and time for proliferation. On the contrary, microbial-reducing steps (e.g. heating >60 °C or drying) diminish the risks. A structured microbial risk assessment of the product manufacturing process including the primary packaging is the best approach to evaluate the microbial growth promoting or reducing steps and to define critical control points (refer to Chapter 2).

The resulting critical control points obtained from the risk assessment can then be part of the process validation. The process validation should demonstrate that the process is under control. As stated in the EU Annex 15, *It is a GMP requirement that manufacturers control the critical aspects of their particular operations through qualification and validation over the life cycle of the product and process.* It is a good practice to include microbiological data during process validation runs not only testing the final product but also by testing product intermediates. For instance, maximum aqueous intermediate hold times may be defined and during process validation these are tested for microbiological purity as supportive evidence.

Aqueous intermediates (e.g. coating solutions for film-coated tablets) have the greatest potential risk of microbial growth and the time during which they are left to stand or processed before a microbial-reducing step should be defined and validated. Further guidance on how to define and validate microbial hold times can be found in Chapter 2.

The US CFR 211.110 text on sampling and testing of in-process materials and drug products actually includes bioburden determination as in-process controls. Whereas for powder-based intermediates with low water activity, in-process bioburden testing may not be necessary, it could be applied for the most growth-promoting solutions (e.g. aqueous granulation or coating solution). For unpreserved nasal spray solutions, testing of bioburden in-process controls before and after microbial reduction or sterilization steps is essential.

1.3.3 Controlled Product Ingredients

Product ingredients (drug substances, excipients, and primary packaging) may be a significant source of microbial contamination risk especially if they are of natural origin. Common excipients of natural origin that often contain a certain bioburden amount are calcium salts, starches, gelatin, acacia, guar gum, dyestuffs, lactose, magnesium stearate, and celluloses. Even if the water activity remains low, bacterial or fungal spores may be found in high numbers in these excipients. The manufacturing processes of the ingredients also impact the resulting bioburden. For instance, synthetically manufactured drug substances are less prone to contamination since manufacturing conditions may be extreme in terms of pH and temperature and antimicrobial inorganic solvents may be used to synthesize/purify the molecule. Nonetheless, as USP <1115> mentions water used in the downstream processes and cleaning may be the most relevant source of contamination. The most critical ingredients in terms of microbiological risk are those that are unprocessed, of natural origin, and with a high water activity.

USP <1115> states that the ingredients are safe and do not pose a risk of infection or toxin if the microbial levels remain in the ranges of those recommended in USP <1111>. In addition, many excipients of natural origin do have microbiological specifications in their respective monograph.

Microbiological acceptance criteria may also be adjusted based on

- the growth-promoting nature of the ingredient
- the quantity of ingredient used in the product
- the risk of proliferation in the manufacturing process
- the microbial-reducing steps in the process

The quality of ingredients may vary from one supplier to another and often pharmaceutical grade quality may be difficult to find on a particular market. It is the manufacturer's responsibility to find and evaluate the most appropriate quality for its use as well as vendors capable of delivering the right quality consistently. This would include initially sending out a quality questionnaire followed by auditing of the vendor.

Actually, the US CFR 211.82 requires that *upon receipt and before acceptance, each container or grouping of containers of components, drug product containers, and closures shall be examined visually for appropriate labeling as to contents, container damage or broken seals, and contamination.* In addition, US 21 CFR 211.84(d)(6) requires that *Each lot of a component, drug product container, or closure with potential for microbiological contamination that is objectionable in view of its intended use shall be subjected to microbiological tests before use.* The release for use may be supported with results obtained by internal QC testing or if the supplier is qualified, on the supplier's certificate of analysis. Further details on the qualification of a supplier or third party can be found in Chapter 17.

The storage of the ingredients prior to utilization should be done in temperature- and humidity-controlled warehouses. The shelf life for each ingredient should be defined and based on experimental data. Materials may be delivered and stored in cotton or paper bags that are permeable to humidity which could lead to microbial proliferation in case they are spoiled with water. Conversely, condensation of water in tight permeable containers that contain hydrated chemicals may occur enabling localized colonization by microorganisms, especially molds. Shipping, transport conditions, and handling should also be evaluated and measures introduced to mitigate risk. A special focus of attention should be given to pallets. Often, wooden ones are used which have a high risk of mold contamination. As soon as they get wet, mold can appear and proliferate. With this, very high level of spores will be produced which can contaminate the material packaging and with the concurrent lock-in into the clean room, a risk of contamination is present.

Non-sterile drug products are protected from moisture or light by their primary packaging (e.g. bottle/cap or blister/foil combination). Verification that the primary packaging combination remains tight throughout the product's shelf life is carried out with stability studies performed under various conditions. Container integrity may be tested directly with physical container closure integrity tests (e.g. dye intrusion test for blister/foil combinations) or indirectly with the products' physical–chemical properties (e.g. water content, impurities, and assay) and bioburden. Also, for non-sterile products with a preservative system, the primary packaging should be evaluated if leachables from the packaging could potentially affect the preservative efficacy.

Further guidance on the risk categorization of pharmaceutical ingredients as well as their testing is described in Chapters 2, 5, and 6.

1.3.4 Controlled Utilities

The utility that represents the most risk to microbial contamination is the process water used as solvent in the non-sterile product formulations or for cleaning of equipment product contacting surfaces. Purified water may contain a high number of Gram-negative bacteria which may be resistant to preservatives in semisolid or liquid formulations. In addition, microbial biofilms may form in the water distribution system piping which could then detach and contaminate the equipment, working surfaces, or the product. The level of microorganisms in water systems can be controlled with a good design of the water distribution system as well as employment of water treatments and monitoring. Water systems are qualified, monitored, and trended at a frequency that is sufficient to ensure that the system is under control and continues to produce water of required quality.

A further utility that might need to be controlled is compressed air or gases used for cleaning/drying or in the manufacturing process itself. In general, compressed air or gases have a low microbial contamination risk. Compressed air, for example, is used to dry equipment after cleaning with water. In this case, after treatment with air, disinfection might follow that will reduce the possible presence of microorganisms on the equipment. However, attention should be drawn to the devices used (e.g. compressed air guns and tubes). If there is residual of water or humidity, microbiological proliferation may occur and thus high levels of microorganisms may be found in the equipment or liberated in the air of the room. Compressed air as well as gases might be used during the manufacturing process, in this case the criticality is higher as contamination of the product is possible. Therefore, the gas or compressed air distribution systems should preferably be microbiologically controlled. In principle, compressed air and gases should have at least the same or better microbiological quality than the ambient air at the point of use.

Furthermore, cleaning and disinfection agents need to be controlled on a regular basis. Although disinfectants are generally considered as microbiologically uncritical, non-sporicidal disinfectant solutions may be contaminated with bacterial spores (e.g. ethanol baths for small equipment parts).

A detailed guidance on utilities can be found in Chapter 7.

1.3.5 Controlled Equipment

Equipment used in the manufacture of non-sterile products may represent a potential source of contamination and they should be designed to ease cleaning, removal of product residuals, and water drainage. As EU EudraLex vol 4 chapter 3 states *Manufacturing equipment should be designed so that it can be easily and thoroughly cleaned. It should be cleaned according to detailed and written procedures and stored only in a clean and dry condition.* Badly designed equipment may result in formation of biofilms in difficult to clean or in areas where water can stagnate. As written in USP chapter <1115>, the preferred material for equipment that is in contact with product is austenitic stainless steel with a roughness average of minimal 15–20 μm.

The equipment cleaning prevents product cross-contamination for non-dedicated equipment and reduces the number of microorganisms to an acceptable level. The cleaning of equipment may be performed automatically (clean-in-place, CIP) or manually by operators. The CIP would be the preferred option since it is a far more controlled process and can be optimized by modifying parameters such as type or concentration of detergent, time of cleaning, water volume, flush rate, etc. However, some equipments are not optimally designed and may contain parts (e.g. dead legs) not accessible for the automatic cleaning. In this case the equipment parts that cannot be automatically cleaned may be demounted and cleaned separately manually in washing room sinks, in baths, or dish washers. This partial disassembly and cleaning of equipment is referred to as clean-out-of-place (COP). In addition, some residuals might first need to be removed manually (scrubbing) before automatic cleaning is applied. In manual cleaning processes, clear written procedures, adequate training, disciplined personnel, and regular oversight are essential. Badly designed manual cleaning procedures may lead to substantial contamination of product contact equipment surfaces (examples of contaminated equipment by inadequate cleaning are described in Roesti (2012)).

Unlike for aseptic filling, sterilization of product contact equipment parts is not required for non-sterile products. Nonetheless, the number of microorganisms should be kept to an acceptable level and no biofilm should be present. In order to reduce the number of microorganisms, a sanitization or disinfection step may be carried out and composed of:

- rinsing with process water heated at 80 °C.
- rinsing with disinfectants (e.g. ethanol/isopropanol). In this case limits must be established for the removal of disinfecting agents used.
- equipment dried in greater than 60 °C chambers to accelerate drying and to prevent growth of mesophilic microorganisms.

Once the cleaning has been performed, the equipment or equipment parts should be completely dried, stored in controlled areas, and be protected from recontamination (e.g. covered by foils). To control the level of microorganisms recontaminating the equipment after cleaning, hold times of cleaned and/or disinfected equipment must be defined and validated.

Certain equipment may not be used regularly or must be stored outside the cleanroom. In such cases a clear procedure must be defined and should be validated. The easiest way is to perform a GMP cleaning after storage as defined in the SOP and validated by the cleaning validation (see below). In such a case, any possible contamination during storage outside the cleanroom will be removed and no specific validation (except the cleaning validation itself) is needed; however, there will be more work, since the equipment must be cleaned twice (once after usage and once after storage). An alternative is to have a validated procedure to store equipment in a warehouse. With the validation it must be shown that there is no contamination risk to the stored equipment over a defined hold time. A popular method is to protect the equipment by wrapping it in plastic. Such type of storage must be validated. Thus, extra validation work may be needed in this case but no additional cleaning after storage would be required. Both procedures are possible and may, for instance, be evaluated by a business case to check which is the more efficient and less time consuming.

Once defined, the cleaning procedure has to be validated. The equipment cleaning validation is essentially carried out to confirm the effectiveness of cleaning procedures that remove residual drug substances thus avoiding cross-contamination of another product manufactured with the same equipment. As per EU EudraLex vol. 4, Annex 15, the risk presented by microbial contamination should be considered in the cleaning validation.

In order to consider a worst-case condition, the cleaning validation should encompass:

- the maximum time between the completion of processing and equipment cleaning (dirty hold time).
- the maximum time between the completion of cleaning and start of processing (clean hold time).

Microbiological results of product-contacting equipment surfaces sampled after the cleaning (carried out at the end of the dirty hold time) and after the clean hold time are used to validate the cleaning procedure.

With new equipment or new cleanrooms and a validated cleaning procedure, microbiological excursion on the surfaces should be controlled. Nonetheless, the older the equipment, the more often excursions might occur. There may be several reasons for this. Examples are:

- **Abrasion of equipment surfaces:** After several years of usage, scratches, fissures, or holes will appear on the surface (even on stainless steel) that can harbor microorganisms and allow them the possibility to survive and proliferate. Polishing or grinding can help to reduce the contamination risk. But after a certain amount of usage, even this will not be enough and the equipment needs to be replaced.
- **Cracked seals:** Something that happens on a regular basis are cracked seals (e.g. silicone sealing used at joints). Here, the best preventive action is to check and change them on a regular basis to avoid recurring microbiological contamination.
- **Cracks in walls and floors:** Especially in older buildings, holes in the floor, cracks in walls or ceilings, or broken silicone joints (around drains, windows, floors, etc.) will develop and must be fixed. QA oversight can be a good preventive tool, where the person in charge checks the facility on a regular basis. An often-encountered issue are holes in the floor. These can easily be fixed using epoxy resins or repair materials. However, in some cases the entire floor needs to be replaced.

1.3.6 Controlled Formulation

During product development, consideration should be given to the susceptibility of non-sterile drug product formulations to microbial growth. Whereas formulations with a low water activity (no microbial growth at a_w of <0.6) such as oral tablets are not at risk, others such as semisolid or liquid products may be susceptible to growth and may require including a preservative system.

Typical preservative systems are listed in Table 1.1 (extracted from Dao *et al.* (2018)).

The effectiveness of the preservative system is evaluated according to USP <51>, Ph. Eur. 5.1.3, and JP 19 or ISO 11930:2019 during the product's development phase. While the pharmacopoeial chapters are intended primarily for pharmaceutical preparations, the ISO norm written for cosmetics is more detailed.

Table 1.1 Examples of preservatives used in non-sterile drug product formulations.

Route of administration	Preservative
Oral	Sodium benzoate, benzoic acid, sodium and potassium, sorbic acid, calcium lactate, paraben (methyl-, ethyl-, propyl-)
Topical	Cetrimide, benzoic acid, thimerosal, imidurea, phenyl salicylate, chlorhexidine, benzalkonium chloride, EDTA, chlorocresol
Nasal	Chlorobutanol, EDTA, benzalkonium chloride, potassium sorbate, chlorocresol, phenylcarbinol

Preservatives are not considered to be a medicinally active component of pharmaceutical formulations, but as excipients. However, an intrinsic effect cannot be excluded, e.g. triggering of allergies or skin irritations in the patient. Therefore, there is an obligation to declare them and whenever possible no preservatives should be used or at least as few as possible.

The pharmacopoeial chapters are explicit in pointing out that the addition of preservatives must not serve as a substitute for production according to good GMP practice. Furthermore, the effectiveness of preservation must be proven over the shelf life.

Thus, with product development the preservative system should be evaluated, for example, on at least three independent batches. The batches selected should be at the lower range of the preservative system concentration so that worst-case preservative concentrations are tested for effectiveness. Additionally, the antimicrobial efficacy test is performed during the shelf life of the product. Here, it is recommended that the test is performed at least at the beginning of the shelf life and at the end. Intermediate time points can be considered but in the authors' opinion are not needed. The company should retest or monitor the preservative system on batches put on shelf life studies (e.g. one batch a year is tested at the beginning and then at the end of its shelf life). The test is, however, not intended as a release test for batch release.

The test as such is well described in the Pharmacopoeia but unfortunately the acceptance criteria among the various Pharmacopoeia are different and therefore a preservative system might comply with the requirements for the United States but not for the EU. Furthermore, especially the interpretation of the results according to the Ph. Eur. can be tricky since two acceptance criteria (A and B) are described, figures for the log reduction are missing and there is no clear definition when growth is present.

Although some test organisms are defined in the pharmacopoeial chapters, further strains might be used depending on the product application or patient group (see e.g. Orth 2007).

Further information on preservation and its efficacy can be found in Orth (2007), Denyer (2007), van Doorne (2007), and Kramer and Assadian (2008).

There are several factors affecting microbial survival and proliferation in a product, these include:

- water activity (relative availability of water in a product for microbial growth)
- pH
- availability of nutrients
- amount of oxygen for aerobic bacteria
- antimicrobial activity of active ingredient or excipient
- inclusion of preservative components

When optimizing formulations, manufacturers can combine different factors to induce a synergistic antimicrobial effect.

The hurdle technology concept may be applied to optimize formulations and to fulfill the increasing consumer demand for preservative-free products. As described in PDA technical report No. 67, microorganisms present in a nonsterile product may be impeded by one or more hurdles (e.g. low pH and low water activity). Adjusting the number of hurdles and overall levels of the hurdles can enhance their effect.

For instance, as per the example in Figure 1.3, increasing the pH from neutral to basic range and lowering the product's water activity would further reduce the risk of microbial proliferation and could possibly reduce the amount or potency of preservatives needed.

Further guidance may be found in Chapter 2.

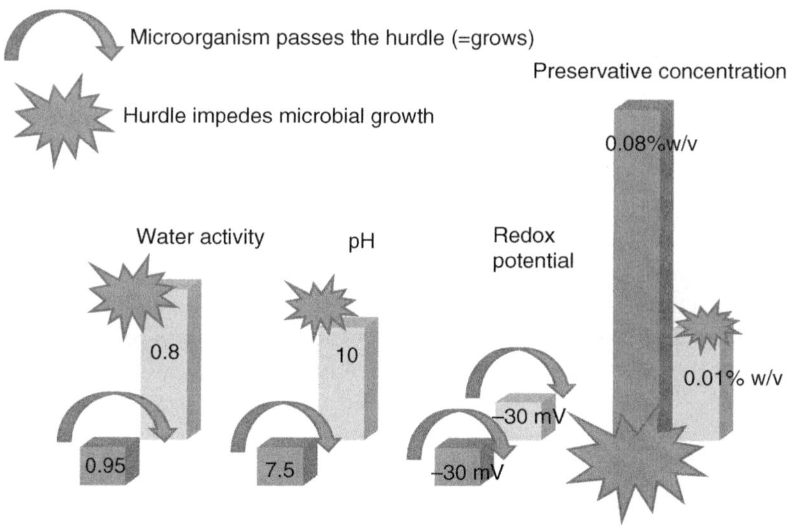

Figure 1.3 Fictive example of hurdle technology applied to a semisolid product. High hurdles reduce the level of microbial contamination thereby increasing drug product safety and quality.

1.4 Conclusion

This chapter gives an overview of the microbiological control strategy for the manufacture of non-sterile drug products or raw materials. The strategy can include the following factors:

- The **facility** with its cleanrooms and adjacent rooms used for production, personnel, materials used, cleaning and disinfection, pest control, and QA oversight. Design, training implementation, and control of these factors will influence the microbiological control of the whole manufacturing process.
- The **procedures** can be addressed by a risk assessment to evaluate and then control the contamination risk. In focus here are validation activities, in-process controls, hold time studies, and growth-promoting conditions.
- The **product ingredients** such as water, excipients, and drug substances need to be controlled to assure an impeccable final drug product. Along with this, the storage of materials (raw materials as well as equipment) should also be addressed.
- The **utilities** – mainly water but also compressed air or gases – used for the manufacturing should be microbiologically evaluated. Also, cleaning and disinfection agents can be a source of contamination.
- The **equipment** will introduce contaminations if not appropriately cleaned, dried, and stored; therefore, special attention should be given by validation approaches or monitoring control. Design and cleaning automation can substantially improve microbiological contamination risk from the equipment.
- The **formulation** of the drug product will have an influence on its quality. Using the hurdle principle as well as preservative factors – already during the development of the product – will facilitate the manufacture of a product with a low microbial contamination risk.

Not all controls described in this chapter need to be strictly applied for each and every type of non-sterile drug product. It is for the reader of this book to define their optimal microbial control strategy and it would depend on the outcome of the comprehensive risk assessment. Nonetheless, if adequate control procedures and monitoring programs are in place, this would provide assurance that the product is safe from a microbiological point of view.

Bibliography

ANVISA, 2013, *Quality Guide for Air Treatment and Environmental Monitoring Systems in the Pharmaceutical Industry*, N. A. o. H. Surveillance, Brasilia.

CFDA, Decree No. 79, 2010, *Good Manufacturing Practice for Drugs (2010 Revision)*, Ministry of Health, China Food and Drug Administration, Beijing.

Dao, H, Lakhani, P, Police, A, Kallakunta, V, Ajjarapu, SS, Wu, KW, Ponkshe, P, Repka, MA, Murthy, SN, 2018, 'Microbial stability of pharmaceutical and cosmetic products', *AAPS PharmSciTech*, vol. 19(1), pp. 60–78.

Denyer, SP, 2007, 'Antimicrobial preservatives and their properties', in Denyer, SP and Baird, RM (eds), *Guide to Microbiological Control in Pharmaceuticals and Medical Devices*, CRC Press, Taylor Francis Group, Boca Raton, 324–348.

EudraLex, 2010, *The Rules Governing Medicinal Products in the European Union, Volume 4, EU Guidelines for Good Manufacturing Practices for Medicinal Products for Human and Veterinary Use, Annex 10, Manufacture of Pressurised Metered Dose Aerosol Preparations for Inhalation*, European Commission, Brussels.

EudraLex, 2014a, *The Rules Governing Medicinal Products in the European Union, Volume 4, EU Guidelines for Good Manufacturing Practices for Medicinal Products for Human and Veterinary Use, Chapter 3: Premises and Equipment*, European Commission, Brussels.

EudraLex, 2014b, *The Rules Governing Medicinal Products in the European Union, Volume 4, EU Guidelines for Good Manufacturing Practices for Medicinal Products for Human and Veterinary Use, Chapter 5: Production*, European Commission, Brussels.

EudraLex, 2015, *The Rules Governing Medicinal Products in the European Union, Volume 4, EU Guidelines for Good Manufacturing Practices for Medicinal Products for Human and Veterinary Use, Annex 15: Qualification and Validation*, European Commission, Brussels.

EudraLex, 2017, *The Rules Governing Medicinal Products in the European Union, Volume 4, EU Guidelines to Good Manufacturing Practice Medicinal Products for Human and Veterinary Use; Annex 1 – Manufacture of Sterile Medicinal Products*, Draft Version, European Commission, Brussels.

FDA, 2004, *Guidance for Industry – Sterile Drug Products Produced by Aseptic Processing – Current Good Manufacturing Practice*, U.S. Department of Health and Human Services, Food and Drug Administration, Rockville, MD.

Goldschmidt, N, Farquharson, G, 2017, 'Understanding cleanliness classifications for life science facilities', *Pharmaceutical Engineering*, vol. 37(2), pp. 38–42.

Goverde, M, 2018, 'Mikrobiologisches monitoring: luft, oberflächen, personal', in Concept Heidelberg (ed), *Gute Hygiene Praxis: Hygiene – Reinigung – Monitoring*, 3rd ed., Editio Cantor Verlag, Heidelberg, 146–227.

Goverde, M, Roesti, D, 2018, *Guidance for Deviation Handling of Microbiological Environmental Monitoring Excursions in Non-sterile Pharmaceutical Manufacturing*, ECA Pharmaceutical Microbiological Working Group, Heidelberg.

ISO 11930, 2019, *Cosmetics – Microbiology – Evaluation of the Antimicrobial Protection of a Cosmetic Product*, International Organization for Standardization, Berlin.

ISO 14644-1, 2015, *Cleanrooms and Associated Controlled Environments – Part 1: Classification of Air Cleanliness by Particle Concentration*, International Organization for Standardization, Berlin.

ISPE, 2016, *Baseline Guide Vol 2: Oral Solid Dosage Forms*, International Society for Pharmaceutical Engineering, Tampa, FL.

ISPE, 2018, *Japan Affiliate: Pest Control Manual*, International Society for Pharmaceutical Engineering, Tampa, FL.

Jimenez, L, 2007, 'Microbial diversity in pharmaceutical product recalls and environments', *PDA Journal of Pharmaceutical Science and Technology*, vol. 61, pp. 383–399.

JP 17th Edition, Chapter <19>, 'Preservatives – Effectiveness Tests', Japanese Pharmacopoeia.

Kramer, A, Assadian, O, 2008, *Wallhäuser – Praxis der Sterilisation, Desinfektion, Antiseptik und Konservierung*, Georg Thieme Verlag, Berlin.

Mexicana NO. NORMA Oficial Mexicana NOM-059-SSA1-2013, *Buenas prácticas de fabricación de medicamentos*, Federal sobre Metrología y Normalización, el Subcomité de Insumos para la Salud 2013, Mexico.

Orth, DS, 2007, 'Principles of preservation', in Denyer, SP and Baird, RM (eds), *Guide to Microbiological Control in Pharmaceuticals and Medical Devices*, CRC Press, Taylor Francis Group, Boca Raton, 309–322.

PDA Technical Report No. 67, 2014, *Exclusion of Objectionable Microorganisms from Non-sterile Pharmaceutical and OTC Drug Products, Medical Devices and Cosmetics*, Parenteral Drug Association.

Ph. Eur. 9th Edition chapter 5.1.3, 'Efficiency of Antimicrobial Preparation', European Pharmacopoeia.

Rieth, M, 2017, *Pharmazeutische Mikrobiologie – Qualitätssicherung, Monitoring, Betriebshygiene*, 2nd ed., Wiley-VCH.

Rieth, M, Krämer, N, 2016, *Hygiene in der Arzneimittelproduktion*, Wiley-VCH.

Roesti, D, 2012, 'Objectionable microorganisms in non-sterile pharmaceutical drug products: risk assessment and origins of contamination', in Moldenhauer, J (ed), *Environmental Monitoring: A Comprehensive Handbook*, vol. 6, PDA/DHI, Washington, DC, pp. 137–158.

Sandle, T, 2012, *The CDC Handbook – A Guide to Cleaning & Disinfecting Cleanrooms*, Grosvenor House Publishing Limited, Tolworth, Surrey.

Seyfarth, H, 2002, 'Mikrobiologisches Monitoring. Teil 1: Notwendigkeit von Umgebungskontrollen/Raumklassifizierung', *La Pharmacie Industrielle*, vol. 64(1), pp.75–80.

Sutton, S, Jimenez, L, 2012, 'A review of reported recalls involving microbiological control 2004-2011 with emphasis on FDA considerations of "objectionable organisms"', *American Pharmaceutical Review*, vol. 15(1), pp. 42–57.

US 21 CFR 211.110, 2018, 'Sampling and testing of in-process materials and drug products', *Code of Federal Regulations*, Title 21, Food and Drug Administration.

US 21 CFR 211.42, 2018, 'Design and construction', *Code of Federal Regulations,* Title 21, Food and Drug Administration.

US 21 CFR 211.46, 2018, 'Ventilation, air filtration, air heating and cooling', *Code of Federal Regulations,* Title 21, Food and Drug Administration.

US 21 CFR 211.56, 2018, 'Sanitation', *Code of Federal Regulations,* Title 21, Food and Drug Administration.

US 21 CFR 211.82, 2018, 'Receipt and storage of untested components, drug product containers, and closures', *Code of Federal Regulations,* Title 21, Food and Drug Administration.

US 21 CFR 211.84, 2018, 'Testing and approval or rejection of components, drug product containers, and closures', *Code of Federal Regulations,* Title 21, Food and Drug Administration.

USP 41-NF 36 chapter <1072>, 2018, 'Disinfectants and Antiseptics', United States Pharmacopeia.

USP 41-NF 36 chapter <1111>, 2018, 'Microbiological Examination of Nonsterile Products: Acceptance Criteria for Pharmaceutical Preparations and Substances for Pharmaceutical Use', United States Pharmacopeia.

USP 41-NF 36 chapter <1115>, 2018, 'Bioburden Control of Nonsterile Drug Substances and Products', United States Pharmacopeia.

USP 41-NF 36 chapter <1116>, 2018, 'Microbiological Control and Monitoring of Aseptic Processing Environments', United States Pharmacopeia.

USP 41-NF 36 chapter <51>, 2018, 'Antimicrobial Effectiveness Testing', United States Pharmacopeia.

Van Doorne, H, 2007, 'Preservative stability and efficacy: formulation influences', in Denyer, SP and Baird, RM (eds), *Guide to Microbiological Control in Pharmaceuticals and Medical Devices,* CRC Press, Taylor Francis Group, Boca Raton, 349–366.

WHO, 2011, *WHO Expert Committee on Specifications for Pharmaceutical Preparations,* WHO Technical Report Series, vol. 961, World Health Organization, Geneva.

ZLG, 2010, *Aide Mémoire 07121105 – Inspektion von Qualifizierung und Validierung in pharmazeutischer Herstellung und Qualitätskontrolle,* Zentralstelle der Länder für Gesundheitsschutz bei Arzneimitteln und Medizinprodukten, Germany.

2

Microbial Contamination Risk Assessment in Non-sterile Drug Product Manufacturing and Risk Mitigation

Tony Cundell

Microbiological Consulting, LLC, Scarsdale, NY, USA

CONTENTS

Pharmaceutical Microbiological Quality Assurance and Control: Practical Guide for Non-Sterile Manufacturing, First Edition. Edited by David Roesti and Marcel Goverde.

2.1 Introduction

As stated in U.S. Pharmacopoeia (USP) General Informational Chapter <1115> *Bioburden Control in Non-sterile Drug Substances and Drug Products,* in terms of microbial contamination risk control, there are two broad categories of drug products: (i) sterile products, in which the bioburden is eliminated using validated sterilization and aseptic processes, and (ii) non-sterile products for which the final product bioburden is controlled to appropriate levels, as governed by the dosage form, based on product attributes, route of administration, and target patient population. Also, this distinction is found in the U.S. Federal Good Manufacturing Practices (GMPs) Regulations 21 CFR 211.113 *Control of Microbiological Contamination (a) and (b).* This book chapter will discuss different approaches to microbial risk assessment and mitigation in non-sterile product formulation, manufacturing process development, and routine manufacture, testing, and release. Sterile products are largely out-of-scope in this chapter and will only be discussed for comparative purposes.

2.2 Regulatory, Compendia, and Industry Guidance

What regulatory guidance is available to help non-sterile drug manufacturers control microbial risk? Key guidance documents available include the USP40/NF35 <1115> *Bioburden Control in Non-sterile Drug Substances and Drug Products,* International Conference on Harmonization (ICH) Q8 *Pharmaceutical Development,* ICH Q9 *Quality Risk Management: Annex 1 – Methods and Tools,* ICH Q10 *Pharmaceutical Quality Management,* and PDA Technical Report No. 54-5 *Quality Risk Management for the Design, Qualification and Operation of Manufacturing Systems.* Note: Another pertinent document is the industry practice PDA Technical Report No. 67 *Exclusion of Objectionable Microorganisms from Non-sterile Pharmaceutical and OTC Drug Products, Medical Devices and Cosmetics* that is discussed in Chapter 11. This chapter will examine these four guidance documents.

USP <1115> *Bioburden Control of Non-Sterile Drug Substances and Products* became official in the Second Supplement to USP37/NF32 on 1 December 2014. This informational chapter was written because the USP Microbiology Expert Committee saw a pressing need for the chapter, as no well-defined regulatory standards or guidance existed for the microbiological/bioburden control of non-sterile pharmaceutical manufacturing environments.

2.3 Putting into Perspective the Microbiological Risk Associated with Non-sterile Products

What Is Risk?

In general, people do not understand risk and take an emotional-driven approach to risk. For example, commercial airline passengers may fear a highly improbable plane crash, discounting the much greater danger of a fatal accident while driving to the airport.

The Dictionary of Contemporary English defines risk as the possibility that something bad, unpleasant, or dangerous may happen. Other more technical definitions are Risk = Frequency (event/time) × Severity or Magnitude (consequences/event) (Sandle 2013) or risk is the combination of the probability of occurrence of harm and the severity of the harm (ICH Q9 2005).

In terms of microbiology, it may be viewed that non-sterile products to some degree were the stepchildren of our industry. A review of industry practices showed that environmental control and monitoring of non-sterile manufacturing ranges from non-existent to programs parallel to those used in aseptic processing. Often, data generated from these latter programs are excessive for the control of the microbiological quality of non-sterile environments in which these products are manufactured, create unnecessary compliance barriers, delaying product release and increase costs.

So, how can we effectively apply the appropriate level of microbial risk control in the manufacture of non-sterile products? To paraphrase the USP chapter, microbial contamination in non-sterile products should be controlled to a level consistent with patient safety, but excessive controls that would add complexity and cost, without a commensurate safety benefit, are not helpful to either the end user of the non-sterile products or the pharmaceutical manufacturer. Therefore, a scientifically pragmatic approach to management of the microbial bioburden in non-sterile products requires consideration of patient risk and the contamination-control objectives required to achieve a practical level of risk management.

While there are many factors that can result in the introduction of microorganisms in non-sterile products, recent data on product failures and recalls indicate that the five factors are the most likely to result in product recalls due to higher than acceptable levels of microbiological content or the presence of objectionable microorganisms. In the opinion of the author, these manufacturing risk factors, in descending order of importance, are: (i) ingredient water; (ii) pharmaceutical ingredients; (iii) process equipment and associated utilities; (iv) manufacturing personnel; and (v) the manufacturing environment.

How can we mitigate these risk factors? In terms of mitigation, control is always superior to product testing. The author strongly believes that bioburden control in non-sterile drug products is achievable by implementing the following controls:

- Procuring pharmaceutical ingredients of high microbiological quality.
- Formulating robust non-sterile products with the lowest possible water activities and effective antimicrobial preservative systems that resist microbial contamination.
- Good bioburden control through sanitary equipment design; sound equipment cleaning; effective cleaning and disinfectant programs; utility management, especially water systems; and personnel hygiene.
- Emphasis on current Good Manufacturing Practices (cGMPs) compliance.
- Risk-based microbial specifications and testing programs.

Chapter 1 elaborates on these different controls.

As way as an example, what life-threatening risks are patients exposed to and how does this compare to the risk of microbial contamination of pharmaceutical drug products?

According to a 2014 U.S. National Vital Statistics Report, the top causes of death annually in the United States are as shown in Figure 2.1.

2.3.1 75 000 Deaths Annually Caused by Infectious Diseases!

As the topic of the chapter concerns the microbial risk associated with non-sterile drug products, we will look more closely at the area of infectious diseases. Estimated causes in descending order of the source of the infectious disease are as follows:

- Community-based infection (50 000 plus deaths)
- Hospital-acquired infection (20 000 plus deaths)
- Food-borne infection (3000 plus deaths)
- Sterile compounding derived infection (10 plus deaths)
- Pharmaceutical drug product derived infection (1 plus deaths)

Although the pharmaceutical industry should not be complacent about microbial contamination and the safety and efficacy of our products, *it is notable that contaminated pharmaceuticals do not make a significant contribution to the frequency of infectious disease.*

Nonetheless, although pharmaceuticals are generally manufactured as suitable for all patient populations, some patient populations have higher risks than the general populations (see Table 2.1) and this should be considered in any risk assessment.

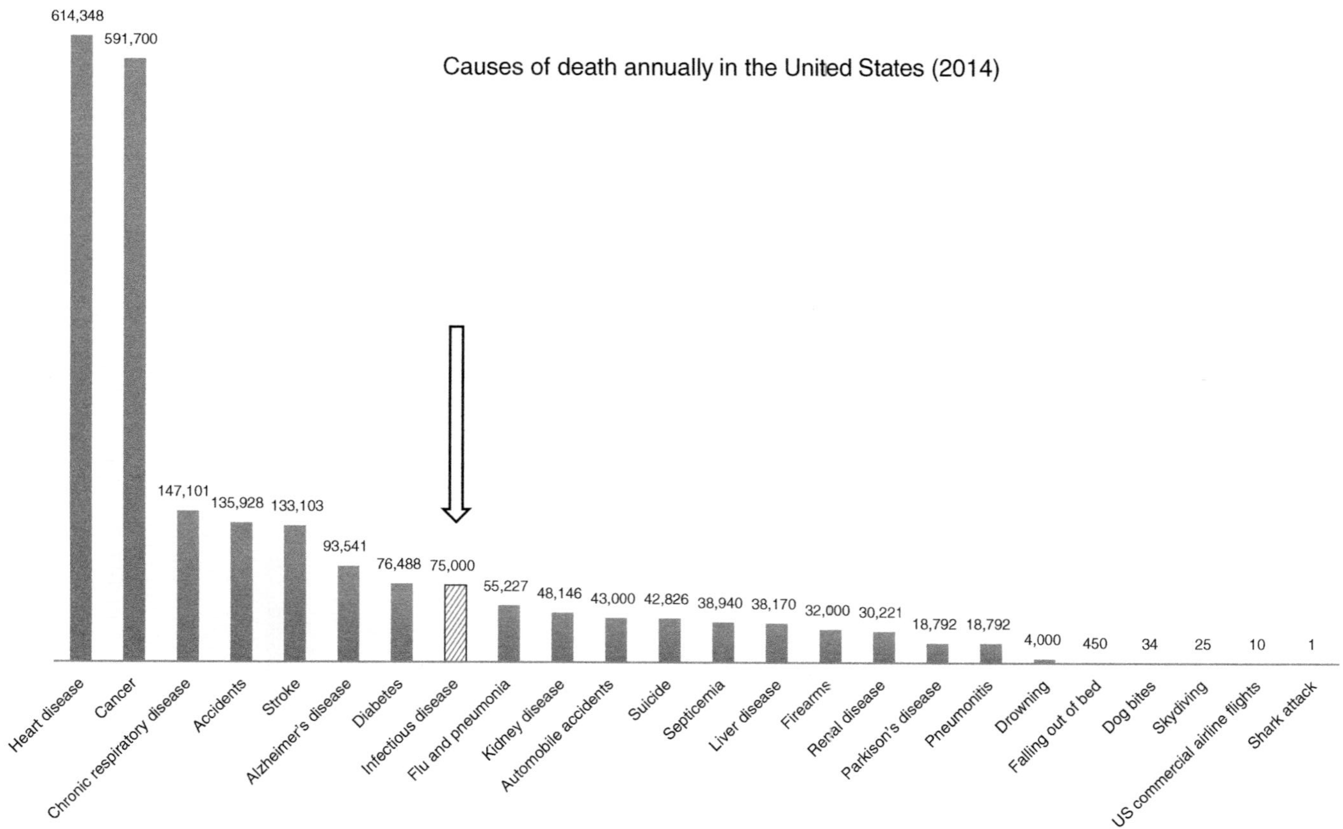

Figure 2.1 2014 U.S. National Vital Statistics Report, the top causes of death annually.

Table 2.1 Patient populations with higher risk of microbial infection.

Patient populations	Medical conditions	Microbiological risks
Chronic medical conditions	Diabetes, cystic fibrosis, alcoholic liver disease, etc.	High microbial counts; objectionable organisms for the dosage form and/or medical condition
Medications, e.g. Proton pump inhibitors (PPI)	Inhibition of gastric acid secretion by PPI	High microbial counts; vegetative bacteria survive the passage through the stomach
Chronic infectious diseases	AIDS and hepatitis	High microbial counts; objectionable organisms for the dosage form and/or medical condition
Substance abuse	Injectable drugs and excessive alcohol consumption	High microbial counts; objectionable organisms for the dosage form
Receiving immunosuppressive treatments	Chemotherapy for cancer, immunosuppression associated with organ transplantation, and corticosteroid use	High microbial counts; objectionable organisms for the dosage form
Special populations	New born, elderly, and pregnant	High microbial counts; objectionable organisms for the dosage form

2.3.2 Susceptibility of Different Patient Populations

Patient population may be classified as healthy, moderately impaired, and immunocompromised. The medical status of the recipients will influence the risks of infection. The number of immunocompromised patients receiving drug products is largely unknown. A recent letter to the Journal of the American Medical Association (JAMA) editors suggested that 4% of US adults self-reported that they have been told at one time by a health professional that they are immune-suppressed. Of these adults surveyed, 2.8% reported current immunosuppression (Harpaz *et al.* 2016).

To this microbiologist, the numbers were revealing and must be considered during risk assessments because microbial specifications do not make any distinction as to medical status for the recipient of a drug product. The default microbiological requirements in USP <1111> have stricter microbial enumeration requirements for more invasive drug products and the absence of specified microorganism requirements are broken down by dosage form. Pharmaceutical product specifically directed toward higher risk patient population may have stricter objectionable microorganism exclusion requirements.

The following may cause immunosuppression, a weakening of the immune system (Lederberg *et al.* 1992):

- Inherited or acquired diseases
- Aging
- Prematurity (neonates)
- HIV infection
- Radiation treatment
- Immunosuppressive mediations for transplantation, chemotherapy, or treatment of autoimmune disease
- Malnutrition
- Pregnancy
- Severe trauma and burns
- Other concurrent infections

2.3.3 Frequency of Drug Product Recall

Recall histories are a useful source of microbial contamination of non-sterile drug products. Based on a survey of US product recalls of non-sterile products published by Sutton and Jimenez (2012), there are around 15–20 recalls annually. Perhaps surprisingly, the major reason for the recalls is the presence of objectionable microorganisms in these products (72%), not exceeding the microbial limit. During a 7-year period from 2004 to 2011, 141 non-sterile drug products were subject to voluntary recall. The recalls by product types were over-the-counter (OTC) drug products (42%), cosmetics and soaps (31%), medical devices (14%), dietary supplements and probiotics (8%), and pharmaceuticals (5%). Because of the strict adherence to cGMPs, it may not be unexpected that pharmaceuticals have the least number of recalls among these product types.

Analysis of the probable cause of the microbial contamination of the non-sterile products by the author suggest that they are in descending order: (i) microbial contamination of water for pharmaceutical purposes; (ii) microbial contamination during the manufacturing process; (iii) failures of preservative systems to protect multiple-use liquid products; (iv) the use of pharmaceutical ingredients with high microbial counts; and (v) improper storage of the products during their shelf life.

The objectionable microorganisms most commonly cited in product recalls are the members of the *Burkholderia cepacia complex*. This topic will be discussed in Chapter 11.

2.4 Risk Assessment Tools

Common risk assessment tools that can be applied in the pharmaceutical industry include Failure Mode and Effects Analysis (FMEA), Failure Mode,

Effects, and Criticality Analysis (FMECA), Fault Tree Analysis (FTA), Hazard Analysis and Critical Control Points (HACCP), Hazard Operability Analysis (HAZOP), and Preliminary Hazard Analysis (PHA).

A key document used by the pharmaceutical industry is ICH Q9: *Pharmaceutical Quality Risk Management* (QRM) *Benefits and Challenges.* QRM can provide a framework for the microbial contamination risk assessment recommending tools like FMEA, FTA, HACCP, cause and effect diagrams, and other statistical tools.

2.4.1 Impact Matrix

A possible approach summarized in Table 2.2 recommended by the author for pharmaceutical manufacturing risk analysis is as follows:

1) Establish three levels of impact (high, medium, and low)
2) Describe the consequences in terms of drug product quality:
   - High – Protracted investigations, chronic product loss, rework, infection outbreaks, product recall, and/or regulatory notification
   - Medium – Investigations, additional testing, and delay to product release
   - Low – Investigations based on available data with minimal delay in product release
3) Describe the consequences in terms of production scheduling, manufacturing, and product release:
   - High – Product deletion to up to three months production delay
   - Medium – Up to one month production delay
   - Low – Up to one week production delay
4) Estimate the frequency of occurrence of product failure:
   - High – Annually to 1 every 10 campaigns
   - Medium – Between 1 and 5 years or between 10 and 50 campaigns
   - Low – 1 event every 5 years or every 51 campaigns
5) Establish the probability to detect product failure:
   - High – If failure occurs it would not be detected until after product release or an intermediate is further processed

Table 2.2 A simple impact matrix for consequences and likelihood of risk.

Risk assessment	Likelihood		
Consequences	1	2	3
3	Medium	Medium	High
2	Low	Medium	Medium
1	Very low	Low	Medium

- Medium – If failure occurs it may be detected by secondary monitoring prior to lot release
- Low – If failure occurs it is readily detected by routine primary monitoring and corrected

2.4.2 Failure Mode and Effects Analysis (FMEA)

FMEA is a risk assessment tool frequently applied in the pharmaceutical industry. The simplest approach is to score (from 1 to 3) in each of the following categories:

- Severity (*S*)
- Frequency of occurrence (*O*)
- Ease of detection (*D*)

Using these criteria, a final FMEA score is the sum of:

$$\text{Severity score} \times \text{Occurrence score} \times \text{Detection score}$$

or

$$\text{RPN} = S \times O \times D$$

where RPN is the risk priority number.

A good example of the application of this tool would be a simple risk assessment of pharmaceutical excipients in terms of their origin. This approach would score animal-derived raw materials as having the highest risk level (18) in terms of both bacterial counts and pathogen presence while synthetic materials would have the lowest risk (2). See Table 2.3 for details.

2.4.3 Hazard Analysis and Critical Control Points (HACCP)

HACCP was first developed to prevent foodborne infection in astronauts by the U.S. National Aeronautics and Space Administration (NASA), the giant

Table 2.3 Risk analysis for excipient based on manufacturing process.

Manufacturing process of excipient	*S*	*O*	*D*	$S \times O \times D$
Synthetic material	2	1	1	2
Semisynthetic material	2	1	2	4
Mineral-derived material	2	2	2	8
Plant-derived material	3	2	2	12
Animal-derived material	3	3	2	18

food company Pillsbury, and U.S. Army Natick Center. This program is now widely used in the food industry to mitigate risk and has applicability to the pharmaceutical industry. HACCP application in our industry is described in ICH Q8 (R2) Pharmaceutical Development.

The seven principles/steps used in HACCP analysis are:

1) Identifying hazards and assessing their severity.
2) Determining the Critical Control Points (CCPs).
3) Establishing control limits.
4) Establishing system to monitor and control CCPs.
5) Establishing corrective actions when a CPP is not under control.
6) Establishing procedures to verify the HACCP system is effectively working.
7) Establishing a documentation and reporting system.

2.4.3.1 Application of HACPP to Tablet Manufacturing

As solid oral dosage forms still comprise around 80% of the pharmaceutical drug products sold, this section will discuss the application of HACCP to tablet manufacturing. The processing steps for the manufacture of a representative film-coated compressed tablet are:

- Procurement of pharmaceutical ingredients
- Warehousing pharmaceutical ingredients
- Batching of the pharmaceutical ingredients
- Blending
- Ingredient water production
- Wet granulation and milling
- Fluid bed drying
- Tablet compression
- Tablet coating
- Packaging
- Distribution

In general, the most critical processing steps in compressed tablet manufacturing with respect to potential microbial contamination are the procurement of pharmaceutical ingredients, ingredient water production, wet granulation and milling, and tablet coating. For example, the holding time of aqueous film coating solutions may be a critical control point due to the ability of bacteria to grow in the solutions that may contain gelatin or other bacterial nutrients. In contrast, fluid bed drying and compression are potentially bioburden-reduction steps due to physical and thermal stress on microorganisms. Refer to examples in Tables 2.4–2.6.

Table 2.4 Procurement of pharmaceutical ingredients.

Manufacturing step	Contamination potential	Preventative measures/CCP	Remarks
Procurement of pharmaceutical ingredients	Low to moderate	Supplier audits Specifying compendial-grade materials Monitoring higher risk materials for microbial counts and absence of pathogens (CCP)	Implement quality agreements with suppliers. Risk analysis based on starting material, ingredient manufacturing process, contribution to the product, and testing histories

Table 2.5 Wet granulation and milling.

Manufacturing step	Contamination potential	Preventative measures/CCP	Remarks
Wet granulation and milling	Moderate	Equipment design Cleaning validation Monitoring purified water used in equipment cleaning and granulation solutions for microbial counts (CCP) Granulation holding time (CCP)	Emphasis on water system and cleaning validation to prevent microbial contamination Water activity measurement may be used to evaluate the ability of the granulation to support microbial growth

Table 2.6 Film coating.

Manufacturing step	Contamination potential	Preventative measures/critical control points	Remarks
Tablet coating	Moderate to high	Equipment cleaning and coating solution holding time (CCP) Incoming microbial testing of ingredients (CCP) Monitor purified water used in coating solutions for microbial counts (CCP)	Water-based coating solutions will support microbial growth. Holding times need to be justified.

How to Set and Justify Holding Times?

Holding time is the time during which a product bulk, intermediate, or ingredient solution can be left to stand without a microbiological risk in the manufacturing process. The microbiological holding times should be defined from the start of a process step where microbial proliferation would occur (e.g. mixing in an aqueous diluent) until the microbial reducing step (e.g. fluid bed drying).

A holding time is first defined with regards to the antimicrobial or growth promotion property of the solution either:

- By performing a microbiological challenge test consisting of inoculating samples of the product with test microorganisms, storing the inoculated samples in the laboratory under defined temperature conditions, and enumerating the microorganisms after a defined storage period. The growth of one or more test organisms within the testing period is defined as an increase of the microbial counts, which is higher than 0.5 $\log_{10}$ units as compared to the start of the experiment (time 0 = spike in product). For example, of an aqueous intermediate, if under standard storage conditions no increase in the microbial count has been observed until and including 12 hours test time, whereas after 24 hours the microbial count is higher than the initial count (growth promotion effect with an increase of the microbial count >0.5 $\log_{10}$ unit), then the maximum holding time at room temperature for the intermediate product is 12 hours.
- Based on a scientific rationale or based on operational requirements without performing a microbiological challenge test.

Once a holding time is set either experimentally or based on a rationale, some companies choose to validate their holding times by manufacturing a minimum one batch and either:

- Testing the solution or product at the end of the holding time, which must comply with defined microbiological requirements.
- Testing the solution or product at the beginning and end of the holding time and confirming that there is no microbial growth. For instance, the number of microorganisms at the end of the holding time should not be more than 0.5 $\log_{10}$ microorganisms from the start of the holding time.

Fluid bed drying has lower risks than the traditional tray drying process because of the extended dry period with tray drying. The mechanism of bioburden reduction in a fluid-bed drying process is: (i) thermal decay and (ii) desiccation (dehydration). The resistance of microorganisms to these stresses are Gram-positive bacteria > yeast and mold ≫ Gram-negative bacteria (Fu and Chen 2006). The tablet compression process, due to the localized temperature and pressure generated, significantly reduces the number of vegetative microorganisms within a blend (Chesworth *et al.* 1977; Blair *et al.* 1991).

2.5 Organizational Risk Management Maturity

It is often informative to assess the maturity of your organization in terms of risk management (Table 2.7). Unfortunately, many pharmaceutical companies are only at the awareness stage of the matrix with employees often being rewarded for their role of fire fighters instead of being true champions of risk management. Obviously this must change, if we want to control risk in the pharmaceutical industry. Although hiring managers and individual contributors with experience in risk management is useful, more important is a cultural change throughout the organization and at every level in the organization. Based on the author's experience in both operations and product development, for risk mitigation emphasis should be placed on formulation and manufacturing process development to ensure the most robust drug products come to the market.

2.6 Hierarchy of Risks

2.6.1 Hierarchy of Risk by Pharmaceutical Ingredient

As described in Cundell (2005), the microbiological quality of the pharmaceutical excipients used to manufacture pharmaceutical and OTC drug products may significantly affect the outcome of individual processing steps and the

Table 2.7 Organizational risk management maturity level.

Risk maturity level	Risk processes	Attitude	Behavior	Skill and knowledge
Skepticism	No formal processes	"Accidents will happen"	"Fear of blame" culture	Unconscious incompetence
Awareness	Isolated use of stand-alone processes	Suspended belief	Reactive, "firefighting"	Conscious incompetence
Understanding and application	Extended use of combined processes	Passive acceptance	Compliance thinking	Conscious competence
Embedding and integration	Risk management embedded in the business	Active engagement	Risk-based decision making	Built-in or internalized competence
Robust risk management	Frequent risk review and improvement	Champion	Innovative and appropriate risk management	Expert

Source: Adapted from Long (2013).

microbiological attributes of the final drug products. Unlike active pharmaceutical ingredients (drug substances) that are manufactured in compliance to GMPs, excipients are purchased from multiple suppliers and in many cases are produced for the food, cosmetics, consumer products, photographic, and paint industries and not specifically for the pharmaceutical industry, so the management of their microbiological quality is less straightforward.

Pharmaceutical ingredients (excipients), especially those of plant and animal origin, and a lesser degree mineral origin, used in product formulation can be a significant source of microbial contaminants and, in the opinion of the author, are the second leading cause of product recalls for microbial contamination after ingredient water. Vendor audits, specifications, testing, package selection, shipping, storage conditions, and expiry dates are all critical in the microbial risk reduction. When manufacturers cannot conduct extensive vendor audits, they should select vendors who have submitted a Drug Master File (DMF) to a competent regulatory authority, implement cGMPs, and comply with USP/NF monographs. Materials that have low water activity, possess high or low pH, are not of natural origin, are inherently antimicrobial, or contain an antimicrobial preservative have a low risk for microbial colonization or proliferation.

To complicate it further, a pharmaceutical ingredient or excipients may be used in a range of pharmaceutical dosage forms, both sterile or non-sterile, or limited to a single dosage form. The material may have a plant, animal, or mineral origin or be chemically synthesized. The starting material may be minimally to extensively processed or chemically modified from a plant, animal, or mineral starting material or produced solely by chemical synthesis. These differences result in different levels of microbial contamination risk.

The origin of the starting material and the manufacturing process will significantly influence the microbiological quality of the excipients. In general, the classification of excipients is by their function in the drug product formulation.

- Binders
- Disintegrants
- Fillers (diluents)
- Lubricants
- Glidants (flow enhancers)
- Compression aids
- Colors
- Sweeteners
- Caking agents
- Buffers
- Preservatives
- Suspending/dispersing agents
- Film formers/coatings
- Flavors
- Printing inks

The different roles of an excipient in a non-sterile product are illustrated in Table 2.8.

What is the relative risk associated with different starting materials and excipients manufacturing processes? As stated earlier, pharmaceutical ingredients may be classified as synthetic, semisynthetic, and derived from plant, animal, or mineral materials.

For excipients, the hierarchy of risk is illustrated in Figure 2.2.

Within this classification, the material may be from slightly to highly processed, with the latter significantly mitigating the risk of microbial contamination (Table 2.9). Details of the starting materials and manufacturing processes of individual excipients may be conveniently obtained from the Handbook of Pharmaceutical Excipients.

Risk assessments can be conducted to determine the potential impact of excipients and measures taken to significantly reduce or even eliminate that risk.

Table 2.8 Common excipients used in compressed tablets.

Excipient classification	Common excipients
Diluent or filler	Lactose, Sucrose, Kaolin, Dibasic calcium phosphate, Calcium sulfate, and Calcium carbonate
Binders	Water, Alcohol, Starch paste, Gelatin solutions, Tragacanth, Sodium alginate, Carboxymethyl cellulose, Polyethylene glycol, and Povidone
Lubricants	Magnesium stearate, Calcium stearate, Talc, Stearic acid, Starch, Mineral oil, Sodium chloride, Sodium benzoate, and Carbowax 4000 or 6000
Disintegrating agents	Corn starch, Methylcellulose, Sodium carboxymethyl cellulose, Alginic acid, Microcrystalline cellulose, and Gums
Sweeting agents	Mannitol, Lactose, Sorbitol, Fructose, Saccharine, and Aspartame
Film coatings	Gelatin and colorants

Figure 2.2 Hierarchy of risk by excipients.

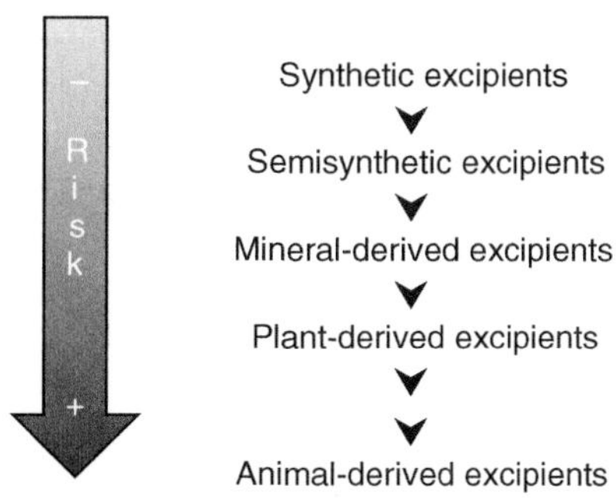

Table 2.9 Classification of excipients by starting material and manufacturing process.

Standing material and manufacturing process	Excipient examples
Synthetic	Povidone (polyvinyl pyrrolidone); Crospovidone (a homopolymer of cross-linked *N*-vinyl-2 pyrrolidone)
Semisynthetic	Captisol (chemically modified cylodextrin); Hydroxypropyl methylcellulose (chemically modified cellulose)
Mineral-derived	Talc (extractive); Dibasic calcium phosphate (processed chemically from a mineral calcium carbonate and phosphoric acid)
Plant-derived	Corn starch; Microcrystalline cellulose; Sucrose
Animal-derived	Lactose (extractive); Magnesium stearate (processed chemically from tallow); Gelatin (purified from bone and hide)

The impact of a single excipient on the microbial content of a drug product will depend on the size of the contribution that the excipient makes to the product formulation, the manufacturing processes, the ability of the product to support microbial growth, and the invasiveness of the route of administration of the drug product.

Clearly, injectable products because of their invasiveness must be sterile while tablets and capsules, in the absence of food-borne pathogens, may contain a moderate bioburden without impacting the recipient of the drug product. Concomitantly, pharmaceutical ingredients contaminated with high numbers of microorganisms or low levels of objectionable microorganisms will impact nasal sprays, vaginal products, topical products, and oral liquids more than tablets and capsules.

A plant, animal, or mineral origin may result in higher microbial counts and the presence of potential pathogens in excipients. Plant and animal materials may be contaminated with fecal matter during cultivation, harvesting, and/or processing. However, animal-derived materials that are subject to significant processing like gelatin may not constitute a microbial risk. This is also true for plant-derived cellulose and cellulose derivatives.

What cGMP regulations apply to the manufacturers of pharmaceutical excipients? Perhaps surprisingly, regulatory agencies are not directly involved in the inspection of excipients manufacturers. Although the cGMP regulations contained in 21 CFR Parts 210 and 211 apply only to finished drug products, Section 501(a) (2)(B) of the U.S. Federal FD&C Act requires that all drug products be manufactured, processed, packaged, and held according to cGMPs with no distinction made between pharmaceutical ingredients and finished products. Excipient manufacturers may file DMF with the U.S. Food and Drug Administration and comply with USP/NF excipient monographs.

Based on the initiative of the Institute of Quality Assurance (IQA), Pharmaceutical Quality Group, and the International Pharmaceutical Excipients Council (IPEC), International Standards for Excipients were published as ISO 9001:2000 *Good Manufacturing Practices Guide for Bulk Pharmaceutical Ingredients*. The standard emphasizes documentation, batch traceability, change control and customer notification, and chemical contamination control. This was subsequently published as the USP general informational chapter <1078> *Good Manufacturing Practices for Bulk Excipients – General Guidance.*

Excipient suppliers should be selected based on their reputation for manufacturing and supplying high-quality pharmaceutical excipients that meet all industry standards. The suppliers should be qualified by audit against a recognized standard such as the 2001 IPEC *Good Manufacturing Practices Guide for Bulk Pharmaceutical Excipients.* Pharmaceutical companies should avoid purchasing excipients from distributors, when the actual excipient manufacturer is unknown, as market forces will determine the quality of the excipients. In general, low prices are reflected by poor quality. In addition to a business contract detailing material grade, price, and delivery schedules, a Quality Agreement should be signed by the excipient supplier and the pharmaceutical company that would emphasis notification of manufacturing changes.

Whenever available, purchase USP-grade excipients. The microbial testing methods for pharmaceutical ingredients are defined in USP Chapter <61> *Microbiological Examination of Non-sterile Products: Microbial Enumeration Tests* and <62> *Microbiological Examination of Non-sterile Products: Tests for Specified Microorganisms.* The microbial enumeration tests are the Total Aerobic Microbial Count (TAMC) and the Total Combined Yeast and Mold Count (TCYMC).

Some USP/NF excipient monographs only list a TAMC requirement. Whereas non-sterile pharmaceutical products are screened using the USP Test for the Absence of Specified Microorganisms appropriate for the dosage form, excipients are typically screened only for the absence of *Escherichia coli*, and less frequently for *Salmonella* spp. if they are unprocessed and have an animal, plant, or mineral origin. USP Chapter <1111> and Ph. Eur. 5.1.4 both named *Microbiological Quality of Non-sterile Pharmaceutical Products* contains the acceptance criteria for the microbial quality of non-sterile dosage forms and pharmaceutical ingredients.

In addition, it is a cGMP requirement to exclude objectionable microorganisms from non-sterile pharmaceutical products. The pertinent sections are 21 CFR 211.113 *Control of microbiological contamination* (a) *Appropriate written procedures designed to prevent objectionable microorganisms in drug products not required to be sterile shall be established and followed.* Furthermore, 21 CFR 211.165 Testing and release for distribution (b) *There shall be appropriate laboratory testing, as necessary, of each batch of drug product required to be free of objectionable microorganisms that may cause infection when given by the route of administration of the drug product and/or cause physicochemical deterioration to the product.*

These regulations emphasize drug products and not excipients. Pharmaceutical manufacturers will experience push back if they attempt to return an excipient because it contains an objectionable microorganism not referenced in a USP/NF monograph.

The question may be asked whether the microbial requirements in USP/NF excipient monographs are technically sound and are risk-based specifications? Sadly they are not. The microbial requirements were never systematically developed, reflect what specifications were included in regulatory submissions, in many cases do not reflect the recommendations in USP <1111>, and are often locked in place by the compendial harmonization process. To illustrate this point, Table 2.10 contains the microbiological requirements for 17 common excipients used in the manufacturing of compressed tablets.

Table 2.10 USP/NF microbial limit requirements of commonly used excipients.

Excipient	Classification	Current compendial testing requirement
Acacia, NF	Processed plant material	Absence of *Salmonella* species
Alginic Acid, NF	Highly purified plant material	TAMC NMT 200 CFU/g; absence of *Salmonella* spp. and *Escherichia coli*
Dibasic calcium phosphate, USP	Processed mineral	None
Croscarmellose sodium, NF	Highly processed plant material	TAMC NMT 1000 CFU/g, TYMC NMT 100 CFU/g; absence of *E. coli*
Crospovidone, NF	Synthetic	None
Gelatin, NF	Processed animal material	TAMC NMT 1000 CFU/g; absence of *Salmonella* spp. and *E. coli*
Lactose monohydrate, NF	Processed animal material	TAMC NMT 100 CFU/g; TYMC NMT 50 CFU/g; absence of *E. coli*
Magnesium stearate, NF	Highly processed animal material	TAMC NMT 1000 CFU/g; TYMC NMT 500 CFU/g; absence of *E. coli* and *Salmonella* spp.
Microcrystalline cellulose, NF	Processed plant material	TAMC NMT 1000 CFU/g; TYMC NMT 100 CFU/g; absence of *Staphylococcus aureus*, *Pseudomonas aeruginosa*, *Salmonella* spp. and *E. coli*
Povidone, USP	Synthetic	None
Sodium starch glycolate, NF	Processed plant material	Absence of *E. coli* and *Salmonella* spp.

Table 2.1 (Continued)

Excipient	Classification	Current compendial testing requirement
Corn starch, NF	Processed plant material	TAMC NMT 1000 CFU/g, TYMC NMT 100 CFU/g; absence of *S. aureus* and *P. aeruginosa* (only when used as an absorbent dusting powder), *E. coli* and *Salmonella* spp.
Starch pre-gelatinized, NF	Processed plant material	TAMC NMT 1000 CFU/g, TYMC NMT 1000 CFU/g; absence of *Salmonella* spp. and *E. coli*
Sucrose, NF	Processed plant material	None
Talc, USP	Extractive	TAMC NMT 100 CFU/g; TYMC NMT 50 CFU/g (intended use for topical administration); TAMC NMT 1000 CFU/g and TYMC NMT 100 CFU/g (intended use for oral administration)
Xanthan, NF	Fermentation	Absence of *Salmonella* spp. and *E. coli*

NMT, not more than, and absence of specified microorganisms in 1 g or ml; TAMC, Total Aerobic Microbial Count; TYMC, Total Combined Yeast and Mold Count.

Some Thoughts on Magnesium Stearate, NF Compendial Microbiological Acceptance Criteria

According to the Handbook of Pharmaceutical Excipients, magnesium stearate, NF is prepared either by the interaction of aqueous solutions of magnesium chloride with sodium stearate or by the interaction of magnesium oxide, hydroxide, or carbonate with stearic acid at elevated temperatures. The starting material stearic acid may be derived from animal fat or cottonseed oil. In the former, stearic acid is manufactured by hydrolysis of fat, which is the continuous exposure to a countercurrent stream of high-temperature water in a high-pressure chamber. The resultant mixture is purified by vacuum steam distillation and the distillates are then separated using selective solvents. Stearic acid may also be manufactured by the hydrogenation of cottonseed and other vegetable oils; by the hydrogenation and subsequent saponification of olein followed by recrystallization from alcohol; and from edible fats and oils by boiling with sodium hydroxide, separating any glycerin, and decomposing the resulting soap with sulfuric or hydrochloric acid. The stearic acid is then subsequently separated from any oleic acid by cold expression.

Based on the review of the manufacturing processes for magnesium stearate, as described above, it is highly unlikely that this excipient will contain a bioburden apart from low levels of molds since magnesium stearate is hydrophobic and will not be susceptible to microbial growth in storage due to the low water *activity (aw = 0.49). Generally, according to USP <1112>, no microbial growth will occur in pharmaceutical excipients at water activities lower than 0.75. Based on these properties and the manufacturing process described above, it is unlikely that magnesium stearate would be contaminated with a Salmonella species and/or E. coli.*

What contribution does this excipient make to the formulation? The level of the lubricant magnesium stearate contained in compressed tablets ranges from 0.5 to 2.5%, so it is a relatively minor constituent in a compressed tablet. Based on the manufacturing process, low water activity, testing history, and low contribution to the dosage form, the author recommends that the microbial limit requirements for magnesium stearate and other pharmaceutical excipients be subject to review and, as justified, as in the case of magnesium stearate, NF, eliminated or modified.

Finally, additional explanatory is provided for excipients testing and risk categorization in Chapter 5.

2.6.2 Hierarchy of Risk by Dosage Form and Processing Steps

A hierarchy of microbial infection risk can be established based solely on the invasiveness of the non-sterile pharmaceutical dosage form. For an infection to occur, the drug product must overcome the natural barriers to microbial infection such as the skin, stomach, and mucosal barriers. The increasing invasiveness would range from solid oral dosage forms that are exposed to gastric and upper intestinal excretions, topical products applied to the skin, to inhalation products that are delivered into the lower respiratory tract. In addition, the drug product attributes that support microbial growth and the medical status of the recipients will influence the risks. Note: Sterile injectable products have the greatest risk of microbial infection due to their parenteral administration but they are outside the scope of this chapter.

The following list (Figure 2.3) provides a hierarchy of broad categories of non-sterile pharmaceutical products with respect to potential risk of microbiological infection to the patient based on the invasiveness of the route of administration, from low to high (after *USP <1115>*).

2.6.2.1 Processing Steps

An individual processing step can potentially increase or decrease the bioburden level of a non-sterile drug product. The unit processing steps for different dosage forms in descending order of microbial contamination risk are described in Table 2.11.

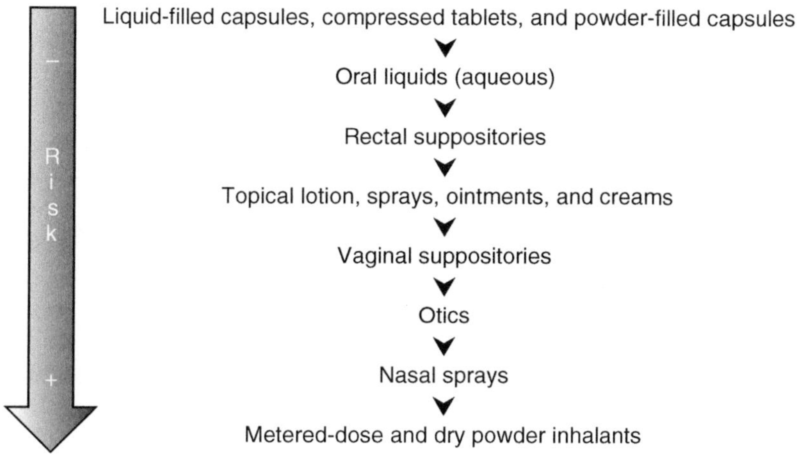

Figure 2.3 Hierarchy of risk by dosage form.

Table 2.11 Summary of unit processing steps in descending order of risk for different dosage forms.

Dosage forms	Unit processing steps in descending order of risk
Compressed tablets	Film coating, wet granulation, tray drying, milling, blending, batching, fluid bed drying, dry granulation, packaging, and compression
Oral liquids	Bulk storage, ingredient water production, mixing, batching, and filling
Lotions, ointments, and creams	Bulk storage, ingredient water production, mixing, batching, and filling
Vaginal and rectal suppositories	Bulk storage, ingredient water production, mixing, batching, and hot filling
Nasal sprays	Bulk storage, ingredient water production, mixing, batching, and filling
Powder dose inhalers	Filling, bulk storage, aerosol storage, mixing, and batching

2.6.2.2 Risk Associated with Different Processing Steps

It is useful to review the microbial risk associated with different manufacturing processes (Table 2.12). Processes associated with sterile product manufacturing are included in the table for comparative purposes only.

Table 2.12 Microbial contamination risk based on dosage form and unit manufacturing operation.

Dosage form	Unit manufacturing operation	Risk rating	Overall risk rating
Biological injectables	Bioreactor – sterilization, inoculum production, and biofermentation	5	4–5
	Recovery and downstream purification	3	
	Cell bank maintenance and viral clearance	5	
	Sterile filtration, aseptic filling, lyophilization, and stoppering and capping	4	
Pharmaceutical injectables	Aseptic filling: traditional cleanrooms	5	4
	Aseptic filling: form-fill-seal or isolator system	3	
	Terminal sterilization	1	
Sterile inhalation solutions	Aseptic filling: traditional cleanrooms	5	4
	Aseptic filling: form-fill-seal or isolator system	3	
	Terminal sterilization	1	
Metered dose Inhalants	Micronization	3	3
	Blending	2	
	Filling/Assembling	1	
Topical liquids, lotions, or creams	Mixing and blending	2	2
	Bulk storage	2	
	Filling	2	
Topical gels or ointments	Emulsification, blending, heating, and Cooling	2	2
	Dispensing	1	
Oral liquids or suspensions	Mixing and blending	2	2
	Filling	1	
Transdermal patches	Dispensing and coating	3	2
	Extrusion	2	
	Coating and drying	1	
	Packaging	1	
Liquid- or powder-filled capsules	Granulation: wet and dry, milling and blending	2	1
	Drying, encapsulation, and packaging	1	

Table 2.12 (Continued)

Dosage form	Unit manufacturing operation	Risk rating	Overall risk rating
Compressed tablets	Granulation: wet and dry, milling and blending	2	1
	Drying, compression, and packaging	1	
	Film coating	2	

Key: 1, little or no risk; 2, little to moderate risk; 3, moderate risk; 4, moderate to high risk; 5, high risk of microbial contamination.

Special Considerations on the Dosage Regime

Usually not considered in microbial risk assessments are the quantity of the dosage form and the duration of the therapy. It is frequently not understood by chemists, pharmacists, and even quality control personnel that microbial testing is based on drug product weight or volume and not potency. For example, the TAMC is measured in colony-forming units per g or ml of the drug product. The weight or volume range of a non-sterile oral dosage may span one to two magnitudes so that the microbial specifications may not directly reflect the microbial challenge to the patient.

Furthermore, the dosage regime may be a short course of treatment such as a five-day antibiotic treatment for an upper respiratory infection versus long-term patient treatment for a chronic illness such as elevated blood pressure or diabetes.

2.6.3 Hierarchy of Risk by Utility System

The contribution of utility system to microbial contamination is depicted in Figure 2.4.

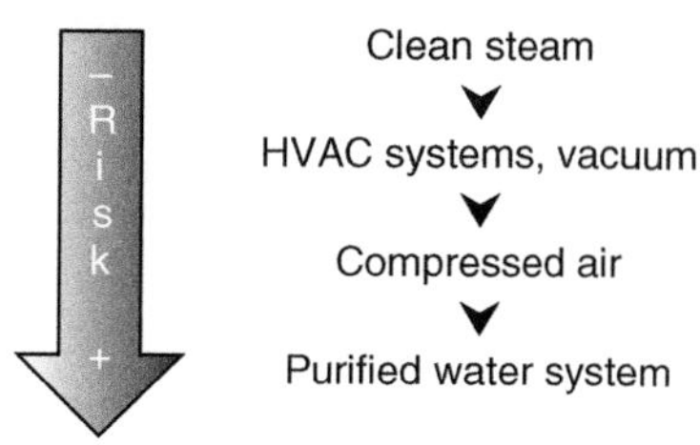

Figure 2.4 Hierarchy of risk by utilities.

2.7 Effect of Product Attributes

2.7.1 What Are the Critical Quality Attributes for a Pharmaceutical Drug?

It is by definition a quality attribute that must be controlled within predefined limits to ensure that the drug product meets its intended safety, efficacy, stability, and performance. The Critical Quality Microbiological Attributes for a non-sterile drug product includes the following:

- Microbial Count, i.e. TAMC and TCYMC (USP <61> and <62>)
- Absence of Specified Microorganisms as mandated by the dosage form as found in USP <1111>
- Absence of Objectionable Microorganisms (21 CFR 211.113)
- Antimicrobial Effectiveness as found in USP <51> (product development only)
- Water activity (product development only)
- Container-closure integrity (product development only)

2.7.2 Role of Formulation in Bioburden Control

The formulation of a non-sterile pharmaceutical drug product will have a profound effect on the ability of microorganisms to survival or growth in a product. The important parameters for bioburden control are pH, osmolality, water activity, redox potential, salinity, alcohol content, and antimicrobial preservative systems.

The reader is referred to a recent review article on the microbial stability of pharmaceuticals and cosmetics for the discussion of preservative systems (Doa *et al.* 2018).

2.7.3 Hurdle Technology Concept

A concept that is useful in the pharmaceutical industry is the hurdle technology promoted by the food microbiologist Leistner (1994) that deliberately combines preservation techniques to establish a series of preservation factors (hurdles) that a microorganism contaminating a product cannot overcome (jump over). These hurdles may include:

- Antimicrobial preservatives
- Extremes of pH
- Water activity
- Redox potential
- Osmolality
- Hot fill

- Delivery system
- Storage temperature

Hurdle technology has been successfully applied to product formulation in the cosmetic industry and is playing a greater role in microbial risk assessment in the consumer health and pharmaceutical industry.

Two Less Well-Known Product Attributes

Water Activity (a_w)

Perhaps the most critical quality attribute is water activity (Cundell and Fontana 2009; Hussong 2009). In general, drug products with water activities less than 0.75 do not support the growth of microorganisms likely to be found in drug products and may be self-preserving. Furthermore, microorganisms cannot develop resistance to low water activities. Generalizations that can be made about water activity include:

- At high a_w, i.e. greater than 0.95, bacteria out compete fungi
- Gram-negative bacteria require higher a_w than Gram-positive bacteria
- Below 0.85 most bacteria will not grow
- Below 0.75 most yeast and mold do not grow
- Between 0.75 and 0.60 only highly specialized microorganisms, not usually found in drug products, will grow
- Below 0.6 no microorganism will grow

Redox Potential

A reduced redox potential is widely used in food packaging employing vacuum packaging; antioxidants or nitrogen blankets to prevent microbial spoilage but is not widely used in pharmaceuticals. The oxidation–reduction or redox potential is defined in terms of the ratio of the total oxidizing (electron accepting) power to the total reducing (electron donating) power. In effect, redox potential is a measurement of the ease by which a substance gains or loses electrons. The redox potential (Eh) is measured in terms of millivolts. A fully oxidized standard oxygen electrode will have an Eh of +810 mV at pH 7.0, 30 °C, and under the same conditions, a completely reduced standard hydrogen electrode will have an Eh of −420 mV.

The major groups of microorganisms based on their relationship to Eh for growth are aerobes, anaerobes, facultative aerobes, and microaerophiles. Generally, the ranges at which different microorganisms can grow are as follows: aerobes +500 to +300 mV; facultative anaerobes +300 to −100 mV; and anaerobes +100 to less than −250 mV. For example, *Clostridium botulinum* is a strict anaerobe that requires an Eh of less than +60 mV for growth; however, slower growth can occur at higher Eh values (Jay 1996).

2.7.4 Concept of Hostility Level

This concept developed by the author collaborating with microbiologist Neil Lewis from Procter and Gamble is similar to the hurdle technology but adds a risk assessment element. Physical, chemical, and microbiological product attributes may influence the survival and/or proliferation of microorganisms within a drug product. As stated earlier, they include water activity, pH, antimicrobial effectiveness, alcohol level, osmolality, redox potential, salinity, and storage temperature. To make a hostility level assessment, a range from level 1 to 5 is established with 1 being the least hostile and 5 the most hostile to the survival of microorganisms. Then a numerical value from 1 to 50 is assigned to each hostility level. The overall hostility level of the formulation will be the sum of the applicable product attribute hostility levels (Table 2.13).

For example, drug product formulated as a compressed tablet as well as syrup to improve dosage of infant and aged populations could be evaluated for its hostility level. The following analysis would be applied to the syrup – with product attributes:

- water activity of 0.85–0.90 (hostility level 20)
- pH 4.5–5.0 (hostility level 20)
- an antimicrobial effectiveness log reduction of greater than 4 logs in 48 hours (hostility level 20)
- sugar concentration greater than 65% (hostility level 50)

This syrup with total hostility level of 110 would have a lower risk of microbial contamination than a formulation with a lower total hostility level.

In general, products with higher hostility levels would have looser bioburden controls while those with lower hostility levels would have tighter bioburden controls such as pharmaceutical ingredients, environmental controls, and reduced microbial testing. Typical physiological requirements of some bacterial pathogen are listed in Table 2.14.

2.8 Emerging Manufacturing Technologies

New manufacturing technologies are being introduced by our industry. Pharmaceutical microbiologist may be unaware of these new technologies and their microbial risk contamination impact. They include jet milling, hot melt extrusion (HME), and continuous manufacturing.

2.8.1 Jet Milling Micronization

The two major types of mills for particle size reduction to 5 μm or less in dry conditions are air-jet fluid energy mills and ball mills. Micronization of drug

Table 2.13 Hostility levels as related to selected product physicochemical and microbiological attribute ranges.

Product attribute	Hostility level 1 (1)	Hostility level 2 (2)	Hostility level 3 (5)	Hostility level 4 (20)	Hostility level 5 (50)
Water activity	0.9–1.0	0.85–0.9	0.8–0.7	0.7–0.6	<0.6
pH	6.5–7.5	5.5–6.5	5.0–5.5	5.0–4.5	<4.5
		7.5–8.5	8.5–9.0	9.0–11.0	>11.0
Antimicrobial effectiveness	None	No increase	>3 logs in 7 days	>4 logs in 48 hours	>5 logs in 24 hours
Hot fill	None	40–50 °C	50–60 °C	60–80 °C	>80 °C
Alcohol level	None	0–5%	5–15%	15–70%	>70%
Propylene glycol level	None	>5%	>10%	>20%	>40%
Sugar level	None	0–20%	20–35%	>35%	>65%
Osmolality	285–310 mosm/kg	310–325 mosm/kg	325–360 mosm/kg	250 mosm/kg	>360 mosm/kg
Redox potential	Aerobes 500–300 mV	Facultative anaerobes 300 to −100 mV			Anaerobes 100 to −250 mV
Salinity	None	0–5%	5–10%	10–20%	>20%
Storage temperature	Ambient	Controlled 20–30 °C	8–15 °C	2–8 °C	Frozen
Single use	Multiple use				Single use
Expiration dating	Days	Weeks	1–3 years	3–5 years	5 years

Table 2.14 Examples of physiological requirements of some common bacterial pathogens.

Organism	Minimum a_w	Temperature range (°C)	Maximum salt (%)	pH range	Oxygen level
Bacillus cereus	0.92	4–55	18	4.3–9.3	Strict aerobe
Campylobacter jejuni	0.987	30–45	1.5	4.9–9.5	Microaerophilic
Clostridium botulinum	0.935	10–48	10	4.6–9.0	Strict anaerobe
Clostridium perfringens	0.97	12–50	8	5.5–9.0	Strict anaerobe
Escherichia coli	0.95	7–49	6.5	4.0–9.0	Facultative anaerobe
Listeria monocytogenes	0.92	−0.4–45	10	4.4–9.4	Facultative anaerobe
Salmonella spp.	0.94	5.2–46.2	8	3.7–9.5	Facultative anaerobe
Staphylococcus aureus	0.83	7–50	25	4.0–10.0	Facultative anaerobe

Source: Adapted from Holt *et al.* (1994).

substances is important in increasing dissolution rates of solid oral dosage forms and inhalation products. Although jet milling exposes the drug substance being micronized to large volumes of high-pressure compressed gas, the use of medical-grade compressed nitrogen and sterile filtration of the jet of nitrogen will mitigate any risk of microbial contamination.

2.8.2 Hot Melt Extrusion

As described in a recent review article by Patil *et al.* (2016), HME is a continuous pharmaceutical process that involves pumping polymeric materials with a rotating screw at temperatures above their glass transition temperature (T_g) and sometimes above the melting temperature (T_m) to achieve molecular level mixing of the active compounds and thermoplastic binders, and/or polymers. This molecular mixing converts the components into an amorphous product with a uniform shape and density, thereby increasing the dissolution profile of the poorly water-soluble drug. Although the dwell time in the extruder-granulator is short, i.e. less than 10 minutes and operating temperature, i.e. 50–160 °C and elevated pressure will eliminate all microorganisms except perhaps the most heat-resistant bacterial spores.

2.8.3 Continuous Tablet Manufacturing

Continuous tablet manufacturing combines the same unit processing steps as batch manufacturing but with a continuous pharmaceutical ingredient feed input and tablet output controlled by real-time in-process analysis and product characterization technologies. Currently the technology is not available for in-process bioburden monitoring of a granulation. Advantages include the ability to move from R&D, to development, scale-up, validation and manufacturing batch size on the same equipment, better process control, less powder segregation, improved flexibility of manufacturing output, and a smaller manufacturing footprint. This approach differs to campaigning of the more familiar batch processes.

The processing steps are drug substance micronization, continuous excipient, drug substance and lubricant feed, blending, granulation, compression or encapsulation, and coating. Equipment should be designed to minimize dead spots where products may build-up. Once the continuous process is successfully started up, it can be continued with adjustments for hours, days, or conceivably weeks to meet drug product sales demand. Little or nothing is known about the microbiological attributes of the tablets or capsule produced by continuous manufacturing. Presumably the passage of dry material through the processing equipment will be to some extent self-cleaning and the more expedition manufacturing would discourage microbial growth. The absence of intermediate holding times between unit processing steps should improve microbiological quality. For example, wet granulations with a water activity level supporting fungal growth would not be held for a sufficient time and storage condition to encourage fungal growth. Furthermore, aqueous ingredients such as granulation and film coating solutions would need specified hold times to limit potent microbial growth. A conservative hold time would be a single manufacturing shift. Alternatively, real-time bioburden monitoring using LASER-induced fluorescence particle monitoring could be applied to mitigate risk of microbial contamination. Antimicrobial preservatives could be added to these processing ingredients but it is unlikely that regulatory agencies would look unfavorably on this option, as they would view it as inconsistent with GMPs.

At the influential Arden House Conference on Continuous Manufacturing, held 16–18 March 2015 in Baltimore, MD, multiple speakers highlighted that process equipment cleaning was a bottleneck in the technology. Running the system empty, in what is called a dry rinse to drive product from the equipment, would be a first step followed by the removal of product contact component parts for out-of-place cleaning. Component cleaning and setup for the next drug product may take two to three days and over a calendar year represent a considerable downtime. The future application of clean-in-place technologies may be possible.

2.9 A Case History

A discussion of this case history highlights many potential risks associated with microbial contamination of non-sterile drug products. In March 2009, the Hong Kong Board of Health reported that four batches of Allopurinol tablets manufactured by a local pharmaceutical company were found to be grossly contaminated with the fungus *Rhizopus microsporus* (>10^3 CFU/g). At least five patients at Queen Mary's Hospital receiving aggressive cancer therapy and treated with anti-gout drug Allopurinol for the common side effect hyperuricemia contracted intestinal mucomycosis and died (Cheng *et al.* 2009). How could compressed tablets, that are considered having a low risk for microbial contamination, be the cause of this fungal outbreak and patient deaths?

The tablets were manufactured at the local facility using a wet granulation that was dried in a tray dryer oven at 50 °C for 4 hours to a water content of 3%. The granulation was then held at 20 °C for up to 14 days prior to tablet compression. A typical formulation is Allopurinol, 100 or 300 mg, corn starch, FD&C Yellow No. 6 Lake (yellow tablets only), lactose, magnesium stearate, and povidone.

A probable source of the mold *R. microsporus* was the corn starch used in the tablet manufacture as it was found to contain 2 CFU of *Rhizopus*/g. Although the ascospores of *R. microsporus* are thermotolerant and would survive 4 hours at 50 °C during tray drying, it appears less likely that a granulation dried to 3% water content and stored at 20 °C for 5–14 days prior to tablet compression would become highly contaminated with *R. microsporus*.

According to the *Handbook of Pharmaceutical Excipients*, corn starch, NF has a water content of 11–14% (water activity 0.6–0.9), which is a water activity range that would support fungal growth. However, corn starch in a granulation dried to 3% has an estimated water activity of 0.1–0.2 and should not support the growth of *R. microsporus*, even if stored at 20 °C for up to 14 days. For the fungus to grow in the tablets, they must have been exposed to an elevated temperature and humidity.

The most clinically important Zygomycetes are in the Order Mucorales (Alvarez *et al.* 2009). Of 190 US clinical isolates tested, the frequency of species were *Rhizopus oryae* (45%), *R. microsporus* (22%), *Mucor corymbifer* (5%), *Rhizomucor pusillua* (4%), *Cunninghamella bertolletiae* (3%), *Mucor indicus* (3%), and *Cunninghamella echinulata* (1%).

The most common site of infection was the sinuses (26%), lungs (27%), and various cutaneous locations (28%).

Rhizopus microsporus is a ubiquitous fungus with the ability to grow at elevated temperatures up to 42 °C and they are often the first fungal invaders of solid substrates. It forms bundles of rosette-forming sporangiophores bearing sporangia and individual ascospores. Perhaps importantly, *R. microsporus* is used in the Asian fermented food tempe which is derived from soya beans.

The optimum conditions for radial growth and biomass dry weight on defined media were temperature 40 °C, water activity 0.995, and ambient air. At $a_w < 0.96$ virtually no growth occurred suggesting fungal growth will not occur in the compressed tablet but could in the wet granulation. Earlier work by Hocking and Miscamble (1995) reported that the minimum a_w for growth of *R. microsporus* on solid media was 0.90.

A review of the evidence by the author of this book chapter is inconclusive as to where *R. microsporus* growth in the Allopurinol compressed tablets occurred. Was it during the storage of the excipient starch, during tablet manufacturing, or due to improper storage of the tablets? The use of a tray dryer is a dated technology that has been largely replaced by fluid bed dryers. The exposure of the wet granulation in the tray dryer or extended storage prior to compression may have encouraged fungal growth. Although solid oral dosage forms, as they are the least invasive non-sterile products, are considered to have the lowest risk of patient infection, cancer patients especially those that are immunologically suppressed due to radiation treatment and/or chemotherapy may be highly susceptible to fungal infection, even from oral administration, and as the result of fungal infection have high mortality rates. This case history justified the lower TCYMC in USP <1111> *Microbiological Quality of Non-sterile Pharmaceutical Products.*

As the strain matching was not used to conclusively confirm the infection was derived from the tablets, it is possible that other environmental factors, e.g. unauthorized food or hospital facility contamination were the source of the infections.

2.10 Conclusions

Microbial contamination risk management tools can be systematically applied to formulation and manufacturing process development, drug manufacturing, and microbial testing and will undoubtedly mitigate risk. The challenge is to ensure an experienced pharmaceutical microbiologist has a place at the table, as microbial contamination risk mitigation is rarely an objective during product development.

Bibliography

Alvarez, E, Sutton, DA, Cano, J, Fothergill, AW, Stichgel, A, Rinaldi, MG, Guarro, J, 2009, 'Spectrum of Zygomycete species identified in clinically significant specimens in the United States', *J. Clin. Microbiol.*, vol. **47**(6), pp. 1650–1656.

Blair, TC, Buckton, G, Bloomfield, S, 1991, 'On the mechanism of kill of microbial contaminants during tablet compression', *Int. J. Pharm.*, vol. 7(3), pp. 111–115.

Cheng, VCC, *et al.*, 2009, 'Outbreak of intestinal infection due to *Rhizopus microspores*', *J. Clin. Microbiol.*, vol. **47**(9), pp. 2834–2843.

Chesworth, KAC, Sinclair, A, Stretton, RJ, Haynes, WP, 1977, 'Effect of tablet compression on microbial content of granule ingredients', *Microbios. Lett.*, vol. **4**, pp. 41–45.

Cundell, AM, 2005, 'Managing the microbiological quality of pharmaceutical excipients', *PDA J. Pharm. Sci. Technol.*, vol. **59**(6), pp. 381–395.

Cundell, AM, 2009, 'Effects of water activity on microorganisms', in Cundell, AM and Fontana Jr, AJ (eds), *Water Activity Applications in the Pharmaceutical Industry*, Davis Healthcare International Publishing, River Grove, IL, pp. 175–204.

Doa, H, *et al.*, 2018, 'Microbial stability of pharmaceutical and cosmetic products' *AAPS Pharm. Sci. Tech.* **19**(1), pp. 60–78.

Fu, N, Chen, XD, 2006, 'Towards a maximal cell survival in convective thermal drying processes' *Food Res. Int.*, vol. **44** pp. 1127–1149.

Harpaz, R, Dahl, RM, Dooling, KL, 2016, 'Prevalence of immunosuppression among U.S. adults, 2013', *JAMA*, vol. **316**(23), pp. 2547–2548.

Hocking, AD, Miscamble, BF, 1995, 'Water relations of some Zygomycetes isolated from food', *Mycol. Res.*, vol. **99**(6), pp. 1113–1118.

Holt, JG, Krieg, NR, Sneath, PHA, Staley, JT, Williams, ST (eds), 1994, *Bergey's Manual of Determinative Bacteriology*, 9th ed., Lippincott, Williams and Williams, Philadelphia, PA, pp. 787.

Hussong, D, 2009, 'Water activity and pharmaceutical manufacturing; a regulatory microbiology perspective', in Cundell, AM and Fontana Jr, AJ (eds), *Water Activity Applications in the Pharmaceutical Industry*, Davis Healthcare International Publishing, LLC, River Grove, IL, pp. 253–257.

ICH Q8 (R2), 2009, 'Pharmaceutical Development, Current Step 4 version'.

ICH Q9, 2005, 'Quality risk management, current Step 4 version 9'.

ICH Q10, 2008, 'Pharmaceutical Quality Management current Step 4, version 4'.

Jay, JM, 1996, *Modern Food Microbiology*, 5th ed., Chapman and Hall, New York, pp. 48–51.

Lederberg, J, Shope, RE, Oaks, SC, 1992, '*Emerging Infections – Microbial Threats to Health in the United States*', Institute of Medicine, National Academy Press, Washington, DC, pp. 294.

Leistner, L, 1994, 'Further developments in the utilization of hurdle technology for food preservation', *J. Food Eng.*, vol. **22**, pp. 421–432.

Long, M, 2013, '*Risk management: regulatory expectations, risk perception and organizational integration*', in Mollah, H, Baseman, H, Long, E (eds), *Risk Management – Applications in Pharmaceutical and Biopharmaceutical*, Wiley, Hoboken, NJ, pp. 49–73.

Parenteral Drug Association, 2014, *Exclusion of Objectionable Microorganisms from Non-sterile Pharmaceutical and OTC Drug Products, Medical Devices and Cosmetics*, PDA Technical Report No. 67.

Parenteral Drug Association, 2017, *Quality Risk Management for the Design, Qualification and Operation of Manufacturing Systems*, PDA Technical Report No. 54-5.

Patil, H, Tiwari, RV, Repka, MA, 2016, 'Hot-melt extrusion: from theory to application in pharmaceutical formulation', *AAPS Pharm. Sci. Tech.*, vol. **17**(1), pp. 20–42.

Sandle, T, 2013 'Risk management and risk assessment for pharmaceutical manufacturing – a contamination control perspective', *Microbiol. Solutions*, vol. **1**, pp. 164.

Sutton, S, Jimenez, L, 2012, 'A review of reported recalls involving microbiological control 2004-2011', *Am. Pharm. Rev.*, vol. **15**(1), pp. 42–57.

USP 40/NF35 <1115>, 2017, 'Bioburden Control in Non-sterile Drug Substances and Drug Products'.

3

Qualification of Microbiological Laboratory Personnel and Equipment

Robert Schwarz

University of Applied Sciences, Wien, Austria

CONTENTS

3.1 Introduction

In any microbiological laboratory not only the equipment needs to be qualified but also the analysts. Since most of the testing is generally not automated, variability between analysts may affect the outcome of the test result. Both equipment and analyst qualification should be addressed:

Pharmaceutical Microbiological Quality Assurance and Control: Practical Guide for Non-Sterile Manufacturing, First Edition. Edited by David Roesti and Marcel Goverde.

- Qualification of laboratory personnel in a microbiological QC environment is crucial linked to the two core topics of GMP: documentation and training.
- Qualification of laboratory equipment in a microbiological QC environment should be based on its impact on the assessment of product quality: direct impact, indirect impact, or no impact.

3.2 Reasons, Requirements, and Strategies for Qualification

3.2.1 Qualification and Re-Qualification of Laboratory Personnel

The QC microbiology lab is part of every pharmaceutical company's manufacturing authorization. Therefore, depending on the market, the corresponding authorization is needed. For example, in the United States or EU the following regulatory requirements are applicable:

- 21 CFR 211.25: *Training in cGMP shall be given on a continuing basis with sufficient frequency to assure that employees remain familiar with cGMP requirements applicable to them.*
- EU GMP Guideline, chapter 2 "Personnel," section 2.11 "Training": *Continuing training should be given, and its practical effectiveness should be periodically assessed.*

The major point regarding education at a university level is that "Pharmaceutical Microbiology" does not exist in any academic syllabus. This concludes that either the practical aspects of it are trained on-the-job or the theoretical aspects can be trained in-house or outhouse.

But, regardless of how intense the training in this topic remains and must be considered when qualifying analysts for performing validated analytic methods, the following problems remain:

- Analyst-induced variability has a major impact on accuracy.
- Analyst-induced bias is a major issue in some types of assay.
- We have developed microbiology acceptance criteria or specifications that ignore the inherent variability of biological systems.

Those points should be addressed in the analyst qualification procedure as well as in the risk assessment for analytical method validation and sufficiently defended to an extent that all measurements were taken to minimize the impact on product quality.

Figure 3.1 shows a general stepwise analyst training concept which also applies to laboratory personnel in a microbiological QC environment.

Each new personnel from analyst to laboratory supervisor must be trained with the appropriate documents defined in the local training program in order to

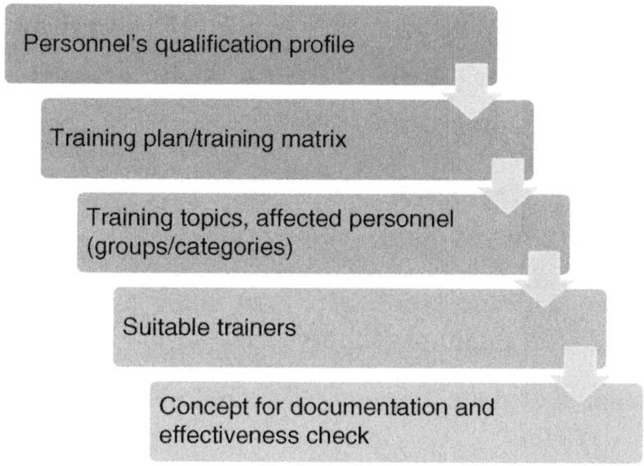

Figure 3.1 Steps of a successful analyst qualification.

accomplish his task. A full understanding of the actual health authority guidelines is required for laboratory supervisors; but for analysts, a basic understanding of these documents is sufficient. All personnel in the microbiology department should however be trained and understand the SOPs concerning good manufacturing practices and health and safety environment requirements in the microbiology department as well as the specific SOPs required to undertake their tasks. As the United State Pharmacopeia (USP) <1117> suggests, following a theoretical training, periodic testing would provide evidence of practical competency of the analyst with regards to general aseptic technique, documentation, housekeeping, or specific test method. For instance, to qualify an analyst for microbial enumeration of Petri plates, following a theoretical training, different agar plates may be prepared with multiple colonies of different types of microorganisms (e.g. bacteria and molds) and the counts of the prepared plates of the analyst to be qualified are compared with those of experienced laboratory personnel.

A periodic vision check, especially for analysts reading test results such as, for instance, enumeration of colonies on agar plates, would also be advisable. This procedure and part of the risk assessments strategy could be comparable to operators performing visual inspection of the final product manually.

The following describes an example of an analyst qualification on a specific test method – the microbiological examination of non-sterile products:

- A representative sample of the different products tested in the laboratory should be considered for the qualification (e.g. hard gelatine capsules, creams, nasal sprays).
- The initial qualification consists of executing under supervision at least one method suitability test run and at least one release test
  - per type of methods used (e.g. pour plate, membrane filtration)

  - with the tests for TAMC/TYMC as well as specified microorganisms
  - with the respective negative controls
- Acceptance criteria:
  - The microorganism recovery rate of the method suitability test should comply with the pharmacopoeia requirements.
  - No contamination of the negative controls.
  - The supervisor must consider the working technique and GMP behavior as adequate.
- The periodic (e.g. annual) requalification consists of performing at least one release and method suitability test run under supervision.

Requalification intervals should be defined in an SOP. Additionally, if a control chart shows a trend or a deviation is based on an analyst failure, additional training and requalification may be necessary. For validated test methods it can be advisable to implement a system for analyst requalification based on specific scenarios. If an analyst has not been performing a test for a long time, he or she must undergo a requalification procedure before analyzing routine samples again. This could be solved with the same time period for every test (e.g. six months) or depending on the test. When this is performed differently for every test, the time period should be based on the complexity of the test and the dependency of it from the analyst. This may cause additional variation which will be reflected in the test's control chart but could prevent those variations effectively in advance.

3.2.2 Equipment Qualification: Which Equipment Needs to Be Qualified in a Microbiological Laboratory?

Equipment qualification including the equipment and devices in a microbiological QC laboratory is a pre-requisite for subsequent validation activities including analytic method validation.

Figure 3.2 shows the dependency of subsequent following validation activities like analytical method validation from laboratory equipment qualification. Today there are equipment used to perform analytic tests independent from analysts. For instance, for bacterial identification tests several steps like incubation, plate reading, and the identification itself are fully automated. Identification itself is based on a comparison of metabolic activity of the isolate and is compared with a database. Therefore, the part of computerized system validation (CSV) in combination with equipment qualification is a major part of the validation activities in a microbiological QC lab and a crucial base for subsequent validations like analytic method validation. The reliability of the analytic method validation is strongly linked to a proper equipment qualification and CSV. Analytic method validation itself is the base for cleaning validation, sterilization validation, and product process validation since the results of analytical tests

Figure 3.2 Validation cascade.

influence the outcome of these validation activities and furthermore influence during routine production of the ongoing-process verification (OPV). In all validation activities mentioned above samples are taken and analyzed. During OPV, the validated state of the processes is periodically checked. Part of this OPV is also taking samples. Depending on the process, its risk analysis, and the results of the testing during the validation, the number of samples can be reduced compared to initial validation. This shows clearly that qualification and validation activities in a microbiological QC lab would impact the monitoring of the manufacturing process as well as the pharmaceutical product's quality attribute and in the end the patient's safety if, for instance, release decisions are based on incorrect data. Therefore, it is important to perform qualification of equipment on a risk-based approach to identify the impact and furthermore the criticality of the equipment used for analytical testing.

3.2.2.1 Equipment Classification According to ISPE

To identify the components that need to be qualified and to determine the extent of the qualification activities a categorization of all the equipment and devices should be performed in a qualification master plan. The evaluation can, for example, be performed according the "Impact Assessment" of international society for pharmaceutical engineering (ISPE) Baseline Vol. 5.

Based on the identification where the specific equipment is categorized, the extent of the workload is defined. Starting at "No Impact" systems do underlie the requirement of "Good Engineering Practice" (GEP) which relies on a proper vendor selection. This is usually in a GMP-driven environment linked to the supplier qualification program. Therefore, a qualified supplier for the microbiological quality control laboratory will follow the requirements of GEP (see also Figure 3.3).

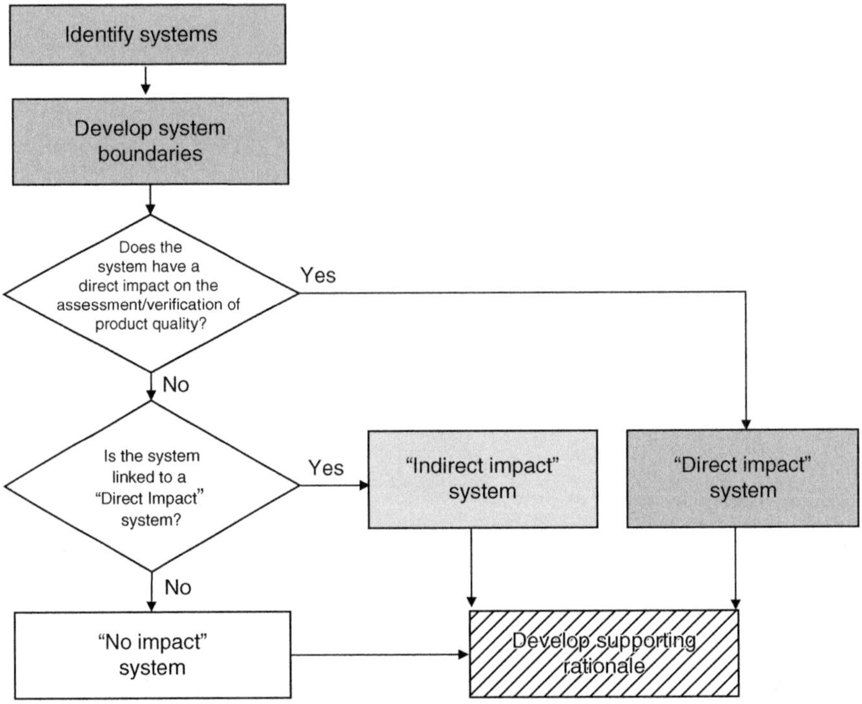

Figure 3.3 Impact assessment according to ISPE Baseline Vol. 5, decision tree.

This categorization should be performed in a proper way including the required framework based on and written assessment in the validation master plan (VMP). For every single object the documented justification is usually part of an implementation risk assessment (Figure 3.4).

Figure 3.4 Categorization defines "working packages."

3.2.2.1.1 Qualification If a qualification is needed it must be performed to fulfill all requirements of the guidelines referred in Section 3.2.3. The most import topics there are the traceability from user requirements and proving the suitability for the intended use.

3.2.2.1.2 Commissioning Commissioning is a systematic approach to the start-up and turnover of facilities, systems, and equipment to end users and ensuring that user requirements and design specifications (DS) are met (according to ISPE Baseline Vol. 5). Activities within this phase may include design reviews, factory acceptance testing, installation verification, and functional testing. Summary reports are generated at the conclusion of commissioning activities and include an overview of the results and any deviations encountered during testing.

3.2.2.1.3 GEP (Based on ISPE "Good Practice Guide: Good Engineering Practice" 2008) Good Engineering Practices (GEPs) consist of proven and accepted engineering methods, procedures, and practices that provide appropriate, cost-effective, and well-documented solutions to meet user requirements and compliance with applicable regulations. GEP underpins activities in the day-to-day operations and forward planning of a pharmaceutical business. The adoption of this methodology leads to a balance of expenditure and activity. In addition, GEP documentation can be leveraged to support verification work.

GEP activities include the following key concepts:

- Project engineering
- Common practices
- Operation and maintenance

The three concepts at the core of most GEP activities include:

- Risk management
- Cost management
- Organization and control

3.2.2.1.4 Critical Instruments According to ISPE Baseline Guide, Vol. 5 Every equipment and all systems that measure or control critical parameters (process/product related) have higher requirements regarding documented testing. This includes calibration and qualification.

Those processes to ensure critical parameters that are heavily depending on manual steps from analysts should be unambiguously described in written protocols (SOPs) and training. Additionally, retraining must be performed regularly.

How to Identify Direct Impact Systems Generally (Modified from ISPE: Engineering Baseline Vol. 5)

1) Is this equipment used to demonstrate compliance with a submitted process?
2) Does the function of the system under routine conditions have direct impact on the product quality?
3) Does a malfunction or inaccuracy of the system have direct impact on the product quality or efficacy?
4) Is the (analytical) result or output from this equipment stored in the batch documentation or used for product release?
5) Does the equipment have product contact?
6) Does this equipment control other critical components without an independent verification?
7) Is the component used to monitor a critical state of a system?

Question 5 would be applicable if samples are drawn and aliquoted directly side-by-side to the production process and parts of the samples are brought back into the product. This could be true at biotechnological processes when using bioluminescence for determining the viability of the production organism.

3.2.2.2 Equipment Classification According to USP

Another possibility for classification of equipment criticality is mentioned in USP <1058> "Qualification of analytical instruments." All equipment or instruments are divided into three different categories; however, it should be noted that the same type of instrument can fit into one or more categories, depending on its intended use.

Group A includes the least complex, standard instruments that are used without measurement capability or user requirement for calibration, such as a magnetic stirrer or vortex mixer. Proper function is ensured by observation, and no further qualification activities are needed for this group.

Group B includes instruments that may provide a measurement or a quantitative condition that can affect a measurement. Examples include a pH meter or an oven. Proper function of instruments in this group may require only some elements of qualification, such as routine calibration, maintenance, or performance checks. The level of qualification will also depend on the criticality of the application. Generally, these instruments may have firmware but not software that is updated by the user.

Group C comprises the majority of analytical instruments and normally includes a significant degree of computerization and complexity, such as high-pressure liquid chromatographs and mass spectrometers. All elements of qualification, including software validation, must be considered to ensure proper functioning of instruments in this group.

Depending on the category, a different level of qualification is needed, thus, in principle, it is the same approach as the one from the ISPE explained above.

3.2.3 Equipment Qualification: How to Qualify Laboratory Equipment

Requirement for equipment qualification is described in various guidelines such as, for instance:

- EU GMP Guideline Annex 11
- EU GMP Guideline Annex 15
- US 21 CFR Part 211
- US CFR Part 11
- USP<1058> Analytical Instrument Qualification
- ICH Q7

In a GMP context, the equipment qualification should demonstrate that the equipment is properly installed and is operating as expected in its specific environment. It includes documented evidence of testing the equipment against acceptance criteria during the different qualification phases. Calibration is part of the qualification procedure and demonstrates that the instruments produce results of defined standards or references within specific limits over an appropriate range of measurements. For equipment or instruments that contain a computerized system, a CSV must be included during the equipment qualification.

The qualification of laboratory equipment or instrument is typically performed in four phases (based on Kavermann (2017); EudraLex Annex 15):

- Design qualification (DQ)
  - Documented evidence if the compliance of the design with GMP is demonstrated.
  - Contains the user requirement specifications (URS; document that specifies what the user expects the equipment or instrument to be able to deliver).
  - Defined material/instrument/test specifications and capabilities.
  - Diagrams and supplier's instructions (e.g. instrument and electrical).
  - Specific microbiology lab requirements (e.g. resistance to disinfection).
  - Supplier assessment and audit report (if applicable).
  - How and when to perform the maintenance of the system.

- Installation qualification (IQ)
  - Documented evidence if the equipment/instrument is properly shipped and installed according to the DS and vendor instructions.
  - Check if all equipment and instrument parts received are correct and undamaged.
  - Verifying shipping and technical documents.
  - Verifies if the instrument is correctly installed and if the power or network/data backup connections are working according to the requirements and assembled correctly.
  - Training by the vendor may be initiated.
- Operational qualification (OQ)
  - Documented evidence that the equipment/instrument operates according to the defined specifications.
  - Equipment/instrument calibration is executed.
  - Functionalities of the equipment/instrument are verified (e.g. testing of sample, pumps, alarms, data management procedures, electronic records and signature functions, temperature mapping in incubators).
  - After a successful OQ, utilization and maintenance SOPs may be written.
- Performance qualification (PQ)
  - Documented evidence that the equipment or instrument performs as expected in the intended operating environment.
  - PQ testing is typically performed under realistic test conditions using system suitability tests or standardization.
  - For microbiological lab equipment/instruments, PQ testing may be consisting of, e.g., recovery of a selected test microorganisms in different sample matrices, incubator temperature mapping with min/max loads or stress tests, method validation for alternative microbiological methods demonstrating noninferiority to compendial methods.

It is a good practice that QA compliance approves each phase prior to starting the next one. After the PQ phase and review/approval of the qualification reports, the instrument is successfully qualified and may be released for GMP testing. It is general practice that the IQ/OQ is performed by the vendor and then the laboratory should evaluate and countersign the documentation.

Once the system has been qualified, it may be re-qualified periodically as defined in a maintenance plan or following a change/update that would impact the qualification status. Periodic intervals are defined based on manufacturer's recommendations, performance experience, use frequency, and local legal requirements.

At first sight, considerations 1 and 8 could be contradictory but they will be combined in the process of grouping (bracketing) described in Annex 15 of EudraLex GMP Guideline where a qualification effort could be reduced to a

WHO Technical Report No. 937, Annex 4

An alternative or supplementary approach is given by this WHO Guideline. It refers to the relation between process parameters and equipment parts as follows:

- For all installation components, subsystems or parameters, critical parameters, and noncritical parameters should be determined.
- If the component comes into direct contact with the product, or if the parameter affects the quality of the drug product. Then, it should be classified as a critical parameter.
- Critical parameters should be qualified.

At the latest during the first qualification – DQ – also timely coordinated with the "design freeze" the classification should be performed. This is crucial because a change in an equipment design may lead to a change in the criticality of a parameter even only if there is a mechanism in place to verify the parameter's output. These tests would be performed during OQ in the function's tests of alarm, failure, and status messages.

For those measuring devices also certain requirements in a GMP-environment like a microbiological QC lab apply:

- Critical measuring devices in a system are identified with a proper risk assessment if not already defined in regulatory documents like a pharmacopoeial method.
- The accuracy of the measure circuit must be known to be considered in the error calculation and to determine the specific measuring range as well as the acceptable range for this specific parameter.

To calculate action limits, the potential measuring error must be subtracted for the upper limit or added to the lower limit from the acceptance limit. For example, temperature range of an incubator acceptance limit: 30–35 °C; accuracy of measuring circuit ±0.5 °C; action limits are 30.5 and 34.5 °C, respectively.

Even though there is – after performing a proper classification of the equipment used in a microbiological QC lab – only a small number of different devices to be qualified, the need for a lean equipment qualification strategy rises throughout the pharmaceutical industry.

reasonable working package without any loss of quality regarding subsequent analytical method validation.

When focusing on this approach, the regulatory requirement and formal background for this type of qualification approach rely on the VMP that must describe how such grouping strategies could be performed.

Considerations for Lean Equipment Qualification (Based on Bieber 2018)

1) No schematic performing of equipment qualification, but risk-based approach where the risk assessment's scope and depth are adjusted to the equipment.
2) Qualification team (all functional roles must always be covered during the whole project):
   - Agree on and execute binding meetings with above functions
   - For qualification documents – no serial approvals, but parallel
   - Functional roles:
     - Engineering
     - Manufacturing
     - Quality assurance
3) Written user requirements based upon process knowledge at the early stage
4) Based on this: URS, critical parameters (incl. ranges and limits)
5) Strict separation between:
   - GMP/GEP
   - Qualification/Commissioning
   - Quality relevant/Technical important
6) Including the vendor (URS/Function specification (FS)/DS):
   - Collaboration/Agreement of URS/Function–design specification
   - Supplier audit
   - Use supplier documentation
7) Determine the required procedures during the qualification phases (qualification matrix)
8) Standardization of tests (i.e. test sheets)

3.3 Critical Aspects of Microbiological Methods

Evolving from classical methods in a university or medical environment, a high number of microbiological test methods are crucially linked to the qualified analyst. Two topics are in focus of the microbiological methods and should be addressed right from the beginning:

- Manual preparation steps that may induce analyst's variability in the final test result.
- Evaluation of test results that may induce analyst's bias.

During implementation of analytical methods these points are evaluated in the necessary risk assessment for each microbiological method required at the latest for analytical method validation. Especially, in conjunction with method robustness those points must be addressed.

As an example, following test and possible solutions to avoid above-mentioned situations are described.

3.3.1 Antibiotic Susceptibility Testing (AST)

For some biotechnological processes, antibiotics are used as part of the culture media. The production organism is genetically modified to be resistant against the used antibiotic, but this is also tested to have supportive data. Especially, in the case when the yield goes down or the concentration of the production organism is low or gets low, this test will be performed to exclude a loss of the production organism's ability to withstand the antibiotic. The test is performed as quantitative bioassays for antibiotics (Figures 3.5 and 3.6).

A crucial aspect in that case could be that there is a prone to subconscious bias as analysts learn to measure the responses into compliance. To avoid this, different solutions may be applicable. Standard procedure would be performing this test with four-eyes principle.

Depending on the pH value and the concentration of the microorganism population a specific diameter for each antibiotic is predefined in the analytic method development as a limit to distinguish if a microorganism is "sensitive" and "resistant" to this antibiotic. In the QC microbiological environment, it has to be taken into consideration that every measuring device, in this case a ruler, has to undergo a calibration program (Figure 3.7).

Compared to the right result, the left result may be hard to evaluate in a manual process (Figure 3.8). First, during analyst training possible sources of a result like in the left should be explained and discussed. Basic microbiological techniques like checking if growth on a plate is a pure culture are required as well as knowledge of the production organism in case of a biotechnological

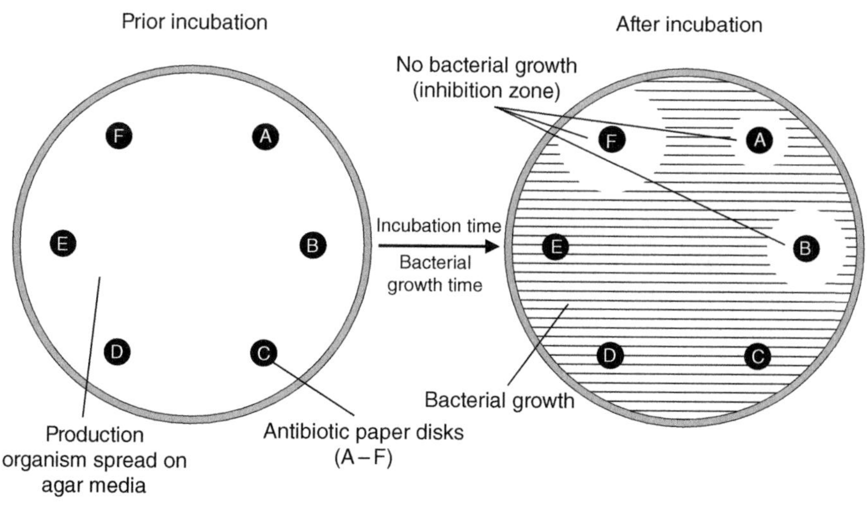

Figure 3.5 Scheme of antibiotic susceptibility testing (AST) prior and after incubation.

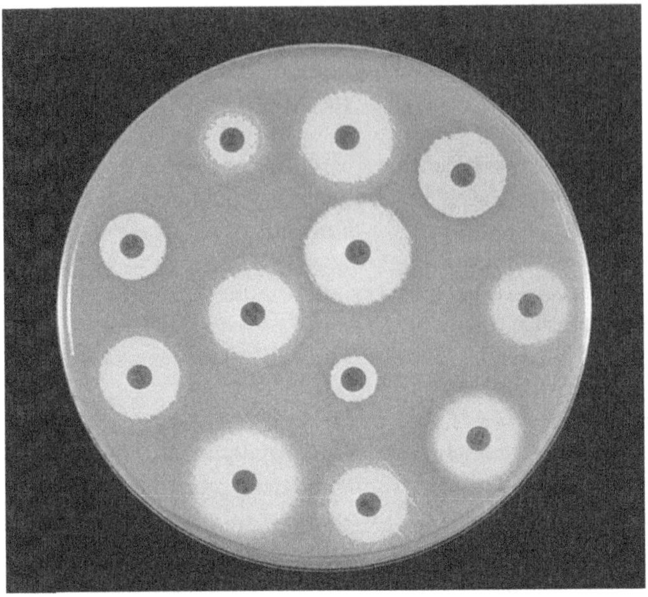

Picture description: *"This 1972 image depicted a Mueller-Hinton agar culture plate that had been used in an antibiotic susceptibility test (AST). Known as the Kirby–Bauer method, each of the small labeled discs, or wafers, contained an antibiotic cocktail. The light halos surrounding each disc, also known as 'reaction zones', represented regions in which the bacteria on the agar's surface did not thrive, due to their sensitivity to the antibiotic that had been soaked into these respective discs. See PHIL 10787 and 10789 for enlargements of regions of this photograph."*
Source: http://www.publicdomainfiles.com/show_file. php?id=13531953614439, this work, identified by www.PublicDomainFiles.com, is free of known copyright restrictions.

Figure 3.6 Agar plate with antibiotic disk after incubation and microbial growth.

production plant. This includes in that specific case also knowledge about genetics, plasmids, and under which conditions a consistent microorganism population show variation.

Based on Figure 3.9 it may even be clearer that a subconscious bias may lead to different results which even differ in a way that one result is a pass (gray dotted circle) or a fail (black circle). In that case the pass says that the production organism incorporated the plasmid that contains the region for developing the specific resistance against the antibiotic which is part of the culture media and should prevent growth of all other microorganisms.

Based on this example another alternative to implementing the four-eyes principle could be to modify the test to perform it as an automated analytic method. This trend to step away from manual processes in the pharmaceutical industry could be seen as "state-of-the-art" to objectivize this and similar tests.

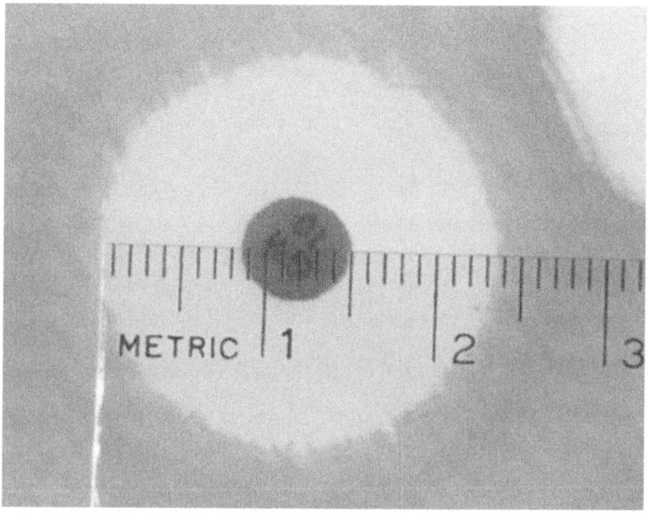

Figure 3.7 Example of evaluating the "reaction zone" (measuring the diameter manually) to identify a "resistant" or "sensitive" result of a specific microorganism against an antibiotic.

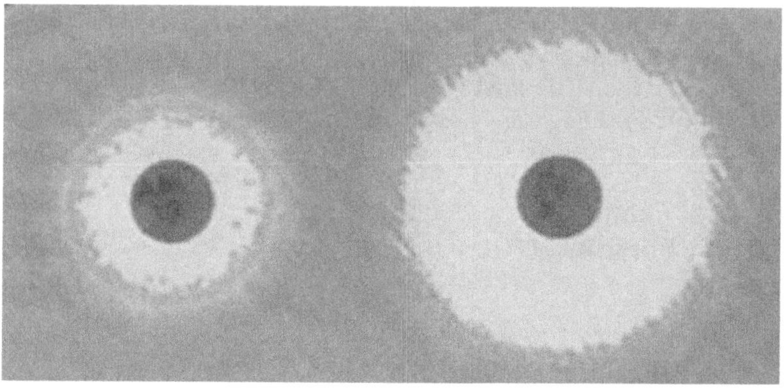

Figure 3.8 Enlargement of Figure 4.6. Comparison of different challengeable antibiotic (AB) disk readings.

For this specific test two versions of automation are the most common ones:

- Liquid medium technique
- Automated reader technique

The first modification is based on an automated dilution of the microbial suspension and adding a liquid antibiotic. After the incubation time the growth is measured via turbidimetry.

The second modification is only a modification in the readout process but the preparation and incubation remain unchanged. This could be an advantage

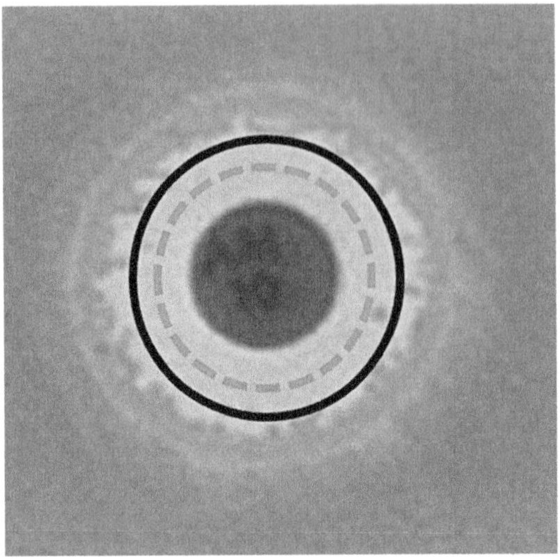

Figure 3.9 Different evaluations lead to results that may comply or may not comply.

because the qualification for the reader and the revalidation effort for the method could be lower compared to the liquid medium technique. Additionally, the traceability of getting the results is higher because the pictures of the plates are stored in the reader and can be transferred and archived on a network server. This allows a re-evaluation if limits are changing. The advantages and disadvantages of an automated reading were also discussed by Le Page *et al.* (2016).

3.4 Practical Examples for Qualification of Laboratory Personnel

3.4.1 Titrations

One of the most often performed techniques in a microbiological QC lab is dilutions – so-called titrations. In contrast to any chemical analysis based on the "nature" of the processes material, the variation when performing a titration is higher.

- An accurate titer for microorganisms as at chemical analysis is a fictional thing
- A standardized technique is a must

Pre-requisites for an accurate titration are calibrated equipment – in this case pipettes. Then, a classical three-step training plan is suitable for qualifying an analyst. Step one is the SOP training. The SOP should cover not only how to perform the method but additionally how to perform self-checks for potential errors during the process that may influence the result. The second step would be to show the analyst to be trained on the method. The method is performed by an experienced and qualified analyst. The third step would be performing the method the first time by the analyst to be trained under supervision. Additionally, the result has to comply in a certain predefined range from the target value of the dilution step. The procedure of the third step could afterwards be implemented as a requalification procedure. Then additionally to the range the frequency should be argued in the risk analysis of the analytical method.

Especially, when it comes to requirements regarding biosafety levels in addition to the theoretical SOP-training, the practical approach with training on the job is an important part of analyst qualification. When handling microorganisms that are not categorized as biosafety level (BSL) 1 or 2, so-called "dry runs" support the analysts under supervision to practice the proper techniques for the dilution steps without the hazardous material.

Depending on the number of tests and the criticality of the test that need low variation in dilution/titration that must be performed alternatively, an automated titration device ("titration robot") could be taken into consideration.

3.4.2 Verification of Spore Count on Biological Indicators

This test is performed as an incoming goods test for biological indicators (BI) for sterilization validation purpose to verify the spore count compared to the certificate of the specific BI lot.

The specification according to USP <55> is that the spore count is not less than 50% and not more than 300% of the label claim. This is often inside the analyst's normal variation when performing dilutions. Therefore, this method can only be performed if the analyst is already qualified to perform titrations. Thus, compared to Figure 3.1, there is additionally a dependency within an analyst's qualification program. Some methods require basic methods as a pre-requisite to perform them properly. This must be addressed and included in the training program of every QC microbiology lab analyst. During the SOP training a short troubleshooting session is advisable. This includes at the latest during this qualification phase basic microbiological knowledge about vegetative microorganism and spores and their importance in pharmaceutical production.

3.4.3 Recovery Rate of Microbiological Swab Sampling

Depending on a pharmaceutical company's strategy, the sampling for environmental monitoring (for further details on this topic see Chapter 8) is sometimes performed by QC staff. The argumentation for this procedure is to achieve less bias during sampling. The effectiveness check of a proper surface cleaning in a pharmaceutical cleanroom is surface sampling with contact plates or swabs during environmental monitoring "at rest." If a manufacturing operator performs this sampling on those surfaces she or he cleaned, this may look like if those persons are checking themselves. There may be a higher incident rate for testing into compliance, even if that occurs subconsciously. On the other hand, especially in high-grade cleanrooms it is better to limit the number of personnel in the critical cleanroom areas to minimize the occurrence of a contamination. Following this argumentation, sampling during routine manufacturing in aseptic production areas is favorably performed by manufacturing operators. To benefit from the advantages of both options, a procedure could be established that during validation or revalidation (e.g. aseptic process simulation) the QC personnel perform the environmental monitoring sampling and during routine manufacturing the operators perform the environmental monitoring. The results will be then compared to show their equivalence.

The recovery testing is performed on sample coupons of those materials that are sampled in monitoring, usually the recovery study consists of

- stainless steel
- glass
- different plastics (e.g. polytetrafluorethylene (PTFE), silicone, and polymethylmethacrylate (PMMA))

In addition to the material, the surface treatment of the coupons should be representative for the surfaces in the production which are part of the monitoring program.

Spiking the surfaces with the microorganism is also crucial because it requires a proper dilution technique. Other challenges like choosing the type of microorganism or dry-out effects that lower the concentration of the microbial suspension are not part of analyst's influence and therefore not focus of analyst's qualification or training.

The test setup of a recovery study itself includes spreading the known concentration of the bacterial suspension on a test surface. This sample is called the spiking sample. It simulates the surface sampled during monitoring. In contrast to a routine sampling the concentration is known based on the proper dilution of the bacterial suspension, the surface area of the coupon, and the coupon's pretreatment (sterilization). The sterilization guarantees that all found microorganisms are only from the spread bacterial suspension. The recovery rate is calculated.

During cleaning process development and validation, it is sometimes of interest to know the bioburden prior to cleaning to get knowledge also about the microbial cleaning performance of the process. It can be compared with swab sampling at environmental monitoring "in operation" (prior cleaning) and "at rest." Usually during cleanroom cleaning this "process capability" is not in focus due to the manual cleaning process and that those surfaces are out of scope of cleaning validation according to Annex 15 of EudraLex GMP guideline.

There are different solutions to address the individual and varying recovery rate for swab sampling between the analysts:

- The analyst-specific recovery is part of the swab sampling documentation and the surface acceptance limit is adjusted by a correction factor.
- The analyst must achieve a minimum recovery (i.e. 50%) to be qualified for swab sampling and the surface acceptance limit is adjusted by a correction factor based on this minimum recovery.
- If the recovery of all analysts exceeds a second level (i.e. 80%), no adjustment factor is necessary.

When focusing on analyst qualification and taking into consideration this qualification as part of a pharmaceutical company's validation program, it would be advisable to perform a risk assessment to justify which procedure is applicable. From the perspective of Annex 15, a "worst-case" approach would be appropriate to use the first option even if an analyst's recovery is significantly above the minimum requirement to prevent single events that after a periodic data review seems to be "calculated into compliance."

Figure 3.10 shows one possible swab sampling method. During method development different solutions could be compared and the sampling method with the best recovery of the specific surface material could be chosen. Nevertheless, it is not advisable to have different swab sampling techniques, i.e. swabbing patterns for different surface materials because this may lead to errors during sampling followed by invalid results or, in a "worst case" situation, that the swab sample will not be analyzed at all.

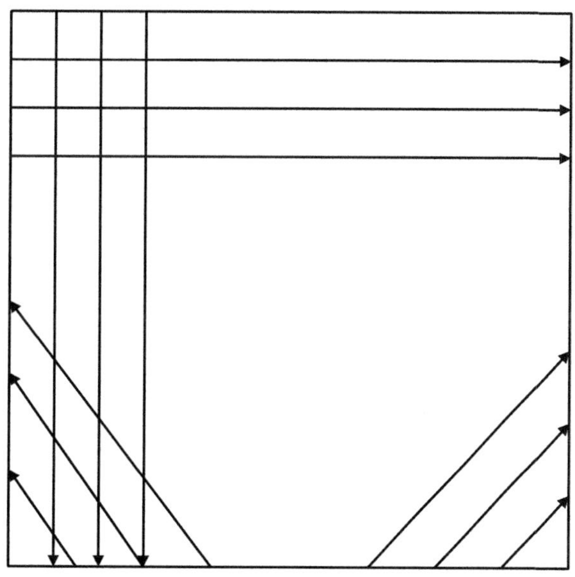

Figure 3.10 Example scheme of swab sampling.

Acknowledgments

I would like thank Ulrich Bieber and Guido Heuwes for teaching me another view on qualification and validation – yet more efficient and comprehensive. Additionally, I would like to show my gratitude to David Roesti to give me the opportunity to participate in this project by believing in my skills and knowledge. Last and most I wish to present my special thanks to Marcel Goverde for his ongoing support and giving me a different view on topics helping me to leave the well-known pathway and entering new roads.

Bibliography

Bieber, U, 2018, 'Modern organization of qualifications', *The Qualification Workshop – What Can a Lean Qualification Look Like?* 13/14 September 2018, Berlin, Germany.

EudraLex, 2011, *Volume 4 – EU Guidelines for Good Manufacturing Practice for Medicinal Products for Human and Veterinary Use – Annex 11: Computerised Systems.*

EudraLex, 2013, *Volume 4 – EU Guidelines for Good Manufacturing Practice for Medicinal Products for Human and Veterinary Use – Part 1 – Chapter 2: Personnel.*

EudraLex, 2014, *Volume 4 – EU Guidelines for Good Manufacturing Practice for Medicinal Products for Human and Veterinary Use – Part 1 – Chapter 6: Quality Control.*

EudraLex, 2015, *Volume 4 – EU Guidelines for Good Manufacturing Practice for Medicinal Products for Human and Veterinary Use, Annex 15: Qualification and Validation.*

ICH Q7, 2000, *Good Manufacturing Practice Guide for active pharmaceutical Ingredients. International conference on harmonisation of technical requirements for registration of pharmaceuticals for human use,* International Council for Harmonisation of Technical Requirements for Pharmaceuticals for Human Use.

ISO/IEC 17025, 2017, *General Requirements for the Competence of Testing and Calibration Laboratories,* International Organization for Standardization.

ISPE, 2008, *Good Practice Guide: Good Engineering Practice,* International Society for Pharmaceutical Engineering.

ISPE Baseline®, 2001, *Pharmaceutical Engineering Guides for New and Renovated Facilities Volume 5 (2001): Commissioning and Qualification 1 (Baseline Guide Volume 5).* International Society for Pharmaceutical Engineering.

Kavermann, H, 2017, 'Lab Layout and Equipment Qualification', *Modern Microbiology Laboratory,* 27–29 June 2017, Prague, Czech Republic.

Le Page, S, Dubourg, G, Rolain, JM, 2016, Evaluation of the Scan® 1200 as a rapid tool for reading antibiotic susceptibility testing by the disc diffusion technique, *The Journal of Antimicrobial Chemotherapy*, vol. 71(12), pp. 3424–3431.

Sutton, S, Singer, D, 2011, Microbiological best laboratory practices, USP <1117> – value and recent changes to a guidance of quality laboratory practice with Don Singer, *American Pharmaceutical Review*, vol. 14(4), pp. 41–47.

US 21 CFR 11, 1997, 'Electronic records; electronic signatures', *Code of Federal Regulations*, Title 21, Food and Drug Administration.

US 21 CFR 211.25, 2018, 'Personnel qualifications', *Code of Federal Regulations*, Title 21, Food and Drug Administration.

US 21 CFR 211.63, 2012, 'Equipment design, size and location', *Code of Federal Regulations*, Title 21, Food and Drug Administration.

USP 39-NF 34 Chapter <1058>, 2016, 'Analytical Instrument Qualification', *United States Pharmacopeia*.

USP 39-NF 34 Chapter <1117>, 2016, 'Microbiological Best Laboratory Practices', *United States Pharmacopeia*.

USP 39-NF 34 Chapter <55>, 2016, 'Biological Indicators', *United States Pharmacopeia*.

WHO, 2006, *Annex 4 – Supplementary Guidelines on Good Manufacturing Practices: Validation*, WHO Technical Report Series, vol. 937, World Health Organization.

WHO, 2010, *Annex 1 – WHO Good Practices for Pharmaceutical Quality Control Laboratories*, WHO Technical Report Series, vol. 957, World Health Organization.

WHO, 2011, *Annex 2 – WHO Good Practices for Pharmaceutical Microbiology Laboratories*, WHO Technical Report Series, vol. 961, World Health Organization.

4

Introduction to Culture Media in Pharmaceutical Microbiology for Non-sterile Products

Marion Louis, Laurent Leblanc, and Félix A. Montero Julian

The Healthcare Business, bioMérieux, Lyon, France

CONTENTS

Pharmaceutical Microbiological Quality Assurance and Control: Practical Guide for Non-Sterile Manufacturing, First Edition. Edited by David Roesti and Marcel Goverde.

4.1 Introduction

The origin of culture media came from the concept defined by Louis Pasteur that an organism causing a disease can be grown outside its original environment. In 1873, Edwin Klebs was the first bacteriologist to separate mixed microbial culture using the dilution method. Then, the botanist Oscar Brefeld worked on the methodology to obtain pure culture on solid media. In 1881, Robert Koch obtained isolated colonies of *Bacillus anthracis* inside gelatin tube and introduced a new culture method called "poured plate." The end of the nineteenth century saw an explosion of the microbiological testing and the development of new culture technics that are still the basics of today's microbiology. All of them are founded on culture media.

A microbiological culture medium has the purpose to support the growth of different kinds of microorganisms, it might contain nutrients, growth factors, salts and minerals, sources of energy (like sugar), and other selective or promoting compounds according to the expected performance. Culture media are of fundamental importance for most pharmaceutical microbiological tests and were integrated early in varied Pharmacopoeias as the main technique to determine the microbial safety of the produced medicine.

The importance of ensuring the quality of drugs is an ancient concern; already in 1518, in England, the requirement of purity for a pharmaceutical compound was officially recognized. Then in 1561, came the title "Pharmacopoeia" (literally, "drug-making") applied for the first time to a book. The development of the practice of the medicine and the variety of drugs conducted to the creation of national Pharmacopoeias; each country compiled their own regulation texts regarding drug manufacturing.

With worldwide globalization came naturally the supranational and international harmonization of the regulatory texts that helped to harmonize the practices in the different countries in the world. Currently, 49 active Pharmacopoeia commissions exist, for 36 Pharmacopoeias worldwide. The last impactful harmonization for non-sterile drugs was implemented in 2009 with the move to harmonize pharmacopoeial microbiological methods for non-sterile products – the main participants being the US Pharmacopoeia (USP), European Pharmacopoeia (Ph. Eur.), and the Japanese Pharmacopoeia (JP) as stated in Table 4.1.

Table 4.1 Harmonized microbial enumeration tests of non-sterile products.

	USP	Ph. Eur.	JP
Microbiological examination of non-sterile products: microbial enumeration tests	<61>	2.6.12	4.05-I
Microbiological examination of non-sterile products: tests for specified microorganisms	<62>	2.6.13	4.05-II
Microbiological quality of non-sterile pharmaceutical products	<1111>	5.1.4	General Information G4

This harmonization became clearly an important need because of the fact that the initial tests described in the USP <61> Microbial Limit Tests before 2009 were not designed to be all-inclusive (detection of all potential pathogens) and the demonstration of the absence of objectionable microorganisms was clearly not the intent of the chapter. This concern was stated by the USP microbiology expert committee at that time in a one-page Stimuli to the Revision Process (USP 1982). Several outbreaks were reported in late 1960 (Price 1984) associated to pathogen-contaminated medications and in 1980, several papers described the capabilities of the *Burkholderia cepacia* to survive in disinfectants (Geftic *et al.* 1979). This situation was a real concern also for the FDA, that the current methods of USP <61> are not able to ensure that any objectionable microorganisms are present in products. Today in 2018, the *B. cepacia* contamination remains a concern (Marquez *et al.* 2017). Additional explanation is given in Chapter 11.

After the harmonization work, the USP chapter <61> Microbial Limit Tests were divided into two chapters: USP chapter <61> Microbial examination of non-sterile products: Microbial enumeration tests, and USP chapter <62> Microbial examination of non-sterile products: Tests for specified microorganisms. The harmonized USP versions are the mirror of the Ph. Eur. chapters 2.6.12 and 2.6.13 and the JP chapter 4.05 as stated in Table 4.1.

These harmonized texts described the microbiological methodologies to attest the safety of the pharmaceutical products. The microbial enumeration test is a simple method to count the number of CFUs in a non-sterile product or raw material. Concerning the growth media used for these methods, the harmonized chapter details two main aspects:

1) Growth promotion of the media (the ability of the media without product to support the growth of low numbers of typical test organisms).
2) Suitability of the enumeration method (the ability of the media to support the growth of low numbers of typical test organisms in the presence of the product).

The main components of the culture media are soybean-casein digest (SCD) agar or broth, potato dextrose agar, and sabouraud dextrose agar or broth. Potential neutralizing agents (glycine, lecithin, polysorbate, thioglycollate, thiosulfate, etc.) may also be added to the media composition.

The USP chapter <62> *Tests for specified microorganisms* specifically described the recommended solutions and culture media dedicated to microbial contamination testing, the ingredients needed for their preparation, and the procedure to verify their suitable properties. The range of culture media proposed varies from liquid (usually named broth) to solid (usually named agar). Thus, medium for the test of bile-tolerant Gram-negative bacteria called Enterobacteria enrichment broth-Mossel, the Violet red blue glucose agar as growth promoting and indicative of *Escherichia coli* and *Pseudomonas aeruginosa*. However, the most prominent change to USP chapter <62> was the enrichment scheme for *Salmonella* and *E. coli*. Formerly, these enteric contaminants were enriched in Lactose broth while the "topical" contaminants (*Staphylococcus aureus* and *P. aeruginosa*) were cultured in SCD broth. The new streamlined method prescribed having all of these "specified microorganisms" enriched in SCD, and the scheme changes significantly after enrichment.

Table 4.2 summarizes the approaches between the ancient and the recent harmonized enrichment approaches for the detection of *E. coli* and *Salmonella* strains.

SCD medium is a highly nutritious multipurpose medium which is used in the preparation of agar plates. Due to the inclusion of both tryptone and soy peptone in the theoretical formulation given as an example in the

Table 4.2 Differences between old and harmonized enrichment approaches.

	USP <62>/Ph. Eur. 2.6.13/JP 35.2	Harmonized USP <62>/Ph. Eur. 2.6.13/ JP 35.2
Escherichia coli	• Lactose	SCD (Soybean-casein digest broth)
	• Streak to MacConkey agar	MacConkey broth
	• If growth with typical morphology, transfer to eosin methylene blue	Streak to MacConkey agar
Salmonella	Lactose	SCD
	Selenite cysteine and tetrathionate	Rappaport Vassiliadis *Salmonella* enrichment broth
	Streak to brilliant green agar, xylose lysine deoxycholate agar, and bismuth sulfide agar	Streak to xylose lysine deoxycholate agar
	Stab streak triple sugar iron agar	Identify presumptive colonies

pharmacopoeial chapters, the medium will support the growth of many fastidious organisms without the addition of serum. In addition, this medium provides amino acids and long-chain peptides as nutritive components for microorganisms, while sodium chloride maintains the osmotic balance.

4.2 Culture Media Challenges and Development

Culture media origins are closely related to clinical diagnostics and in order to meet pharmaceutical requirements, the challenge of culture media was to adapt them to these different expectations.

While culture media are now very well described in the regulations, their development and production have several specificities that could impact the non-sterile product results.

In this chapter, we will review the main requirements and expectations for the development and the control of culture media in order to meet the best performances and guarantee patient safety.

4.3 Importance of Culture Media for Patient Safety

Safety is always taken for granted, but a lot goes into ensuring the safety of non-sterile products. Reliable culture media in terms of production and development indirectly support product safety which helps to protect the patient. Consequently, expectations on culture media performances, robustness, and reliability must be excellent while guaranteeing the maximum convenience regarding the use and storage. Therefore, numerous parameters have to be examined and taken into account for the development of highly performing culture media, as shown in Figure 4.1.

4.4 Culture Media Are all Different

Culture media compositions dedicated to pharmaceutical controls are described in pharmacopoeial chapters. However, the performances of culture media from different manufacturers can be highly different as they depend on several parameters such as the raw materials used to make the products (origin, purity, and variability), the manufacturing process (media temperature and cooling parameters), and the development.

Pharmacopoeias provide the minimum expectations to reach for the different parameters (performances, selectivity, and physical–chemical parameters); nonetheless, culture media manufacturers can strengthen the expectations required.

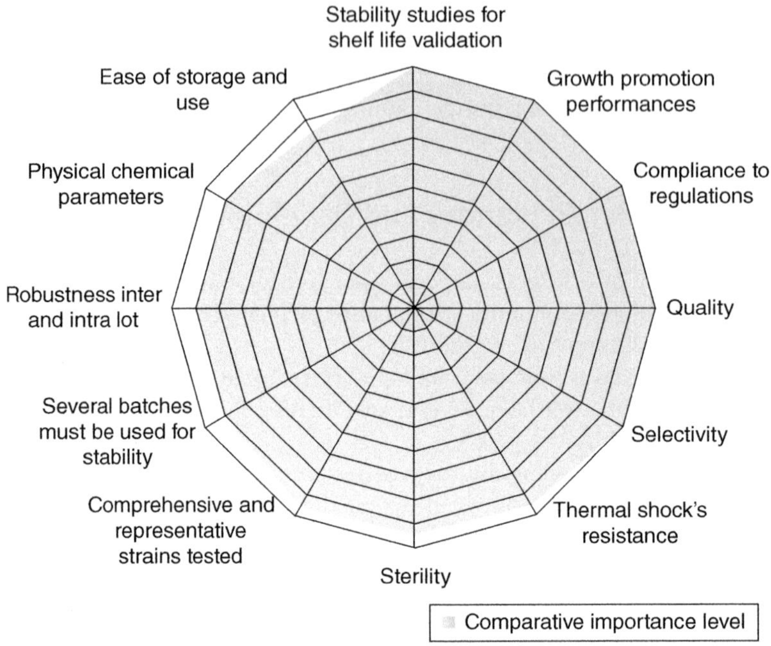

Figure 4.1 Example of product requirements for a culture media and their comparative importance level (from 0 to 10).

The general notice chapter of the Pharmacopoeia mentions that it may be necessary to modulate the concentration of some ingredients to meet optimal performances. Thus, the whole development of culture media to meet performance and quality expectations is extremely important and results from a delicate balance between the raw material selection, formulation, and production parameters.

4.4.1 Importance of Raw Materials

4.4.1.1 Origins

Culture media contain all the elements that microorganisms need to grow efficiently. All culture media are made with the same basic compounds:

- A carbon source (commonly provided by sugars). Also named carbohydrates, these products are produced industrially from fruits or vegetables (e.g. maize, rice, and wheat) which are digested by enzymatic hydrolysis to create saccharides.
- A source of amino acids and nitrogen (provided by different kinds of peptones). Sourced from an animal or vegetal origin, peptones are the products of a chemical (acid) or enzymatic (e.g. pepsine and trypsine) hydrolysis of

proteins. These proteins arise from muscles, organs, milk (casein), etc., for animals and from soya, maize, wheat, etc., for plants.

- Different salts and buffer (chemical products). Salts are important to provide the microorganisms essential elements for their growth. The main salt found in various formulations is NaCl, but others can be added to support the development of sensitive strains such as $MgCl_2$, K_2HPO_4, or $CaCl_2$. Buffers regulate the pH of the media and allow to maintain the optimal pH condition for microbial growth.
- Demineralized water.
- Other kinds of ingredients are also introduced in the formulations to give the media several properties. For solid media, the addition of a gelling molecule such as agar, for example, provides the properties required for Petri dish usage, and for selective media the supplementation of selective agent segregates resistant microorganisms that are enabled to grow.

4.4.1.2 Selection

The formulation of culture media is established after extensive work on the ingredient selection. Not all raw materials are suitable for all applications and even with the same origin, they can display a huge difference. As an example, casein is the main peptone used in culture media and depending on the origin of the milk and the different treatments it underwent, the growth quality can be dramatically different.

4.4.1.3 Variability and Controls

The majority of the compounds making culture media are from a biological origin (for example, algae for the agars or animal origins for the peptones) and are, by definition, subjected to natural variability. Environmental conditions impact directly the quality of the material. Drought, flooding, or natural climate variability will lead to adaptation of the metabolisms of the organisms and will modify the final performance of the ingredients that are manufactured. The variability of these ingredients impacts directly the quality of the culture media and, to guarantee for each batch of culture media optimal performances, a modulation of their concentrations can be necessary.

The control of this variability is at the heart of the knowledge of culture media manufacturers. Each batch of raw material is evaluated and, according to its quality, its concentration is adjusted to get the perfect level of performance. In addition, the interaction of the different components in the final formulation is performed. It may happen that some interactions between ingredients occur and prevent the usage of the product in a specific formulation but allows the usage in another. This is only noticed with mastering each ingredient that the reproducibility and robustness of the culture medium can be assured. For instance, Figure 4.2 shows adjustments of the peptone concentrations from different batches to reach the targeted performance defined for the culture medium lot release.

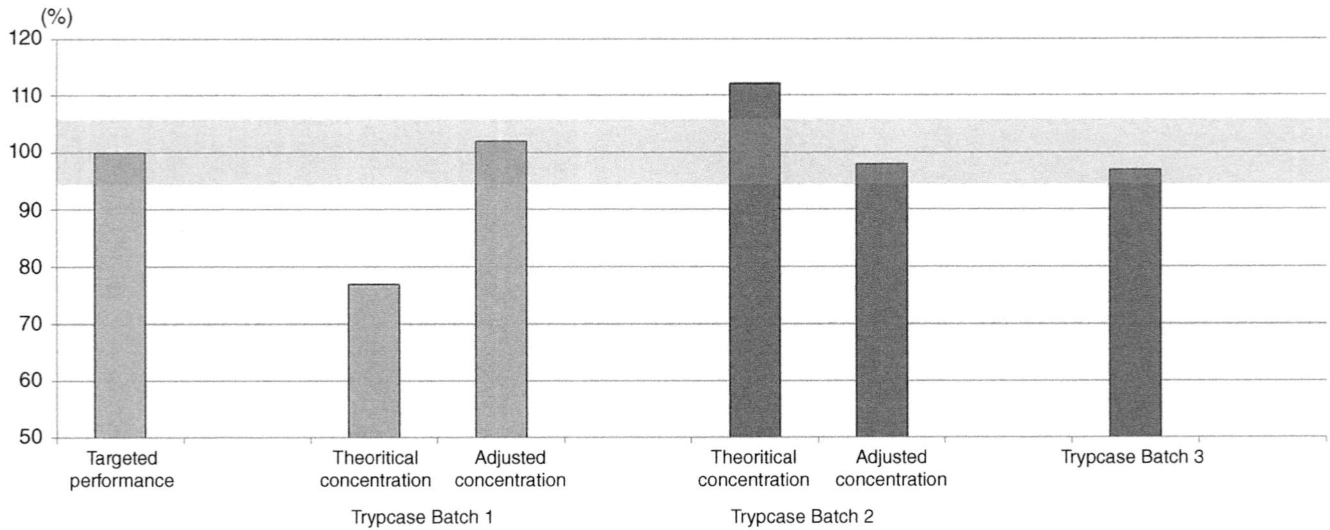

Figure 4.2 Example of adjustment of the trypticase concentration to adapt the global medium performance at the required specification.

4.4.2 Manufacturing Process

The industrial production of culture media requires a stable and validated process. The homogeneity of the production is one of the most important parameters. Mixing, sterilizing, and pouring are important steps of the process that are validated to deliver the good quality media. The formulations are made with sensitive products that must be handled with care all along the production to avoid any degradation of their quality. Some products are thermosensitive and must be introduced in the medium after sterilization, others are light sensitive and must be protected from the light. Each medium has its own manufacturing challenges and then a dedicated and validated process (e.g. heating temperature, F_0-sterilization, and cooling conditions). However, they all follow the overall same core process as described in Figure 4.3.

4.4.3 Development of Culture Media

The product development process has a strong impact on the quality and performances of culture media.

The way culture media have been designed and validated, and how their performances have been challenged during the whole shelf life including critical parameters such as thermal shocks, growth promotion tests with wild strains in addition to the pharmacopoeial test organisms, etc., provide the level of reliability of the manufacturer and its products.

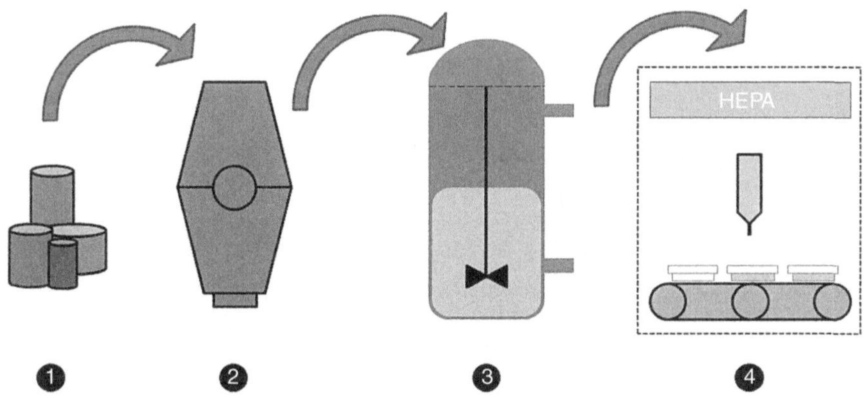

Figure 4.3 Main manufacturing steps for a culture medium: Step 1: The raw materials are selected, tested, and released by the quality control (QC). Step 2: The ingredients are blended and homogenized together in a thin powder. Step 3: The powder is mixed with water and sterilized in a tank. Non-thermosensitive liquid additives may be added with this step. Step 4: Pouring of the medium in its final container (it can be a Petri dish, a tube, or a flask). During this step, thermosensitive additives may be added in an aseptic environment (HEPA filter, ISO 5). Additional packaging steps follow the manufacturing of the medium and permit the protection and the shipment of the product to the final user.

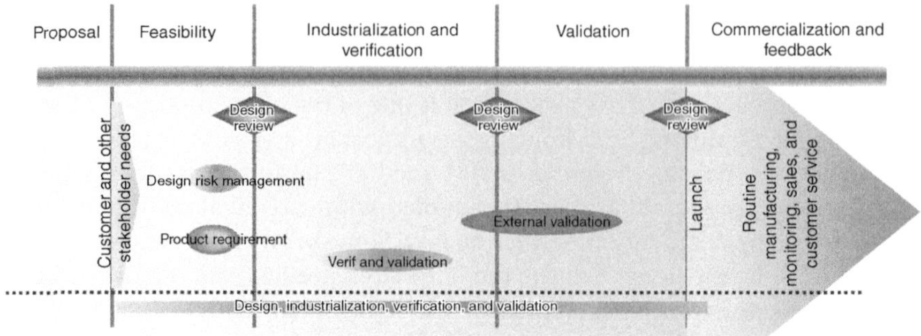

Figure 4.4 Example of a product development process.

Culture media manufacturers should implement a defined process of product development under quality assurance to guarantee robustness of the media performances and processes with regular committee controls. Such design controls are notably described in ISO 9001 and ISO 13485 "Medical devices – Quality management systems" (§7.3 design and development) and 21CFR Part 820 "Quality System Regulation" (subpart C §820.30 design control) (Figure 4.4).

Such processes strengthen the reliability of the culture media for the long term as the aim for end users is finally to validate the media once and for all and to avoid any possible switches to other suppliers or culture media references due to quality/performance issues.

4.4.4 Stability Studies

All prepared media should be labeled and used before the expiration date. The expiry date or shelf life of the culture medium should be validated under the routine production conditions which also include packaging, transportation, and storage. The performances of a culture medium must be guaranteed during its entire shelf life. Those performances should be defined on several parameters of key importance to ensure the right level of microbial detection in product control or environmental monitoring.

Therefore, during the development of a culture medium product, several important requirements must be assessed to meet the expected performances and robustness during the entire shelf life of the product. Here are some examples of parameters (not exhaustive) that need to be evaluated at each time point to define the shelf life of a culture medium:

- Regular growth promotion of a wide panel of microorganisms described in the different Pharmacopoeias and also with house isolates representative of the pharmaceutical environments.

- Regular neutralization properties of the culture medium to ensure either good neutralization of residual disinfectants on surfaces or preservatives in product formulations.
- Regular test and challenge of the inhibitory properties for selective media.
- Regular physical–chemical points of control, such as moisture/exudation, media shrinkage/cracks on the plates, pH, and agar strength.
- Inter batch robustness: several batches must be monitored for stability.
- Intra batch robustness to confirm homogeneity of performances within the batch (beginning, middle, and end of the batch).

Prior to initiating the stability studies, the validation batches should be split into two groups: one group undergoing a thermal shock treatment (see section 4.4.4.1) and the other group being stored at the final storage temperature. The thermal shock cycle should mimic the shipment conditions of culture media in different parts of the world, and should also be correlated to the final temperature storage conditions. Indeed, specific storage conditions such as flexible storage temperatures between 2 and 25 °C imply a different thermal shock profile compared to culture medium traditionally stored at 2–8 °C. During the stability studies, as long as the culture medium meets the performance expectations at each time point, then the shelf life evaluation goes on, until two consecutive time points on the thermal shock-treated batches with lower performances than the non-thermal shock-treated batch are detected.

The end of the shelf life is defined when two successive data points do not meet the performance expectations on the different items that are followed during the whole shelf life. The objective is to prevent any false negative results and avoid any consequences for product quality and patient safety. A schematic representation of the stability studies is represented in Figure 4.5.

4.4.4.1 Thermal Shocks

Culture media are submitted to potential temperature variations since their production release until their final use and must be robust enough to provide homogeneous performance results whatever the location of the final user is. The logistic channel is extremely important to control.

Considering this requirement, stability studies are performed by manufacturers to evaluate culture media performances until final delivery to pharmaceutical sites. These stability data must be gathered as part of the quality management and product development process to qualify the shelf life.

During the development, thermal stress should be applied to mimic changing conditions that could occur from shipment, storage conditions, and changing temperatures related to weather.

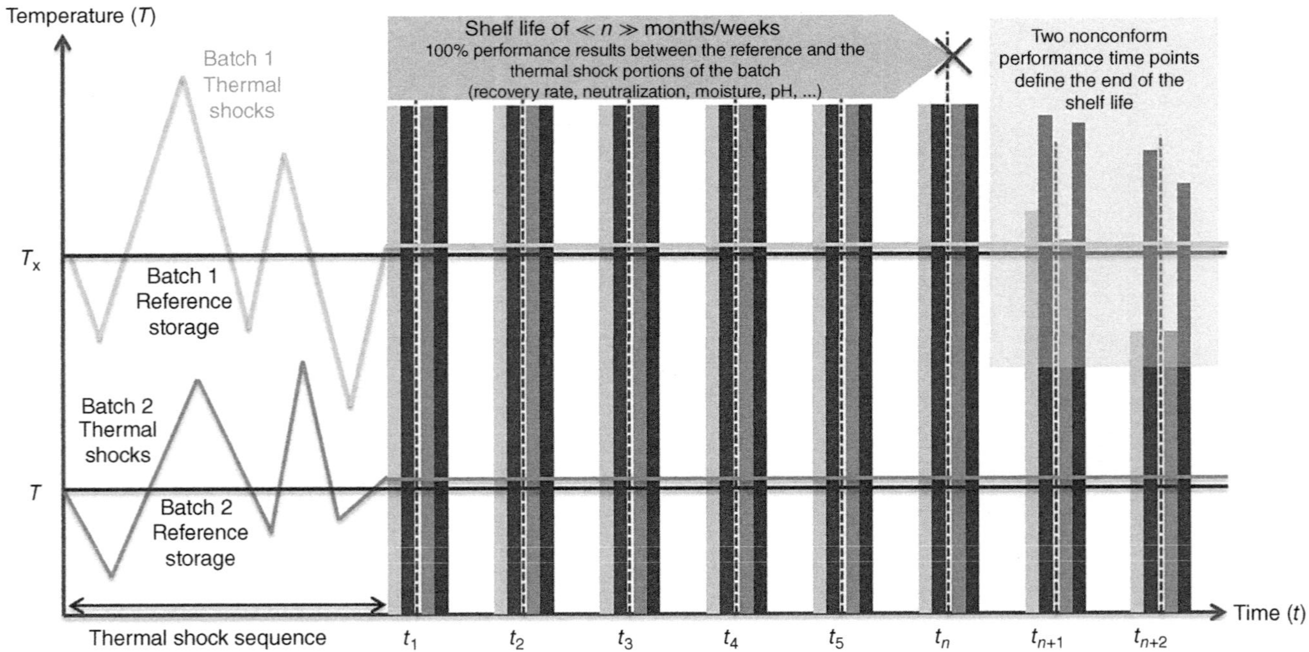

Figure 4.5 Example of a stability study of different culture media batches including thermal shocks sequence.

4.5 Innovation in Regard to Culture Media

The general notice chapter of the Pharmacopoeia mentions that it may be necessary to modulate the concentration of some ingredients to meet optimal performances. While still conforming to the culture medium-described formula, it gives to culture media manufacturers a short window for the improvement of culture media in order to offer the best performances and add innovation as the pharmaceutical environment is still evolving.

4.5.1 Objectionable Organisms Recall

Pharmaceutical environment keeps evolving and continuous culture medium innovation is important to strengthen the control of non-sterile products.

For example, a recent summary of drug product recalls has revealed that more than 70% of the non-sterile product recalls between 2004 and 2011 were for objectionable microorganisms and not exceeding microbial limits (Sutton and Jimenez 2012).

In terms of the microorganisms implicated, 54% were for *B. cepacia*, 15% for some *Pseudomonas* including *P. aeruginosa*, 21% for unspecified fungi, 11% for members of the family Enterobacteriaceae, and 9% for *Bacillus cereus*.

Burkholderia cepacia poses a special risk for manufacturers of health and personal care products. Actually, at the time of writing this chapter, the USP is drafting a chapter <60> *Microbiological Examination of Nonsterile Products – Tests for Burkholderia cepacia Complex* for a specified microorganism testing of *B. cepacia* complex (USP PF 2018). Some innovative media have been developed to target and detect *B. cepacia* complex. For example, a specific chromogenic culture media for the detection of *B. cepacia* complex were developed as a general-purpose medium used for total aerobic microbial counts (TAMC) with an additional chromogenic substrate that enables the specific detection of *P. aeruginosa* and *Burkholderia* species. The development of such media clearly strengthens the non-sterile product controls and patient safety. It indeed brings additional peace of mind in knowing that the sample does not harbor high-risk objectionable organisms like *B. cepacia*.

Manufacturers will now have the possibilities to follow recommendations for the detection and to use specific media allowing the growth and easier detection of these species. Additional explanation is written in Chapter 11.

Culture media needs to be regularly improved and adapted to the evolution of the environment and patient healthcare in order to help non-sterile manufacturers to anticipate new needs.

4.5.2 Increase Media Flexibility and Ease of Use

Culture media composition used to control the non-sterile products are described in regulations in terms of theoretical formulation, production, and physical–chemical parameters which leaves little place for innovation to take place. Thus, manufacturers of culture media are cutting their developments to provide non-sterile manufacturing sites more flexibility and ease of use for their controls such as higher shelf life to reduce quality control release tests or extended range of temperature storage.

4.6 Quality Controls

4.6.1 Quality Release Test Performed by the Culture Media Manufacturer (External Provider or In-house Media Manufacturer)

Each batch of culture media produced must be controlled in order to detect any manufacturing issues that could have occurred such as raw material variability, sterilization or autoclaving issue (see Section 4.7), or personnel error during the QC manipulation.

These batches should be submitted to a stringent quality control that must include physical–chemical parameters testing (pH and aspect of the media), growth promotion test with the pharmacopoeial strains, inhibitory properties for selective media, and absence of microbial contamination.

The growth promotion and selectivity tests reflect at least the requirements from the Pharmacopoeia in terms of quality-controlled strains, inoculum size, incubation time, and temperature (Table 4.3).

For the nutritive media, the following acceptance criteria must be achieved according to the Pharmacopoeia (e.g. Ph. Eur. 2.6.12):

- **Liquid media:** Clearly visible growth, comparable to that previously obtained with a previously tested and approved batch of medium.
- **Solid media:** The growth obtained must not differ by a factor greater than two from the calculated value for a standardized inoculum. For a freshly prepared inoculum, the growth of the microorganisms is comparable to that previously obtained with a previously tested and approved batch of medium.

For the selective media, the following acceptance criteria must be achieved according to the Pharmacopoeia (e.g. Ph. Eur. 2.6.13, see Table 4.4):

- **Test for growth-promoting properties, liquid media:** Inoculate a portion of the appropriate medium with a small number (not more than 100 CFU) of the appropriate microorganism. Incubate at the specified temperature for not more than the shortest period of time specified in the test. Clearly visible growth of the microorganism comparable to that previously obtained with a previously tested and approved batch of medium occurs.

Table 4.3 Requirements for the Growth Promotion Test according to Ph. Eur. chapter 2.6.12.

		Growth promotion	
Microorganisms	**Preparation of test strain**	**Total aerobic microbial count**	**Total yeasts and mold count**
Staphylococcus aureus such as: ATCC 6538 NCIMB 9518 CIP 4.83 NBRC 13276	Casein soybean digest agar or casein soybean digest broth 30–35 °C 18–24 h	Casein soybean digest agar and casein soybean digest broth ≤100 CFU 30–35 °C ≤3 days	–
Pseudomonas aeruginosa such as: ATCC 9027 NCIMB 8626 CIP 82.118 NBRC 13275	Casein soybean digest agar or casein soybean digest broth 30–35 °C 18–24 h	Casein soybean digest agar and casein soybean digest broth ≤100 CFU 30–35 °C ≤3 days	–
Bacillus subtilis such as: ATCC 6633 NCIMB 8054 CIP 52.62 NBRC 3134	Casein soybean digest agar or casein soybean digest broth 30–35 °C 18–24 h	Casein soybean digest agar ≤100 CFU 30–35 °C ≤5 days	–
Candida albicans such as: ATCC 10231 NCPF 3179 IP 48.72 NBRC 1594	Sabouraud dextrose agar or Sabouraud dextrose broth 20–25 °C 2–3 days	Casein soybean digest agar ≤100 CFU 30–35 °C ≤5 days	Sabouraud dextrose agar <100 CFU 20–25 °C <5 days
Aspergillus brasiliensis such as: ATCC 16404 IMI 149007 IP 1431.83 NBRC 9455	Sabouraud dextrose agar or potato dextrose agar 20–25 °C 5–7 days or until good sporulation is achieved	Casein soybean digest agar and casein soybean digest broth ≤100 CFU 30–35 °C ≤3 days	Sabouraud dextrose agar <100 CFU 20–25 °C <5 days

- **Test for growth-promoting properties, solid media:** Perform the surface-spread method, inoculating each plate with a small number (not more than 100 CFU) of the appropriate microorganism. Incubate at the specified temperature for not more than the shortest period of time specified in the test.

Table 4.4 Growth promoting, inhibitory and indicative properties of media from Ph. Eur. chapter 2.6.13.

	Medium	Property	Test strains
Test for bile-tolerant Gram-negative bacteria	Enterobacteria enrichment broth-Mossel	Growth promoting	*Escherichia coli* *Pseudomonas aeruginosa*
		Inhibitory	*Staphylococcus aureus*
	Violet red bile glucose agar	Growth promoting + indicative	*E. coli* *P. aeruginosa*
Test for *E. coli*	MacConkey broth	Growth promoting	*E. coli*
		Inhibitory	*S. aureus*
	MacConkey agar	Growth promoting + indicative	*E. coli*
Test for *Salmonella*	Rappaport Vassiliadis *Salmonella* enrichment broth	Growth promoting	*Salmonella enterica subsp. enterica* serovar Typhimurium or *S. enterica* subsp. *enterica* serovar Abony
		Inhibitory	*S. aureus*
	Xylose lysine deoxycholate agar	Growth promoting + indicative	*S. enterica* subsp. *enterica* serovar Typhimurium or *S. enterica* subsp. *enterica* serovar Abony
Test for *P. aeruginosa*	Cetrimide agar	Growth promoting	*P. aeruginosa*
		Inhibitory	*E. coli*
Test for *S. aureus*	Mannitol salt agar	Growth promoting + indicative	*S. aureus*
		Inhibitory	*E. coli*
Test for clostridia	Reinforced medium for clostridia	Growth promoting	*Clostridium sporogenes*
	Columbia agar	Growth promoting	*C. sporogenes*
Test for *Candida albicans*	Sabouraud dextrose broth	Growth promoting	*C. albicans*
	Sabouraud dextrose agar	Growth promoting + indicative	*C. albicans*

Growth of the microorganism comparable to that previously obtained with a previously tested and approved batch of medium occurs.

- **Test for inhibitory properties, liquid or solid media**: Inoculate the appropriate medium with at least 100 CFU of the appropriate microorganism. Incubate at the specified temperature for not less than the longest period of time specified in the test. No growth of the test microorganism occurs.
- **Test for indicative properties:** Perform the surface-spread method, inoculating each plate with a small number (not more than 100 CFU) of the appropriate microorganism. Incubate at the specified temperature for a period of time within the range specified in the test. Colonies are comparable in appearance and indication reactions to those previously obtained with a previously tested and approved batch of medium.

Additional parameters can be tested such as specific strains included to the growth promotion test, aspect of media, weight, labeling, etc., to strengthen the release control of the batches.

Furthermore, to the pharmacopoeial strains, non-sterile manufacturers might include in-house and objectionable organisms during the growth promotion test once they have been selected to have a particular interest for the company (e.g. according to risk analysis, already found during environmental monitoring).

4.6.2 Quality Control Test Performed on Ready-to-Use Culture Media Purchased from External Manufacturers

As described previously, manufacturers of culture media are responsible to release conform culture media that are controlled according to at least pharmacopoeial specifications for each batch manufactured. Moreover, they also need to determine the resistance of the culture media during development by applying variations of thermal shocks in order to simulate the shipment.

However, despite all these controls and securities, non-sterile manufacturers should control themselves the conformity of the culture media they receive from an external provider. This control should occur for each batch number and for each delivery in order to specifically address any potential issues that could have occurred during the specific shipment of those batches. The thermal shock sequences tested during the culture media development cannot be exactly the same as the ones submitted in reality and in some cases, some products could be submitted to higher or lower temperatures than the ones tested.

As a consequence, batches of culture media must be controlled by the user according to the pharmacopoeial requirements before use in order to leverage this potential risk.

The controlled batch to release must be controlled according to a *reference batch* and the recovery rate must be between 50 and 200% for solid media.

"Reference batch" could be interpreted in different ways and could impact the recovery rate finally obtained for the controlled batch to be released. The counts of this "reference batch" could mean:

- Counts obtained on a previous validated batch but not in parallel of the controlled batch to release.
- Counts obtained on a previous validated batch and in parallel of the controlled batch to release.
- Counts of the controlled batch from the certificate of analysis of the culture media supplier.

What is important to leverage during the growth promotion test for a new batch of media is the limitation of the discrepancies in test conditions. Generally, the best conditions are to compare batches (batch to release and reference batch) from the same media (same reference) and to respect the same incubation conditions and same inoculum. In most of cases, the controlled batch to release and the reference batch are from the same reference number and are tested in parallel with the same inoculum preparation and incubation conditions.

For example, the counts of a batch of selective agar can hardly be compared to the ones of a nutritive agar as the media compositions are totally different with different uses and objectives. Selective agar contains inhibitory agents that could slightly impact the growth of strains.

Moreover, two batches of the same media should be controlled according to the same conditions to avoid non conform recovery rate due to possible discrepancies concerning the incubation conditions and the inoculum concentration. In the Pharmacopoeia, the only recommendation regarding inoculum is that it must be not more than 100 CFU for a nutritive as well as selective media so when the growth promotion test is not performed at the same time for the controlled batch to release and the reference batch, a batch can be rejected due to inoculum preparation discrepancies while the batch remains conform.

In some cases, it is not possible to test a batch to release with a previous validated batch in parallel as the reference batch is expired. In this case, the value obtained previously on this reference batch can be used to calculate the recovery rate but users should respect as close as possible the conditions to be able to have reproducible and reliable results.

4.6.3 Importance of the Quality Control Strains

Pharmacopoeial strains are used to challenge the fertility properties of the culture media. The recovery obtained is compared to previous validated batches and allows to release or reject the batch.

As a consequence, the strain calibration has an important impact on the culture media batch release. If the strain inoculum is not well calibrated or if the number of CFU aliquoted on culture media is lower than expected, it could lead to reject a conform batch (leading to all the investigation and recontrols necessary).

The possible reasons explaining a low recovery could be:

- Variability of the strain stability that is not adequately validated.
- Defect of homogeneity of the strain solution during preparation and inoculation (e.g. vortex failure and pipetting error from the operator or from a non-calibrated pipette).
- Storage conditions not respected or thermal shocks that could lead to excessive stress of the microorganisms.
- For calibrated strains: when the inoculum is not precise and accurate enough.

It is then recommended to use calibrated, precise, and accurate inoculum with robustness and homogeneity to have trustful results and evaluate the risk to have discrepancy recovery rate due to the strain heterogeneity of their inoculum.

4.6.4 Outsourcing Strategy and How to Perform an Audit at a Growth Media Partner

Ready-to-use media has become more important than in-house prepared media, especially explained by the high constraints that need to be respected during the development, selection of raw materials, and manufacturing parameters to guarantee high and consistent performances. Developing and producing culture media needs a specific expertise and regulatory authorities are used to question and challenge this approach in order to determine the correct management of the culture media performances that have a key impact on the pharmaceutical product to release.

Moreover, using ready-to-use culture media in many cases allows to save technician time and money. As a consequence, pharmaceutical strategy for culture media consists in mostly outsourcing this critical and specific approach. In this case, auditing growth media suppliers is necessary to strengthen the outsourcing strategy for culture media. Usually, the frequency of a routine audit is between two and three years. A paper audit can be initially proposed in order to provide all the important information (certification, accreditation, and quality management system) before going directly to the site.

Usually the audit can begin with a presentation of the company, the description of the audit objectives and scope, followed by a visit of the manufacturing plant according to the normal flow of production: from the reception of raw materials until the preparation to the delivery in order to cover the whole flow of the culture media produced. The quality control laboratory can also be audited to control their procedures, methods of testing, and Good Laboratory Practices.

Audits can then be focused on the quality management, production, facilities and equipment, packaging and labeling, and laboratory and validation.

Within the scope of the audit, the following elements can be considered in order to leverage the capability of the supplier to develop, produce, and ship the culture media under the coverage of a quality system management.

4.6.4.1 Quality System

- Change control: how the culture media manufacturer manages their changes
- Deviation and out of specification management
- Complaints follow-up: trends and actions
- Investigations and CAPA: following complaints or deviations
- Training of the personnel
- Raw material reception and controls
- Shelf life validation: performances data all along the shelf life and stability studies
- Storage and transportation validation
- Batch manufacturing documentation
- Traceability of the product from the raw material reception to the production of the end finished product
- Qualification of suppliers

4.6.4.2 Facilities and Equipment

- Master Validation Plan
- Equipment qualification
- Equipment change control and revalidation
- Autoclave revalidation program
- Calibration
- Maintenance
- Environmental monitoring program: procedures, trends, and follow-up

Regular routine audits allow to strengthen the management of culture media by the manufacturer in order to better understand their organization, procedures, and production, but also to leverage the production of culture media that is part of the pharmaceutical products release control. Pharmaceutical industries and culture media manufacturers should work closely in order to agree on the criticism the media could have for the pharmaceutical site, and to reinforce their controls to provide the more adapted culture media solution.

4.7 Culture Media Troubleshooting

Culture media are made from biological components and consequently are exposed to a natural variability. In addition, depending on all the different manufacturing parameters (e.g. filling and cooling temperature and autoclave), transportation and storage conditions, and potential manipulation error, the media or the results obtained on the media may be questioned.

To leverage these elements, it is recommended to launch a robust validation or to use ready-to-use media for which manufacturers have already done all the evaluations and validations.

Hereafter, different examples of problems encountered with growth media are illustrated.

4.7.1 Temperature Storage Issues

Culture media are recommended by their manufacturers to be stored within a certain range of temperatures during its shelf life. Outside this tolerance, performances of the media may be affected.

4.7.1.1 Storage Conditions Below 2 °C

Generally, culture media are recommended to be stored at a temperature superior to 2 °C to avoid any performance drifts. Negative temperatures may freeze the media and have an impact on the agar network (see Figure 4.6) that can affect the performances (e.g. fertility and inhibitory properties).

Thus, media are not recommended to be stored below 2 °C, and a specific attention must be paid to the storage location as the temperature can decrease below 2 °C close to the fridge wall where the cooling unit is located. In this area, culture media may be subjected to negative temperatures.

The storage of the culture media in their original box can avoid being directly in contact with these negative temperatures.

4.7.1.2 Exceptional Excursions of Temperature During Shelf Life

If culture media are subjected to isolated excursions of temperature, their performances may be impacted. Thus, the impacted media are generally not used.

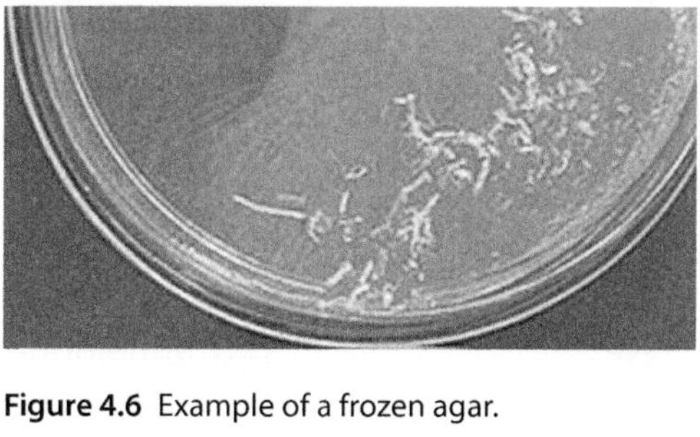

Figure 4.6 Example of a frozen agar.

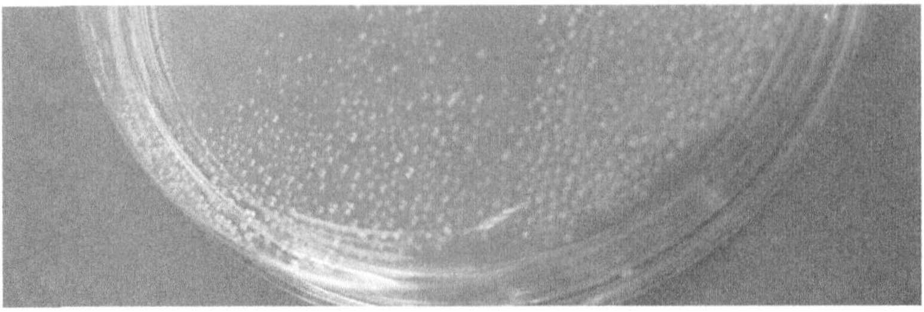

Figure 4.7 Example of water condensation.

If users really insist in using the concerned media, a deviation must be opened and the manufacturer must be contacted to have their recommendations on the media performance.

4.7.2 Water Condensation and Excessive Moisture

Water is an important part of the culture media composition and in some occasions, some water condensation could be observed on the lid or on the surface of the culture media.

This condensation is generally resulting from thermal shocks that could have occurred between the filling and the cooling of the media (humidity is condensed on the lid, as shown in Figure 4.7) or shipment/storage of the media.

In little quantity, this condensation does not have any impact on performances or microbiological state of the media but could result in user discomfort (difficulty to manipulate the plate or in some occasions, the water at the surface of the media could result in the swarming of some bacteria).

The difficulty results in the determination of which condensation level is acceptable for the pharmaceutical industries that use these media to release their products.

4.7.3 Fertility Issues

When a batch of media is rejected due to fertility failure, investigation must be initiated to understand the origin of the defect. Investigation could follow a 5 M (Ishikawa, see Figure 4.8) methodology to determine the root cause and address the correct CAPA.

Preliminary verification can consist in confirming that the media was not expired at the time of the control and that it did not experience defects such as dehydration that could have decreased the growth-promoting properties.

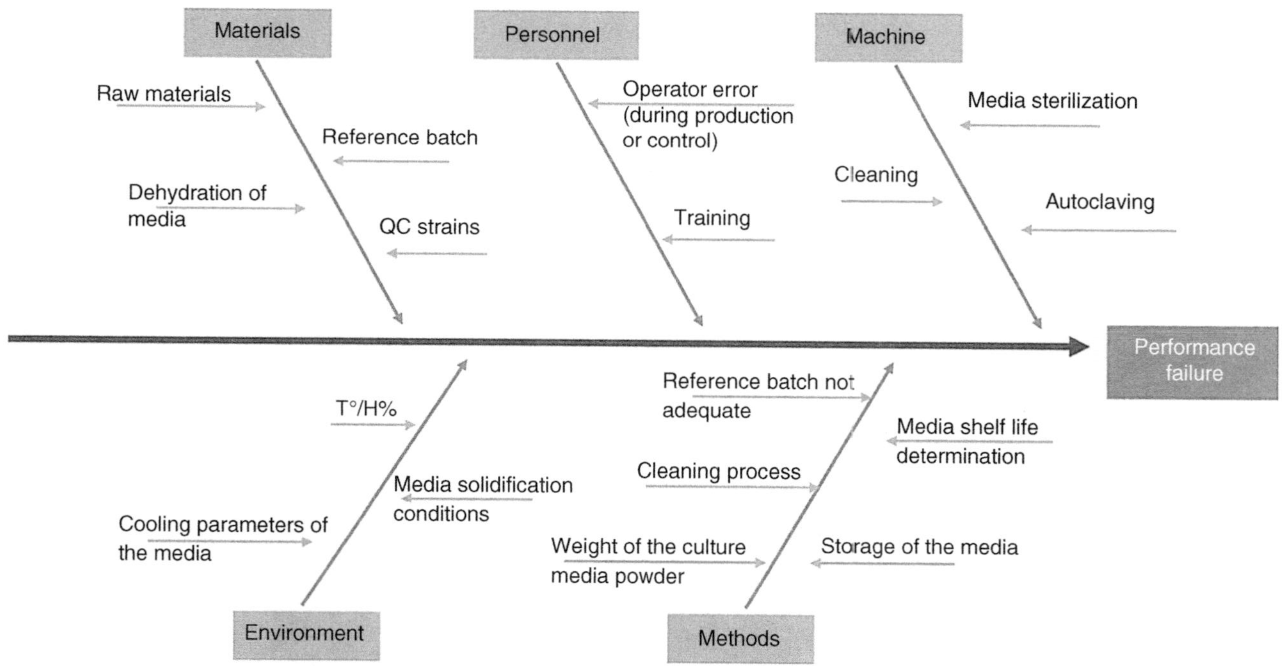

Figure 4.8 Example of Ishikawa methodology for a growth promotion failure investigation.

Specific attention must also be paid to the whole production process of the culture media as critical factors may impact directly the fertility of the media such as the raw material origins and variability, the sterilization of the media (temperature and time), and the autoclaving parameters.

Raw materials are from biological origins, and could be variable from one batch to another one. This variability could be minor without any impact on the performances, but in other cases, it could affect the growth properties of the media. In this case, robust developments of the media to determine the raw materials' origin and their possible variability are really important to avoid missing any growth defect for some strains not detected only during quality control.

Sterilization and autoclaving steps could also have an impact on the media fertility properties. These parameters must be set accordingly and controlled if any fertility deviation is detected.

The methods used for the quality control must also be investigated to leverage any potential false negative result due to manipulation error.

4.7.4 Crystals in Xylose Lysine Deoxycholate (XLD) Culture Medium

Crystals are known defects sometimes visible on XLD culture medium that could be mixed up with a contamination as shown in Figure 4.9. These crystals are the results of the crystallization of the salt raw materials and do not alter the performances of the culture media.

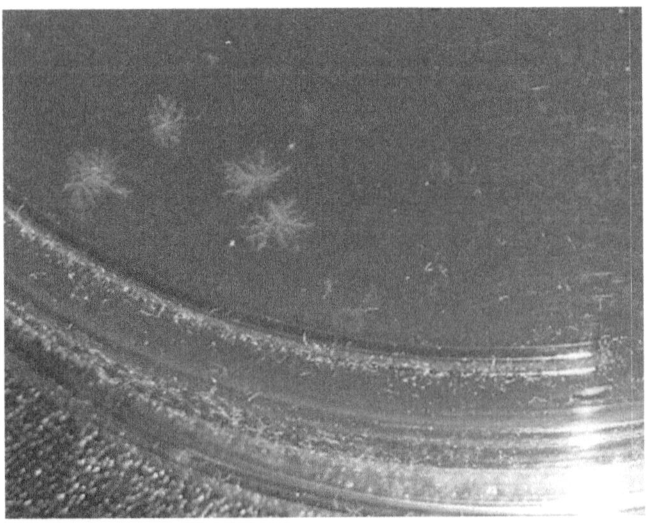

Figure 4.9 Crystal on XLD agar.

4.8 Conclusion

Culture media play a key role in the quality control of the pharmaceutical industry and thereby in patient safety. Their importance is more and more highlighted, from their development (requirements and expectations, and selection of raw materials), production until their final control and use, they must provide reliable results, allowing non-sterile manufacturers to be confident in the microbiological quality of products released on the market.

Moreover, due to the continuous evolution of the patient environment and safety needs, culture media must be adapted and should continuously be innovative. As described in this chapter, the development and manufacture of culture media is a complex and sophisticated process that manufacturers of culture media must recognize for its user value. Manufacturers should work closely with non-sterile manufacturers to respond to their needs to the benefit of the patient.

Bibliography

Geftic, SG, Heymann, H, Adair, FW, 1979, 'Fourteen-year survival of *Pseudomonas cepacia* in a salt solution preserved with benzalkonium chloride', *Appl. Microbiol.*, vol. 37(3), pp. 505–510.

ISO 13485, 2016, '*Medical Devices – Quality Management Systems*', International Standards Organization.

ISO 9001:2015, 2015, '*Quality Management System*', International Standards Organization.

JP 17th Edition, Chapter <4.05>, 2017, 'Microbial Limit Test' Japanese Pharmacopoeia.

Marquez, L, Jones, KN, Whaley, EM, Koy, TH, Revell, PA, Taylor, RS, Bernhardt, MB, Wagner, JL, Dunn, JJ, LiPuma, JJ, Campbell, JR, 2017, 'An outbreak of Burkholderia cepacia complex infections associated with contaminated liquid docusate', *Infect. Control Hosp. Epidemiol.*, vol. 38(5), pp. 567–573.

Ph. Eur. 9th Edition chapter 1, 2018, 'General Notices', European Pharmacopoeia.

Ph. Eur. 9th Edition chapter 2.6.12, 2018, 'Microbiological Examination of Non-sterile Products: Microbial Enumeration Tests', European Pharmacopoeia.

Ph. Eur. 9th Edition chapter 2.6.13, 2018, 'Microbiological Examination of Non-sterile Products: Test for Specified Microorganisms', European Pharmacopoeia.

Price, J, 1984, 'Establishing microbiological specifications for pharmaceutical not required to be sterile', *Pharm. Manuf.*, pp. 26–29.

Sutton, S, Jimenez, L, 2012, 'A review of reported recalls involving microbiological control 2004-2011 with emphasis on FDA considerations of "objectionable organisms"', *Am. Pharm. Rev.*, vol. 15, pp. 42–57.

US 21 CFR 211.820, 2018, 'Quality System Regulation' subpart C §820.30 design control.
USP, 1982, 'Microbial contamination of sterile and nonsterile articles, with special reference to *Pseudomonas cepacia*', *Pharmacopeial Forum*, vol. 8(4), p. 2239.
USP 41-NF 36 chapter <61>, 2018, 'Microbiological Examination of Nonsterile Products: Microbial Enumeration Tests', United States Pharmacopeia.
USP 41-NF 36 chapter <62>, 2018, 'Microbiological Examination of Nonsterile Products: Tests for Specified Microorganisms', United States Pharmacopeia.
USP PF 29(5), n.d., In-Process Revision: '<61> Microbial Examination of Nonsterile Products: Microbial Enumeration Tests', *Pharmacopeial Forum*.
USP PF 29(5), n.d., In-Process Revision: '<62> Microbial Examination of nonsterile products: Test for Specified Microorganisms', *Pharmacopeial Forum*.
USP PF 44(5), 2018, In-Process Revision: '<60> Microbiological Examination of Nonsterile Products – Tests for Burkholderia cepacia Complex'.

5

Microbiological Examination of Non-sterile Final Dosage Forms and Raw Material Including Acceptance Criteria and Testing Frequency

David Roesti

Novartis Pharma Stein AG, Stein, Switzerland

CONTENTS

Pharmaceutical Microbiological Quality Assurance and Control: Practical Guide for Non-Sterile Manufacturing, First Edition. Edited by David Roesti and Marcel Goverde.

5.1 Microbiological Acceptance Criteria

5.1.1 Final Dosage Forms

The microbiological acceptance criteria of non-sterile drug products are generally based on the tripartite harmonized informational chapters of the Ph. Eur./USP and JP and are composed of a quantitative total aerobic microbial count and total yeasts/molds count as well as the absence of specified microorganisms (see Table 5.1). Even if these chapters are in the informational section of the respective pharmacopoeia, these acceptance criteria are generally followed. In addition, the US 21CFR 211.165 requires that *There shall be appropriate laboratory testing, as necessary, of each batch of drug product required to be free of objectionable microorganisms* which leaves a wide area for interpretation and would increase the selection of specified microorganisms to be included in the microbiological examination of non-sterile products. Actually, in a PDA 2013 survey (PDA 2013), more than 85% of respondents from pharmaceutical, over-the-counter drug products, medical devices, and cosmetics answered that they were performing microbiological testing according to the harmonized pharmacopoeia chapters and 32% were screening for the absence of additional objectionable microorganisms.

Specified or Objectionable Microorganisms?

The specified microorganisms are the ones that are listed in the respective Pharmacopoeia chapters. This is generally considered as a minimum expectation to fulfill based on the route of administration. However, this list is not exhaustive and the product manufacturer has also to take into account other microorganisms that are potentially objectionable for their product. Actually, the pharmacopoeia writes that in addition to the microorganisms listed, the significance of other microorganisms should be evaluated. Pharmacopoeia also writes that where warranted, a risk-based assessment of the relevant factors is conducted by personnel with specialized training in microbiology and in the interpretation of microbiological data. If the level or the type of the detected contaminating microorganisms cannot exclude a hazard for the product or the recipient, the release of a drug product should be evaluated and justified very carefully even if the acceptance criteria are not exceeded.

The PDA TR-67 writes that *When a drug manufacturer that markets nonsterile pharmaceuticals is trying to comply with the FDA mandate delineated in 21 CFR 211.84, CFR 211.113 and CFR 211.165, it must be clear that those products are "free of objectionable microorganisms."* PDA TR-67 also recommends that product testing and decisions about which products will or will not be routinely tested should be based on risk and that *in addition to drug manufacturers, excipient and API manufacturers need to ensure that their materials are free of objectionable microorganisms, particularly dosage forms whose pharmaceutical ingredients are included in the formulation*. The risk is dependent on the nature of the product, manufacturing process, and route of administration. For instance, for solid oral dosage forms with a low water activity (<0.75) a special screening for objectionable microorganisms may not be necessary.

For products such as topicals, liquid oral solutions, inhalants, or nasal solutions where there may be a concern for microbiological contamination, isolates from the plate counts, as well as enrichment testing, should be identified and evaluated by a risk assessment.

Already some health authorities have indicated that only the absence of the pharmacopoeia-listed specified microorganisms may not be sufficient in some dosage forms. For example, the Australian Therapeutic Goods Administration (TGA) consider that all Pseudomonad bacteria (and not just *Pseudomonas aeruginosa*) are objectionable organisms in a non-sterile medicine that is intended for topical use. In addition, FDA have already expressed their concern with the presence of *Burkholderia cepacia* in non-sterile, water-based drug products and specifically for inhalation products, liquid oral solutions, as well as topicals (Torbeck *et al.* 2011; FDA 2017).

If a type of microorganism is considered objectionable for the product in question following a risk assessment and if during manufacturing the controls in place cannot guarantee its absence, the routine release testing would need to include this microorganism an additional absence of specified microorganism test. Risk assessment and definitions of objectionable microorganisms are further described in detail in Chapter 11.

Other country-specific Pharmacopoeia have also similar acceptance criteria limits (e.g. British, Indian, or Korean Pharmacopoeia) as the harmonized chapters even if some minor differences do arise (e.g. Chinese Pharmacopoeia) as indicated in Table 5.1.

According to Ph. Eur. 5.1.4, USP <1111>, and JP G.4, when an acceptance criterion for microbiological quality is prescribed, it is interpreted as follows:

Table 5.1 Acceptance criteria of final dosage forms based on harmonized Ph. Eur. 5.1.4, USP<1111>, and JP G4 and additional criteria based on ChP 1107.

Route of administration	TAMC	TYMC	Specified microorganism
Dry powders and capsules for inhalation	10^2 CFU/g	10^1 CFU/g	Absence in 1 g: *Pseudomonas aeruginosa* *Staphylococcus aureus* Bile-tolerant Gram-negative bacteria
Oromucosal, gingival applications	10^2 CFU/g or ml	10^1 CFU/g or ml	Absence in 1 g, or 1 ml *P. aeruginosa* *S. aureus* ChP In addition, absence of *Escherichia coli* in 1 g, 1 ml or 10 cm^2
Cutaneous application	10^2 CFU/g or ml	10^1 CFU/g or ml	Absence in 1 g, or 1 ml *P. aeruginosa* *S. aureus*
Nasal application	10^2 CFU/g or ml	10^1 CFU/g or ml	Absence in 1 g, or 1 ml *P. aeruginosa* *S. aureus* ChP In addition, absence of *E. coli* in 1 g, 1 ml or 10 cm^2
Auricular application	10^2 CFU/g or ml	10^1 CFU/g or ml	Absence in 1 g, or 1 ml *P. aeruginosa* *S. aureus*
Preparations for vaginal use	10^2 CFU/g or ml	10^1 CFU/g or ml	Absence in 1 g, or 1 ml *P. aeruginosa* *S. aureus* *Candida albicans*
Transdermal patches/ Transdermal therapeutic systems	10^2 CFU/ system	10^1 CFU/ system	Absence in 1 system: *P. aeruginosa* *S. aureus*
Aqueous preparations for oral use	10^2 CFU/ml	10^1 CFU/ml	Absence in 1 ml *E. coli* ChP In addition, absence of *Salmonella* in 10 g or 10 ml for chemical and biological preparation containing mineral and raw materials of animal or vegetal origin

Table 5.1 (Continued)

Route of administration	TAMC	TYMC	Specified microorganism
Nonaqueous preparations for oral use	10^3 CFU/g or ml	10^2 CFU/g or ml	Absence in 1 g or ml: *E. coli* ChP In addition, absence of *Salmonella* in 10 g or 10 ml for chemical and biological preparation containing mineral and raw materials of animal or vegetal origin
Rectal use	10^3 CFU/g or ml	10^2 CFU/g or ml	No additional requirements ChP Absence of *S. aureus* and *P. aeruginosa* in 1 g or 1 ml
Ph. Eur. Special provision for oral dosage forms containing raw materials of natural (animal, vegetal, or mineral) origin for which antimicrobial pretreatment is not feasible and for which the competent authority accepts TAMC of the raw material exceeding 10^3 CFU/g or CFU/ml	10^4 CFU/g or ml	10^2 CFU/g or ml	Not more than 10^2 CFU of bile-tolerant Gram-negative bacteria in 1 g or 1 ml Absence of *Salmonella* in 10 g or 10 ml Absence of *E. coli* in 1 g or 1 ml Absence of *S. aureus* in 1 g or 1 ml
Ph. Eur. Special provision for premixes for medicated feeding stuffs for veterinary use using excipients of plant origin for which antimicrobial treatment is not feasible.	10^5 CFU/g or ml	10^4 CFU/g or ml	Not more than 10^4 CFU of bile-tolerant Gram-negative bacteria in 1 g or 1 ml Absence of *E. coli* in 1 g or 1 ml Absence of *Salmonella* in 25 g or 25 ml

10^1 CFU: maximum acceptable count = 20;
10^2 CFU: maximum acceptable count = 200;
10^3 CFU: maximum acceptable count = 2000, and so forth.

Mathematically this interpretation is not correct. Yet, the authors of the harmonized chapters wanted to include the variability of microbiological testing within the limits. Indeed, it is considered that counts within a factor 2 range are within the typical variability of microbiological testing. Therefore, for a 10^3 CFU acceptance criteria, a maximum acceptable count of $2 \times 1000 = 2000$ CFU is acceptable *as long as no objectionable microorganisms are present* (evaluation via a risk assessment, see text above and Chapter 11). Some health authorities may not accept this interpretation and require that the mathematical interpretation is strictly used (e.g. 10^3 CFU = 1000 CFU). The manufacturer should then adapt the interpretation based on the local regulations in which the product is marketed.

What to Do in Case a Product Is Marketed in Multiple Countries with Different Acceptance Criteria?

The manufacturing site may either:

- Test the final dosage form with microbiological limits based on the Pharmacopoeia with the most strictest acceptance criteria. This option is especially relevant for bulk testing of the final dosage form prior to packaging.
- Release the product based on the local specification of the manufacturing site and retest in the country in which the product is sold according to the local market legislation.
- Cross-validate the different test methods to show equivalence and test only one method. This option is also relevant for bulk testing.
- Test each different packaged product as per the local registration of the country in which the product is sold. This is the option requiring the most resources for the site.

5.1.2 Raw Materials

For raw materials such as drug substances and excipients, the harmonized informational chapters of the Pharmacopoeia (Ph. Eur. 5.1.4, USP <1111>, JP G.4) define general acceptance criteria for Microbiological Quality of Nonsterile Substances for Pharmaceutical Use with a total aerobic microbial count (TAMC) limit of 10^3 CFU/g or ml and TYMC of 10^2 CFU/g or ml.

Individual monographs may be more stringent or require the absence of additional specified microorganisms. For example, the USP monograph for lactose monohydrate requires a TAMC of not more than 1×10^2 CFU/g and TYMC of not more than 5×10^1 CFU/g and includes the absence of *Escherichia coli* in 1 g. There are nonetheless not many individual monographs which contain requirements for microbiological testing.

The ICH Q6A writes that for microbial limits of new drug substances that total count of aerobic microorganisms, the total count of yeasts and molds, and the absence of specific objectionable bacteria (e.g. *Staphylococcus aureus, E. coli, Salmonella, P. aeruginosa*) should be tested and determined using pharmacopoeial procedures. *The type of microbial test(s) and acceptance criteria should be based on the nature of the drug substance, method of manufacture, and the intended use of the drug product.* Raw materials which are extracted or manufactured from material of natural origin may be highly contaminated with microorganisms. Examples of such raw materials are calcium salts, starches, gelatin, acacia gum, guar gum, dyestuffs, lactose, magnesium stearate, and celluloses. Spore-forming microorganisms generally predominate since the water activity of such excipients is very low and bacterial spores would better survive desiccation steps during the excipient's production. Also molds may proliferate in hygroscopic excipients. Furthermore, if the storage conditions do not prevent humidity from penetrating the raw material bags, microorganisms could develop to a critical level.

Actually some companies have also the policy to define the acceptance criteria of the raw materials based on the product for which these materials will be used. For instance, if the excipient is intended for a product used as a cutaneous application, the excipient should also be free of *S. aureus* and *P. aeruginosa*.

If the same excipient is used for different product categories, the incoming (inbound) test would need to cover the acceptance criteria of the different products. For instance, if lactose is used in a powder for inhalation, this lactose would also be tested for the absence of *P. aeruginosa, S. aureus*, and Bile-tolerant Gram-negative bacteria. In addition, where deemed necessary and based on a risk assessment, additional objectionable microorganisms may also be screened in raw material testing since the specified microorganisms of the pharmacopoeia may not include other potential objectionable microorganisms (e.g. *B. cepacia*).

Inbound testing of raw materials may not be required if the material is not specifically susceptible to a microbial contamination, storage conditions are under control, and purchased from a certified supplier.

5.1.3 Internal Out of Expectation (OOE) Limits

In addition to the quantitative specifications, internal out of expectation (OOE) limits may be defined. The OOE limit is served to flag unexpected high counts that may indicate a drift from controlled manufacturing conditions. If this limit is exceeded multiple times it may indicate an adverse trend and would require investigation to determine the impact on the product's quality or patient safety as well as the root cause of contamination and corresponding actions. Exceeding the OOE limit would not necessarily lead to product rejection since the limit is below the specification but this decision is dependent on the investigation and the type of microorganisms identified in the product. If the outcome of the OOE investigation is to release the excipient, API, or drug product, the investigation report should preferably be supplemented with a risk assessment.

If a sufficient amount of data is available, the OOE level could be defined based on historical results (refer to Chapter 10) and if not enough data are available (e.g. for products in development), the OOE level can be kept at a proportion of the specification level (e.g. 30 or 50%).

5.2 Testing Frequency

5.2.1 Final Dosage Forms

The testing frequency of non-sterile products generally depends on the criticality of the route of administration, susceptibility of the product to enable microbial growth (e.g. water activity and preservatives), the composition (e.g. amount of excipients of natural origin), local regulatory requirements, the manufacturing process (e.g. amount of available water after granulation), the verification of the suitability of the test method, the analytical results from the "antimicrobial effectiveness test" as well as the history of the product (refer to Table 5.2 for examples).

The ICH Q6A advises that the drug product should be tested unless its components (e.g. *raw materials incl. water*) are tested before manufacture and the manufacturing process is well controlled via validation of decontamination processes which have been proven effective.

A simplified decision tree is available in the ICH Q6A (#8 Microbiological attributes of non-sterile products) to justify the testing frequency. According

Table 5.2 Examples of critical and noncritical factors for microbial contamination.

Critical	Noncritical
Product containing raw materials from natural origin (animal, vegetal, and mineral), especially if unprocessed or weakly processed	Synthetic raw materials
Water activity >0.75	Water activity <0.6
Wet granulation and aqueous coating	Dry granulation and tableting
pH value 6–8	pH value <3 or >10
Long holding times of unpreserved aqueous solutions at room temperature (e.g. >24 hours)	Short hold times (e.g. <8 hours)
Inhalation, nasal sprays	Oral and rectal application
High water activity formulation without preservative	Semi-solid with experimentally proven preservative system

to the ICH Q6A, microbiological quality testing may be reduced based on a scientific justification taking into account the following factors:

- Microbiological quality of excipients and APIs
- Manufacturing process validation
- Intrinsic microbiological properties (ingredients, process, etc.)
- History of good microbiological quality

Table 5.3 provides examples of testing frequencies based on the criticality of the final dosage forms.

5.2.1.1 Which Frequency to Set by Skip-lot Testing?

The testing frequency is typically based on the susceptibility of the product to become contaminated, the maximum tolerable interval for which no results are available and operative/financial reasons (*even if this should not be opposed to quality*!). For instance, a testing frequency of every tenth lot may be selected because in case of deviations, the number of lots between the last good result and the failed one is not too large and the lots in between may be retested during the deviation investigation. Obviously, testing every hundredth lot is not relevant considering the low probability of detecting contamination due to the low sample volume and reduced testing frequency. Some cases may justify a frequency based on a time period and not on amount of batches (e.g. monthly or quarterly testing); for instance, in cases where a very high number of lots or products are produced on the same manufacturing line.

In order to support the decision to reduce the testing frequency with data, it is recommended that the first lots are tested before switching to skip-lot testing (e.g. first 10 or 20 lots).

Table 5.3 Suggested testing frequencies based on product route of administration.

Route of administration	Testing frequency
Dry powders and capsules for inhalation	Lotwise
Oromucosal, gingival applications	Lotwise
Cutaneous application	Lotwise or skip-lot (if well preserved)
Nasal application	Lotwise or skip-lot (if well preserved)
Auricular application	Lotwise or skip-lot (if well preserved)
Preparations for vaginal use	Lotwise or skip-lot (if well preserved)
Transdermal patches/Transdermal therapeutic systems	Lotwise or skip-lot (if well preserved)
Aqueous preparations for oral use	Lotwise or skip-lot (if well preserved)
Nonaqueous preparations for oral use	Skip-lot or no testing
Rectal use	Skip-lot or no testing

FDA Exception to Skip Lot

Whereas the tripartite ICH Q6A guideline was also signed by the FDA, since end of 2013, some FDA assessors pointed out that skip-lot testing does not comply with 21 CFR 211.165(a) and (b).

> Sec. 211.165 Testing and release for distribution:
>
> (a) For **each batch** of drug product, there shall be appropriate laboratory determination of satisfactory conformance to final specifications for the drug product, including the identity and strength of each active ingredient, prior to release. Where sterility and/or pyrogen testing are conducted on specific batches of short-lived radiopharmaceuticals, such batches may be released prior to completion of sterility and/or pyrogen testing, provided such testing is completed as soon as possible.
> (b) There shall be appropriate laboratory testing, as necessary, of **each batch** of drug product required to be free of objectionable microorganisms.

Indeed, according to the new interpretation of the FDA, if acceptance criteria for microbial enumeration tests including specified microorganisms are written in the company's product release specification, they should be tested batch-wise. However, FDA are also of the opinion that, microbial limits testing may be **omitted** from the product release specification provided adequate upstream microbiological controls are established and documented.

FDA recommends omitting the microbial limits specification by addressing several points (Metcalfe 2013):

- Identification and justification of critical control points in the manufacturing process that could affect microbial load of the drug product.
- Describe microbiological monitoring and acceptance criteria for the critical control points that have been identified. Verifying the suitability of the testing methods for the drug product. Conformance to the acceptance criteria established for each critical control point should be documented in the batch record in accordance with 21 CFR 211.188.
- Activities taken when microbiological acceptance criteria are not met at control points should be described.
- The results of microbial limits testing performed on exhibit or stability batches of the drug product should be provided.
- Microbial limits testing should be performed at least at the initial stability testing time point.

A fictive example of justification omitting microbial limits testing is provided below:

Rationale for Omitting Microbial Limits Testing of Product A Film-coated Tablet

Composition and Manufacturing Process of Product A Film-coated Tablet
Product A film-coated tablet (FCT) is composed of the drug substance XX, and the excipients XY, XZ, and YY. Every incoming batch of the excipients used in the manufacture of Product A FCT are analyzed for microbiological quality.

The experimentally determined water activity of Product A FCT is less than 0.6 which would not enable microbial proliferation since according to USP <1112>, a water activity of more than 0.6 is necessary to enable microbial growth.

The manufacturing process of Product A FCT can be summarized as follows:
- *add description of the manufacturing process*

According to the manufacturing process described above, microbial reducing steps take place during process steps *process step 1* and *process step 2*. Microbial enhancing steps are *process step 3* and *process step 4*. For these two critical steps, a holding time for the granulation solution and the coating suspension is defined as maximum X hours based on an experimental study and has been validated in the manufacturing process.

Microbiological Control and Monitoring
The microbiological control program at facility X for the manufacturing Product A FCT is based on applicable health authority guidelines and cGMPs. The control and monitoring strategy is highlighted below and all procedures are defined in SOPs. If a result from the monitoring activities exceeds the defined levels, an investigation is performed including a product impact assessment if necessary. Therefore, the drug product cannot be released until any deviations of the monitoring activities are decided and closed.

Microbiological contamination is controlled based on the following procedure:
- *List all microbial controls*

Microbiological contamination control in the manufacturing environment for Product A FCT is monitored according to the following program:
- *List the monitoring program*

Historical Microbiological Testing of Product A FCT
- *List the historical data (at least 20 results)*

Conclusion
Based on the above rationale, we propose not to perform the release microbiological examination test for the final dosage form of Product A FCT and this test will be omitted from the specifications.

5.2.2 Drug Substances and Excipients

If an excipient or a drug substance has microbiological requirements in their corresponding pharmacopoeia monographs, then they ought to be tested lotwise (or evaluated based on a certificate of analysis for a qualified supplier) as incoming control. For other drug substances and excipients, the ICH Q6A also provides guidance on selecting the testing frequency as described in decision tree #6: microbiological quality attributes of drug substance and excipients.

Another approach is to use a decision table to determine testing frequency of drug substances and excipients with different categories (see example Table 5.4) or risk assessment tools such as HACCP or FMEA (refer to Chapter 2).

Table 5.4 Example of raw material testing frequency decision table.

Name of drug substance/excipient	Product route of administration	Natural origin?	Water activity	Growth-promoting or antimicrobial properties?	Historical data	Testing frequency
Excipient 1	Inhalation	No	0.5	None	Always comply with the acceptance criteria	Lotwise since inhalation application
Excipient 2	Oral	Yes	0.4	None	Always comply with the acceptance criteria	Skip-lot testing
Excipient 3	Oral	Yes	0.4	None	A few out of specification cases	Lotwise due to historical data
Excipient 4	Topical	No	0.8	Growth promoting	Growth but comply with acceptance criteria	Lotwise due to growth-promoting properties
Drug substance 1	Topical	No	0.2	Antimicrobial	Always comply with the acceptance criteria	Skip-lot
Drug substance 2	Oral	No	0.2	Antimicrobial	Always comply with the acceptance criteria	Skip-lot or no testing
Drug substance 3	Inhalation	No	0.4	Antimicrobial	Always comply with the acceptance criteria	Skip-lot testing

Additional criteria to Table 5.4 may include, for instance, quantity of the material used in the manufacturing process, what is the final proportion in the final drug product, as well as growth-promoting or microbial-reducing steps during the manufacturing process in which the raw material is used.

The raw materials' inbound testing frequency may be reduced to annually (as part of the supplier verification program) if the supplier is certified and delivers a certificate of analysis with the required microbiological testing. Supplier certification should encompass the following considerations with regards to microbial contamination (not exhaustive):

- Source of material and raw material manufacturing process
- Quality performance of the material over a determined trial period
- Result of audits performed by the purchasing company as well as results of health authority inspections

5.3 Procedure if Microbial Growth Occurs in Routine Testing

If the specification is exceeded or if an adverse trend is acknowledged, then the microorganisms have to be identified and a deviation investigation has to be initiated. Deviation investigations are described in Chapter 12 (Figure 5.1).

For the more critical non-sterile drug products APIs or excipients (e.g. topicals, inhalants or nasal solutions, growth promoting products), even if the counts remain within expectation or specification levels, it is a good practice to identify the recovered microorganisms and assess if they are objectionable or not. Criticality may, for instance, depend on microbiological controls and monitoring, antimicrobial or growth-promoting steps during the manufacturing process, the nature of the product, the final dosage form's route of administration, and the historical results (refer to Table 5.2).

5.4 Sampling

The harmonized Pharmacopoeia chapters on microbiological examination of non-sterile products require a standardized volume of sample (see Table 5.5).

For typical release testing, the sample size should cover also volume needed for additional specified microorganism tests and a remaining volume for retests in case of deviations. This means that for instance at least 30 g for a nonaqueous preparation for oral use.

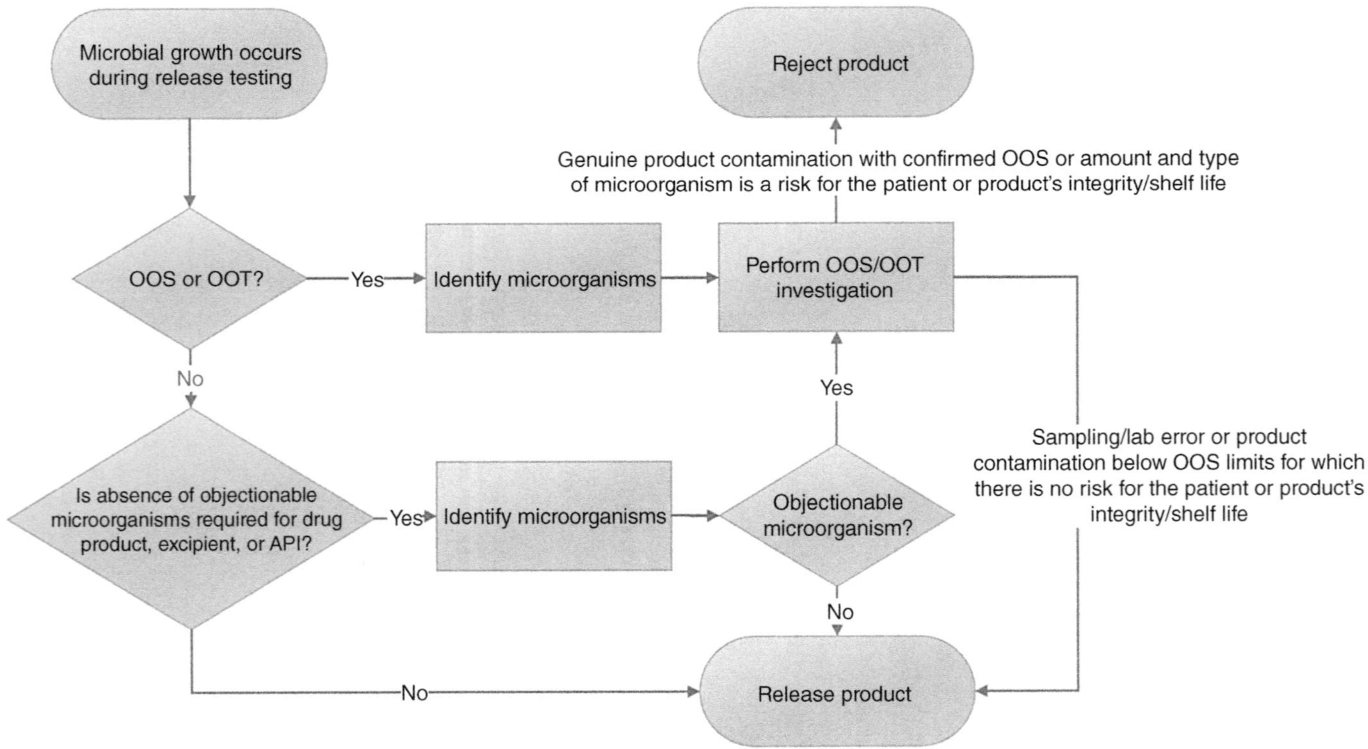

Figure 5.1 Example of procedure if microbial growth occurs. OOS, out of specification; OOT, out of trend; API, active pharmaceutical ingredient.

Table 5.5 Standard sample volume for microbiological examination of non-sterile products.

Type of product	Sample size
Solid product/drug substance/excipient	10 g
Liquid product/drug substance/excipient	10 ml
Solids or liquids in aerosol form	10 Containers
Transdermal patches	10 Patches
Oromusosal films	10 units
Drug substance when amount per dosage unit ≤1 mg	Not less than the amount present in 10 dosage units or 10 g or 10 ml of the product
Drug substance with limited sample quantity (<1000 ml or 1000 g)	1% of the batch
Products where the total number of entities in a batch is less than 200	2 units
Products where the total number of entities in a batch is less than 100	1 unit

Test Bulk or Packaged Final Dosage Form?

USP <610> supports testing of unpackaged drug products in case their design renders the sampling process difficult and prone to extraneous contamination risk. Typical products would be low-content inhaled and nasal drug products. In these cases, testing may be performed on the bulk product, i.e. final dosage formed that is not finally packaged instead of the finally packaged final dosage form.

When a drug product is manufactured for many countries, it may be valuable to test the final formulated product at the bulk level prior to packaging instead of testing each packaged product. This option should be approved by the health authorities during registration filing as some local legislations might not accept testing on bulk level.

Nonetheless, if testing is carried out on the bulk product alone, the primary packaging should be tested for microbial purity and the filling and packaging should be carried out under a microbiologically controlled area with a validated process. Testing should be performed on the packaged product if this is not the case.

Statistically, with a 10 g sample size the probability to detect a microbial contamination is very low. This is also due to the fact that microbiological contamination in a product is generally not homogeneously distributed and aggregates of microorganisms might be located in individual spots. The JP G4 actually recommends that in general, a mixture of samples must be randomly taken from at least different three portions of a same batch. Portions of the beginning, middle, and end of the batch campaign would be preferable to cover the length of the campaign.

Actually, for bulk drug product or for raw materials testing, the sampling should also cover different bags/containers. In this case, a typical rule is to apply the following formula:

$$n = \sqrt{N} + 1$$

This means if for excipient A you have received 22 separate bags from one batch, at least 6 bags need to be sampled. The test may be performed on a pooled sample containing an even amount of product/material from each container/bag.

Even if the products are considered non-sterile, extraneous contamination should be prevented during sampling, i.e. with trained operators wearing special gowning (e.g. Tyvek suite), gloves (e.g. nitrile or latex) that have been disinfected (e.g. with ethanol 70%) using either sterile or disinfected utensils and in sample bags or bottles that are sterile. For products with the tightest microbiological acceptance criteria the use of an additional sampling cabinet would be recommended. Low-level contaminations of non-sterile products through bad aseptic praxis can lead to avoidable deviations caused by contamination during sampling (Roesti 2012).

For most non-sterile products, a sample hold time does not need to be defined since microbial proliferation is not expected due to the nature of these products (e.g. low water activity and preservatives). In some cases where microbial proliferation would occur (e.g. in aqueous intermediates), then a sample hold time (=time from which the sample is taken until time start of analysis) should be defined. Failure to follow such sample hold time would lead to a different number of microorganisms that would not represent the conditions at the time of sampling. Generally, for growth-promoting solutions, a sample hold time of maximum 2 hours at room temperature or 24 hours at 2–8 °C is sufficient without additional validation.

5.5 Nutrient Medium Controls

Self-prepared nutrient medium may be submitted to a test for pH value, absence of microbial contamination, and growth-promoting properties before use. The pH value of ready prepared medium (= purchased medium) could also be based on the quality certificate of a certified supplier but as per USP <61>/Ph. Eur. 2.6.12, each batch must be also tested for growth-promoting properties.

If the growth medium control does not meet the requirements, the deviation procedure must be followed.

5.5.1 pH Value

The pH value should be checked after and the nutrient media must have cooled down to room temperature before measurement.

5.5.2 Absence of Microbial Contamination

The absence of microbial contamination is ensured on at least one container or filled agar plate of each batch. It is recommended that the incubation period at the respective medium-specific conditions lasts the maximum time defined (e.g. at least five days for soybean–casein digest agar [SCDA] and seven days for sabouraud dextrose agar [SDA]).

5.5.3 Growth Promotion Tests

Growth promotion of nutrient media used in microbial enumeration tests is described in detail in the USP <61> and follows the following procedure (refer to Table 5.6):

- Spread/inoculate the test organisms separately, at a concentration of ≤100 CFU, on/into the medium to be examined. The test strains must be obtained from an official culture collection (e.g. American Type Culture Collection [ATCC]). The microbial test inoculums may be either bought from a qualified supplier or prepared in-house according to a standard procedure.
- Incubate at the defined temperature for not longer than the minimum incubation time of the microbial enumeration tests.

  After growth, the following acceptance criteria must be fulfilled:
  - Liquid media: Clearly visible growth, comparable to that previously obtained with a previously tested and approved batch of medium.
  - Solid media: The growth obtained must not differ by a factor greater than 2 from the calculated value for a standardized inoculum. For a freshly prepared inoculum, growth of the microorganisms is comparable to that previously obtained with a previously tested and approved batch of medium.

Soybean–Casein Digest Broth (SCDB)

Pancreatic of casein	17.0 g
Papaic digest of soybean	3.0 g
Sodium chloride	5.0 g
Dibasic hydrogen phosphate	2.5 g
Glucose monohydrate	2.5 g
Purified water	1000 ml

Soybean–Casein Digest Agar (SCDA)

Pancreatic digest of casein	15.0 g
Papaic digest of soybean	5.0 g
Sodium chloride	5.0 g
Agar	15.0 g
Purified water	1000 ml

Sabouraud Dextrose Agar (SDA)

Dextrose	40.0 g
Mixture of peptic digest of animal tissue and pancreatic digest of casein (1 : 1)	10.0 g
Agar	15.0 g
Purified water	1000 ml

Table 5.6 Growth promotion test of agar and liquid media for TAMC and TYMC.

	Growth promotion test	
Microorganism	**TAMC**	**TYMC**
Staphylococcus aureus (e.g. ATCC 6538)	SCDA or SCDB ≤100 CFU 30–35 °C ≤3 days	–
Pseudomonas aeruginosa (e.g. ATCC 9027)	SCDA or SCDB ≤100 CFU 30–35 °C ≤3 days	–
Bacillus subtilis (e.g. ATCC 6633)	SCDA or SCDB ≤100 CFU 30–35 °C ≤3 days	–
Candida albicans (e.g. ATCC 10231)	SCDA or SCDB ≤100 CFU 30–35 °C ≤5 days	SDA ≤100 CFU 20–25 °C ≤5 days
Aspergillus brasiliensis (e.g. ATCC 16404)	SCDA or SCDB ≤100 CFU 30–35 °C ≤5 days	SDA ≤100 CFU 20–25 °C ≤5 days

Nonselective SCDA and SDA are used as growth media for the membrane filtration, pour plate, and surface spread method whereas nonselective SCDB is used for the Most-Probable-Number (MPN) method. PDA TR-67 also suggests to use blood agar or litmus milk agar for the screening of objectionable microorganisms where appropriate.

I Have on My Plate Enumerated 102 CFU, Is This Critical in Comparison to the Requirement of ≤100 CFU?

For growth promotion tests some health authority inspectors or compliance officers have a very strict interpretation of the ≤100 CFU requirement. They would expect that if the microbial counts exceed 100 CFU, that the test is considered invalid and that a deviation must be opened to investigate the exceeding number.

This is a far too strict interpretation of the USP's requirement since it does not take into account the typical variability of microbiological testing. With a target inoculum of 80-100 CFU it is probable that some counts may exceed 100 CFU. To reduce the risk of exceeding 100 CFU some testing laboratories use target inocula of 30-50 CFU. USP FAQ to the USP <61> chapter provides additional explanation to the interpretation of the acceptance criteria: *For example with an inoculum of 100 CFU, acceptable counts are: 100/2 = 50 CFU to 100 × 2 = 200 CFU. The factor is introduced to take account of the variability of the method.* Therefore, exceeding slightly the 100 CFU is microbiologically not relevant.

Key in the growth promotion test is that the same inoculum culture and inoculum volume is used when comparing two different media lots.

Growth promotion of nutrient media used in the absence of specified microorganism tests is described in detail in the USP <62> and the procedure depends on the medium property to be evaluated (refer to Table 5.7):

- **Growth-promoting properties** – Inoculate the medium with not more than 100 CFU. Incubate for not more than the shortest period of time specified in the test, i.e. if a period of 18–72 hours is specified, growth must occur within 18 hours.
  - Liquid media: Clearly visible growth (comparable to that of a previously tested and approved batch of medium).
  - Solid media: Growth of microorganisms (comparable to that of a previously tested and approved batch of medium).
- **Indicative properties** – Inoculate the solid medium with not more than 100 CFU. After incubation for a period within the specified range of the test, the colonies must be comparable in appearance and indication reactions to those obtained with a previously tested and approved batch of medium.
- **Inhibitory properties** – Inoculate the solid/liquid medium with at least 100 CFU. After incubating for not less than the longest period of time specified in the test, no growth of the test-organisms should occur, i.e. if a period of 18–72 hours is specified in the test, no growth may occur after 72 hours.

Table 5.7 Growth promotion test of agar and liquid media for the detection of specified microorganisms.

Medium	Properties	Test strains
Test for bile-tolerant Gram-negative bacteria		
Enterobacteria Enrichment Broth-Mossel (EEB)	Growth-promoting	*Escherichia coli* (e.g. ATCC 8739) *Pseudomonas aeruginosa*
	Inhibitory	*Staphylococcus aureus*
Violet Red Bile Glucose Agar (VRBGA)	Growth-promoting and indicative	*E. coli* *P. aeruginosa*
Test for *E. Coli*		
MacConkey Broth (MCB)	Growth-promoting	*E. coli*
	Inhibitory	*S. aureus*
MacConkey Agar (MCA)	Growth-promoting and indicative	*E. coli*
Test for *Salmonella*		
Rappaport Vassiliadis Salmonella Enrichment Broth (RVSEB)	Growth-promoting	*Salmonella enterica* (e.g. ATCC 14028)
	Inhibitory	*S. aureus*
Xylose Lysine Deoxycholate Agar (XLDA)	Growth-promoting and indicative	*S. enterica*
Test for *P. aeruginosa*		
Cetrimide Agar (CA)	Growth-promoting	*P. aeruginosa*
	Inhibitory	*E. coli*
Test for *S. aureus*		
Mannitol Salt Agar (MSA)	Growth-promoting and indicative	*S. aureus*
	Inhibitory	*E. coli*
Test for *Candida albicans*		
Sabouraud Dextrose Broth (SDB)	Growth-promoting	*C. albicans*
Sabouraud Dextrose Agar (SDA)	Growth-promoting and indicative	*C. albicans*

Shelf Life Validation of Nutrient Medium

During shelf life, the growth medium must keep its growth-promoting and physical–chemical properties. In order to validate a shelf life, the following approach may be followed:

At the beginning of the shelf life, test the pH of the medium and verify the growth promotion of a selection of microorganisms (e.g. compendial test strains and relevant in-house isolates) as described above. Three replicates per microorganisms at a minimum should be used. Store the media for a period of time determined as shelf life (e.g. six months) under the storage conditions that will be used. After storage, retest the pH and inoculate the medium using the same selection of microorganisms as for the beginning of shelf life. In parallel, inoculate a freshly prepared medium with the same microbial inoculum used for the six months stored medium.

As acceptance criteria, the pH must be within the defined range (e.g. 7.0–7.4) and the microbial recovery on the stored medium of each test strain should not differ by a factor 2 of the recovery of the freshly prepared medium. In addition, when pooling the data, the overall microbial recovery should not be significantly different between stored and freshly prepared medium. For this evaluation use either a classical statistical hypothesis test (e.g. 2 sample *t*-test if data are normally distributed) or a non-inferiority test.

The media should then not be used beyond the expiry date determined using the shelf life validation.

In order to comply with continuous process verification standards, it is also advisable to check the validity of the shelf life periodically (at least once per year) by carrying out growth promotion tests when a batch reaches the end of shelf life and compare the microbial recovery to a fresh medium batch.

5.6 Test Method Overview

Microbiological examination of non-sterile dosage forms is generally executed according to the methods described in the harmonized chapters of the Ph. Eur. 2.6.12/2.6.13, USP <61>, <62>, and JP 4.05 I/II.

The tests should be carried out aseptically under conditions designed to avoid accidental contamination of the product to be examined. To this end the tests are typically executed under a Unidirectional Air-Flow cabinet. The cabinet's testing environment should be tested periodically for air and surfaces using either settle plates or active air monitoring and contact plates to demonstrate that microbiological levels remain low (Figure 5.2).

Receipt of the samples and the test results (raw data), as well as their evaluation and interpretation are to be documented according to cGMP rules.

For the quantitative microbial enumeration testing, the membrane-filtration method, the plate-count methods, or the MPN method can be employed. The test method to be applied depends on the physical/chemical (e.g. solubility and miscibility) and the antimicrobial properties of the product. If possible, the membrane filtration method is to be preferred to the plate-count methods and the latter to the MPN method (Table 5.8).

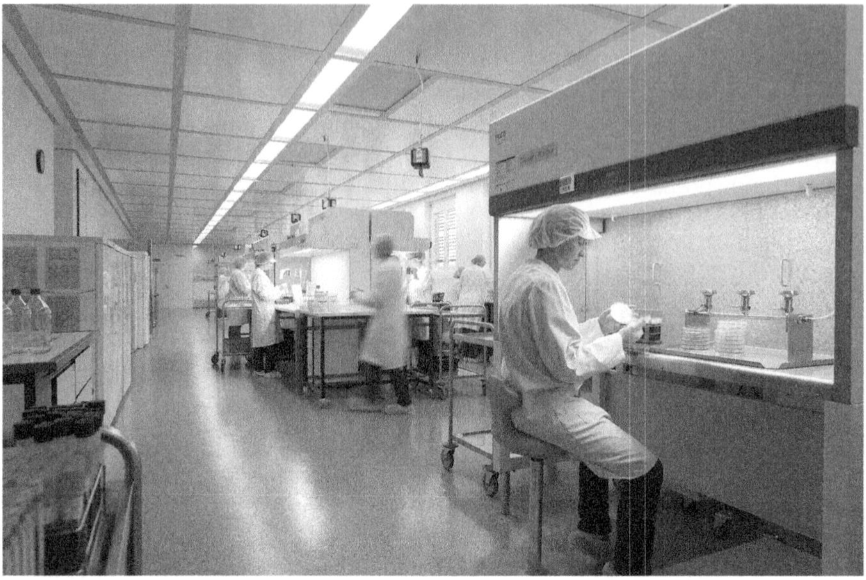

Figure 5.2 Example of laminar flow hoods within a microbiological laboratory. *Source:* Courtesy of BAV Institut GmbH Offenburg, Germany.

Table 5.8 Overview of test methods based on the nature of the product.

Nature of product	Method of choice
Aqueous, water-soluble, and oily products	**Membrane filtration** with subsequent incubation of the filters on appropriate, solid nutrient media *Whenever possible, use the membrane filtration method since it has a higher sensitivity.*
Products insoluble in water and fatty products	**Plate-count procedures** Transfer of the suspended or emulsified materials into (pour-plate) or onto nutrient media (surface-spread) which contain agar or **Enumeration using dilution series (MPN)** Transfer of the suspended or emulsified materials into series of tubes with liquid nutrient medium. *Whenever possible, use the pour-plate method.*

For the qualitative absence of specified microorganisms, the diluted or undiluted product to be examined is directly transferred into the enrichment media. After the specified incubation time, at a given temperature, subcultures are cultivated on selective agar plates or selective liquid medium.

5.7 Verification of the Suitability of the Method

If the pharmacopoeia compendial test method is used, it is not necessary to carry out an extensive analytical test method validation where many different validation criteria have to be fulfilled (e.g. limit of detection, accuracy, precision, etc.). Nonetheless, it should be verified that microorganisms present in a product can be recovered with the test method used and that the product does not inhibit microbial growth or affects the performance or reading of the method.

The method suitability is verified:

- With each new product or dosage form and/or its corresponding API or excipient.
- When there is a change in the test procedure.
- When there is a change in the product which could have an influence on the test result or microbial growth.

The suitability of the methods is checked with and without the presence of the product, preferably by three independent experiments. Some companies also check the suitability on three different batches to cover batch variation.

For very early phases of development, where not enough product is available for a suitability test, some companies use a standard test method (e.g. using several inactivators and dilutions) based on the product's composition without having executed a suitability test.

A bracketing approach can be used when performing method suitability for a drug product with various dosage strengths when the highest and lowest dosage strengths differ only by their API content. Placebos should be tested separately.

To verify testing conditions, a negative control is performed using the chosen diluent in place of the test preparation. There must be no growth of microorganisms. A failed negative control would require an investigation.

5.7.1 Sample Preparation

The verification of the method enables to determine the suitable dilution factor and type of diluent used. Typical diluents are buffered sodium chloride–peptone solution pH 7.0, phosphate buffer solution pH 7.2, or SCDB. The sample preparation detailed below is based on USP <61> and the author's experience:

- Aqueous or water-soluble products
  Dissolve or dilute 10 g or 10 ml of the product to be examined in the diluent to the appropriate dilution level (e.g. 1 : 10 with 10 ml or 10 g product in 90 ml diluent). The characteristics of the product may necessitate the use of

larger volumes. If necessary, adjust the pH value to pH 6–8. If appropriate or needed, prepare further dilutions using the same diluents.

- Nonfatty products, insoluble in water
  Suspend 10 g or 10 ml of the product to be examined in the diluent to the appropriate dilution level (e.g. 1 : 10). Help suspend e.g. by vigorously shaking in a suitable container which contains glass beads, or by constant stirring for not longer than 30 minutes. If necessary, homogenize the suspension mechanically by means of a mixer or other apparatus. The characteristics of the product may necessitate the use of larger volumes. A suitable surface-active agent such as polysorbate 80 may be added to assist the suspension of poorly wettable substances. If necessary, adjust the pH value to pH 6–8. Prepare further dilutions using the same diluent.
- Fatty and oily products
  Homogenize 10 g or 10 ml of the product to be examined with the minimum required amount of sterile polysorbate 20 or polysorbate 80 or another non-inhibitory surface-active agent. As an alternative, sterile-filtered isopropyl myristate may be used for homogenization. Heat if necessary to not more than 40 °C, or in exceptional cases, to not more than 45 °C for a short period.

  Mix carefully at this temperature (40 °C). Then, if necessary, add a diluent pre-warmed to not more than 40 °C to make a 1 : 10 dilution. Maintain the temperature at 40 °C for the shortest period necessary for the formation of an emulsion and in no case for more than 30 minutes.

  Note:
  Only sterile-filtered isopropyl myristate may be used since hydrolysis products with antimicrobial activity are formed on heating the substance.
- Liquids or solids in the form of aerosols
  Disinfect the outside of 10 dosing units by wiping with a sterile paper cloth soaked in 70% isopropyl alcohol. Spray the entire contents via a sterile spraying head into a sterile glass bottle with glass beads. Shake the dosing units well prior to and during spraying. When no more aerosol is coming out, punch a hole in the side of the container using suitable tongs, in order to enable the aerosol to evaporate completely. Then, using flat tongs and applying gentle pressure, cut open the units in the upper quarter of the container height. Pipette 5 ml of diluent (e.g. buffered sodium chloride – peptone + 5% polysorbate 80), pre-warmed to 40 °C, into each dosing container. While pipetting, rotate the container in order to thoroughly wet it inside. Then, transfer the whole volume of all 10 dosing units to the bottle containing glass beads and shake well.
- Transdermal therapeutic systems (TTS)
  Disinfect the outside of 10 patches with sterile paper cloths soaked in 70% isopropyl alcohol and remove the cover sheets using sterile forceps. Stick each patch onto sterile gauze and transfer to a sterile glass bottle with a magnetic bar. Then add e.g. 500 ml of diluent and stir on a magnetic stirrer for at least 30 minutes.

5.7.2 Method Suitability for Microbial Enumeration Tests

The diluted product (starting with a 1 : 10 dilution) and the diluent chosen (as a positive control) are inoculated with the test microorganisms specified in Table 5.9, a level of not more than 100 CFU. Each test strain is examined separately. In some particular cases, it may be necessary to use additional test strains as those provided in the corresponding pharmacopoeia (e.g. *B. cepacia* for inhalation products).

5.7.2.1 Membrane Filtration

Prior to use, sterilize the filtration units. The filters must have a diameter of approximately 50 mm and a nominal pore size of not more than 0.45 μm. It is not specifically required in the harmonized pharmacopoeia to carry out the

Table 5.9 Test microorganisms and incubation conditions for determining the suitability of the microbial enumeration test based on USP<61>and Ph. Eur. 2.6.12.

Microorganism	Total aerobic microbial count	Total combined yeasts/molds count
Staphylococcus aureus	SCDA MPN SCDB ≤100 CFU 30–35 °C ≤3 days	–
Pseudomonas aeruginosa	SCDA MPN SCDB ≤100 CFU 30–35 °C ≤3 days	–
Bacillus subtilis	SCDA MPN SCDB ≤100 CFU 30–35 °C ≤3 days	–
Candida albicans	SCDA ≤100 CFU 30–35 °C ≤5 days MPN not applicable	SDA ≤100 CFU 20–25 °C ≤5 days
Aspergillus brasiliensis	SCDA ≤100 CFU 30–35 °C ≤5 days MPN not applicable	SDA ≤100 CFU 20–25 °C ≤5 days

test in duplicate. Some testing laboratories nevertheless duplicate the determination as part of the inoculum variability is then covered.

Typically dilute the product as indicated in sample preparation, add the appropriate microbial suspensions, and filter the equivalent of 1 g or 1 ml product containing the microbial suspension. The microbial suspension is added in the product dilution. The volume of the microbial suspension should not exceed 1% of the volume of diluted product. Filter immediately, and rinse the membrane filter with an appropriate volume of diluent. The volume of rinsing liquid can be adjusted and depends on the properties of the sample to be examined (e.g. antimicrobial activity) but should not exceed 5 × 100 ml. Excessive rinsing may result in microorganisms being rinsed through the filter. In the case of oily products, the rinsing liquid may contain a surface-active agent such as polysorbate 20 or 80 at a concentration of not more than 5%.

After rinsing, take hold of the filters by the edge using sterile forceps and transfer with the contaminated side facing upwards onto the appropriate agar medium. Incubate the plates upside down (with the lid facing downwards) and carry out the determination in duplicate. The incubation is not more than 3 days at 30–35 °C for bacteria and not more than 5 days at 20–25 °C for yeasts or molds.

Example of a procedure for a 1 : 10 dilution.

- Weigh in a sterile glass bottle 10 g or 10 ml of the filterable product (Figure 5.3).
- Add 90 ml of the suitable diluent (e.g. buffered sodium chloride–peptone solution). Mix and/or let dissolve the product.

Figure 5.3 Weighing of a cream in a sample container placed on a precision balance. Source: Courtesy of BAV Institut GmbH Offenburg, Germany.

- In parallel, fill a control bottle with 100 ml diluent (e.g. buffered sodium chloride–peptone solution).
- Add in each of 5 containers (1 per test microorganism) 10 ml of the dissolved product.
- Spike each container with the diluted product and control bottle with 0.1 ml of a microbial inoculum that contains 3,000–10,000 CFU/ml test microorganism (→microbial suspension volume = 1%).
- Filter 10 ml of the inoculated product diluent immediately (→1 g of product containing 30–100 CFU microorganisms are filtered).
- Rinse 1–3 times with 100 ml of buffered sodium chloride–peptone solution.
- Transfer the filter on the SCDA or SDA agar plate.
- Carry out the same procedure for the spiked control containing only diluent and test microorganisms and for the negative control containing only diluent.
- Incubate for bacteria not more than 3 days at 30–35 °C and for yeasts and molds not more than 5 days at 20–25 °C.

5.7.2.2 Plate Count Methods

The plate count methods are composed of the pour plate and the surface spread method whereas the pour-plate method is generally preferred since the microorganisms are better distributed than the surface plate method and there is less mechanical/desiccation stress. Plate count methods are executed in duplicate plates per test condition.

5.7.2.2.1 Pour Plate Typically dilute the product as indicated in sample preparation, add the appropriate microbial suspensions. The volume of the microbial suspension should not exceed 1% of the volume of diluted product. Then, pipette 1 ml of the product dilution containing the microbial suspension into Petri dishes and add approximately 20 ml of liquefied agar at a temperature less than 45 °C. The incubation is not more than 3 days at 30–35 °C for bacteria and not more than 5 days at 20–25 °C for yeasts or molds.

Example of a procedure for a 1 : 10 dilution:

- Weigh in a sterile glass bottle 10 g or 10 ml of the filterable product.
- Add 90 ml of the suitable diluent (e.g. buffered sodium chloride–peptone solution). Mix and/or let dissolve the product. In parallel, fill a control bottle with 100 ml diluent (e.g. buffered sodium chloride–peptone solution).
- Transfer 10 ml of the product dilution in a suitable recipient (e.g. sterile beaker).
- Inoculate the 10 ml aliquot with not more than 100 μl of each microbial suspension (concentrated at a level of 3,000–10,000 CFU/ml).
- Pipette 1 ml of the spiked product dilution into empty sterile Petri plates and add the appropriate volume (~20 ml for 9 mm Petri plates) of liquefied agar-based medium (at a temperature of not more than 45 °C). The final product concentration in the Petri plate is 0.1 g. Repeat this step to have duplicate results (Figures 5.4 and 5.5).

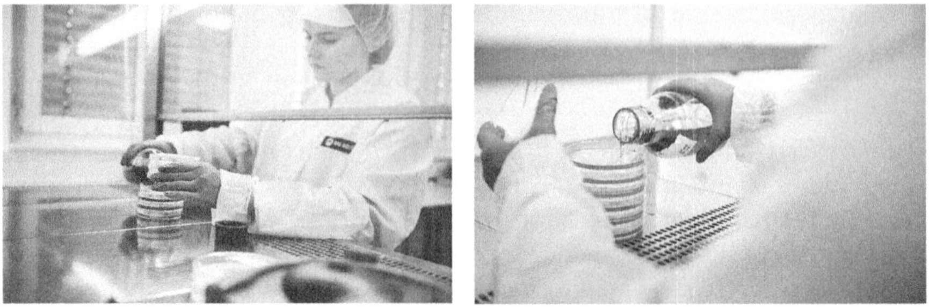

Figure 5.4 Pour plating medium into Petri dishes. *Source:* Courtesy of BAV Institut GmbH Offenburg, Germany.

Figure 5.5 Solidified pour-plated agar plates in the incubator. *Source:* Courtesy of BAV Institut GmbH Offenburg, Germany.

- Allow the agar medium to solidify and incubate the plates upside down (with the lid facing downwards).
- Carry out the same procedure for the spiked control containing only diluent and microorganisms.
- Pipette 1 ml of the diluent containing no microorganisms and pour liquid agar for the negative control.

5.7.2.2.2 ***Surface Spread Method*** The test procedure is the same as for the pour plate apart that instead of pipetting the inoculated production dilution or diluent control in an empty Petri dish, 0.1 ml quantities of the dilutions are pipetted on the surface of an agar medium plate and then spread using a Drigalski spatula.

Incubate the plates upside down (with the lid facing downwards). Carry out the determination in duplicate. The incubation period is not more than 3 days at 30–35 °C for bacteria and not more than 5 days at 20–25 °C for yeasts or molds.

Enumeration and Requirements for the Membrane Filtration and Plate Count Methods
After incubation, the colonies on the duplicate plates of the product dilutions and controls are enumerated. The mean count of the product dilutions divided by the mean count of the controls should not differ by a factor greater than 2 (50–200%).

Examples:

Product A
By 1 : 10 dilution
Pour plate product #1 = 64 CFU
Pour plate product #2 = 48 CFU → mean 56 CFU
Pour plate control #1 = 79 CFU
Pour plate control #2 = 67 CFU → mean 73 CFU

$$\frac{\text{Mean product}}{\text{Mean control}} \times 100 = \frac{56\,\text{CFU}}{73\,\text{CFU}} \times 100 = 0.77 \times 100 = 77\%$$

Interpretation of result: For product A, a 1 : 10 product dilution is suitable.

Product B
By 1 : 10 dilution
Pour plate product #1 = 24 CFU
Pour plate product #2 = 15 CFU → mean 19.5 CFU → 20 CFU*
Pour plate control #1 = 50 CFU
Pour plate control #2 = 69 CFU → mean 59.5 CFU → 60 CFU*

$$\frac{\text{Mean product}}{\text{Mean control}} \times 100 = \frac{20\,\text{CFU}}{60\,\text{CFU}} \times 100 = 0.33 \times 100 = 33\%$$

Interpretation of result: For product B, a 1 : 10 dilution is not suitable and the method verification must be repeated either by increasing the product dilution or by adding neutralization agents as explained in Section 5.7.4.

In the quantitative tests TAMC and TYMC, the calculation of the product-specific detection limit is based on the lowest dilution which yielded a recovery rate of a factor 2 for a 1 g, 1 ml, or 1 system product basis. For example, if a

* Mean is 19.5, respectively, 59.5 that may be rounded to the above decimal since in microbiology half colonies do not exist.

recovery rate of at least 50% is obtained for a 1 : 10 product dilution in the TAMC pour-plate suitability test, then the detection limit for the TAMC test is 10 CFU/g or ml.

For filtration, generally the equivalent of 1 g or ml or the product is filtered and the detection limit is then 1 CFU/g or ml.

If the recovery rate is too low in the 1 : 10 dilution, but sufficient in the 1 : 100 dilution, the detection limit is 100 CFU/g or ml. If the recovery rate of microorganisms is not sufficient, additional dilutions can be tested (e.g. 1 : 20, 1 : 50, 1 : 100, etc.) or more or larger plates can be used (e.g. 4 or 20 plates or 100 ml Petri dishes). If possible, the detection limit of the method should lie below the product specification by a factor of at least 10. The dilution level should not exceed the product specification level.

To demonstrate acceptable microbial recovery from the product, the lowest possible dilution factor of the prepared sample must be used for the test. Where this is not possible due to antimicrobial activity or poor solubility, further appropriate protocols must be developed. If inhibition of growth by the sample cannot otherwise be avoided, the aliquot of the microbial suspension may be added after neutralization, dilution, or filtration (see Section 5.7.4).

5.7.2.3 Most Probable Number Method

The MPN method is reserved for the determination of the TAMC in situations where the membrane filtration or plate-count methods are inapplicable for chemical/physical reasons. The MPN method may also be used to enumerate swarming bacteria on agar media. The MPN method is generally not suitable for the determination of the total combined yeasts/molds count since unreliable results are obtained for the enumeration of molds. However, the author is of the opinion that using an appropriate protocol (for example, higher tube volume, longer incubation periods, other incubation temperatures), molds can be also enumerated.

The choice of the dilutions to be taken depends on the properties of the product to be examined (e.g. gelation ability and antimicrobial activity) and the given specifications.

- Prepare a series of at least three serial 10-fold dilutions of the product. The first dilution (1 : 10) is 1 g or 1 ml of the product to be tested in 9 ml diluent (e.g. buffered sodium chloride–peptone). The 1 : 10 dilution is also spiked with the 100 µl of test microorganism. The microbial suspension concentration is 300–1000 CFU/1 ml, thus the 1 : 10 dilution contains 30–100 CFU microorganisms. The second (1 : 100) is 1 ml of the first dilution in 9 ml diluent. The third is 1 ml of the second dilution in 9 ml diluent (Figure 5.6).
- Repeat the preparation but using 1 ml of diluent in the first dilution as control.

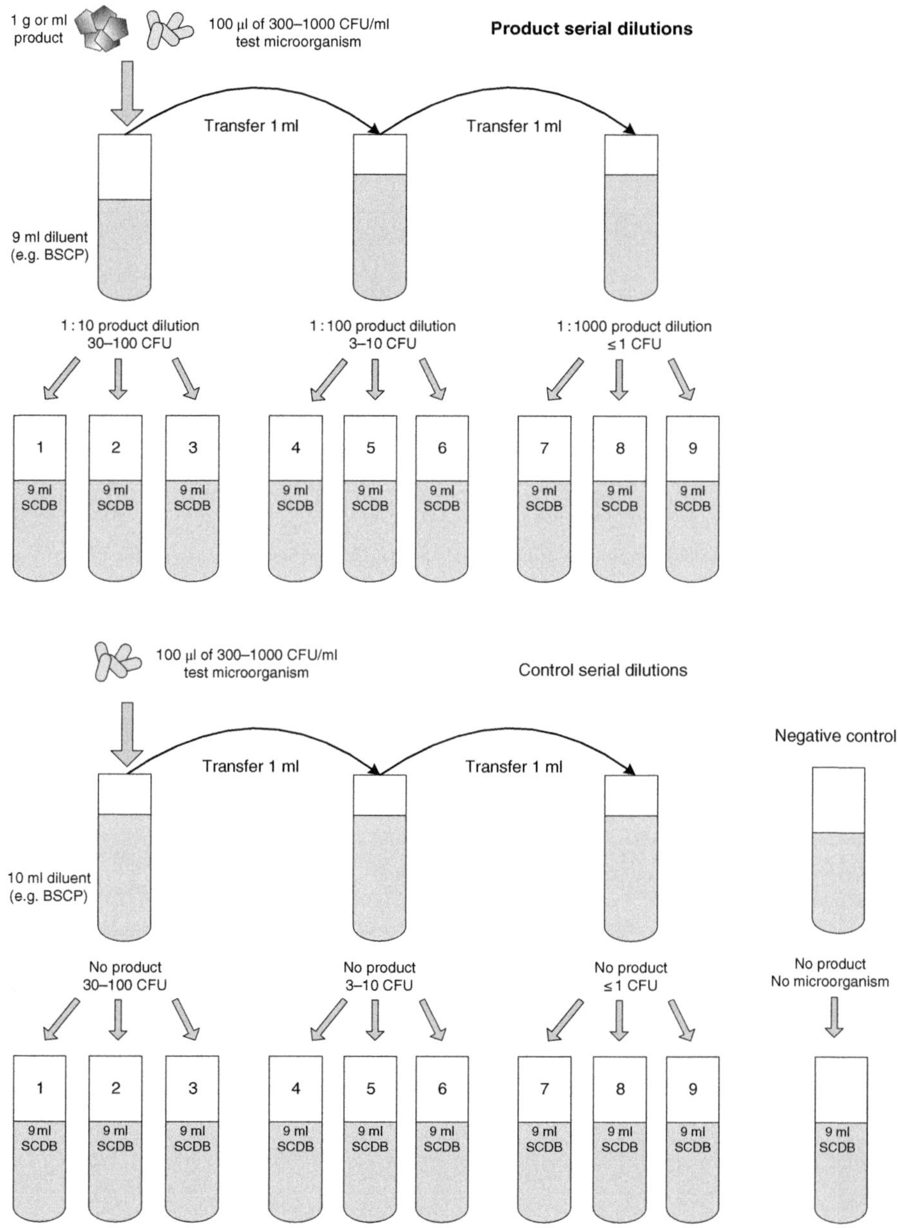

Figure 5.6 Schematic overview of the MPN serial dilutions.

- Prepare 9 ml quantities of SCDB to a series of 19 test tubes. Arrange the tubes in 3 sets of 3 tubes +1 tube for the negative control.
- Into tubes 1–3, pipette 1 ml quantities of the 1 : 10 product. Into tubes 4–6, pipette 1 ml quantities of the 1 : 100 dilution; into tubes 7–9, pipette 1 ml quantities of the 1 : 1000 dilution.
- Repeat the same for the control without product.
- Add in a final tube 1 ml quantities of the diluent as a negative control.
- Incubate all the tubes at 30–35 °C for 3–5 days.

Depending on the properties of the product (antimicrobial effect), it is necessary to carry out the analysis in Erlenmeyer flasks and to increase the incubation volumes to e.g. 100 ml of SCDB (dilution effect).

If, due to the nature of the product, the results of the test are inconclusive or doubtful (turbidity due to the material), remove material from each of the test tubes after the incubation period and prepare subcultures on solid SCDA medium. Incubate the subcultures at 30–35 °C for 24–48 hours.

5.7.2.2.3 Enumeration and Requirements for MPN Method Determine the most probable number of microorganisms per g or ml of product according to Table 5.10. The calculated value from the product dilution must be within 95% confidence limits of the results obtained with the control.

Table 5.10 Most-probable-number values of microorganisms.

Three test tubes at each level of dilution				
Observed combinations of numbers of tubes showing growth in each set				
0.1 g or ml per test tube (1 : 10)	0.01 g or ml per test tube (1 : 100)	0.001 g or ml per test tube (1 : 1000)	Most probable number per g or ml	95% confidence limits
0	0	0	<3	0–9.4
0	0	1	3	0.1–9.5
0	1	0	3	0.1–10
0	1	1	6.1	1.2–17
0	2	0	6.2	1.2–17
0	3	0	9.4	3.5–3.5
1	0	0	3.6	0.2–17
1	0	1	7.2	1.2–17
1	0	2	11	4–35

Table 5.10 (Continued)

Three test tubes at each level of dilution				
Observed combinations of numbers of tubes showing growth in each set				
0.1 g or ml per test tube (1 : 10)	0.01 g or ml per test tube (1 : 100)	0.001 g or ml per test tube (1 : 1000)	Most probable number per g or ml	95% confidence limits
1	1	0	7.4	1.3–20
1	1	1	11	4–35
1	2	0	11	4–35
1	2	1	15	5–38
1	3	0	16	5–38
2	0	0	9.2	1.5–35
2	0	1	14	4–35
2	0	2	20	5–38
2	1	0	15	4–38
2	1	1	20	5–38
2	1	2	27	9–94
2	2	0	21	5–40
2	2	1	28	9–94
2	2	2	35	9–94
2	3	0	29	9–94
2	3	1	36	9–94
3	0	0	23	5–9423
3	0	1	38	9–10438
3	0	2	64	16–181
3	1	0	43	9–181
3	1	1	75	17–199
3	1	2	120	30–360
3	1	3	160	30–380
3	2	0	93	18–360
3	2	1	150	30–380
3	2	2	210	30–400
3	2	3	290	90–990
3	3	0	240	40–990
3	3	1	460	90–1980
3	3	2	1100	200–4000
3	3	3	>1100	

For example:

Interpretation of result: For product A, the MPN value "93" is within the 95% confidence limits of the control (16–181). The method for product A is suitable (Table 5.11).

Interpretation of result: For product B, the MPN value "20" is within the 95% confidence limits of the control (17–199). The method for product B is suitable (Table 5.12).

Interpretation of result: For product B, the MPN value "20" is outside the 95% confidence limits of the control (30–380). The method for product C is **not** suitable (Table 5.13).

Table 5.11 Example of MPN calculation product A.

0.1 g per tube (1 : 10 dilution)	0.01 g per tube (1 : 100 dilution)	0.001 g per tube (1 : 1000 dilution)	MPN	95% lower and upper confidence limits
Product dilution				
3	2	0	93	18–360
Control				
3	0	2	64	16–181

Table 5.12 Example of MPN calculation product B.

0.1 g per tube (1 : 10 dilution)	0.01 g per tube (1 : 100 dilution)	0.001 g per tube (1 : 1000 dilution)	MPN	95% lower and upper confidence limits
Product dilution				
2	0	2	20	5–38
Control				
3	1	1	75	17–199

Table 5.13 Example of MPN calculation product C.

0.1 g per tube (1 : 10 dilution)	0.01 g per tube (1 : 100 dilution)	0.001 g per tube (1 : 1000 dilution)	MPN	95% lower and upper confidence limits
Product dilution				
2	0	2	20	5–38
Control				
3	2	1	150	30–380

5.7.3 Suitability of the Test Method for Absence of Specified Microorganisms

The suitability of the test method is verified according to the Pharmacopoeia chapters (e.g. USP <62>) and enables to determine the detailed test parameters (e.g. medium, dilutions, rinsing volumes, and exact incubation times) to ensure adequate recovery of specified microorganisms. The product is prepared as described in the testing of products for specified microorganisms using the shortest incubation period possible (refer to Section 5.8.3). There are cases where applying strictly this rule would end up having analysts having to work in the middle of the night to evaluate growth. For instance, the suitability test of the absence of *E. coli* would require a first incubation of not more than 18 hours, a second of not more than 24 hours, and a third of not more than 18 hours (e.g. start incubation 4 p.m., end reading 4 a.m.).

This absurd timeline in terms of organization could be overcome either by incubating much less than the recommended hours (but with a risk of failing the suitability test due to lesser recovery of the test microorganism) or on the contrary extending the time of incubation at one of the enrichment steps. For instance, for the absence of *E. coli*, the first incubation could last slightly less than 24 hours and the other incubation periods as recommended (e.g. start incubation 4 p.m., end reading 10 hours). However, in this case, it must be ensured that during routine product testing, the minimum incubation time of 24 hours for the first incubation is followed. The test strain is added at the time of mixing in the prescribed growth medium (e.g. SCDB) at a concentration of not more than 100 CFU. Each test strain is added individually. The control is composed of the diluent without product.

The test procedure is considered as valid if the specified microorganisms are detected with the expected indicative reactions, as listed in Table 5.14.

Note: If other specified microorganisms are used, the enrichment or selective media may differ from the pharmacopoeia examples. For example, to test for the absence of *B. cepacia*, a pre-enrichment in SCDB for 24–48 hours at 30–35 °C followed by selection in a *B. cepacia* selective agar (e.g. Oxidation–fermentation polymixin bacitracin lactose agar, *B. cepacia* selective agar, or Pseudomonas cepacia agar) for 18–72 hours at 30–35 °C may be applied. Actually, at the time of writing this chapter, the USP is drafting a chapter ⟨60⟩ *Microbiological Examination of Nonsterile Products – Tests for Burkholderia cepacia Complex* for a specified microorganism testing of *B. cepacia complex.* This new USP chapter was available in the US PF 44 (5) for public commenting until the end of 2018.

Table 5.14 Indicative reactions of specified microorganisms in the method suitability test.

Medium	Reactions	Test strains
Test for bile-tolerant, Gram-negative bacteria		
VRBGA	Growth of colonies with reddish precipitate	*Escherichia coli, Pseudomonas aeruginosa*
Test for *E. coli*		
MCA	Growth of red, generally non-mucoid colonies sometimes surrounded by a reddish precipitation zone	*E. coli*
Test for *Salmonella*		
XLDA	Growth of red colonies, with or without black centers	*Salmonella enterica*
Test for *P. aeruginosa*		
CA	Growth of mostly greenish colonies	*P. aeruginosa*
Test for *Staphylococcus aureus*		
MSA	Growth of yellow/white colonies surrounded by a yellow zone	*S. aureus*
Test for *Candida albicans*		
SDA	Growth of white, yeast-like colonies	*C. albicans*

5.7.4 Examples of Procedures in Case the Method Suitability Fails

In cases where the microbial recovery is not sufficient to demonstrate that the method is suitable, modification of the test procedure may be carried out. These include:

- Increasing the product dilution. For instance, if microbial recovery by a 1 : 10 dilution is too weak, then increasing the product dilution to 1 : 20, 1 : 50, or 1 : 100 may reduce the antimicrobial activity. However, the dilution level should not be too excessive. It would not be acceptable to have dilutions higher than the acceptance criteria (e.g. 1 : 1000 dilution for an acceptance criterion of not more than 100 CFU/g).
- Increasing the rinsing volume for the membrane filtration method. Nonetheless, the rinsing should not be too excessive (e.g. not more than 5×100 ml) to avoid that microorganisms are eventually rinsed through the membrane or a loss of filter retention capacity/integrity.
- Addition of neutralizing agents to the diluents or culture medium. In general, the diluents or nutrient media are supplemented with polysorbates (e.g. Tween 20 or Tween 80) as well as soybean lecithin. Other neutralizing agents

Table 5.15 Common neutralizing agents and neutralizing method for interfering substances.

Interfering substance	Potential neutralizing method
Alcohols	Dilute
Aldehydes	Dilute Glycine Thiosulphate
Bis-biguanides	Lecithin
EDTA	Mg^{2+} or Ca^{2+} ions
Glutaraldehydes	Sodium bisulphite
Halogens	Thiosulphate
Iodine	Polysorbate
Parabens	Lecithin Polysorbate
Phenols	Dilute
Quaternary ammonium compounds	Lecithin Polysorbate
Mercurials	Sodium bisulphite Thioglycollate Thiosulphate
Sorbates	Dilute

that may be used are described in Table 5.15 (list not exhaustive). If neutralizing agents are used, their absence of toxicity for microorganisms should be demonstrated (e.g. by carrying out a blank with neutralizer and without product).

For membrane filtration, if the antimicrobial activity cannot be eliminated after having modified the procedure as described above, the microbial suspension may be added in the rinsing liquid of the final rinsing step.

For the pour-plate method, if the antimicrobial activity cannot be eliminated after having modified the procedure as described above, the microbial suspension may be added directly in the Petri dish without contacting the product dilution before the agar-based medium is poured.

Another possibility if the suitability cannot be shown is the use of an alternative method. Instead of the pour-plate method, the membrane filtration or MPN method might give complying results. Other methods such as ATP Bioluminescence or SimPlate (etc.) would be an alternative, but would require validation prior to use.

If the antimicrobial activity in relation to the microorganism tested cannot be neutralized in a given product, it can be assumed that the inhibited microorganism will not be present in the product. Nevertheless, routine testing of this product with the highest dilution and/or suitable inactivators compatible with microbial growth and the specific acceptance criterion should be performed in order to verify the antimicrobial activity as well as possibly to detect other test strains not included in the method suitability test.

5.8 Microbiological Examination of Non-sterile Products

5.8.1 Microbial Enumeration Tests: Membrane Filtration and Plate Count Methods

The microbiological examination of non-sterile products is executed according to the method that was confirmed as suitable. A negative control may be performed during product testing to demonstrate the freedom of microbial contamination in the nutrient media and diluents and the efficiency of the aseptic measures and materials used in the test. The negative control should preferably cover all steps of the testing procedure. It should consist of preparing the dilution but replacing the diluent used (e.g. if 10 g of product is diluted in 90 ml diluent, the negative control consists of 100 ml diluent). A failed negative control must require an investigation.

5.8.1.1 Membrane Filtration

The filters are generally composed of mixed cellulose esters and have a diameter of approximately 50 mm and a nominal pore size of not more than 0.45 μm. Usually 10 ml of the dilution representing 1 g or 1 ml of the product are filtered under a vacuum. After filtering, rinsing with the volume and type of rinsing liquid determined in the method suitability test is performed. After rinsing, take hold of the filters by the edge using sterile forceps and transfer with the contaminated side facing upwards onto the appropriate agar medium. Incubate the plates upside down (with the lid facing downwards).

5.8.1.2 Pour Plate Method

Test the sample based on the method confirmed in the method suitability test. Typically pipette 1 ml of the product dilution into Petri dishes (e.g. Ø 9 or 14 cm) and add the appropriate volume of liquefied agar-based medium (e.g. SCDA and SDA) at a temperature of not more than 45 °C. Allow the agar medium to solidify and incubate the plates upside down (with the lid facing downwards). Carry out the determination in duplicate.

5.8.1.3 Surface Spread Method

Prepare the sample using the method that has been shown to be suitable. Use Petri dishes of normal size (9–10 cm Ø) with solid nutrient medium. Typically spread at least 0.1 ml quantities of the product dilution on the surface of the medium. Incubate the plates upside down (with the lid facing downwards). Carry out the determination in duplicate. Incubation times and the way to calculate the results are the same as described for the pour-plate method.

5.8.1.4 Incubation

TAMC: Incubate the agar-based medium (e.g. SCDA) at 30–35 °C for 3–5 days.
TYMC: Incubate the agar-based medium (e.g. SDA) at 20–25 °C for 5–7 days.

5.8.1.5 Counting

According to the harmonized pharmacopoeia chapters for TAMC all colony-forming units (CFUs) including yeasts and molds on SCDA must be enumerated and for TYMC all CFUs on SDA including bacteria. If the growth of bacteria is expected in SDA and this might result in exceeding the total yeast and molds counts acceptance criteria, SDA may be also supplemented with antibiotic (e.g. chloramphenicol) to limit bacterial proliferation.

The Chinese Pharmacopoeia recommends daily reading of the plates. The author would not recommend daily reading since it is not necessary in most cases as microbial bioburden remains low in final formulated products and daily reading would cause momentary shifts in the incubation temperature, risk of cross contamination (especially if molds are growing), and unnecessary laboratory work.

The author would rather recommend for products with an expected high bioburden, where it might be difficult to count the colonies due to overlapping of the individual types of microorganisms, to count the cultures before the end of the defined incubation period at an appropriate interval.

After counting the colonies, the number of microorganisms per g or ml of product are calculated, assuming that for every visible colony there exists one CFU. Calculate the arithmetic mean of the colony counts from the duplicate plates of the given dilution. The colony counts are rounded up to the next whole number (worst case) before being multiplied by the dilution factor.

For membrane filtration if growth occurs on multiple dilutions, the dilutions that contain 25–200 bacterial and yeast colonies and/or 10–50 mold colonies per filter are preferably taken into account. If more than one dilution has growth within the optimal range, the worst-case result should be reported. For the pour-plate or surface spread method since the reading area is larger, the dilution where 25–250 bacterial and yeast colonies and 8–80 mold colonies per plate are preferably taken into account.

If No Growth Occurs Do I Report 0 CFU?

No, since the final results depend on the detection limit and product dilution.

In the pour-plate method a 1 : 10 dilution means that 0.1 g/ml of the product was pipetted in the empty Petri dish before medium was poured. This means that considering the acceptance criteria based on a 1 g basis, at least 10 microorganisms would need to be present in the ml of the 1 : 10 dilution so that at least 1 colony is detected. Therefore, the detection limit is 10 CFU/g. If during routine testing no colonies are present on the nutrient media the final result is not 0 CFU/g but less than 10 CFU/g to take into account this limit of detection.

In the membrane filtration method if 1 g of the product is filtered, the result with no growth is less than 1 CFU/g.

Example 5.1

Assuming that 20 and 12 colonies are counted on the filters from a 1 g product dilution:

$$\frac{20+12}{2}=16\times1=16\,\mathrm{CFU/g}$$

Example 5.2

Assuming that 12 and 13 colonies are counted on the plates from a 1 : 10 product dilution (equivalent 0.1 g). In this case the colony counts are rounded to the highest number:

$$\frac{12+13}{2}=13\times10=130\,\mathrm{CFU/g}$$

Example 5.3

For a product with 2 dilution series, the following should occur. Assuming that 35 and 26 colonies are counted on the plates from dilution 1 (1 : 10), and 2 and 4 colonies on the plates from dilution 2 (1 : 100):

$$\frac{35+26}{2}=31\times10=310\,\mathrm{CFU/g}$$

The 1 : 100 dilution is not taken into account since it is less precise.

Example 5.4
Assuming that no colonies are counted on the plates, the count per g or ml of sample is not 0 but has to take into account the detection level based on the lowest tested dilution. In case of a 1 : 10 dilution with no growth, the result would then be:

$$< 10\,\text{CFU/g}$$

5.8.2 Microbial Enumeration Tests: MPN Method

5.8.2.1 Enumeration by Means of Serial Dilutions (MPN Method)

Prepare a series of 10 tubes containing 9 ml of SCDB or the medium determined in the method suitability test. Arrange the tubes in 3 sets of 3 tubes +1 tube for the negative control. Into tubes 1–3, pipette 1 ml quantities of the product dilution, dissolved or homogenized in the ratio 1 : 10. Into tubes 4–6, pipette 1 ml quantities of the 1 : 100 product dilution; into tubes 7–9, pipette 1 ml quantities of the 1 : 1000 product dilution; and into tube 10, pipette 1 ml of the diluent as a negative control.

Incubate all the tubes at 30–35 °C for 3–5 days. No growth of microorganisms must occur in the negative control.

If as demonstrated during method suitability, due to the nature of the product, the results of the test are inconclusive or doubtful (turbidity due to the product), remove material from each of the test tubes after the incubation period and prepare subcultures on solid agar medium. Incubate the subcultures at 30–35 °C for 24–48 hours.

Determine the most probable number of microorganisms per g or ml of product according to Table 5.10.

For example, if the growth combination is 3-1-1, the resulting most probable number is 75 CFU/g.

5.8.3 Test for Specified Microorganisms Procedure

The detailed test parameters (e.g. medium, dilutions, rinsing volumes, and exact incubation times) are established in the product-specific method suitability test. Tests are generally based on the harmonized compendial chapters as described below. An additional negative control composed of e.g. 1 ml of a diluent solution may be used during testing.

Examples of procedure for the testing of specified microorganisms adapted from USP <62> are written below.

Bile-Tolerant, Gram-Negative Bacteria

Inoculate 10 g or 10 ml of the product directly into a volume of SCDB corresponding to the volume determined in the test for suitability of the method. Whenever possible, inoculate 90 ml of SCDB with 10 g or 10 ml of the product. Homogenize the mixture and incubate at 20–25 °C for a time sufficient to resuscitate the bacteria but not sufficient to encourage multiplication of the organisms (usually two hours but not more than five hours).

Mix the container and transfer 10 ml of the resuscitation culture or the quantity of the contents corresponding to 1 g or 1 ml into 90 ml of EEB medium. Then, incubate at 30–35 °C for 24–48 hours.

Subculture onto VRBGA and incubate at 30–35 °C for 18–24 hours.

The growth of red colonies with a reddish precipitate indicates the presence of bile-tolerant, Gram-negative bacteria.

The product passes the test if there is no growth.

In addition, the USP<62> also describes a quantitative test for bile-tolerant Gram-negative bacteria.

Pseudomonas aeruginosa

Mix 10 ml of the prepared product corresponding to 1 g or 1 ml, into 90 ml of SCDB or into another suitable volume of SCDB determined in the test for suitability of the method. Incubate at 30–35 °C for 18–24 hours.

Subculture onto CA and incubate at 30–35 °C for 18–72 hours.

If growth occurs, the presence/absence of *P. aeruginosa* must be confirmed by suitable biochemical tests or identification systems.

The product passes the test if there is no growth of *P. aeruginosa*.

Staphylococcus aureus

Mix 10 ml of the prepared product corresponding to 1 g or 1 ml into 90 ml of SCDB or into another suitable volume of SCDB determined in the test for suitability of the method. Incubate at 30–35 °C for 18–24 hours.

Subculture onto MSA and incubate at 30–35 °C for 18–72 hours.

Yellow/white colonies surrounded by a yellow zone indicate the presence of *S. aureus*. The presence/absence of *S. aureus* must be confirmed by suitable biochemical tests or identification systems.

The product passes the test if there is no growth of *S. aureus*.

Escherichia coli

Mix 10 ml of the prepared product corresponding to 1 g or 1 ml into 90 ml of SCDB or into another suitable volume of SCDB determined in the test for suitability of the method. Incubate at 30–35 °C for 18–24 hours.

Mix well and transfer 1 ml of the SCDB culture into 100 ml of MCB. Incubate at 42–44 °C for 24–48 hours.

Subculture onto MCA and incubate at 30–35 °C for 18–72 hours.

If growth occurs, the presence/absence of *E. coli* must be confirmed by suitable biochemical tests or identification systems.

The product passes the test if there is no growth of *E. coli*.

Salmonella

Inoculate 10 g or 10 ml of the product directly into 90 ml or into another volume of SCDB corresponding to the volume determined in the test for suitability of the method. Incubate at 30–35 °C for 18–24 hours.

Add 0.1 ml of the SCDB culture to 10 ml of RVSEB and incubate at 30–35 °C for 18–24 hours.

Subculture onto XLDA and incubate at 30–35 °C for 18–48 hours.

Growth of well-developed, red colonies with or without black centers indicates the presence of salmonellae. The presence/absence of *Salmonella* species must be confirmed by suitable biochemical tests or identification systems.

The product passes the test if there is no growth of *Salmonella* species.

Candida albicans

Mix 10 ml of the prepared product corresponding to 1 g or 1 ml into 90 ml of SCDB or into another suitable volume of SDB determined in the test for suitability of the method. Incubate at 30–35 °C for 3–5 days.

Subculture onto SDA and incubate at 30–35 °C for 24–48 hours.

If growth occurs, the presence/absence of *C. albicans* must be confirmed by suitable biochemical tests or identification systems.

The product passes the test if there is no growth of *C. albicans*.

Clostridia

Prepare a sample using a 1 in 10 dilution (with a minimum total volume of 20 ml) of not less than 2 g or 2 ml of the product to be examined. Divide the sample into 2 portions of at least 10 ml. Heat one portion at 80 °C for 10 minutes, and cool rapidly. Do not heat the other portion.

Selection and subculture – Use 10 ml or the quantity corresponding to 1 g or 1 ml of the product to be examined of both portions to inoculate suitable amounts (determined as described under Suitability of the Test Method) of Reinforced Medium for *Clostridia*. Incubate under anaerobic conditions at 30–35 °C for 48 hours. After incubation, make subcultures from each container on Columbia Agar, and incubate under anaerobic conditions at 30–35 °C for 48–72 hours.

Interpretation – The occurrence of anaerobic growth of rods (with or without endospores) giving a negative catalase reaction indicates the presence of *Clostridia*.

This is confirmed by identification tests. The product complies with the test if colonies of the types described are not present or if the confirmatory identification tests are negative.

If growth occurs on selective media, then the presence of microorganisms should be identified for confirmation. Even if the pharmacopoeia allows classical biochemical ID methods, the author recommend the use of genetic identification or identification systems that can identify to the species level (refer to Chapter 9).

5.8.4 Method Transfer

The methods used to perform the microbiological examination of non-sterile products are sometimes transferred from one analytical laboratory to another either located in the same organization or to an external contract laboratory.

Before transferring methods, it should be first ensured that the receiving site is working according to the same cGMP standards as the receiving site and in case of external laboratories that it has been appropriately qualified for testing (refer to Chapter 17). For transfer of analytical laboratory methods typically parallel testing is performed on the same product and results are directly compared with a tolerable acceptable difference. In microbiology, parallel comparison is nonsense. First of all, microbiological methods are far too variable and microorganisms are not homogenously distributed in the product. Secondly, if the product in question has overall good microbiological quality, the result would be 0 counts for the sender site and 0 counts for the receiving site which does not demonstrate anything.

Therefore, the author would suggest that the receiving site executes at least one method suitability test run using the same method as the sender site but with its own medium and test strains and laboratory conditions. If microbial recovery is within the acceptance criteria (factor 2), then the sender site is qualified to execute the test. A transfer protocol must be written by the sender site, a transfer report by the receiving site, and both must be approved by the quality assurance department of both sites.

An alternative is that the contract laboratory that executes the test carries out its own method suitability test with a different test method on at least three independent experiments.

If the receiving site fails to reproduce sufficient recovery, then a deviation investigation should be initiated and led by the sender site. The investigation should cover comparisons of nutrient media used and type and preparation of test strain inoculum since these two elements may affect significantly microbial recovery. If the receiving site constantly fails to recover an adequate number of microorganisms even after remediation actions have been implemented, then it would not be qualified to execute the tests and an alternative solution must be found.

5.9 Elements to Consider for Raw Data Sheets

Pharmaceutical companies are working under cGMP rules that imply also demonstration of correct test execution according to a validated test method with qualified equipment and material by a qualified analyst and verification.

Therefore, the raw data sheet that contains test execution and results should contain at least the following elements:

- Name of the analyst that has executed the test. If several analysts were implied in the testing, they should sign under each part of the test that they were involved in (e.g. weighing, dilution, pouring, filtering, results reading).
- Date and time of analysis. For procedures depending on time (e.g. incubation time), the time from start and end of the procedure should be written.
- Reference to the suitable method used (e.g. Method suitability report number or SOP). Alternatively, the method procedure may be described directly in the worksheet.
- Qualified equipment used (e.g. incubator, weighing balance, laminar flow hood, water baths).
- Dilutions used and resulting calculation for final results.
- Results. They should cover the final counts for TAMC and TYMC as well as biochemical methods or identification methods used for confirmation of specified microorganisms as well as the name of the species identified.
- Outcome of results. For instance, test complies or does not comply.
- Review and approval from supervisor and/or quality assurance.

Data integrity of GMP data generated when executing microbiological examination of non-sterile products is explained in Chapter 12.

Acknowledgments

The author would like to thank the BAV INSTITUT GmbH, Offenburg, Germany, for generously providing the photos of the microbiological laboratory included in this chapter.

Bibliography

Barlasov, J, Sutton, S, Jakober, R, 2014, Recovery of stressed (acclimated) *Burkholderia cepacia* complex organisms, *Am. Pharm. Rev.*, vol. 17(3), pp. 16–24.

ChP Chapter 1105, 'Microbiological Examination of Non-Sterile Products: Microbial Enumeration Test', 2015.

ChP Chapter 1106, 'Microbiological Examination of Non-Sterile Products: Test for Specified Bacteria', 2015, Chinese Pharmacopoeia.

ChP Chapter 1107, 'Specification for Microbiological Limit of Nonsterile Pharmaceutical Products', 2015, Chinese Pharmacopoeia.

Da Silva, LVF, Velloso, LDF, Bento, CN, Gytin, E, Tateno, AF, Levi, JE, Rodrigues, JC, Ramos, SRTS, 2002, 'Use of selective medium for *Burkholderia cepacia* isolation in respiratory samples from cystic fibrosis patients', *Rev. Inst. Med. Trop. Sao Paulo*, vol. 44(4), pp. 203–208.

FDA, 2017, 'FDA advises drug manufacturers that *Burkholderia cepacia* complex poses a contamination risk in non-sterile, water-based drug products', https://www.fda.gov/drugs/drug-safety-and-availability/fda-advises-drug-manufacturers-burkholderia-cepacia-complex-poses-contamination-risk-non-sterile (accessed 20 July 2019).

ICH Q6A, 2009, 'Specifications: Test Procedures and Acceptance Criteria for New Drug Substances and New Drug Products: Chemical Substances', International Conference on Harmonisation of Technical Requirements for Registration of Pharmaceuticals for Human Use.

JP 17th Edition, Chapter <4.05>, 'Microbial Limit Test', 2017, Japanese Pharmacopoeia.

JP 17th Edition, Chapter G4, 'Microorganisms-Microbial Attributes of Non-sterile Pharmaceutical Products', 2017, Japanese Pharmacopoeia.

Mahenthiralingam, E, Urban, TA, Goldberg, JB, 2005, 'The multifarious, multireplicon *Burkholderia cepacia* complex', *Nat. Rev. Microbiol.*, vol. 3, pp. 144–156.

Metcalfe, J, 2013, 'Regulatory Review Perspectives for Non-Sterile Drug Products', PDA 8th Annual Global Conference on Pharmaceutical Microbiology.

PDA Survey, 2013, *Objectionable Microorganisms for Nonsterile Pharmaceutical, Consumer Health, Medical Devices, Dietary Supplement, and Cosmetic Products*, Parenteral Drug Association.

PDA Technical Report No. 67, 2014, *Exclusion of Objectionable Microorganisms from Nonsterile Pharmaceuticals, Medical Devices, and Cosmetics*, Parenteral Drug Association.

Ph. Eur. 9th Edition chapter 2.6.12, 'Microbiological Examination of Non-sterile Products: Microbial Enumeration Tests', 2018, European Pharmacopoeia.

Ph. Eur. 9th Edition chapter 2.6.13, 'Microbiological Examination of Non-sterile Products: Test for Specified Microorganisms', 2018, European Pharmacopoeia.

Ph. Eur. 9th Edition chapter 5.1.4, 'Microbiological Quality of Non-sterile Pharmaceutical Preparations and Substances for Pharmaceutical Use', 2018, European Pharmacopoeia.

Roesti, D, 2012, 'Objectionable microorganisms in non-sterile pharmaceutical drug products: risk assessment and origins of contamination', in J. Moldenhauer (ed.), *Environmental Monitoring – A Comprehensive Handbook*, vol. 6, PDA/DHI: Washington, DC, pp. 137–158.

Sutton, S, Jimenez, LA, 2012, Review of reported results involving microbiological control 2004-2011 with emphasis on FDA consideration of "objectionable organisms"., *Am. Pharm. Rev.*, vol. 15(1), pp. 1–12.

TGA Guidance on Therapeutic Goods, 'Order No. 77 Microbiological Standards for Medicines', 2008.

Torbeck, L, Raccasi, D, Guilfoyle, D, Friedman, R, Hussong, D, 2011, '*Burkholderia cepacia*: this decision is overdue', *PDA J. Pharm. Sci. Technol.*, vol. 65(5), pp. 535–543.

US 21 CFR 211.113, 'Control of Microbiological Contamination', 2012.
US 21 CFR 211.165, 'Testing and Release for Distribution', 2012.
USP 41-NF 36, 'General Chapters: <610> Alternative Microbiological Sampling Methods for Nonsterile Inhaled and Nasal Products', 2018, United States Pharmacopeia.
USP 41-NF 36 chapter <1111>, 'Microbiological Examination of Nonsterile Products: Acceptance Criteria for Pharmaceutical Preparations and Substances for Pharmaceutical Use', 2018, United States Pharmacopeia.
USP 41-NF 36 chapter <1112>, 'Application of Water Activity Determination to Nonsterile Pharmaceutical Products', 2018, United States Pharmacopeia.
USP 41-NF 36 chapter <61>, 'Microbiological Examination of Nonsterile Products: Microbial Enumeration Tests', 2018, United States Pharmacopeia.
USP 41-NF 36 chapter <62>, 'Microbiological Examination of Nonsterile Products: Tests for Specified Microorganisms', 2018, United States Pharmacopeia.
USP PF 44(5) In-Process Revision: <60>, 'Microbiological Examination of Nonsterile Products – Tests for *Burkholderia cepacia* Complex', 2018, United States Pharmacopeia.

6

Microbial Requirements and Testing of Primary Packaging

Marcel Goverde

MGP Consulting GmbH, Binningen, Switzerland

CONTENTS

Pharmaceutical Microbiological Quality Assurance and Control: Practical Guide for Non-Sterile Manufacturing, First Edition. Edited by David Roesti and Marcel Goverde.

6.1 Introduction

To test for microbial contamination, non-sterile products are normally controlled according to USP <61> and <62> or Ph. Eur. 2.6.12 and 2.6.13 (see Chapter 5). Acceptance criteria for these products are given by USP <1111> or Ph. Eur. 5.1.4. In order to be able to comply with these microbial requirements, the raw materials must show a low bioburden, and the production rooms, equipment, and processes must comply with GMP guidelines. Furthermore, the staff must be qualified and receive regular training in hygienic behavior. With all the measures for comprehensive microbiological quality assurance, the primary packaging that comes into contact with the preparation must not be forgotten.

Often the non-sterile product is tested on the level of the bulk and in such cases the bioburden from the primary packaging is not assessed on this level. Even if the packed final drug product is controlled for microbial contamination and the results exceed the specification, it is not clear whether the contamination comes from the product (e.g. the bulk), was introduced during the packaging process, or comes from the packaging material itself. Thus, it makes sense to test the drug product on the bulk level and also to test primary packaging material, and to control the packaging process, e.g. by means of environmental monitoring and cleaning validation. This chapter will deal with testing of primary packaging.

6.1.1 Definition of Primary Packaging

In their Guidance for Industry, the FDA (1999) defines packaging material as follows: *A primary packaging component means a packaging component that is or may be in direct contact with the dosage form.*

Examples of primary packaging materials for non-sterile dosage forms are:

- Containers
- Tubes
- Pipettors
- Droppers
- Foils
- Blister-pack material
- Bags
- Plugs
- Caps

In general, these packaging materials are made of different components, which are often glass, metal, and plastics. Some of these components are specifically defined in Ph. Eur. chapter 3 "Materials for containers and containers" but without microbial specifications:

- Plasticized poly(vinyl chloride)
- Polyolefin

- Polyethylene with or without additives
- Polypropylene
- Poly(ethylene-vinyl acetate)
- Silicon
- Non-plasticized poly(vinyl chloride)
- Polyethylene terephthalate
- Glass
- Metal, especially aluminum
- Cotton

The primary packaging of pharmaceutical dosage forms must be designed, produced, and processed in a way that does not change the safety, identity, strength, quality, and purity of the medical product beyond its specification (Rieth and Krämer 2016, Figure 6.1). From a microbiological point of view, the most important functionality is the protection it offers from contamination as well as preventing ingress of moisture that may lead to subsequent spoilage (Figure 6.1). Firstly, highly contaminated primary packaging might directly contaminate the drug product. This could impair the therapeutic effect or adversely affect the galenic stability of the preparation; or the microorganisms, especially if they are pathogenic, could directly harm the patient. Secondly, the packaging material has the functionality of protecting the drug product from contamination from the outside. Only properly designed, properly used packaging that is suitable for the preparation and delivered in good quality can fully meet this task (Hecker 1992). Furthermore, if moisture can enter the packed drug product, the

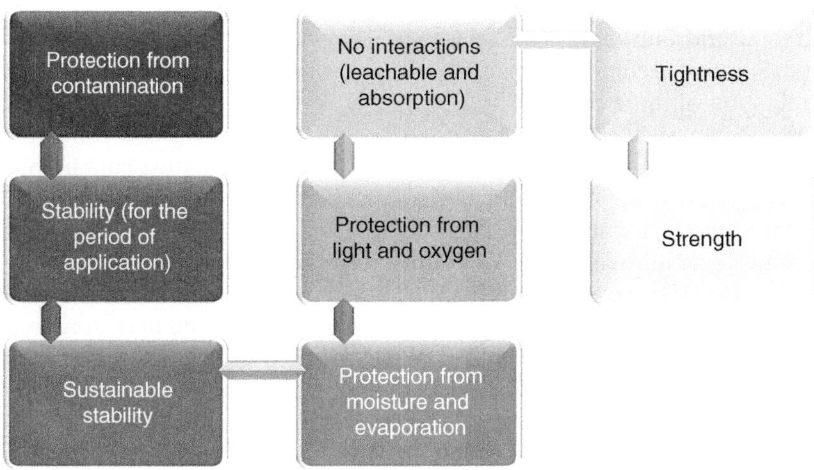

Figure 6.1 Summary of the different functionalities of primary packaging materials with focus on microbial contamination, i.e. descending functionality from a microbiological point of view. *Source:* Adapted from Rieth and Krämer (2016).

water activity might increase to a level where microbial growth, especially molds, is possible and will spoil the product, putting the patient at unnecessary risk.

The microbial contamination introduced through primary packaging needs to be addressed for certain products such as liquid dosage forms or inhalers. The following list of non-sterile-drug-product application forms attempts to prioritize the criticality for the patient if microbial contamination is introduced into the drug product by the primary packaging:

1) Non-preserved liquid products (e.g. nasal spray and inhalation products)
2) Non-sterile products with high water activity ($a_w > 0.75$)
3) Non-preserved creams
4) Transdermal patches
5) Preserved creams
6) Capsules filled (preserved) with liquid
7) Suppositories
8) Dry products (e.g. tablets and capsules in powder form or inhalers)

6.1.2 Microflora in Packaging Materials

Basically, when packaging material is manufactured under controlled conditions (GMP or ISO standards) it will have a very low bioburden due to the high temperature used during production (e.g. melting of glass, metal, or plastic). Only during the subsequent process (handling of finished primary packaging, storage, and shipment, especially when cardboard is used), might microbial contamination be reintroduced to the material (Payne 2007).

Several studies have been published on the microbial testing of primary packaging. Krüger (1970) examined different primary packaging materials (polystyrene and aluminum tubes, glass and polyethylene containers, and PVC blisters). Some of them (e.g. polystyrene tubes for oral tablets) had a microbial contamination of more than 1000 CFU/unit in 60–85% of the samples tested. This high contamination was probably due to the packaging and transport conditions from the supplier since the tubes showed optical dust contamination on receipt. On the other hand, in the same study, very low bioburden was found, for example, for aluminum tubes.

High microbial numbers (>10^4 CFU/unit) were found in cotton used as primary packaging (Dacarro *et al.* 1987). Such high numbers can be confirmed by the authors' experience where similar contamination was found in cotton used as an overlayer or protection for tablets in containers (e.g. vitamins). This high level of contamination is not surprising since cotton is a natural product (protective case around the seeds of the cotton plants), which when untreated can have a high bioburden.

The most exhaustive published study is probably that by Payne (2007) based on the data from Negretti (1981). The data of this study are summarized in Table 6.1. The highest contamination rates were found for cap liners (96%),

Table 6.1 Summary of the data from Negretti (1981).

Type of container	Number examined	Number contaminated	Number of containers contaminated with				
			Bacilli	Mold	*Escherichia coli*	*Pseudomonas aeruginosa*	*Staphylococcus aureus*
Glass bottles	1000	546	308	103	6	4	25
Plastic bottles	1000	948	466	686	19	0	40
Metal tubes	1000	897	521	248	11	0	64
Droppers	500	468	187	189	10	0	44
Cap liners	500	482	257	151	8	0	14
Blisters	200	67	42	9	0	0	4

Source: Adapted by Payne (2007).
The following further species were found: *Enterococcus faecalis, Klebsiella pneumoniae, Serratia* spp., *Enterobacter* spp., *Yersinia* spp., *Citrobacter* spp., *Proteus* spp., and *Hafnia* spp.

plastic bottles (95%), droppers (94%), and metal tubes (90%). The lowest levels of contamination were found for blisters with 34%. It is interesting to note that several potential objectionable microorganisms such as *Escherichia coli, Staphylococcus aureus, Pseudomonas aeruginosa,* or *Klebsiella pneumoniae* were also found. However, the containers had a generally low level of contamination, i.e. only in 2.5% of the cases was the CFU number per container above 50.

A further study was published by Bouska *et al.* (1980). Here, several labs tested a total of approximately 2000 glass bottles, 11 different types or volumes from 5 different suppliers. In 41.1% of the cases no microorganisms were detected and in 71–100% of the tests 0–9 CFU/unit were found. For 4 bottle types, more than 100 CFU/unit were found in 0.9%, 1.7%, 1.7%, and 6.5% of the tests. The latter bottle (a 50 ml bottle from supplier A) showed a generally relatively high contamination rate, thus this one could be seen as an out-of-trend (OOT) supplier or bottle type. In only one bottle type was the highest detected CFU value above 1000 CFU/unit (again the 50 ml bottle from supplier A), all the others were below 1000 CFU/unit. In total, 98.3% of the bottles tested showed a value below 100 CFU/unit. From a qualitative point of view, the isolates found were identified and in only two tests was *S. aureus* found, in one case *Streptococcus* group D and in another case a coliform bacterium. In conclusion, these results show generally very low bioburden in glass bottles used for pharmaceutical preparations.

Hecker (1992) tested glass containers used for oral liquids and found results comparable to Bouska *et al.* (1980), i.e. the bioburden was found to be very low.

For plastic containers, foils, and dosing aids, the bioburden was also in general below the internal acceptance criteria. They found aerobic spore-forming bacteria or molds when the contamination came from the environment and typical human-related bacteria (*Micrococcus* or *Staphylococcus*) when human interventions during the production, packaging, or transport were involved. Pathogenic or objectionable microorganisms were found in only very rare cases (Hecker 1992).

These findings can be confirmed by the authors' experience. During all the testing seen so far, only in rare cases was relevant contamination detected (see Section 6.7 for some examples).

The microbiological testing data from January 2016 to July 2018 for the primary packaging of a diverse portfolio of non-sterile products from a pharmaceutical manufacturer was analyzed. In total, 1417 primary packaging tests were performed. Table 6.2 gives an overview of the amount tested and the batches that were rejected or investigated. From the six rejected batches five were for inhalers, which have very tight specifications (10 CFU/inhaler), and two of these five cases were rejected because the CFU number could not be evaluated (swarming bacteria). In the other 3 cases, 11 CFU/unit were found and the root cause was a contamination introduced by the supplier. The last case was the cover of a container, which showed agar plates overgrown with molds. Here, the investigation showed that the real CFU number was within specification, but it was classified as OOT. In Table 6.3 the distribution of the CFU numbers found for total aerobic microbial count (TAMC) and molds is summarized for the different categories of primary packaging. These data show the high microbiological quality of primary packaging. In 99.6% of the cases, the CFU number was less than 10 CFU/unit. The 10 CFU/unit specification limit applies only to the inhaler, while higher numbers (mostly 100 CFU/unit) are acceptable for the other materials.

6.1.3 Antimicrobial Packaging

Active packaging is a packaging system that possesses attributes beyond basic barrier properties. This is achieved by adding active ingredients to the

Table 6.2 Summary of the number of accepted and rejected batches tested per year.

Year	Accepted batches	Rejected or OOT batches
2016	590	5
2017	564	1
2018	257	0

Remark: one of the rejected or OOT (out of trend) batches could be released after further investigation (see text for details).

Table 6.3 Summary of the number of CFU found for each group of primary packaging (see text for details).

	Bags/others		Bottles		Seals		Foils		Inhalers		Total	
CFU	TAMC	Mold	TAMC	Mold	TAMC	Mold	TAMC	Mold	TAMC	Mold	TAMC	Mold
0	153	206	9	35	4	13	628	665	383	457	**1177**	**1376**
1	60	26	19	4	9	0	30	4	68	5	**186**	**39**
2	10	0	3	0	0	0	6	1	3	0	**22**	**1**
3	6	0	0	0	0	0	1	0	2	0	**9**	**0**
4	2	0	0	0	0	0	2	0	0	0	**4**	**0**
5	0	0	2	0	0	0	2	0	1	0	**5**	**0**
6	0	0	4	0	0	0	0	0	0	0	**4**	**0**
7	0	0	2	0	0	0	0	0	0	0	**2**	**0**
8	1	0	0	0	0	0	1	0	0	0	**2**	**0**
11	0	0	0	0	0	0	0	0	2	0	**2**	**0**
25	1	0	0	0	0	0	0	0	0	0	**1**	**0**
>200	0	1	0	0	0	0	0	0	3	0	**3**	**0**
Total		233		39		13		670		462		1417

For example, for *Bags/others* for TAMC in 153 tests 0 CFU were found and for 206 tests 0 CFU molds. TAMC, total aerobic microbial count.

packaging system or using an actively functional polymer (Han 2003). These systems are able to inhibit or kill microbial contaminants in the packed product. To the author's knowledge, antimicrobial packaging is not applied in pharmaceutical packaging development. In contrast, in the food industry, the use of specific antimicrobial active substances in packaging materials (Table 6.4) has been established for several years. Its primary goals are (i) safety assurance, (ii) quality maintenance, and (iii) shelf-life extension (Han 2003), similar to the functionality of primary packaging for drug products, as mentioned above. It is likely this approach is not taken in the pharmaceutical industry as a lot of drug products do not promote microbial growth per se and, where this is the case, an antimicrobial preservation system (e.g. parabens, benzoic acid, propylene glycol, and ethanol) is added to the drug. However, the use of antimicrobial active substances in the primary packaging rather than in the drug product itself could be an attractive option.

Another possibility to reduce the risk of microbial contamination in the packaged product and thereby increase drug product safety is hurdle technology. This technology is described in the PDA TR 67 (2014) for its application in the pharmaceutical industry, especially in non-sterile manufacturing, but was developed by the food industry (e.g. Han 2003). With hurdle technology, barriers are introduced to the process to reduce microbial contamination risks. In manufacturing, hurdles might be high temperature, low water activity, high acidity, reduced redox potential, etc. For the primary packaging system, hurdles would be physical barriers, moisture barriers, or antimicrobial ingredients in the packaging (see Figure 1.3 in Chapter 1) that in combination would reduce microbial contamination or its proliferation.

Table 6.4 Summary of some antimicrobial active substances.

Antimicrobial active substance	Example	Packaging
Organic acids	Benzoic acid, paraben, sorbate	PE, LDPE, styrene
Enzymes	Lysozym, EDTA, glucose oxidase	PVAL
Bactericides	Nisin	PE, HPMC
Fungicides	Benomyl, imazalin	Ionomer, PE, LDPE
Polymers	Chitosan	Paper, LDPE, nylon
Natural extracts	Grapefruit juice, silver, eugenol	LDPE, nylon, paper
Gas	Ethanol, hinokitiol	Silica gel, plastic films

Source: Adapted from Han (2003).
HPMC, hydroxypropyl-methylcellulose; LDPE, low density polyethylene; PE, polyethylene; PVAL, polyvinyl alcohol (water-soluble films).

6.2 Guidelines and Literature

The pharmaceutical industry requires specifications, regulations, and quality control requirements, which clearly dictate how the product or the process needs to be controlled. Without such regulations, they often face the problem that they do not know how to produce or control the product. The same is true for the microbiological control of primary packaging. Indeed, there are a lot of regulatory or guiding documents on primary packaging in regard to usage, composition, etc., but hardly anything on microbiological quality. In the current section some regulatory texts are summarized and interpreted according to the author's views on microbiology.

Searching the online version of the European Pharmacopoeia (Ph. Eur.) for the word "containers" in combination with "microbiology," only 2 hits appear: Ph. Eur. chapters 2.6.7 and 5.1.6. However, both chapter do not give any information on the microbiological control of containers or primary packaging. There are several interesting Ph. Eur. chapters on physical aspects of containers (see Table 6.5): however, none of them gives us any information for microbiological testing.

Further guiding regulatory documents are summarized in Table 6.5. In most of these documents no microbiological requirements or information on microbiological testing are given.

In EudraLex (2015) chapter 5 "Production" in paragraph "Packaging material," it is mentioned that primary packaging should be controlled in a comparable way to starting materials. Here, the question is what kind of controls apply to starting materials, especially with regard to microbiology. Furthermore, in Annex 5 of the EudraLex (2004), which talks about the sampling of starting materials and packaging materials, again no microbiological requirements for microbiology are found.

Two paragraphs from the Code of Federal Regulation (CFR 2017a, 2017b) are cited in Table 6.5, where the microbiological testing of containers can be interpreted, at least to a certain degree. From the FDA there are two Guidance for Industry documents, which mainly refer to parenteral products (see Table 6.5, FDA 1999, 2004). However, in the author's opinion, these could also be applied to primary packaging materials used for non-sterile manufacturing.

In chapter 3.2.2 of Annex 9 of the WHO (2002), some information is given for routine testing of packaging materials and containers. Here, it is clearly stated that microbiological tests are also needed (Table 6.5). It is interesting to note that the EMA document from 1998 is quoted here (EMA 1998). In the new version of this document no microbiological requirements for packaging material are mentioned (EMA 2005).

In contrast, a passage from a German drug law (AMG 2017) is very clear: in § 55 (8) in conjunction with § 13 of the regulations governing the operation of

Table 6.5 Overview of regulatory and guiding documents for primary packaging in regard to microbiological information or requirements.

Documents	Chapters	Content	Microbiology
Ph. Eur.	1.3	General aspects of containers used for pharmaceutical products	No microbiological requirements
Ph. Eur.	3.1	Materials used for the manufacture of containers	No microbiological requirements
Ph. Eur.	3.2	General requirements for container which are or may be in direct contact with the product	No microbiological requirements
CFR	211.80(a)	*There shall be written procedures describing in sufficient detail the receipt, identification, storage, handling, sampling,* ***testing****, and approval or rejection of* ***components and drug product containers and closures****; such written procedures shall be followed.*	There is no direct link to microbiology but the question is why should microbiology not be part of container testing?
CFR	211.84(d)	*Each lot of a* ***component, drug product container, or closure*** *that is liable to microbiological contamination that is objectionable in view of its intended use* ***shall be subjected to microbiological tests before use***	Here it is open to the reader to interpret whether microbiological testing is needed or not. In the author's opinion, it must at least be evaluated whether a risk is present
FDA – Aseptic Guide	VI. Components and containers/closures	*It is important to characterize the microbial content (e.g. bioburden, endotoxin) of each component that could be contaminated and establish appropriate acceptance limits*	This guideline is for aseptic processing and not for non-sterile manufacturing, thus not applicable; however, in terms of the content it is also reasonable for non-sterile manufacturing
FDA – Guidance for Industry	Container Closure Systems	The product needs to be protected from microbiological contaminations	As with the Aseptic Guide, this guideline mainly refers to parenteral products

EudraLex	5.40 and 5.48	Mentions the control of primary packaging	No clear link to microbiology, however, it is mentioned that primary packaging should be given attention comparable to that given to starting materials
EudraLex	Annex 8	Aspects of sampling of packaging material	No microbiological requirements
WHO – Annex 9	3.2.2 Testing program	Here, microbiological testing is mentioned for packaging material and containers	No further details for testing are given (e.g. which kind of test and which acceptance criteria)
EMA	Guideline	Guideline on Plastic Immediate Packaging Materials	No microbiological requirements
AMG & ApBetrO & BAK	AMG: § 55 (8), ApBetrO: § 13	Primary packaging must protect the medicinal product from physical, microbiological, and chemical changes	Microbiological testing of primary packaging is mentioned in the guidance to this paragraph
DIN/EN/ISO 15378:2018-04	Primary packaging material for medicinal products	During manufacturing of primary packaging material, contamination risk needs to be prevented and controlled	Again, what is needed from a microbiological point of view is open to interpretation, especially since the pharmaceutical industry is not normally directly responsible for manufacturing the primary packaging
EN ISO 11737-1:2006	Annex 1–3	Guidance and method details for the bioburden evaluation of medical devices	Several approaches for bioburden testing, sample size, data interpretation, validation of the test method (for further details, see Box 6.3)

pharmacies (ApBetrO 2017) it is stated that primary packaging should protect the product from microbiological contaminations. Furthermore, in the guidelines to this document it is very clearly stated that primary packaging material must be subjected to microbiological examination (BAK 2016b).

Finally, in the DIN/EN/ISO 15378:2018-04 it is stated that during manufacturing of primary packaging material, contamination risk needs to be prevented and controlled. However, this implies no real control at the level of the customer (which applies to non-sterile manufacturing of drug products). For the testing and validation of microbiological testing methods, the ISO 11737 gives some methodological guidance mainly to be applied for medical devices but which can also be used for primary packaging (for further details, see Box 6.3).

In summary, although clear requirements for the microbiological testing of primary packaging are not given, it is quite obvious that testing can be anticipated with at least some of the text summarized in Table 6.5.

6.3 Acceptance Criteria and Testing Frequency

Clear acceptance criteria, which are often used as specifications, are defined for non-sterile pharmaceutical preparations and substances for pharmaceutical use; for example, in the informative USP chapter <1111> and Ph. Eur. chapter 5.1.4, and for microbiological attributes of non-sterile nutritional and dietary supplements in USP chapter <2023> or for herbal medicine in Ph. Eur. chapter 5.1.8. The harmonized acceptance criteria from USP <1111> and Ph. Eur. 5.1.4 are summarized in Table 6.6. It is very important to mention that the interpretation of these acceptance criteria allows a factor of 2. Therefore, if a drug product has an acceptance criterion of 10^1 CFU/g, its maximum acceptable count is 20 CFU/g. Furthermore, as well as the specific microorganisms defined, the significance of other microorganisms should be evaluated in terms of use of product, nature of the product, method of application, intended recipient, use of immunosuppressive agents, and presence of disease, wounds, and organ damage. If some objectionable microorganisms are found, the product should be rejected (for details on objectionable microorganisms, see Chapter 11).

6.3.1 Examples of Acceptance Criteria

One possible approach for acceptance criteria for primary packaging would be to apply the criteria from Table 6.6. This means 1000 CFU are used for a non-aqueous preparation for oral use for TAMC, and by using 1000 CFU/g for the product itself, in the worst case the product has 2000 CFU/g in total and is therefore still within the acceptance criterion of Table 6.6 using the factor 2 stated above. However, this is not really the intention of the factor 2

Table 6.6 The harmonized acceptance criteria for non-sterile pharmaceutical preparations and substances for pharmaceutical use according to USP <1111> and Ph. Eur. 5.1.4.

Route of administration	TAMC (CFU/g or CFU/ml)	TYMC (CFU/g or CFU/ml)	Specified microorganisms
Nonaqueous preparations for oral use	10^3	10^2	Absence of *Escherichia coli* (1 g or 1 ml)
Aqueous preparations for oral use	10^2	10^1	Absence of *E. coli* (1 g or 1 ml)
Rectal use	10^3	10^2	–
Oromucosal use, gingival use, cutaneous use, nasal use, auricular use	10^2	10^1	Absence of *Staphylococcus aureus* (1 g or 1 ml) Absence of *Pseudomonas aeruginosa* (1 g or 1 ml)
Vaginal use	10^2	10^1	Absence of *P. aeruginosa* (1 g or 1 ml) Absence of *S. aureus* (1 g or 1 ml) Absence of *Candida albicans* (1 g or 1 ml)
Transdermal patches (limits for one patch including adhesive layer and backing)	10^2	10^1	Absence of *S. aureus* (1 patch) Absence of *P. aeruginosa* (1 patch)
Inhalation use (special requirements apply to liquid preparations for nebulization)	10^2	10^1	Absence of *S. aureus* (1 patch) Absence of *P. aeruginosa* (1 patch) Absence of bile-tolerant Gram-negative bacteria (1 g or 1 ml)
Substances for pharmaceutical use	10^3	10^2	–

TAMC, total aerobic microbial count; TYMC, total combined yeast and mold count.

assumption and this also raises the question: Which unit is used to calculate the CFU for primary packaging. By 25 cm^2 for a foil or per unit for a container? This approach would in any case allow quite high numbers for the primary packaging. But since the unit is not clearly defined, defending these acceptance criteria in applications, audits, or inspections could be quite tricky.

Another very reasonable approach was published by Hecker (1992) and later by Seyfarth (2003). Here, as a rule of thumb, 10% of the acceptance criteria for the primary packaging from Table 6.6 are used. In Table 6.7 the approach of Hecker (1992) and Seyfarth (2003) are summarized. They defined the acceptance criteria for their primary packaging per unit (e.g. bottles), per g (for cotton and foam), and per 100 cm^2 for foils. This is indeed a very pragmatic approach, however, for bottles in particular, the differences in volume can be tremendous and therefore a special criterion might be needed. This is partially addressed in Table 6.7 by differentiating between two volumes. Furthermore, the question arises whether a criterion for specified microorganisms is needed or not. Hecker (1992), for example, tests for the absence of *E. coli*, *Salmonella*, *P. aeruginosa*, and coagulase-positive *Staphylococcus* per unit. For containers used for topical medications, it is also necessary to show the absence of Enterobacteriaceae. Seyfarth (2003) does not outline an approach for specified microorganisms.

A more recent publication of possible acceptance criteria for primary packaging material was published by Rieth (2017). He defined a very low level of bioburden and for only three different materials:

- Packaging for solid dosage forms: 10 CFU/g or ml of packaging content
- Packaging for liquid/pasty dosage forms: 1 CFU/ml of packaging content
- Packaging for plugs, closures, and covers: 10 CFU/unit (product contacting surface)

These are more stringent than those presented in Table 6.7, but again the question of the size of the primary packaging arises.

Two user examples of acceptance criteria for primary packaging materials are given in Tables 6.8 and 6.9. For that in Table 6.8, each grown colony was

Table 6.7 Summary of the acceptance criteria for primary packaging according to Hecker (1992) and Seyfarth (2003).

Containers	TAMC	TYMC
Bottles of oral medications		
≤50 ml	max. 100 CFU/unit	max. 10 CFU/unit
>50 ml	max. 1000 CFU/unit	max. 100 CFU/unit
Bottles and tubes for topical medications		
≤50 ml	max. 10 CFU/unit	max. 5 CFU/unit
>50 ml	max. 100 CFU/unit	max. 20 CFU/unit
Closures and application devices	max. 100 CFU/unit	max. 10 CFU/unit
Cotton and foam rubber	max. 1000 CFU/g	max. 100 CFU/g
Foils	100 CFU/100 cm^2	10 CFU/100 cm^2

Source: Adapted from Hecker (1992) and Seyfarth (2003).
TAMC, total aerobic microbial count; TYMC, total combined yeast and mold count.

Table 6.8 User example of acceptance criteria for primary packaging.

Packaging material	TAMC	Indicator microorganism
Container up to 200 ml	25 CFU/unit	Absent in tested unit
Container 201–2000 ml	100 CFU/unit	Absent in tested unit
Container >2000 ml	200 CFU/unit	Absent in tested unit
Other materials	25 CFU/unit	Absent in tested unit
Foils	100 CFU/100 cm^2	Absent on 100 cm^2

The following species were defined as indicator microorganisms: *Escherichia coli*, *Salmonella* spp., *Staphylococcus aureus*, *Pseudomonas aeruginosa*, and Enterobacteriaceae. TAMC, total aerobic microbial count.

Table 6.9 User example of acceptance criteria for primary packaging.

Packaging material	TAMC	TYMC
Foils for inhalation or patches	10 CFU/100 cm^2	2 CFU/100 cm^2
Inhaler	10 CFU/unit	2 CFU/unit
Foils for oral, solid dosage forms	100 CFU/100 cm^2	10 CFU/100 cm^2
Tubes, bottles (≤50 ml)	10 CFU/unit	2 CFU/unit
Tubes, bottles (>50 ml)	100 CFU/unit	10 CFU/unit
Bottles for ointments (≤50 ml)	10 CFU/unit	2 CFU/unit
Bottles for ointments (>50 ml)	100 CFU/unit	10 CFU/unit
Bottles for tablets (≤50 ml)	100 CFU/unit	10 CFU/unit
Bottles for tablets (>50 ml)	1000 CFU/unit	100 CFU/unit
Cap or cover	100 CFU/unit	10 CFU/unit

TAMC, total aerobic microbial count; TYMC, total combined yeast and mold count.

additionally identified to exclude any objectionable microorganisms that were mainly defined as those specified in Ph. Eur. chapter 5.1.4: *E. coli*, *Salmonella* spp., *S. aureus*, *P. aeruginosa*, and Enterobacteriaceae (bile-tolerant, Gram-negative bacteria). The second example (Table 6.9) does not check for any specified or objectionable microorganisms but more primary packaging types are differentiated and not only TAMC is checked but also TYMC.

6.3.2 Calculation Approach to Define Acceptance Criteria

As shown in Section 6.3.1, several quite pragmatic approaches exist to define the microbiological acceptance criteria for primary packaging materials. In this section a more scientific approach is presented where the criterion is defined according to the route of application and the surface of the primary packaging in contact with the amount of drug. Two examples are given in Boxes 6.1 and 6.2.

This approach calculates the amount of the drug product that comes into contact with primary packaging material. This is very important since, for

Box 6.1 Calculation Approach for TAMC Testing of a Primary Packaging Material (in This Case an Aluminum Foil) in Relation to the Foil Surface and the Tablet Size. The Tablet Is Orally Applied

- Weight per tablet: 500 mg
- Thus, two tablets weigh 1 g and according to Table 6.6 the acceptance criterion for TAMC is 10^3 CFU/g
- The following foil surface area comes into contact with 1 tablet: approximately 12 cm^2, thus for 2 tablets this size is 24 cm^2
- 24 cm^2 roughly corresponds to the surface of a standard contact plate
- Therefore, with one contact plate the contribution of the foil to 1 g of the oral tablet is tested
- Taking the 10% of the acceptance criterion stated in Table 6.6, the maximal acceptable contribution to 1 g drug product would be 10^2 CFU
- Consequently, a threshold value of **10^2 CFU/culture plate or 25 cm^2** would be the acceptance criterion for this primary packaging

Box 6.2 Calculation Approach for TAMC Testing of a Primary Packaging Material (in This Case a Plastic Tube) in Relation to the Tube's Surface and the Volume of the Gel. The Gel Is Cutaneously Applied

- The tube contains 30 g gel
- According to Table 6.6 the acceptance criterion for TAMC is 10^2 CFU/g
- The inner surface of the tube is approximately 60 cm^2, i.e. 2 cm^2 come into contact with 1 g of the gel
- A contact plate has a surface of approximately 25 cm^2
- Taking 10% of value stated in Table 6.6, the acceptance criterion would be 10^1 CFU/g gel
- Consequently, 10^1 CFU per 2 cm^2, i.e. approximately **125 CFU/contact plate** or 25 cm^2

Alternatively:

- 30 g may contain 30×10^2 CFU
- Consequently, 60 cm^2 may contain 3×10^2 CFU (using the 10% rule)
- Therefore, the threshold is **3×10^2 CFU/tube**

example, an individually packed tablet will have much more contact with the primary packaging compared to tablets in a container.

With this approach a product-specific acceptance criterion is needed for each primary packaging. This will generate a lot more work than the approaches with a fixed level; however, once carried out this approach is much more specific, and the calculation itself is quite easy. Probably the bigger challenge is its application in the testing laboratory. In the lab, for example, several dozen foils are tested at the same time without necessarily knowing to which product they belong. Thus, the biggest challenge in the author's opinion is not the calculation but the correct assignment of the primary packaging to the right drug product.

6.3.3 Comparison with Food Industry

Clearer primary-packaging acceptance criteria are given for the food industry. Table 6.10 summarizes some examples. When these figures are compared with those detailed above, it can be seen that they are quite stringent, which makes sense as food products have higher potential for microbial proliferation, which is normally not the case for drug products.

6.3.4 Testing Frequency

For the testing frequency the main question is, if each batch (or lot) needs to be tested (lot-wise testing) or if a lower testing frequency (e.g. every fifth or tenth

Table 6.10 Some examples of acceptance criteria for primary packaging used in food industry.

Packaging material	Parameter	Standard value	Basis	Methods (e.g.)
Foils	TAMC TYMC Enterobacteriaceae	≤2 CFU/100 cm^2 ≤1 CFU/100 cm^2 Absent/100 cm^2	IVLV	Pour-plate or contact method
Cup, bowls (≤500 ml)	TAMC TYMC Enterobacteriaceae	≤10 CFU/100 ml ≤1 CFU/100 ml Absent/100 ml	IVLV	Rinsing method
Disposable container for milk	TAMC Coliforme	≤50 CFU/100 ml Absent/100 ml	IDF, FDA	Rinsing method

Source: Adopted from Beckmann (2010).
CFU, colony forming units; FDA, US Food and Drug Administration; IDF, International Dairy Federation; IVLV, Industrievereinigung für Lebensmitteltechnologie und Verpackungen; TAMC, total aerobic microbial count; TYMC, total combined yeast and mold count.

batch, so-called skip-lot testing) is acceptable. To reduce the testing frequency from lot-wise to skip-lot testing a risk-based approach should be implemented. Examples for such risk-based approaches to define the testing frequency for non-sterile drug products as well as drug substances and excipients are given in ICH Q6 (EMA 2000). Using the decision tree #6 *Microbiological quality attributes of drug substances and excipients* as basis, already with the first step (*Is the drug substance/excipient capable of supporting microbial growth or viability?*) the decision tree ends with the statement: *Provide supporting data. Microbial limits acceptance criteria and testing may not be necessary.* To provide supporting data, for example, the first 5 or 10 batches of the primary packaging material is tested and if the results do comply with the acceptance criteria, then a skip-lot testing (or no testing at all?) is possible. Such an approach is quite straight forward and easy. An alternative would be to look at the production process of the primary packaging to check for any microbiological contamination risks (i.e. low risk is present if there are microbiological inhibitory steps such as high temperature, no usage of water, microbiological-controlled environment, and packaging process, storage, and transportation conditions). In such cases a reduced testing can be implemented (for further details on process risk assessment, see Chapter 2 or USP chapter <1115> or PDA TR No. 67).

However, in certain cases when the drug product is a product with a high risk for the patient, e.g. inhalers used for inhalation products, a lot-wise testing is preferably implemented.

6.4 Test Methods

Specific methods for testing primary packaging for pharmaceutical use are, as far as the author knows, not described in any guiding or regulatory document but there are some descriptions in the book by Rieth (2017). However, in general, the methods used for microbiological examination of non-sterile products (Ph. Eur. 2.6.12 and 2.6.13 or USP <61> and <62>) can be applied. Using these methods, no method validation is needed since these are pharmacopoeial methods, i.e. in Chapter 1 in General Notices of Ph. Eur. it is stated that *the test methods given in monographs and general chapters have been validated in accordance with accepted scientific practice and current recommendations on analytical validation. Unless otherwise stated in the monograph or general chapter, validation of the test methods by the analyst is not required.* With this background it is easier to use these methods instead of developing further methods that need to be validated. However, some adaptation might be needed, which will be discussed hereafter. The main question that arises with these adaptations is whether they will change the method in such a way that a validation (e.g. according to Ph. Eur. 5.1.6 or USP <1223>) is needed or whether a suitability test or a more sophisticated suitability test will suffice. Finally, it may also be questioned if a suitability test as such is needed since primary packaging are generally inert with limited release of leachable chemicals and therefore no inhibition of microbial growth in nutrient media is expected. Thus, it might also be a possibility to skip the suitability test provided a scientific justification is in place.

Interestingly, ISO 11737-1:2006 gives guidance and detailed methods for the determination of bioburden as used in the validation and monitoring of medical devices to be sterilized. This approach could in general – at least for some type of primary packaging – be applied (for further details, see Box 6.3).

Box 6.3 Some Details on the Approaches from ISO 11737-1:2006

The ISO describes in the three annexes some possible approaches for the enumeration of bioburden on medical devices and its validation (suitability test). They provide two general ideas: (i) repetitive treatment of a sample product (i.e. that the product is several times tested for microbial contamination with the same method) or (ii) product inoculation with known levels of microorganisms followed by quantitative assessment of the extent of recovery using a specific method (e.g. rinse and membrane filtration or agar plating). For the second approach the question arises how comparable this is to the natural situation, i.e. how representative is such an inoculation compared to the natural bioburden on a product.

To remove microorganisms from the product, treatments used may either consist of rinsing (or elution) together with some form of physical force or direct surface sampling. A surfactant may be used to enhance the recovery. As **physical forces in combination with rinsing** the following methods are outlined: stomaching, ultrasonication, shaking, vortex mixing, flushing, blending, and swabbing. For **direct sampling** contact plating, agar overlaying or most probable number is mentioned.

For **validation using the repetitive recovery** approach the product to be tested is sampled several times (e.g. a surface is tested using contact plates until there is no significant increase in the accumulated number of microorganisms recovered). One possibility to know how many microorganisms are left on the surface is the coating of the surface with molten recovery medium. Using contact plates on surfaces the recovery with the first plate has been shown to be approximately 50% (BUT for laboratory inoculated surfaces with known microbial suspensions and not natural inoculated surfaces, see Section 6.4.3). In such a case a correction factor should be implemented.

Furthermore, **validation using product inoculation** is outlined in the ISO norm. The selection of the microorganisms is challenging since the use of vegetative microorganism is difficult in practices because loss of viability can occur on drying. Therefore, the use of bacterial spores is recommended. Microbial inoculation has limitations such as encrustation, adhesion or non-adhesion of the suspension, clumping, and variation in the level of the inoculum, and these limitations should be considered when inoculating the product.

For **data interpretation,** interesting is also the approach to not only look for an exceedance of a specific limit but also to investigate changes of bioburden level on the material which would be some kind of trending. Here, control charts (Shewhart charts) are mentioned (see also Chapter 10). Limits should be based on historical data.

6.4.1 Membrane Filtration

Many primary packaging materials can be tested using the membrane filtration method described in Ph. Eur. 2.6.12 or USP <61>. A short protocol on how such testing for containers (bottles, vials, tubes, etc.) can be performed is summarized here:

1) Sample a predefined number of containers (e.g. 40 units).
2) Fill each container with a washing solution (e.g. buffered sodium–chloride–peptone solution pH 7.0 or phosphate buffer pH 7.2 or casein soya digest broth as described in Ph. Eur. 2.6.13 or USP <62>). In general, the addition of e.g. 0.2% polysorbate 80 is recommended for better wetting of the material surface.
3) Shake the units for 10–15 minutes or leave them for 30 minutes at room temperature to release the microbes on the surface.
4) As a negative control, the same amount of washing solution is used to show the absence of microorganisms in the material used and to demonstrate aseptic handling.
5) The washing solution is pooled or directly filtered as for the compendial method. Usually the entire solution can be filtered: however, depending on the acceptance criterion it might be helpful to filter only part of the solution or to filter different volumes (e.g. 1 ml, 10 ml, 100 ml, and rest of solution).
6) If necessary, rinse, for example, with 100 ml rinsing solution.
7) The filter is then added to casein soya digest agar (CASO) or sabouraud dextrose agar (SDA) plates.
8) Incubate at the required temperature (30–35 °C for 3–5 days for TAMC, 20–25 °C for 5–7 days for TYMC).
9) After incubation, enumerate the CFU per plate and calculate the CFU/unit.

A similar procedure can be used for small covers, stoppers, etc. In this case a predefined number of units are added to between 500 and 1000 ml of the washing solution and then the same procedure is applied from point 3 onward as described above. Using this approach, it should not be forgotten that the outside or the side that will not come into contact with the product is being tested.

Usually, this should be no problem but in some cases it might matter and then another method or approach is needed.

For testing of specified microorganisms, see Section 6.4.4.

Box 6.4 Membrane Filtration Approach for Big Containers

Testing of big containers (e.g. 20 l or 60 l) is carried out using the contact plate approach (see Section 6.4.3) or the membrane filtration approach (see Section 6.4.1). For the latter, the container is filled with 1–2 l of washing solution and the container is closed using its lid. Then the container is shaken, e.g. by rolling the container on the floor until the entire inner surface is wetted with the washing solution.

6.4.2 Pour-Plate Count

For large covers or if a blister needs to be tested, the pour-plate count method can be used. In his book, Rieth (2017) describes the following approach (adapted):

1) Add a sterile filter paper soaked with sterile water to a Petri dish or a box.
2) The units to be tested are placed on the filter paper.
3) Pour agar into the units (e.g. cover or blister) as described in the pharmacopoeial chapters.
4) As a negative control pour some agar into sterile Petri dishes.
5) Close the Petri dish or box to avoid desiccation.
6) Incubate at the required temperature (30–35 °C for 3–5 days for TAMC, 20–25 °C for 5–7 days for TYMC).
7) After incubation, enumerate the CFU per unit.

An alternative would be to place the primary packaging (if possible) directly in a big Petri dish; however, the membrane filtration method is likely more accurate and easier to carry out.

To detect specified microorganisms, the use of swabs (see Section 6.4.4) is likely to be the easiest approach.

6.4.3 Contact Plates and Swabs

Although contact plates are not described in the chapter referred to above, it is an established method and the one referred to in some regulatory or guiding documents (e.g. USP <1116>; FDA 2004; EudraLex 2008; ANVISA 2013). Contact plates are very easy to use, however, they also have their limitations (e.g. the recovery rate is at ~50%; USP <1116> and Goverde 2018). The contact plate method can be used for the testing of foils and also for bigger containers.

If the surface to be tested is not smooth enough, swabs can be used instead of a contact plate (Goverde *et al.* 2017).

Details on the technique for contact plates, dip slides, or swabs are given in several norms (e.g. ISO 18593 or DIN 10113-3). In general, the contact plate filled with a suitable agar medium is pressed against the surface with approximately 500 g for 5–10 seconds. Some experiments were performed on artificially inoculated surfaces with in-house isolates to test for a difference in recovery between 10, 5, and 1 second. In these experiments, no significant difference could be found (Goverde 2018; Berchtold unpublished data). However, pressing for 10 seconds is an established and reasonable time.

The culture medium used, as for the other methods, is normally CASO agar (Casein Soya Digest Agar). But if testing for yeasts and molds, SDA (Sabouraud Dextrose Agar) can also be used. When testing for specified microorganisms, swabbing could well be the best approach. After swabbing the surface, the swab is added directly to the enrichment broth as described in Ph. Eur. 2.6.13 or USP <62>.

The incubation is normally performed as described above according to the compendial microbiological chapters.

6.4.4 Test for Specified or Objectionable Microorganisms

To test for specified microorganisms, it is possible to use the methods described in Ph. Eur. 2.6.13 or USP <62> using the membrane filtration or the swab technique described in Sections 6.4.1 and 6.4.3. In this case the primary packaging is washed, the washing solution filtrated, and the filter then added aseptically to the first enrichment broth. Alternatively, the material is swabbed and the swab is then added to the first enrichment broth. Further steps are the same as for product testing (see Chapter 5).

If the primary packaging should be free of objectionable microorganisms, again the same approach can be applied as for product testing (see Chapter 11). Thus, each single colony found either by enumeration or specified testing is identified and evaluated for its criticality by means of risk assessment.

Another possibility that is often used is that no specific test for specified microorganisms is defined. Instead, each colony found in the enumeration test, as for the objectionable approach, is identified to exclude specified or objectionable microorganisms. However, with such an approach a relevant deviation from the pharmacopoeial method may occur and therefore some validation that the specified microorganisms can be detected might be needed.

6.5 Suitability Test

Probably one of the most important activities for QC testing is to show the suitability of the test method for recovering microorganisms in the presence of the product. Although primary packaging material is normally quite inert and no inhibition of microorganisms can be expected, this should be shown by means of a suitability test. It is probably not necessary for each individual type or article number of primary packaging material to be tested, i.e. grouping (e.g. all foils made out of aluminum) could be justified.

The main problem with primary packaging is that direct inoculation of the surface of the material – as for most nonaqueous products – is difficult since the inoculated microorganisms (e.g. *P. aeruginosa*) die off very fast with the drying process. Using spores only (e.g. *Bacillus subtilis* or *Aspergillus brasiliensis*) is likely to give reasonable results for the direct inoculation approach with a drying step (see also Box 6.3).

Thus, as for product testing (e.g. tablets), the product itself is not spiked but the agar used or the washing solution containing the primary packaging. This approach is elaborated for the membrane filtration method (Section 6.5.1) and the contact plate approach (Section 6.5.2).

It must be clearly stated that some of the primary packaging might have a certain bioburden. In such cases the material is treated before starting the suitability test (e.g. by sterilization or disinfection with 70% EtOH) to avoid cross contamination. If this is not possible, the calculation needs to be adjusted by using another control to establish the natural bioburden CFU per tested unit.

6.5.1 Suitability Test Approach for the Membrane Filtration Method

The points below describe an approach to testing primary packaging according to the membrane filtration method (e.g. bottles, covers, and inhalers):

1) Enough containers are prepared to analyze the amount tested in the routine (e.g. 20 units) for each microorganism.
2) Test strain solutions with a defined CFU number are prepared. The test strains from Ph. Eur. 2.6.12 or USP <61> are recommended. In certain cases, further species (e.g. *Burkholderia cepacia*) might be advisable.
3) When bottles are being tested (as described in Section 6.4.1), an inoculated test solution is prepared. The washing solution described above is inoculated with max. 100 CFU of the test strain and the 20 containers are filled equally with this solution. This procedure is repeated for each individual test strain.
4) As a positive control, the same amount of test solution as for the primary packaging testing is prepared and inoculated with the test strain (max. 100 CFU). A single positive control is prepared for each strain.
5) As a negative control, the washing solution is tested in a comparable way but without any strains or primary packaging added.
6) The filled containers as well as the positive and negative control are kept at room temperature at least as long as for routine testing (e.g. >30 minutes), but not longer in order to avoid microbial proliferation.
7) The entire solution is pooled for each strain or directly filtered for each experimental approach.

8) If necessary, rinse, for example, with 100 ml washing solution.
9) The filter is then added to CASO or SDA agar plates.
10) Incubate at the required temperature (30–35 °C for max. 3 days for TAMC, 20–25 °C for max. 5 days for TYMC).
11) After incubation, enumerate the CFU per plate.
12) A factor of 2 as the acceptance criterion as described in the compendial chapters is used. This means the recovery of the tested solution from the primary packaging must show at least 50% of the CFU number of the positive control for the corresponding strain. Furthermore, the negative control must not contain any microorganisms.
13) In certain cases, the non-toxicity of the washing solution needs to be shown by comparing the CFU number of the positive control with the CFU number of the test strain applied.

6.5.2 Suitability Test for the Contact Plate Approach

When using the contact plate approach (e.g. for foils), we encounter once again the problem that the primary packaging material cannot be inoculated directly. An indirect approach is also possible here:

1) The material (e.g. foil) is sampled with a contact plate. Two plates are used for each strain.
2) The two contact plates used for each strain are each inoculated with 100 μl of a test strain suspension with max. 100 CFU/100 μl.
3) As a positive control per strain two unused contact plates are inoculated in the same way.
4) For the negative control, 100 μl of a sterile buffer or water are used instead of a test strain solution.
5) The test strains from Ph. Eur. 2.6.12 or USP <61> are recommended. In certain cases, further species (e.g. *B. cepacia*) might be advisable.
6) Incubate at the required temperature (30–35 °C for max. 3 days for TAMC, 20–25 °C for max. 5 days for TYMC).
7) After incubation, enumerate the CFU per plate.
8) A factor of 2 as described in the compendial chapters is used as the acceptance criterion. This means the CFU recovery of the tested primary packaging must show at least 50% of the CFU number of the positive control for the corresponding strain. Furthermore, the negative control must not show any microorganisms.

6.5.3 Pour-Plate Method and Specified Microorganisms

Comparable to the two approaches mentioned in Sections 6.5.1 and 6.5.2, a suitability test for the pour-plate method can be performed where the test strain solution with 100 CFU/100 μl is added directly to the agar when it is poured. The approach then continues as for the compendial method.

For specified microorganisms using the membrane filtration method, either the test strain is added to the solution for filtering (direct approach) or the test strain is added to the first enrichment broth (indirect approach). In general, the direct approach should be used where possible. When testing for specified microorganisms, the suitability test will differ significantly depending on whether the pour-plate or the contact plate approach is used. The easiest way is to inoculate the agar or the contact plate with the corresponding specified microorganisms under investigation. However, in this case no enrichment and selection are used.

6.6 OOS Procedure

Although the microbiological quality of the primary packaging material is very good and only in rare cases does an exceedance of the requirements occur (see Section 6.1.2 for some examples), every company needs a clear procedure in its SOP on how to handle any results that do not comply. In general, the procedure described in Chapter 12 can also be applied to the primary packaging. However, some further questions might arise.

First of all, when there is an exceedance with primary packaging, clarification is required of whether it is a deviation or an out of specification (OOS). In general, the latter appears when clear specifications for the material were defined, which are approved by the authorities. If testing is regarded more as a form of monitoring – for example, when it can be shown by a risk-based approach that the primary packaging material or even the product with which it is packed is usually not at risk – this then points to a deviation. Indeed, this is only a question of wording, since the procedure as such is quite similar (for details, see Chapter 12). In Figure 6.2 a general procedure for a deviation is depicted that also applies to an OOS.

An event happens, which in our case would be the exceedance of the microbiological level or detection of a specified or objectionable microorganism. A discrepancy (deviation or OOS) is then opened either in a software system or on paper. The latter must of course be data integer. Then, the investigation starts by checking for a sampling error (which for primary packaging can happen) or a lab error (this would correspond to a level 1 OOS investigation). If this is not the case, then a thorough investigation is initiated (this would correspond to the level 2 OOS procedure). As for any exceedance, the real root cause should be found so that the relevant corrective and preventive actions (CAPAs) can be defined. After implementation of the relevant CAPAs, their efficacy should be checked after a defined time period. Finally, the case is challenged by QA. If the case is complete, scientifically sound and well documented,

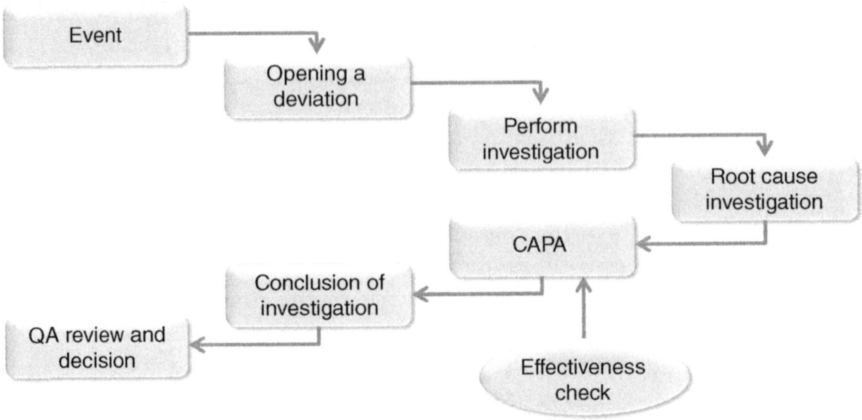

Figure 6.2 Simplified procedure for handling a deviation or an out of specification for primary packaging material. CAPA, corrective action and preventive action.

the case is summarized and approved by QA. The product can then be released for use or must be rejected.

The following list gives some possible checkpoints for handling a primary packaging material that exceeds the microbial requirements:

- Put batch on hold
- Identification of microorganism
- Investigate for a sampling or lab error
- Resampling material and retesting of samples
- Clarify the intended use of the primary packaging material
- Root cause investigation
- Risk assessment for the batches released so far
- Possible treatment of the primary packaging material (e.g. washing, heat treatment, irradiation, and fumigation)
- Contact with or complaint to the manufacturer
- Microbiological testing of the final packaged product
- Higher testing frequency
- Decision on approval or rejection of the batch

6.7 Examples of OOS or OOE Cases

In the following subsections, three examples of a discrepancy are described. The approaches were adapted but not harmonized to show the different approaches used by the investigating party.

6.7.1 Example 1: OOE of a Metal Container (1 l) Used for API Storage

This report contains the following chapters:

(a) Summary and results
(b) Lab investigation
(c) Further investigation with CAPAs
(d) Decision information
(e) Attachments

(a) Summary and results

Specification	TAMC: 100 CFU/unit; alert level 30 CFU/unit
	Absence of specified microorganisms in 4 units
Original sample	TAMC: 42 CFU/unit (identification of *Bacillus* species)
	Absence of specified microorganisms in 4 units
Retest	Not possible since there are no more samples
Resampling	TAMC: 25 CFU/unit (identification of *Bacillus* species)
	Absence of specified microorganisms in 4 units
Conclusion	The current batch did not exceed the specification, but an alert level exceedance is present. Resampling showed comparable results slightly under the alert level with the same species found in the original sample.
	The containers are used for API storage. However, the API powder is packed into sterile plastic bags, which are then packed into the metal containers. Therefore, the metal container itself is not in direct contact with the product (and therefore not really primary packaging). However, it was decided as a contamination control risk to test the containers as primary packaging.
	The root cause of the contamination is probably due to the metal container being packed in untreated paper. Further preventive action to change this procedure was initiated and is being followed by the Quality Management (QM) department.
Recommendation	The batch can be released for its intended purpose.
Reported results	TAMC: 42 CFU/unit
	Absence of specified microorganisms in 4 units
	Comment: Identification of *Bacillus* species, see deviation PPM-18-002

(b) Investigation

Lab error: The correct procedure was used and followed. The negative controls showed no contamination. The analyst remarked that the cans were wrapped in (probably untreated) paper.

Sampling error: No sampling error was found to be the root cause (see attachment II).

(c) Further investigation with CAPAs (Table 6.11)

Table 6.11 Tabular summary of further investigation with CAPAs.

Action	Comment
Information to QA and production	See attached email to Mr. X and Ms. Y from 16 June 2018 (attachment III)
Retest and resampling	Results see chapter a) and attachment I
Initiation of a task force with experts	Meeting with QA, person responsible for product QC and production on 18 June 2018. See attachment IV for meeting minutes
Decision for initiation of an investigation report	No need, it is only an exceedance of the alert level
Trend of historical data	The current product was tested for the first time; however, similar products from the same supplier had been previously tested. All results were below the alert level, but microbial growth appeared in most samples (see attachment V)
Increased sampling needed?	The product will be tested lot-wise for the next five batches
Root cause	The most probable root cause is the container being wrapped in untreated paper. Further investigations are ongoing and will be reported by the QM department
Further actions	• Lot-wise testing • QM department to follow up on the case • The container must be disinfected and wiped with 70% EtOH before use
Recommendation of task force	Since it is only an alert level exceedance and the can is not in direct contact, the batch can be released. However, all cans in this batch and also the following batches must be disinfected until higher quality is implemented

(d) Decision information

Product category	Primary packaging (not in direct contact with the drug product/substance)
Supplier	Company XYZ
Assessment of identified isolates	All species found are Gram-positive, spore-forming rods, which are typically found in the environment. They are not human pathogens.
Product risk assessment	The containers are not in direct contact with the product. The contamination is low relative to the amount of product packed in the containers.
Historical data	It is a new product, however, comparable cans from the same supplier showed comparable contamination (see attachment V).

(e) Attachments

(I) Copy of results with identification of isolates (12 pages)
(II) Record of sampling with statement from team leader
(III) Email to production and QA
(IV) Meeting minutes of task force from 18 June 2018
(V) Historical data of product

6.7.2 Example 2: Possible OOS for a Foil Used for Inhalation Products

Initial Situation

During routine testing of a foil used to pack capsules used for inhalation, one out of four plates could not be enumerated due to spreading bacteria. The other three plates showed no microbial growth (0 CFU/plate). As the testing method, four contact plates were pressed on the foil in a laminar air flow (LAF) cabinet (for details on testing method, see SOP-12345). Incubation was at 30–35 °C. The specification is 10 CFU/100 cm^2.

Level 1 Investigation

The level 1 investigation found no lab error. The following points were investigated:

- Contact plates were within the expiry date and the growth promotion test complied.
- The negative control showed no microbial growth.
- Other samples tested during the same session showed no microbial growth.
- The environmental monitoring of the LAF during the last three months did not show any conspicuous results. No spore-forming bacteria were found during this time period.
- The analyst was qualified.
- The interview with the analyst did not highlight any special issues; however, she mentioned that the foil is sampled by the supplier and sent to the quality control lab packed in cardboard.
- The incubator used showed no conspicuous results during the last environmental testing. It ran within the requested requirements.
- The isolate was identified as *Paenibacillus glucanolyticus*, a Gram-positive spore-forming species, which is known to spread.

Level 2 Investigation

Retest: With a level 2 investigation, the original foil was retested. In addition, the foil roll was internally sampled; i.e. an analysis was performed internally on the foil roll in an air-flow cabinet by a qualified person and not on the QC-specific samples from the supplier. The foil was protected using a sterile plastic bag during transport to the lab. All the samples complied with the requirements without microbial contamination.

Historical data: Other batches tested in the past had detected microbial contamination that was mostly within the requirements (especially with spore-forming bacteria). With the last out of expectation (OOE) discrepancy, it was mentioned that the QC samples provided by the supplier might get contaminated during the transport due to the use of cardboard (see deviation #1231 and #1250). Cardboard is known to have microbial contaminants, especially spore-forming bacteria.

Investigation

The supplier stated that there were no special issues during the sampling, preparation, and transport of the samples for QC testing. As with the last deviation (#1250), the supplier confirmed that they have no microbiological precautions to sample and pack the foil for QC samples. However, the samplers are trained and sample in an ISO 8 environment directly after production of the foil. The sampler wears cleanroom clothes and a hair net but no gloves and no face masks. Furthermore, the cardboard used for transport packaging is untreated. At the customer's the sample is checked into SAP and sent to the corresponding QC lab for testing.

The cardboard used to pack the samples was tested using contact plates. These results showed microbial contamination with different bacterial species (including spore-forming, Gram-positive rods).

Root cause

From the investigation it can be concluded that no microbial contamination above the specification is present. An OOS was opened since one plate could not be enumerated due to the presence of spreading bacteria. However, the investigation highlighted a potential sampling improvement. Since the QC test samples are sampled at the suppliers under noncontrolled microbiological conditions, contamination of the sample is possible, especially since the samples are packed in untreated cardboard.

Risk Assessment and CAPAs

A risk assessment was written for the batches of the foil already released. Compliant results and the product characteristics (low water activity, patient population, and no objectional microorganism detected) indicate that there is no microbial contamination risk for the packed drug products.

In a special experiment it could be proven that 1 CFU of *P. glucanolyticus* is able to spread on the agar plates used for testing.

With a former OOS for the same foil, the supplier was asked to change its sampling procedure. So far, sampling has not undergone improvement, and the same procedure is in place. In the future, therefore, as a preventive measure, QC samples will be sampled by the customer in a specific LAF cabinet by trained personnel, and the samples will be wrapped in a sterile plastic bag for transportation.

Furthermore, other contact agar plates are evaluated to see if they can reduce the swarming effect of *P. glucanolyticus* (this preventive action is ongoing and will be reported in a separate test report; GMP-#R-900-02).

QA Decision

Initial testing identified the contamination as *P. glucanolyticus*, a spore-forming, Gram-positive, swarming bacterium. Due to the swarming, enumeration of the agar plate was not possible and as a worst case an OOS discrepancy was initiated. With retesting and resampling, it could be shown that no contamination above the specification is present. Therefore, it can be assumed that the batch complies with the specification. The initial results are invalidated and the results of the retest and resampling are reported.

However, investigation of the current and a former discrepancy showed that there is a contamination risk with the sampling procedure by the supplier (sampling under noncontrolled microbial conditions and packaging of the samples in cardboard). As a preventive measure, in the future sampling will be performed under aseptic conditions by trained personnel by the customer.

From a microbiological point of view, the current batch can be released.

Attachments:

A1: Copy of raw data from first testing
A2: Email to production and QA
A3: Details on lab error investigation
A4: Copy of results of retest and resampling
A5: Historical data of product
A6: Information from the supplier
A7: Microbiological testing of cardboard used for packaging
A8: Risk assessment for batches already released
A9: Copy of test report that 1 CFU of *P. glucanolyticus* is swarming
A10: Memo for the addition of internal foil sampling (including training record of sampler)
A11: Copy of test plan for improving contact plates concerning swarming bacteria (GMP-#P-900-02)

6.7.3 Example 3: Contaminated Inhaler

QA Summary of the Discrepancy

XY inhalers made by YZ were tested for microbial contamination according to SOP-1234 using 20 units. The results showed an OOS with 18 CFU/inhaler (specification is 10 CFU/unit). The lab investigation showed no lab or sampling error (lab investigation is documented in LIR-2018-009). A further investigation was performed using another 20 inhalers from the same sampling as well as another 20 inhalers from the same batch but with a new sampling by the same person and under QA oversight. According to the QA oversight no error was observed during sampling. QC testing showed comparable contamination for both samples.

Testing identified Gram-positive, human-related cocci (*Staphylococcus epidermidis, Micrococcus luteus*), and Gram-positive, environment-related rod (*Bacillus thuringiensis, Bacillus* spp.) contaminants. Since the supplier is supposed to send the inhalers for gamma irradiation to decontaminate them, no such contamination should be possible. After contacting the supplier, it was mentioned that the inhalers had been reprocessed due to a possible particle issue after gamma irradiation. This means that all the inhalers were checked individually by operators at the supplier's in a noncontrolled environment without using any protective measures after irradiation.

Thus, the root cause of the current deviation is a wrong procedure by the supplier by recontaminating the inhalers after gamma irradiation due to another nonmicrobial-related issue.

As a corrective action, the inhaler will be rejected since a second gamma irradiation is not validated. As a preventive measure, the technical agreement with the supplier will be adjusted to avoid any physical contact after gamma irradiation by using a seal to show that the irradiated packed material has not been opened.

Batch decision: Reject.

Risk for other batches: Since every batch is tested for microbiology and all batches during the last 12 months did not show any exceedance of the action or alert level, no contamination or patient risk is present.

6.8 Conclusion

The microbiological testing of primary packaging material is not clearly regulated; however, it appears reasonable that it should be tested with predefined acceptance criteria. Several regulatory or guiding documents mention the testing of primary packaging material but it is not clear if this also applies to microbiology. However, it is the author's opinion that from a cGMP point of view primary packaging material needs to be checked to a certain degree for microbiological contamination as this is part of the overall microbial contamination control strategy. Thus, all primary packaging should be evaluated by a risk-based approach for the criticality of the drug product (or raw material) and the intended patient population. In certain cases, it needs to be tested batchwise or skip-lot-wise. No testing might even be acceptable if its impeccable quality can be proven by risk assessment.

For drug products as well as substances for pharmaceutical use, clear acceptance criteria are given by the Pharmacopoeia. No such recommendations are available for primary packaging material. There are some publications that suggest acceptance criteria and the current chapter provides some customer examples. However, it is the author's opinion that probably the most appropriate approach would be to calculate the microbiological level based on the area of the primary packaging material that comes into contact with the amount of drug product. Calculation examples are given in the present chapter.

The methods used should be as proximate as possible to the ones given by the Pharmacopoeia for non-sterile drug products. Additionally, verification that microorganisms can be adequately recovered from primary packaging (method suitability test) should preferably be conducted. Here, a pragmatic approach to inoculation should be applied since the direct inoculation of primary packaging is not feasible in a robust and reproducible way.

Whenever there is an exceedance of the acceptance criteria, a clear procedure for handling such discrepancies must be defined in a SOP.

The microbiological testing or evaluation of primary packaging material used for drug products as well as substances used for pharmaceutical use should be part of the entire microbiological quality control strategy. This is the only way to assure the impeccable microbiological quality of any drug product.

Bibliography

AMG, 2017, *Gesetz über den Verkehr mit Arzneimitteln. Arzneimittelgesetz – AMG.* Bundesministeriums der Justiz und für Verbraucherschutz, pp. 1–151.

ANVISA, 2013, *Quality Guide for Air Treatment and Environmental Monitoring Systems in the Pharmaceutical Industry.* ANVISA Brasília.

ApBetrO, 2017, *Verordnung über den Betrieb von Apotheken. Apothekenbetriebsordnung – ApBetrO.* Bundesministeriums der Justiz und für Verbraucherschutz, pp. 1–27.

BAK, 2016a, *Leitlinie der Bundesapothekerkammer zur Qualitätssicherung: Prüfung und Lagerung der Primärpackmittel.* Bundesapothekerkammer, pp. 1–4.

BAK, 2016b, *Kommentar zur Leitlinie der Bundesapothekerkammer zur Qualitätssicherung: Prüfung und Lagerung der Primärpackmittel.* Bundesapothekerkammer, pp. 1–8.

Beckmann, K, 2010, 'Chemische, mikrobiologische und sensorische Methoden der Analyse von Packmitteln.' *Analytica 2010, QS und Analytik in der Molkereiindustrie, 24.03.2010.*

Bouska, T, Dalla Torre, M, Felds, P, Hecker, W, Spingler, E, Züger, A, 1980, 'Kontrolle von Leerbehältern anhand einer mikrobiologischen Studie mit Gläsern für orale Liquida.' *Pharm. Ind.*, vol. 42, pp. 1195–1198.

CFR – Code of Federal Regulation Title 21 Sec. 211.80, 2017a, *General requirements.*

CFR – Code of Federal Regulation Title 21 Sec. 211.84, 2017b, *Testing and approval or rejection of components, drug product containers, and closures.*

Dacarro, C, Bonferoni, MC, Montanari, L, Mazza, P, 1987, 'Microbiological analysis of pharmaceutical purified cotton', *Farmaco Ed. Prat.*, vol. 42, pp. 17–27.

DIN 10113-3:1997-07, 1997, *Determination of Surface Colony Count on Fitment and Utensils in Foodareas – Part 3: Semiquantitative Method with Culture Media Laminated Taking up Equipment (Squeeze Method).* Deutsches Institut für Normung.

DIN/EN/ISO 15378:2018-04, 2018, *Primary Packaging Material for Medicinal Products – Special Application of ISO 9001:2015 According to the Good Manufacturing Practice (GMP).* Deutsches Institut für Normung.

EMA (ed.), 1998, 'Plastic primary packaging materials', in *The Rules Governing Medicinal Products in the European Union. Vol. 3. Guidelines on the Quality, Safety and Efficacy of Medicinal Products for Human Use.* Brussels, European Commission, pp. 75–82.

EMA, 2000, *ICH Topic Q 6 A. Note for Guidance Specifications: Test Procedures and Acceptance Criteria for New Drug Substances and New Drug Products: Chemical Substances.* CPMP/ICH/367/96.

EMA, 2005, *Guideline on Plastic Immediate Packaging Materials.* CPMP/QWP/4359/03, EMEA/CVMP/205/04.

EudraLex, 2004, *The Rules Governing Medicinal Products in the European Union. Vol. 4. EU Guidelines to Good Manufacturing Practice Medicinal Products for Human and Veterinary Use. Annex 5: Manufacture of Immunological Veterinary Medical Products.* European Commission.

EudraLex, 2008, *The Rules Governing Medicinal Products in the European Union. Vol. 4. EU Guidelines to Good Manufacturing Practice Medicinal Products for Human and Veterinary Use; Annex 1 - Manufacture of Sterile Medicinal Products (Corrected Version).* European Commission.

EudraLex, 2015, *The Rules Governing Medicinal Products in the European Union. Vol. 4. EU Guidelines for Good Manufacturing Practice for Medicinal Products for Human and Veterinary Use. Chapter 5 "Production".* European Commission.

FDA, 1999, *Guidance for Industry: Container Closure Systems for Packaging Human Drugs and Biologics.* Food and Drug Administration.

FDA, 2004, *Guidance for Industry: Sterile Drug Products Produced by Aseptic Processing – Current Good Manufacturing Practice.* Food and Drug Administration.

Goverde, M, 2018, 'Mikrobiologisches Monitoring: Luft, Oberflächen, Personal', in *Gute Hygiene Praxis: Hygiene – Reinigung – Monitoring,* Concept Heidelberg (ed). 3. überarbeitete und erweiterte Auflage 2018. Editio Cantor Verlag.

Goverde, M, Willrodt, J, Staerk, A, 2017, 'Evaluation of the recovery rate of different swabs for microbial environmental monitoring', *PDA J. Pharm. Sci. Technol.,* vol. 71, pp. 33–42.

Han, JH, 2003, 'Antimicrobial food packaging', in *Novel Food Packaging Techniques,* R. Ahvenainen (ed.). CRC Press, pp. 56–65.

Hecker, W, 1992, 'Bestimmung der mikrobiellen Kontamination von Primärpackmitteln', *Pharm. Ind.,* vol. 54(12), pp. 1042–1046.

ISO 11737-1:2006, 2006, *Sterilization of Medical Devices – Microbiological Methods – Part 1: Determination of a Population of Microorganisms on Products.* International Organization for Standardization.
ISO 18593:2016-11, 2016, *Microbiology of the Food Chain - Horizontal Methods for Sampling Techniques from Surfaces Using Contact Plates and Swabs (ISO/DIS 18593:2016),* German and English version prEN ISO 18593:2016. International Organization for Standardization.
Krüger, D, 1970, 'Die Rolle des Verpackungsmaterials im Rahmen der Produktionshygiene', *Pharm. Ind.,* vol. 32, pp. 1051.
Negretti, F, 1981, 'Findings on the microbiological characteristics of pharmaceutical containers', *Boll. Chim.-Farm.,* vol. 120, pp. 193–201.
Payne, DN, 2007, 'Microbiological ecology of the production process', in *Guide to Microbiological Control in Pharmaceuticals and Medical Devices,* Denyer, SP, Baird, RM (eds.). CRC Press, pp. 51–68.
PDA, 2014, 'Exclusion of Objectionable Microorganisms from Nonsterile Pharmaceuticals, Medical Devices and Cosmetics', Technical Report No. 67, 2014, ISBN: 978-0-939459-70-4. Parenteral Drug Association.
Ph. Eur. Chapter 1 (9th Edition), 'General notices'.
Ph. Eur. Chapter 2.6.7 (9th Edition), 'Mycoplasmas'.
Ph. Eur. Chapter 2.6.12 (9th Edition), 'Microbiological examination of non-sterile products: microbial enumeration tests'.
Ph. Eur. Chapter 2.6.13 (9th Edition), 'Microbiological examination of non-sterile products: Test for specified micro-organisms'
Ph. Eur. Chapter 3 (9th Edition), 'Materials for containers and containers'.
Ph. Eur. Chapter 5.1.4 (9th Edition), 'Microbiological quality of non-sterile pharmaceutical preparations and substances for pharmaceutical use'.
Ph. Eur. Chapter 5.1.6 (9th Edition), 'Alternative methods for control of microbiological quality'.
Ph. Eur. Chapter 5.1.8 (9th Edition), 'Microbiological quality of herbal medicinal products for oral use and extracts used in their preparation'.
Rieth, M, 2017, *Pharmazeutische Mikrobiologie.* 2. Auflage. Wiley-VCH Verlag.
Rieth, M, Krämer, N, 2016, *Hygiene in der Arzneimittelproduktion.* Wiley-VCH Verlag.
Seyfarth, H, 2003, 'Mikrobiologische Qualität von Arzneimitteln, Teil II: Ausgangsstoffe, Wasser, Primärpackmittel', *Swiss Pharma,* vol. 25(10), pp. 5–18.
USP 40 Chapter <1111>, 'Microbiological examination of nonsterile products: acceptance criteria for pharmaceutical preparations and substances for pharmaceutical use'.
USP 40 Chapter <1115>, 'Bioburden control of nonsterile drug substances and products'.
USP 40 Chapter <1116>, 'Microbiological control and monitoring of aseptic processing environments'.
USP 40 Chapter <1223>, 'Validation of alternative microbiological methods'.
USP 40 Chapter <2023>, 'Microbiological attributes of nonsterile nutritional and dietary supplements'.

USP 40 Chapter <61>, 'Microbiological examination of nonsterile products: microbial enumeration tests'.

USP 40 Chapter <62>, 'Microbiological examination of nonsterile products: tests for specified micro-organisms'.

WHO Technical Report Series, No. 902, 2002, *Annex 9 – Guidelines on packaging for pharmaceutical products*. World Health Organization.

7

Utilities Design and Testing

Tim Sandle

Bio Products Laboratory, Hertfordshire, United Kingdom

CONTENTS

Pharmaceutical Microbiological Quality Assurance and Control: Practical Guide for Non-Sterile Manufacturing, First Edition. Edited by David Roesti and Marcel Goverde.

7.1 Introduction

To fully understand how different microbial hazards can affect the quality of a non-sterile product and how microbiological control can be achieved, a risk based approach is required (see also Chapters 1 and 2). To achieve this, the health status of the patient and the types of contaminants are need to be understood (following Sutton's classic schema) (Sutton 2006). In addition, any contamination-control assessment needs to factor in the potential sources of contamination as presented from the environment. With non-sterile products, temperature and humidity are extrinsic factors that affect microbial growth (Denyer and Baird 1990), while the intrinsic factors limiting growth in most formulations include low water activity, extreme pH levels, and the presence of antimicrobial compounds (Bloomfield and Baird 1996). Hence, the risk from the external environment, as influenced by a "utility," is highly dependent upon the type of product processed under a given set of conditions.

While the association between contamination in a product and contamination from the environment is not necessarily causative, with the manufacture of non-sterile products microbial risks can arise from critical utilities like water, from improperly cleaned equipment, from the surrounding process environment, as well as from people (Charnock 2004; Tyski 2011). Many such contributing factors are assessed in the chapters that make up this book; the focus of this chapter is with one of those factors – critical utilities.

Underpinning many of the activities of pharmaceutical production are "utilities," for manufacturing cannot take place without controlled sources of air or water, and many different types of utilities are used in non-sterile product manufacturing. Water systems, clean air provided into working areas, compressed air, and nitrogen are just some examples of the utilities that provide the backbone to pharmaceutical operations. Each of these can be source of microbial contamination and, when not controlled, each is capable of affecting the quality of the product. This chapter reviews the primary utilities used in non-sterile pharmaceutical manufacturing, with a focus on good design principles and the testing undertaken to verify the status of the utility. As with other aspects of pharmaceutical development, utilities should be seen as part of the overall "Quality by Design" strategy that forms the basis of ICH Q8 (2009). In addition, each utility should undergo a form of qualification before it is accepted for routine use; and each should be subject to periodic testing, either directly (as with sampling water) or indirectly (as with environmental monitoring to assess the operational capabilities of the controlled space).

The chapter undertakes the review of utilities in two parts. Section 7.2 looks at what the different types of utilities are and discusses how they are designed and how they are controlled. This is examined within the context of Good Manufacturing Practice (GMP). Section 7.3 of the chapter looks at operational issues that affect critical utility performance from the microbiological contamination perspective.

7.2 Defining, Developing, and Maintaining Utilities

7.2.1 What Are Utilities?

The word "utilities" is somewhat tenuous: it can refer to the provider (or producer) of a utility or to the utilities themselves. By utilities, in the general sense, this means gas, electricity, telephony, water services, and so on. In the pharmaceutical context, and from the perspective of GMP, what is of concern are the so-termed "critical utilities." Here, the operative word "critical" refers to those utilities that can affect the safety, identify, potency-strength, quality, or purity of the product. This is generally taken to be water of pharmaceutical quality (including water used for the cleaning of equipment), air as supplied into the clean workspace (controlled environment), and services like compressed gases. From outside the microbiological perspective, critical utilities may also include backup systems (to address events like a major power outage or data capture), waste disposal systems, and drainage. These nonmicrobial impacting systems are not addressed in this chapter.

There are a number of fundamentals that need to be considered with any GMP critical utility. These are (Hart 2005):

- Good design principles: The design of an appropriate utility needs to be efficient and also abide with defined GMP requirements. GMP requirements will vary according to the product under manufacture.
- GMP-compliant design of equipment and utilities: A GMP-compliant design of equipment is the basis for fulfilling the technical requirements and specifications of the utility. Design will include safety, operational reliability, and measures to protect the product from adulteration.
- Validation and qualification: Each utility and any associated item of equipment must be qualified; in turn, qualified equipment and validated processes

as the prerequisites for producing pharmaceutical quality. The approach taken to validation can be risk based (Chao 2005).

- Routine operation: During operation a number of important parameters relating to a given utility need to be met (such as with water, a permitted maximal value for microbial counts). The lifetime of a utility can be enhanced, and the prospect of a successful revalidation achieved, through a robust preventive maintenance program together with regular calibration. This forms an essential part of the pharmaceutical quality system. The approach to calibration, as with validation above, can be risk based. That is, with an understanding of the life cycle of the equipment or utility together with the performance of other comparable systems.
- Quality assurance aspects: Many parts of the quality assurance system impact upon utilities. Of importance are system changes (which need to be captured through change control) and deviations, especially those that have or might have a direct impact on the pharmaceutical material produced. The types of changes and deviations of greatest importance are those that can impact upon the validated status of a utility. Quality assurance can also be supported through regular system audits.

7.2.2 Good Design Principles

The installation of a utility should follow good design principles. These principles offer a framework and a mind-set to achieve acceptable functionality while meeting stringent tests of "fitness for purpose" in relation to pharmaceutical facilities. Moreover, good design principles also form part of GMP. This means constructing a framework for quality assurance to ensure that products are consistently produced and controlled by the application of appropriate standards for their intended use (MacGregor and Bruwer 2008). Included as part of good design is the necessity to understand the process overall, which is often through the construction of process flow diagrams (Yang and Cui 2004). The primary factors to include in process flows relating to non-sterile manufacturing are (Somma 2007):

- Material: Incoming raw materials – packaging components and product sampling. Materials can be designated as work in progress or as finished goods.
- Personnel: Detailing change facilities (and gowning requirements), manufacturing (operations and quality assurance), materials management, support services (maintenance), administration, and quality control.
- Cleaning of equipment and facility: Note needs to be taken of dirty equipment staging, cleaning and disinfection, inspection, assembly, validation, and equipment-part storage.
- Waste material, including liquids, solids, and trash: such as methods of neutralization or sterilization; holding times; removal and disposal of water, together with recycling.

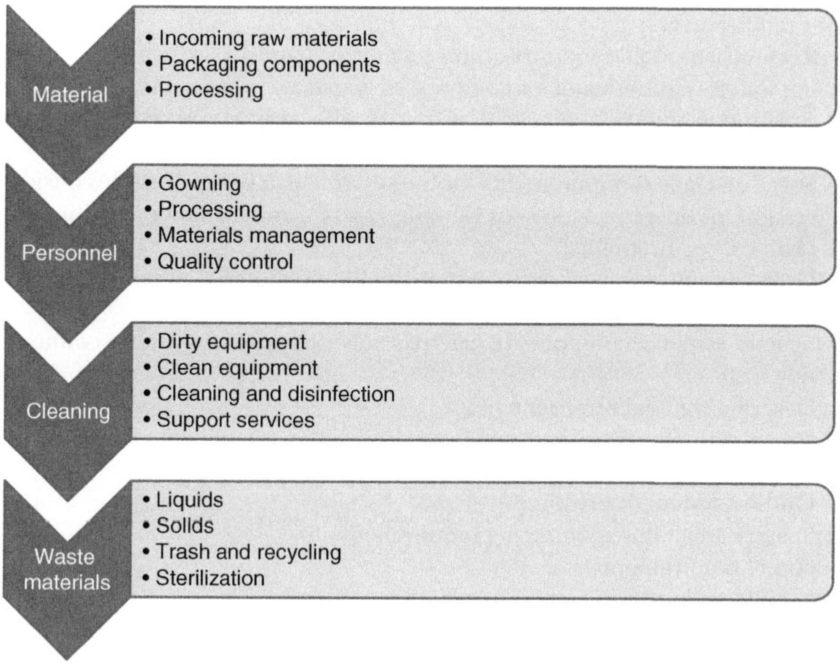

Figure 7.1 Primary process flows for non-sterile manufacturing.

The above steps are illustrated in a process flow diagram, as set out in Figure 7.1.

Included in such analysis should be the steps taken to avoid cross contamination, such as between clean and dirty equipment. For each of these factors, utilities will play a part in affecting the final outcome and potential final quality of the product.

The basis of "good design" begins with a User Requirement Specification (URS). The URS specifies what the user expects the utility to do and the standards that need to be met. Sometimes the items listed in the URS are differentiated as "mandatory," "desirable," "optional," and "future enhancements." The URS document allows potential suppliers to tender for and, if successful, to plan what to design to. The document also acts as a guide to planning costs, setting timetables, pinpointing milestones, considering the testing needed to verify acceptance, and so on (Chu *et al.* 2017).

For example, a URS for a pharmaceutical water system might include:

- Process requirements:
  a) Capacity questions, such as: How much water is needed? What is peak load? How many user points? What will be the usage priority?
  b) Quality questions, such as: What water quality does the system need to produce (both chemical and microbiological)? References should be made to pharmacopeia requirements.

- User requirements:
  a) Review of available support utilities (such as quantity and type of incoming water; requirements for compressed air, steam, power, and drainage).
  b) Treatment required for feed water to the system. Assess sampling requirements.
  c) Space and layout requirements for the system installation. Include consideration for operator access for sampling and for maintenance activities.
- Mechanical requirements:
  a) Detail piping standard(s) to be met (including material of construction, surface finish, and absence of dead-legs).
  b) Specify sanitary components (such as valves, pumps, tri-clamp connections, etc.)
  c) Describe welding requirements.
  d) Assess number of air breaks in floor drains, to prevent contamination.
  e) Add piping slope requirements.
  f) Outline system drainability.
  g) Specify any inline monitoring requirements.
- Electrical requirements.
- Water system sanitization requirements, e.g. ozone, heat.
- Automation requirements:
  a) Details of automation hardware required.
  b) Software Design Specification (SDS)
  c) Controller/human machine interface (HMI) function.
  d) Data capture and storage requirements.
  e) 21 CFR Part 11 compliance statement.
- Documentation requirements, such as:
  a) As-built piping and instrumentation diagrams.
  b) Panel layout and electrical diagrams.
  c) Isometric piping diagrams.
- Validation requirements:
  a) Validation plan.
  b) Equipment qualification (IQ, OQ, and PQ).
  c) Ongoing monitoring, including chemical and microbial expectations.
- Health and safety
- General maintenance requirements.

Once the utility and its provider have been selected, the validation process should be assessed and inbuilt into the subsequent steps. This approach is laid out in EU GMP Annex 15 (EudraLex 2015). This Annex describes the principles of qualification and validation that are applicable to the facilities, equipment, utilities, and processes used for the manufacture of medicinal products. A specific requirement with this regulation is that any planned changes to the facilities, equipment, utilities, and processes, which may affect the quality of the product, are formally documented and the impact on the validated status or control strategy assessed.

7.2.3 Validation Master Plan

The user must define the approach to validation at the start of the project. A good way to capture this intent and the main areas to be covered is by constructing a Validation Master Plan (VMP). The VMP document delineates the validation program that will subsequently be executed. VMPs can be for specific utilities or equipment; or they can cover an entire facility. VMPs can include some or all of the following, depending upon the project remit:

- scope
- responsibilities
- facility description
- building and plant layouts
- controlled environments
- storage areas
- personnel and material flow
- water and solid waste handling
- utilities like water systems
- ventilation and air-conditioning system
- clean steam, compressed air, gases, and vacuum system.

Also considered are the types of manufacturing equipment, building management systems (BMS's), and products to be validated; together with process validation and cleaning validation requirements (Ocampo *et al.* 2007).

7.2.4 Qualification of Utilities

Documents stemming out of the VMP include a design qualification (DQ), installation qualification (IQ), operational qualification (OQ), and performance qualification (PQ). Each of these serves a different, but developing, purpose (Aleem *et al.* 2003):

- DQ: This demonstrates that the equipment or utility meets the written acquisition specification.
- IQ: This describes the preinstallation detail that the equipment or utility plus its component parts and location are fit for the purpose and satisfy the objectives of the user to carry out the intended function to expected standards. In addition, the IQ qualifies the equipment or utility is installed correctly with tubing, fittings, syringes, and valves connected appropriately.

- OQ: This refers to the qualification that needs to be accrued out following installation to verify the performance criteria presented by the manufacturer. The OQ thus confirms if the instrument satisfies its agreed metrics and that it is valid in the working environment.

  OQs vary with different types of utilities. Some examples are:
  - Water For Injection Systems: 14 days of consecutive sampling. Results can be used to set monitoring limits.
  - Purified Water Systems: Typically, 14 days consecutive sampling. Results can be used to set monitoring limits.
  - Clean Steam Systems: Typically, three days sampling.
  - Gases/Clean Dry Air Systems: Typically, three days of sampling.
  - Heating, ventilation, and air conditioning (HVAC) system: Sampling "at rest" for particulates. In-operation state often extended to minimum of five days for microbiological counts.
- PQ: This refers to the test or validation protocol carried out by the user with device in operation for either actual product batches or simulation. A set of tests provides documentary evidence that the instrument or utility is maintaining the agreed values.

  PQs, as with OQs, vary with different types of utilities. Some examples are:
  - Water for Injection Systems: 30–35 days of consecutive sampling (typically divided into 25–30 days of standard operational sampling plus 5–10 days for a shutdown recovery test). This should be followed by a 1 year review of the data, with sampling typically performed weekly.
  - Purified Water Systems: Typically two to four weeks consecutive sampling is required for the first phase of the PQ. This is followed by sampling at a routine frequency (established on the basis of risk), with data reviewed after one year to complete the second phase of the PQ.
  - Clean Steam Systems: Typically 14 days sampling.
  - Gases/Clean Dry Air Systems: Typically seven days of sampling.
  - HVAC system: Sampling "in operation" for particulates. In-operation state often extended to minimum of five days for microbiological counts.

After the completion of each satisfactory qualification stage, a format release for the next step in the qualification and validation should be made as a written authorization.

7.2.5 Upgrading, Reconstructing, and Renovating Utilities

At some point in time a utility will either need significant upgrade or repair, or it will need to be replaced. Each of these activities should be risk based and attempted under change control. If manufacturing is intended to continue while an upgrade takes place, then routine processing requires protecting and special measures need to be taken in order to protect operations. The basis of protection will include understanding the process and points of risk (a risk assessment tool like Hazard Analysis Critical Control Points can be useful for this purpose) (WHO 2009). As part of the controls, additional segregation and cleaning and disinfection may be required. In addition, any external personnel involved also need to be controlled; here, a degree of protection can be introduced through access control. The approach taken should be documented and full training given.

Alterations and upgrades must go through a change control system and changes will require risk assessment. The risk assessment should take account of any impact upon the validated parameters of the system being altered and should conclude as to whether additional validation is required.

Major changes that would require requalification of the utilities distribution system would for instance be:

- Capacity increase or decrease of the distribution system (modification in the number of points of use as well as water loops).
- Upgrade of the old system.
- Major additions to the system.
- Change of sanitization method for a water system.
- Rebalancing exercises for cleanrooms.
- Replacement of treatment modules by a nonidentical module.
- Change of the feed water quality grade or origin.
- Certain types of software upgrades.

Some organizations elect to undertake physical and procedural requalification and maintenance irrespective of whether critical parameters have been affected. This is to ensure that the utility remains in the qualified state within defined time intervals.

The approach for requalification typically consists of the tests that have previously demonstrated that the utility is operating correctly. This tends to be on a smaller scale (such as, with a water system, seven days of sampling). Should data indicate less satisfactory results than the whole system, qualification should be repeated.

7.2.6 Outsourcing

The design, operation, or renovation of a utility may be outsourced to a third party (see also Chapter 17). The extent to which this happens depends upon the degree of specialism within the organization (and perhaps the criticality of the time-to-response should a utility be off-line). Where outside contractors are used, these personnel should be controlled. Control will either come under the departmental client or the auditing department. This control mechanism should include a process for selecting and auditing suppliers, using a risk-based approach. This may involve seeking case studies from other firms that have used a proposed supplier. In addition, the training records of contractors should be reviewed and approved in advance of any works being undertaken. Before starting work, each of the staff supplied by the contractor should undergo induction within the facility for health and safety and for GMP.

This section of the chapter has provided a framework for designing, qualifying, and maintaining utilities. The second part of this chapter focuses on specific types of utilities.

7.3 Review of Critical Utilities

7.3.1 Compressed Air and Gases

Compressed gas is a general term for gas stored or held under pressure that is greater than atmosphere. Compressed gases are used at different stages of the pharmaceutical manufacturing process. Applications include weighing stations process line; use of gas to maintain an inert atmosphere above a liquid or powdered product inside a storage tank, silo, reactor, process equipment, or other vessel; use of liquid nitrogen for the preservation of biological samples; use of inert gas to pressurize new, repaired, or modified tanks, pipelines, and vessels; and use of inert gas to displace air and contaminants from storage tanks. Furthermore, compressed gases such as air, nitrogen, and carbon dioxide are deployed in operations involving purging or overlaying.

Compressed gas sampling for microorganisms is an important part of contamination control assessment (Sandle 2013b). While sampling is important, the method of sampling can hinder by the design of the gas system, where sampling is not easily conducted in an aseptic manner, or by the design of the air-sampling instrument. This section reviews the important aspects of compressed air sampling for microbiological assessment and looks at possible sources of contamination, should microorganisms be recovered.

Purity is a factor that needs to be maintained with compressed gas; hence, the gas should be supplied oil free. Purity overall is achieved through a combination of filtration, purification, and separation. The process of creating the compressed gas can additionally introduce water vapor; thus, a process must be in place to remove water vapor before the gas is expelled into a critical zone like a cleanroom. Compressed gas is typically discharged from the compressor hot and it will contain water vapor. Temperature is reduced by using a post-compressor cooler and, as the gas condenses, the water vapor and other impurities can be removed. The risk of water vapor is particularly high with compressed air, which is drawn into a compressor via the atmosphere. Atmospheric air contains a high proportion of water vapor (that is water in a gaseous form). Water removal is achieved through a combination of filtration and dehumidification.

Where air is drawn in from the outside, the process of drawing in air also introduces microorganisms, which require filtering out. The level of filtering depends upon whether "sterile" air is required (absence of viable microorganisms) or air with a low bioburden.

Compressed gas can be supplied at source either sterile or non-sterile. Sterility, where required such as with an inhalation product, is achieved through the use of a bacterial retentive membrane filter (0.2 μm pore size). Where a sterile-filtered gas is required, it is important that the sterilizing grade filter is maintained dry for condensate in a gas filter will most probably cause blockage or lead to microbial contamination. Risks of condensate are controlled by heating and use of hydrophobic filters (to prevent moisture residues in a gas supply system). Filters should also be changed periodically. As part of ongoing quality control, filters must be integrity tested at installation and at end of use.

7.3.1.1 Compressed Gas Standards

Although national standards bodies have guidance documents for compressed air sampling, and reference is made within FDA and EU GMPs, the general approach and requirements for compressed gases are set out in a multipart ISO standard: ISO 8573. This standard consists of the following parts (ISO 2010):

- Part 1: Contaminants and purity classes
- Part 2: Test methods for aerosol oil content
- Part 3: Test methods for measurement of humidity
- Part 4: Test methods for solid particle content
- Part 5: Test methods for oil vapor and organic solvent content
- Part 6: Test methods for gaseous contaminant content
- Part 7: Test method for viable microbiological contaminant content
- Part 8: Test methods for solid particle content by mass concentration
- Part 9: Test methods for liquid water content

Part 1 outlines the required purity classes based on the concentration of particles and level of impurities. The potential "impure" contaminants for compressed air, which can affect whether a required purity class is met, include:

- Particles (such as dirt, rust, and pipe scale), with particles assessed by size. For example, as a result of the mechanical compression process, additional impurities may be introduced into the air system. Generated contaminants include compressor lubricant, wear particles, and vaporized lubricant. Furthermore, fittings and accessories can contribute to particles.
- Water (in both vapor and liquid forms). Water is typically assessed by vapor pressure dew point. This is the temperature at which the air can no longer "hold" all of the water vapor which is mixed with.
- Oil (including aerosol, vapor, and liquid forms).

Table 7.1 Maximum number of permitted particulates used in the pharmaceutical industry for compressed gas.

	Particle size limits per m^3		
ISO 8573 class	**0.1–0.5 μm**	**0.5–1.0 μm**	**1.0–5.0 μm**
1	≤20'000	≤400	≤10

With purity, many parts of the pharmaceutical industry will use class 1 compressed gas based on the maximum number of permitted particulates. The particle limits are summarized in Table 7.1.

A separate standard exists for the production of compressed air. This is ISO 12500, a four-part standard:

- ISO 12500-1:2007 – Filters for compressed air – Test methods – Part 1: Oil aerosols
- ISO 12500-2:2007 – Filters for compressed air – Test methods – Part 2: Oil vapors
- ISO 12500-3:2009 – Filters for compressed air – Test methods – Part 3: Particulates
- ISO 12500-4:2009 – Filters for compressed air – Methods of test – Part 4: Water

With ISO 12500 there are no specific microbial testing requirements.

7.3.1.2 Microbial Survival in Compressed Gases

Although compressed gas and air systems are relatively harsh environments, they can aid microbial survival if there are available nutrients. The availability of nutrients is dependent upon the purity of the gas and airline. Nutrients suitable for metabolizing by microorganisms include water and oil droplets. Another factor that can affect survival is temperature, especially where temperatures are warmer (Stewart *et al.* 1995).

In addition to vegetative cells, bacterial spores are well equipped to survive the harsh environmental conditions. Spores are resistant to the types of temperature ranges and moisture levels found within compressed gas lines. Another risk exists with biofilm, where microbial communities can potentially form and develop through attachment to air lines and tubing.

Although these risk factors exist, typically no microorganisms would be expected to be recovered from compressed gas lines. Research has shown that many microorganisms can survive and multiply in pressurized systems up to 10 bar and some are at least able to recover after being pressurized. However, at 160 bar pressure upwards, survival rates are very low. Where low-level counts are recovered, these require investigation. More often the source is adventitious contamination, although a fault with the compressed air line cannot be ruled out.

Although microbial contamination of compressed air or gas is a rare event, incidents can occur. Sources of contamination include:

- Source of the air of gas. With air, this is intake air from surroundings (which can contain oil, dirt/dust and moisture/water vapor, microorganisms).
- Piping distribution systems. Piping distribution and air storage tanks, more prevalent in older systems, will have contaminant in the form of rust, pipe scale, mineral deposits, in addition to bacteria.
- Bacterial retentive filter. The filter may become blocked, lose its integrity, or become wet.
- Compressor failure. The compressor itself can create a contaminated environment. For example; the compressor's prefilters can become overloaded with dust and lint, causing the filter to cease functioning properly.
- Sample valve. The point-of-use sample valve may not be designed correctly or become faulty.

7.3.1.3 Microbiological Requirements

Microbial content itself does not influence the gas purity class assigned, although the standards recommend that microbial levels are assessed. Acceptable microbial numbers are subject to a separate assessment; with such an assessment is based on an interpretation of GMP.

It is to note that the health authority guidelines cited in Table 7.2 have been written for sterile medicinal products. Nonetheless, a consensus is that the microbiological quality of the gas must be at least as good as the cleanroom air quality in which the process is taking place. Note that for EU GMP Grade A/ISO 14644 class 5 areas, the microbial count would then be less than 1 CFU/m^3 and the particle levels conform to the area at rest ≤3520 particles/m^3. Many companies, however, do not monitor gas or compressed air used in grade A or B areas since these are sterile filtered as close as possible to the point of use. In such cases a filter integrity test is executed in lieu of microbiological monitoring.

Compressed air sampling should form part of an environmental monitoring program, along with cleanroom assessments. The program should take into account air points to be tested. This could be every point, points considered to be of greater risk (such as product contact), or representative points along a loop. The frequency of testing must also be considered, and this too would need to tie into risk.

While there is an argument, as set out here, for the testing of compressed air where there is product contact there is less of a consensus over the testing of nitrogen. While nitrogen gas can be used to dispense or transfer most fluids from storage, the ISO standard has no specific microbial testing requirements

Table 7.2 Comparison of microbiological requirements in different standards/health authority guidelines.

Guideline	Requirement
ISO 8573 Compressed air – Part 7: "Test method for viable microbiological contaminant content" (2003)	As indicated above, compressed gas requires assessment against a number of parameters, including particles and viable microorganisms. The part of the standard used for making assessments is ISO 8573 Compressed air – Part 7: *Test method for viable microbiological contaminant content*
IPSE Good Practice Guide – Process Gases (2011)	Table 7.1 of the guide indicates that particle counts (both viable and inert) should *Typically equal to the at rest condition of the area served*
2004 FDA Aseptic Filling Guidance document (2004)	*A compressed gas should be of appropriate purity (e.g. free from oil) and its microbiological and particle quality after filtration should be equal to or better than that of the air in the environment into which the gas is introduced*
New EU Annex 1 draft	*Compressed air and gases that come in direct contact with the product/container primary surfaces should be of appropriate chemical particulate and microbiological purity, free from oil and must be filtered through a sterilizing filter at the point of use. Where used for aseptic manufacturing, confirmation of the integrity of the final sterilization filter should be considered as part of the batch release process*

and very few microorganisms, of the types common to pharmaceutical manufacturing environments, would be likely to survive. On this basis a risk-based justification could be made not to perform nitrogen gas testing.

The user will need to determine whether each compressed gas line requires testing and the frequency of testing. Certainly all product contact compressed gases should be assessed. A sampling plan should also consider, and adapt to, the following:

- Cleanroom grade
- Type of product manufactured
- Increased or reduced production schedules
- Seasonal changes
- Equipment changes and modifications
- Replacement of hardware or filters and dryers
- Inactivity of system

Table 7.3 proposes microbiological requirements and minimum testing frequencies for gas or compressed air microbiological monitoring.

*These test frequencies can be lowered, based on risk assessment.

With the action levels set out in Table 7.3, the levels achieved in a practical setting will probably be far lower. The user should therefore set alert limits

Table 7.3 Example of minimum microbiological requirements for gas monitoring.

Grade	Testing frequency quantitative (volumetric)[a]	Action level[b] Number of viable aerobic organisms in gases (CFU/m^3)
C	Monthly	100
D	Quarterly	200
Defined for inhalation product manufacturing	Quarterly	200
Defined for solid oral dosage product manufacturing	Quarterly	500

[a] Testing frequency should be based on a risk assessment.
[b] Additional alert levels to be defined based on the water system performance and historical data.

based on an historical review of the data and use these limits for trending purposes. Here, limits setting is not dissimilar to approaches used for setting environmental monitoring alert levels.

7.3.1.4 Sampling

When sampling compressed air for microorganisms, it is important that the air is depressurized and that the flow rate is controlled. Control of the flow rate is important to ensure that a cubic meter of air is sampled within the required sampling time (this time will be instrument dependent). If the air sampler takes 36 minutes to capture a cubic meter of air, then it will be sampling at 1 ft^3/min. An external regulator will be needed to bring the flow rate down to the sampling rate of instrument. This is assessed using a flow meter. Pressure reduction to atmospheric conditions is of great importance and knowing the flow allows the agar exposure time to be assessed, so that one cubic meter of air is sampled.

It is also important that isokinetic sampling of the air occurs and that air velocity is reduced until it is within the range of the sampler as identified by the manufacturer. This is not only necessary for obtaining the correct sample size but also impacts on the possibility of microbial survival. The level of impact stress has been shown to affect microbial recovery on agar and be dependent upon the impaction velocity of the cells into the agar as well as the design and operating parameters. Due to the fact that any microorganisms present are transported under pressure and then suddenly released into atmospheric conditions, they may be damaged by the immediate expansion of the gas and the resulting shearing forces.

The head of the instrument and any attachments should ideally be sterile before use, to avoid contamination. The culture medium used with the instrument should have been tested for growth promotion and, as for environmental monitoring, has been validated as suitable for gas-viable air monitoring. With most samplers the head will be autoclavable. Some users disinfect the tubes and hoses used to connect the sampler with a disinfectant like 70% isopropyl alcohol. This is mentioned as an option in the ISO standard, although this is erroneously described as "sterilization." Where a disinfectant is used it is important to run the air through the sampler without any agar plate in place; this is necessary to evaporate the disinfectant and to remove any residues. The presence of disinfectant could potentially lead to a "false negative."

With sampling, the sample inlet is connected to the compressed gas line and air is directed over an agar plate or strip. The method works by compressed gas, under reduced pressure, called "partial flow," is forced over the surface of an agar plate. Any microorganisms are impinged onto the surface of the agar.

The sampling time should be sufficient in order to sample one cubic meter of the agar. After sampling, the agar plate or strip is removed and incubated within a microbiology laboratory. At the end of incubation, the agar is examined for colony forming units. Incubation can be for aerobic or anaerobic organisms, or both. The extent to which either is present should be based on initial validation and by taking into account whether such organisms pose a patient risk, should they end up being transferred into the end product.

If colony forming units are recovered, these should be assessed against the appropriate levels defined (e.g. see Table 7.2). It is good practice to identify the contaminants recovered especially if the counts exceed the levels; the identification may provide important information as to the origin of the bacteria that would support a deviation investigation.

7.3.1.5 Instrumentation for Sampling

The type of instrument recommended in the ISO standard is a "slit-sampler, a type of impaction air tester," although alternative samplers can be used, if justified. With a standard impactor sampler, air is drawn through a sampling head via a pump or fan and accelerated, usually through a perforated plate (sieve samplers), or through a narrow slit (slit samplers). This process creates a laminar flow through the sampler head. Hence, the air sampler should be fitted with a diffuser capable of maintaining laminar flow conditions. This is necessary so that particles pass through the sample head in a controlled flow.

The velocity of the air is determined by the diameter of the holes in sieve samplers and the width of the slit in slit samplers. When the air strikes the collection surface on the agar plate, it makes a tangential change of direction. This causes any suspended particles to be thrown out by inertia, impacting onto the collection surface. When the correct volume of air has been passed through the sampling head, the agar plate can be removed and incubated (Sandle 2011).

When selecting a suitable sampler, three parameters should be checked and evaluated. These requirements are undertaken by the instrument provider, given the specialist equipment required. These are (Sandle 2010):

- The physical efficiency of the sampler. This is the relative efficiency of the sampler in collecting particles over a range of sizes. Physical efficiency is measured against membrane filtration sampling and the d_{50} value assessed. The d_{50} is the aerodynamic diameter, above which the collection efficiency of the impactor approaches 100%. Knowing the d_{50} value gives an indication of the sizes of particles likely to be collected by the sampler for the d_{50} is equivalent to particle size at which 50% of the particles are collected, and 50% pass through the sampler because that are too small to impact (Hinds 1982).
- The biological efficiency. This is the relative efficiency of the sampler in collection of microorganisms on a surface so that they remain viable and can be counted post-incubation. Biological efficiency is compared with an established reference sampler. This is selected according to the manufacturer. Assessment of air samplers involves the use of a controlled microbial population passed into a nebulizing chamber.
- The flow rate of the sampler. With all sample sizes, the flow rate of air through the sampling head is critical to the accuracy of the result.

In terms of culture media, sampler models available either collect air samples onto contact plates (55 or 84 mm diameter), standard Petri dishes (90 mm diameter), or onto agar strips.

Outside of these requirements, the ideal device should be portable to permit sampling throughout a series of cleanrooms. Devices should also be cleanable and resistant to common cleanroom disinfectants. The ideal material of construction is stainless steel.

7.3.1.6 Culture Medium Used and Incubation Conditions

An appropriate agar must be selected. An example is tryptone soya agar, which is a generally nutritious medium designed to recover a range of bacteria and fungi. A key factor to take into account is whether the process of sampling leads to undue desiccation of the agar, rendering any recovered microorganisms unable to grow on the agar due to depletion of growth nutrients. This will be affected by the flow rate, type of compressed gas, and the model of air-sampler, together with the type of culture medium. A risk will remain that microbial cells will become damaged by mechanical stress during the sampling process and lose viability. These factors should be evaluated through a study (Morring *et al.* 1983).

An example of such study is provided below.

As plates are exposed to air over time, they undergo a loss of weight due to desiccation. The degree of weight loss varies depending upon the environment in which the plate is exposed and the instrument used. The desiccation of the agar may be detrimental to the viability of any microorganisms which may have settled onto the surface of an exposed agar plate. The process of desiccation can be considered of in terms of total water loss or by reduced access to moisture due to the formation of a “skin layer” onto the agar surface.

The purpose of the validation assessment is to show whether the plates retain the ability to support microbial growth after the maximum exposure time (which is defined by the air-sampler run time).

When designing a validation test protocol to examine the impact of weight loss, there are a number of factors to consider. These include:

- The type of culture medium.
- The use of neutralizers in the culture medium (this may or may not be a factor depending on the application of the plates).
- The air sampler type and model.
- The hydration state of the medium and the impact of this upon the rate of desiccation.
- The metabolic and physical state of any microorganism that may be deposited onto the plate surface.
- The length of the exposure time.
- The type of gas used (multiple studies will need to be run for different gases, such as for nitrogen and compressed air).

There are also different approaches to take when designing when and how agar plates will be assessed. These are captured in the three options below.

Approach 1:
Should plates be exposed first in the air sampler with the maximum exposure time and then inoculated with the microorganism?

The disadvantages with this approach are that:

- Additional moisture could be available from the culture medium which could skew the obtained result.
- The use of laboratory prepared strains is not representative of the environmental flora. The use of environmental isolates may provide a greater degree of robustness because they will have adapted to have survived in adverse environmental conditions. However, once environmental isolates have been cultured in the laboratory, they arguably become laboratory cultures and phenotypically different from environmental flora.
- A further issue with this approach is that running samplers with plates and inoculating them is a greater challenge because a microorganism was more likely to be deposited onto the surface of a settle plate during exposure. Furthermore, at the end of the sampling time, plates will have undergone maximum weight loss, whereas inoculating the plates at the start of the incubation could have resulted in microorganisms being carried from the surface with moisture loss (because the microbial population applied would be as a suspension), which would not have been an accurate challenge.

Approach 2:
Should plates be inoculated and then exposed?

The disadvantage with this approach is that:

- This approach does not assess the ability of the plate to recover microorganism at the end of the exposure time.

Approach 3:
Should plates be run first, store in the process area until collected (to assess sample hold times prior to incubated), then incubated for the maximum incubation time, and then inoculated?

Generally, this is the best approach for capturing all data variables.

In terms of suitable microorganism to use, a representative Gram-positive rod, Gram-positive coccus, Gram-negative rod, yeast-like fungus, and filamentous fungus should be used together with, as is the regulatory preference, isolates from the user's manufacturing environment.

- *Bacillus subtilis* (ATCC 6633)
- *Candida albicans* (ATCC 10231)
- *Staphylococcus aureus* (ATCC 6538)
- *Aspergillus brasiliensis* (ATCC 16404)
- *Pseudomonas aeruginosa* (ATCC 9027)
- Environmental isolates (two or more)

In addition, the agar medium should be removed from the sampler as quickly as is practicable and transferred to the required incubator. This is to avoid the culture medium from drying out or deteriorating.

With the incubation conditions selected, the time and temperature should be suitable for the recovery of mesophilic microorganisms from ambient air temperature, particularly Gram-positive organisms given that such bacteria are better equipped to survive in dry environments (Moissl-Eichinger *et al.* 2012). The typical requirement is to look for mesophilic bacteria and fungi (those that would grow across the temperature range 20–35 °C). However, anaerobic or microaerophilic organisms may need to be considered if such organisms are likely to be present and/or pose a risk to the product. Here, some users would elect to use one representative temperature whereas others would elect to use a two-step incubation regime, such as (Sandle 2014a):

- 20–25 °C for 3–5 days, followed by
- 30–35 °C for 3–5 days.

The selected incubation time should be based on growth promotion studies. If certain microorganisms are considered a problem, alternate incubation times, conditions, or culture media can be considered.

7.3.1.7 Reporting Requirements

When reporting the results from a compressed air sampling session, in relation to microbial counts, the following information is advised in the ISO 8573 standard:

1) Whether the compressed airline was "sterile" or "non-sterile" (something of importance should non-sterile processing occur in the same facility as sterile processing)
2) The date of sampling
3) The date of measurements
4) The location of the sample

In addition to this, the result should be added, expressed as colony forming units per cubic meter of air (CFU/m^3).

7.3.1.8 Bacterial Endotoxin and Compressed Gases

The ISO 8573 standard has the option of sampling compressed air for bacterial endotoxin. Such testing remains relatively uncommon and it is only necessary should the compressed gas have a direct product contact and where there is a concern with Gram-negative bacteria. In most cases there should be no likelihood of endotoxin being present, especially in the context of non-sterile manufacturing.

The sampling method for bacterial endotoxin is tricky and inexact. Either colonies are examined for Gram-negative bacteria, and assessment is made about endotoxin risk; or the *compressed air* is passed through pyrogen-free water.

7.3.2 Cleanrooms and Controlled Environments

Most non-sterile pharmaceuticals are processed under controlled environmental conditions, but not often in classified cleanrooms. This is because the contribution of people to microbiological contamination of most non-sterile products is considered relatively minor and the contribution from the air is, in the most part, negligible. There is a GMP consideration for some liquids and inhaled products to be processed in a controlled environment so as to minimize microbiological contamination, and this is entirely justified on a risk basis (this might stem from, for example, the FDA "Pharmaceutical cGMPs for the 21st Century" document which requires an environmental risk assessment). However, for the vast majority of non-sterile products, the environment contributes little risk to the product. Risk can be kept at a low level through good HVAC design and by demonstrating control by a level of environmental monitoring. These issues are addressed below.

7.3.2.1 Heating, Ventilation, and Air Conditioning (HVAC)

The space within which non-sterile processing occurs will either be a controlled environment or a cleanroom. Where a cleanroom is used, this is typically assessed according to EU GMP Grade D (where the air quality maximal value is 200 CFU/m^3) for the stricter cases or Grades E/F for the less stricter cases. The use of Grades E and F is mentioned in the WHO GMP Guidelines for HVAC systems intended to support non-sterile environments. Here, Grade E is described as being applicable to semisolid and liquid dosage forms (with an air quality maximal value of 500 CFU/m^3) and Grade F for the manufacture of tablets, capsules, and coated tablets (air quality maximal value of 800 CFU/m^3). By being controlled, there is an expectation that the particulate levels will be within an acceptable range. The control of particulates relates to the permitted number of particles that ingress into the room space, the motion that the particles follow within the space (there are particles that either enter or which are generated from people or machinery), and the rate at which particles in the air are removed from the room.

Environmental cleanliness is determined by several factors:

- The quantity of air introduced into the space.
- The effectiveness of air distribution through the space.
- The effectiveness of the removal of air contaminants.

The ingress and flow of particles is a consequence of the (HVAC) system, especially the filtration part which will most likely be via a high efficiency particulate air (HEPA) filter. The types of filters required for different applications depend on the quality of the ambient air and the return air (where applicable) and also on the air change rates. The HVAC will also control room temperature and humidity levels as required for specific processes. Thus, HVAC encompasses the processes and technology required to maintain the required set of physical conditions (WHO 2016).

Air filtration and air change rates should be set to ensure that the defined clean area condition is attained. The air change rates should be determined by the manufacturer and designer, taking into account the various critical parameters using a risk-based approach, considering the area condition required, and whether a specific room cleanliness condition is in fact required together with whether the room condition is rated for an "at rest" condition or an "operational" condition. The type of equipment in the room and whether the equipment generates high levels of particles is also of importance when considering what is expected in terms of cleanroom air.

Other factors to account for with the design are:

- The quality and filtration of the supply air.
- Particulates generated by the manufacturing process.
- Particulates generated by the operators.
- Configuration of the room and air supply and extract locations.
- Having sufficient air to achieve containment effect and to clean up the area.
- Having sufficient air to cope with the room heat load.

- Ensuring there is sufficient air to balance extract rates.
- Ensuring sufficient air to maintain the required room pressure differential.

Directional airflow within production areas should assist in preventing contamination. This is normally a turbulent airflow. Airflows should be planned in conjunction with operator locations, so as to minimize contamination of the product by the operator. Unidirectional airflow (UDAF) should be used for weighing booths or sampling booths to provide operator and product protection and should also have a slight air inflow from the room to enhance containment. This can be demonstrated by smoke airflow pattern tests, or other appropriate tests.

The high-pressure differential between the clean and less clean zones should be of sufficient magnitude to ensure containment and prevention of flow reversal, but should not be so high as to create turbulence problems. Where appropriate, temperature and relative humidity should be controlled, monitored, and recorded, where relevant, to ensure compliance with requirements pertinent to the materials and products and provide a comfortable environment for the operator where necessary.

7.3.2.2 Validation

The design, installation, and commissioning of the HVAC system requires the use of a specialist contractor (to take note of sometimes overlooked factors like the variation of solar radiation density outside of the pharmaceutical facility which can affect the extent that control is maintained during the summer months). Factors to consider include the building materials, the air infiltration rate, the type of energy sources, ease of engineering access, and so on.

The validation of the system can be undertaken in-house or by using an independent company. The qualification phase is important, especially if it can impact upon the quality of the product (in terms of its stability, such as the consequence of the ambient air temperature being out of range) or microbial survival (which could arise from particulate ingress with survival being promoted through optimal temperatures).

7.3.2.3 Monitoring and Control

To control the HVAC, many pharmaceutical facilities will use a computer-based BMS, where hardwire functions to bridge any gap autonomously between desired and actual environmental conditions. BMS's are capable of monitoring the following:

- HVAC/mechanical systems
- Temperature
- Humidity
- Electrical power distribution
- Security and building access

- Surveillance
- Life safety systems (fire alarms, gas leaks, seismic, etc.)
- Lifts and elevators
- Lighting

A BMS consists of a personal computer connected to a range of distributed data acquisition modules. The data acquisition modules provide a method of connecting to room-mounted sensors to monitor the above parameters like temperature and differential pressure. Each room within the facility would be monitored for each required parameter, with alarm limits applied to notify users of "out of limits" operation.

Whether or not a BMS is in place, the following parameters, once appropriate ranges have been decided, enable an assessment of process area environmental control:

- Temperature
- Relative humidity
- Supply air quantities for all diffusers
- Return air or exhaust air quantities
- Room air change rates
- Room pressures (pressure differentials)
- Room airflow patterns
- Unidirectional flow velocities
- Containment system velocities (if applicable)
- HEPA filter penetration tests
- Room particle counts
- Room cleanup rates
- Microbiological air and surface counts where appropriate

The frequency at which these parameters should be assessed will come from a risk assessment, based on the potential risk to the processed product.

7.3.2.4 Environmental Monitoring to Show Environmental Control

Periodic monitoring of airborne microorganisms in the air and on surfaces is helpful in assessing overall levels of control. Particle monitoring can be used should an investigation warrant this (e.g. helping to identify a source of microbial contamination). Before instituting a microbiological environmental monitoring program into a non-sterile facility, the following questions can help to structure a monitoring regime:

- What are you looking for?
- Where will you monitor, how and how often?
- What is the relationship, if any, between environmental monitoring data and patient risk?
- How much is unacceptable and why?

- What is acceptable and why?
- What action will you take if results are high?
- How will you assess the effectiveness of that action?

Depending upon the answers to these questions these responses can inform about the nature of monitoring. Environmental monitoring is addressed in Chapter 8.

7.3.2.5 Energy Efficiency

The topic of energy efficiency is of concern to many manufacturers and, given that airborne contamination risks are often lower with non-sterile products compared with sterile products, "over-specification" of design needs to be avoided. HVAC systems are costly to operate and energy dependent, not least because air is used to condition the room environment (which requires the temperature and humidity of the air to be altered). The way the air is conditioned is either by the direct heating of air (passing air over a heat exchanger) or through the use of heated or chilled water; or through both, as is the case with an air handling unit (AHU). The AHU is a secondary part of the HVAC system (in contrast to the boiler and chiller, which forms the primary part of the HVAC system). Each of these different processes requires considerable amounts of energy to be used.

Taking temperature control as an example, the temperature difference between ambient air and air inside a process room causes heat to flow across the building envelope. In winter, heat is lost to the outside requiring the HVAC system to supply a high heat input via an external energy source. In turn, this can cause the air to become too dry, causing operator discomfort; to avoid this humidification is required, which adds further to the energy burden (Wijeysundera 2016). Humidification of air occurs by adding moisture to an air stream that is flowing steadily, using psychrometric modeling.

7.3.3 Water Systems

Water is potentially a major source of microbiological contamination, as a poorly designed and controlled water system can contain high numbers of microorganisms, especially Gram-negative organisms which may be less susceptible to the killing effect of chemical preservatives. Thus, where water is a key formulation constituent or process component, its control is of crucial importance.

All microorganisms require water to grow. Many non-sterile formulations have very low levels of available water; either because they are dry or solid (tablets, capsules, powders, and so on), they are water free (ointments), or they have formulation components which reduce the amount of water available to microorganisms (so-called humectants). The dry or solid products may have intermediate manufacturing steps that may also promote growth of microorganism such as wet granulation and water-based coating solutions. If the holding time of these intermediates is not controlled, microbial proliferation may occur until developing levels of microorganisms that may provoke product degradation or result in amounts exceeding the product's final specification levels. The products which contain substantial amounts of water (or intermediates and additives which do) constitute a significant microbiological threat. Thus, oral liquids, topical liquids, creams, semisolids, etc., constitute a potential microbiological risk. This is why so many of these products are formulated to contain a chemical preservative agent, the efficacy of which is established during development and confirmed periodically on commercial lots. Thus, the nature of the formulation should be considered as part of the overall microbiological risk assessment in relation to the risks posed by water.

There are many grades of water used in the pharmaceutical industry. Water for manufacturing may be potable mains water, water purified by ion-exchange, reverse osmosis or distillation or Water for Injection. Water systems are a critical component in controlling bioburden in the non-sterile manufacturing environment. Water is used in manufacturing, cleaning, and rinsing.

With non-sterile product production, purified water is most commonly used. The microbiological limit applying to purified water is normally not more than 100 CFU/ml as defined, e.g. by the Ph. Eur. Monograph Aqua Purificata.

7.3.3.1 Types of Water

7.3.3.1.1 Potable Water Potable water may be used for some pharmaceuticals, but perhaps more so for cosmetics and toiletries. In the pharmaceutical industry potable water is deemed good enough for cleaning purposes (e.g. walls and floors in non-sterile units). The degree to which cleaning water must be microbiologically controlled is a function of where it is to be used, what products and equipment it is being used in association with, and of the volumes to be used. Potable water has a microbiological specification in line with national requirements (the WHO indicates 500 CFU/ml, although many countries set lower limits) and the absence of *Enterobacteriaceae*. However, the quality of water may vary both from time to time and from place to place. In the same territories water authorities will not allow water to be used directly from the mains but insist on break tanks. This represents a significant source of potential contamination because on prolonged storage microorganisms either settle out or attach themselves to the storage vessel surfaces and grow as living biofilms. This results in the so-called "bottle effect," whereby bacterial growth and activity are substantially enhanced through growth as a biofilm (due primarily to increased nutrient trapping and concentration) as opposed to a free floating (planktonic) lifestyle. The intermittent throughput of the storage tank ensures that, unless treated, the contents serve as a source of infection.

7.3.3.1.2 Deionized Water Deionized water is used extensively in the manufacture of tablets, syrups, suspensions, creams, lotions, and for washing of all manufacturing equipment. It is prepared by passing potable water through anion and cation exchange resin beds to remove the ions. Any bacteria present in the mains water will therefore be present in the deionized water. Deionization beds are prone to contamination because they must be protected from the corrosive potential of chlorine which acts as a bacteriostat in potable water. Those beds that are not regenerated frequently with strong acids or alkali are often heavily contaminated. Consequently, there is a lot of emphasis on the development of new resins that are able to resist microbial contamination (Song *et al.* 2017).

7.3.3.1.3 Purified Water Produced by Reverse Osmosis The process of producing water by reverse osmosis involves forcing water by an osmotic pressure through a semipermeable membrane which acts as a molecular filter. Solubles dissolved in the water are impeded and those with a molecular weight in excess of 250 do not diffuse at all. In this manner microorganisms, and pyrogens, are removed, resulting in sterile water being produced. Contamination may, however, occur in the storage vessel on the distribution system if they are not kept free from microorganisms. Care must also be taken to disinfect the membrane at regular intervals. This interval will be determined by the results of regular sampling but will probably be of the order of once per month, depending on use.

7.3.3.1.4 Distilled Water (WFI) Distilled water is very high-quality water, similar in standard to reverse-osmosis water, if produced by a still designed to prevent the entrainment of water droplets. As it leaves the still, distilled water is sterile, but its microbiological quality can deteriorate quickly as a result of a fault in the cooling system, the distribution system, or incorrect storage conditions. The flora of contaminated distilled water is usually Gram-negative bacteria (commonly *Pseudomonas* spp. or *Methylobacterium* spp.), often as a pure culture. Owing to its high cost, distilled water is usually used only for parenteral manufacture either as an ingredient or as a pyrogen-free rinsing agent for product contact surfaces. It can on occasion be used in the formulation of oral and topical pharmaceutical products where a low bacterial count is needed.

7.3.3.2 Good Water System Design

The specifications for water purification equipment, storage, and distribution systems should take into account the following (Sandle 2016):

- The risk of contamination from leachates from contact materials.
- The adverse impact of adsorptive contact materials.
- Hygienic or sanitary design.

- Corrosion resistance.
- Freedom from leakage.
- Configuration to avoid proliferation of microbiological organisms.
- Tolerance to cleaning and sanitizing agents (thermal and chemical).
- The system capacity and output requirements.
- The provision of all necessary instruments, test, and sampling points to allow all the relevant critical quality parameters of the complete system to be monitored.

The storage and distribution system should be considered as a key part of the whole system, and should be designed to be fully integrated with the water purification components of the system. The storage and distribution system should be configured to prevent recontamination of the water after treatment and be subjected to a combination of online and off-line monitoring to ensure that the appropriate water specification is maintained.

With materials and design (Zoccolante 2005):

1) Corrosion resistance. PW, HPW, and WFI are highly corrosive. To prevent failure of the system and contamination of the water, the materials selected must be appropriate, the method of jointing must be carefully controlled, and all fittings and components must be compatible with the pipework used. Appropriate sanitary specification plastics and stainless-steel materials are acceptable for pharmaceutical systems. When stainless steel is used it should be at least grade 316L. The system should be passivated after initial installation or after modification. When accelerated passivation is undertaken, the system should be thoroughly cleaned first, and the passivation process should be undertaken in accordance with a clearly defined documented procedure.
2) Smooth internal finish. Once water has been purified it is susceptible to microbiological contamination, and the system is subject to the formation of biofilms when cold storage and distribution is employed. Smooth internal surfaces help to avoid roughness and crevices within the WPU system. Crevices are frequently sites where corrosion can commence. The internal finish should have an arithmetical average surface roughness of not greater than 0.8 μm arithmetical mean roughness (Ra). When stainless steel is used, mechanical and electropolishing techniques may be employed. Electropolishing improves the resistance of the stainless-steel material to surface corrosion.
3) Jointing. The selected system materials should be able to be easily jointed by welding in a controlled manner. The control of the process should include as a minimum, qualification of the operator, documentation of the welder setup, work-session test pieces, logs of all welds, and visual inspection of a defined proportions of welds.
4) Where heat exchangers are employed to heat or cool water within a system, precautions should be taken to prevent the heating or cooling utility from

contaminating the water. The more secure types of heat exchangers of the double tube plate or double plate and frame configuration should be considered. Where these types are not used, an alternative approach whereby the utility is maintained and monitored at a lower pressure than the water may be considered. Where heat exchangers are used they should be arranged in continually circulating loops or sub-loops of the system to avoid unacceptable static water in systems. When the temperature is reduced for processing purposes, the reduction should occur for the minimum necessary time. The cooling cycles and their duration should be proven satisfactory during the qualification of the system.

5) Elimination of ball valves will help to address microbial contamination in water systems (Collentro 2011).

7.3.3.3 Microbial Control of Water

Water treatment equipment, and storage and distribution systems used for purified water should be provided with features to control the proliferation of microbiological organisms during normal use, as well as techniques for sanitizing or sterilizing the system after intervention for maintenance or modification. The techniques employed should be considered during the design of the system and their performance proven during the commissioning and qualification activities. Good design is necessary in order to minimize the chances of biofilms developing. Biofilms form when there is the opportunity for microorganisms in the water to attach onto surfaces. Adherence is enhanced by many species within the community secreting a polysaccharide coating which is "slime like" and very adhesive. The function of the coating is to encourage the attachment of other bacteria, to trap nutrients, and to provide a degree of protection. Biofilm formation in water systems is often a product of poor design. This may relate to the type of material used for the pipework, the finish of the pipework, the diameter of the pipe, the velocity of the circulating water, or the presence of dead-legs (bends in the pipe where the water velocity slows) (Sandle 2013a).

Thus, the important points to note are (Sandle 2014b):

1) Maintenance of continuous turbulent flow circulation within water distribution systems reduces the propensity for the formation of biofilms. The maintenance of the design velocity for a specific system should be proven during the system qualification and the maintenance of satisfactory performance should be monitored. During the operation of a distribution system, short-term fluctuations in the flow velocity are unlikely to cause contamination problems provided that cessation of flow, flow reversal, or pressure loss does not occur.
2) The system design should ensure the shortest possible length of pipework.
3) For ambient temperature systems, pipework should be isolated from adjacent hot pipes.

4) Dead-legs in the pipework installation greater than 1.5 times the branch diameter should be avoided.
5) Pressure gauges should be separated from the system by membranes.
6) Hygienic pattern diaphragm valves should be used.
7) Pipework should be laid to falls to allow drainage.
8) The growth of microorganisms can be inhibited by:
   a) ultraviolet radiation sources in pipework;
   b) maintaining the system heated (guidance temperature 70–80 °C);
   c) sanitizing the system periodically using hot water (guidance temperature >70 °C);
   d) sterilizing or sanitizing the system periodically using superheated hot water or clean steam; and
   e) routine chemical sanitization using ozone or other suitable chemical agents. When chemical sanitization is used, it is essential to prove that the agent has been removed prior to using the water. Ozone can be effectively removed by using ultraviolet radiation.

Many options are available for treating and improving water quality. The use of any particular method depends on what is causing the microbial deterioration, the source of the problem, the quality of the water required and the volume to be treated, and the type of distribution system. The design of a system is influential on the size of the microbial populations and the ability of the user to remove them. Dead-legs, long pipework runs to taps, undrainable pipes, and U-bends all create microbiological problems once installed.

Three methods are routinely used for treating water, namely chemicals, filtration, and UV light. Chemical treatment (e.g. sodium hypochlorite and chlorine gas) is applicable to raw, mains water, but can also be used to treat distribution systems of water produced by distillation, deionization, and reverse osmosis. The concentration of the chemical used will vary depending upon the location of the water in the distribution system. Membrane filtration, using a 0.22 μm porosity-filter, is useful where the usage is moderate and a continuous circulation of water can be maintained. Thus, with the exception of that drawn off for use, water is continually being returned to the storage tank and refiltered. In principle, filtration works well but is relatively expensive for high throughputs because the filters may need regular changing to prevent blockages and "grow through." For this reason, use of 0.22 μm filters as a means of controlling contamination in waters used directly for product manufacture is frowned upon. In essence, filters should only be used prior to the distribution process. Ultraviolet radiation (254 nm) is used for the disinfection of water of good optical clarity, and works particularly well in a recirculating system where water flows over a multiple lamp system. Caveats are that penetration of UV light into water is small, and any dead bacteria present in the system will further hinder penetration.

7.3.3.4 Monitoring Water

In terms of testing, USP <1231>, Water for Pharmaceutical Purposes notes that: ...*water systems need to be operated and maintained in a controlled manner that requires that the system be validated to provide assurance of operational stability and that its microbial attributes be quantitatively monitored against established alert and action levels that would provide an early indication of system control.*

With non-sterile processes the primary concern is with monitoring the water for bioburden and screening water for the presence or absence of objectionable microorganisms, based on a risk review of the product and its intended use (there is not normally any requirement for endotoxin testing). For microbiological monitoring of total aerobic microbial counts in water, the traditional categorization is that there are two basic forms of media available: "high nutrient" and "low nutrient." Those media traditionally categorized as high-nutrient include plate count agar, soybean casein digest agar, and M-heterotrophic plate count agar. These media are intended for the general isolation and enumeration of heterotrophic or copiotrophic bacteria. As alternative to high-nutrient media, low-nutrient media, such as R2A agar and National Water Research Institute agar have a larger variety of nutrients than the high-nutrient media. These low-nutrient media were developed for use with potable water due to their ability to recover a more nutritionally diverse population of microorganisms found in these environments. The use of R2A is prescribed for microbiological testing of water in the Ph. Eur. monograph on purified water and WFI and recommended in the JP chapter G8 Water Quality Control of Water for Pharmaceutical Use.

The plates are generally incubated within a temperature range of 30–35 °C and for a minimum time of 5 days. For the selection of microbiological culture media and the incubation conditions used for monitoring of water and clean steam, some companies support their decision by validation studies. For such studies, it is recommended to perform growth promotion tests using a large selection of typical in-house water-related isolates and/or to perform a study comparing counts of different sampling points in production areas (thus the original microflora and metabolic state) during initial qualification of new water distribution systems. The optimal test method is membrane filtration, either using general or specific agars.

If water is used in a component of the product formulation, the microbiological specification for water is typically based on the Pharmacopoeia monographs. Whereas for the monitoring, some companies apply in addition to the specification level, an action and alert level on historical data or a fraction of the water specifications as per USP <1231> (three-tier approach); other companies apply a two-tier approach, i.e. a specification and action level or an action level and alert level.

Considering the variability of microbiological counts and based on the fact that, in general, the water specification is not a final product critical quality attribute or critical in process control (IPC), in general, is an action level based on the Pharmacopoeia requirements and an alert level based on the water system performance sufficient. For water-based products without or containing a weak preservative system, the above criteria would not apply and tight specifications would need to be introduced.

As an example, Table 7.4 provides examples of microbiological requirements and testing frequency may be applied for water microbiological testing. Note that physical requirements have not been included but should be based on the relevant GMP guidelines and Pharmacopoeia.

During the water system qualification phase, an increased testing frequency should be performed in order to provide supportive data of the water controls and provide a better assessment of the water microflora, especially verifying that objectionable microorganisms are absent.

In addition to the requirements defined in Table 7.4, the absence of objectionable microorganisms may be included. The selection of objectionable microorganisms is based on different criteria such as, for instance, the water system, the product manufacturing process, type of product, risk of proliferation in product, patient risk, and health authority requirement. Generally *P. aeruginosa*, *Burkholderia cepacia*, and *Enterobacteriaceae* are typical representatives of a water system quality for pharmaceutical manufacturing. With objectionable organism testing for the presence of *B. cepacia* complex organisms has become a noteworthy regulatory concern. This is because these

Table 7.4 Example of microbiological/biological requirements of water used in pharmaceutical manufacturing.

Type of water	Testing frequency[a]	Action level[b] Total aerobic microbial count	Action level Endotoxin
Potable water	Monthly	500 CFU/ml	Not required
Purified water	Weekly	100 CFU/ml	Not required
Water for injection	Daily	10 CFU/100 ml	0.25 EU/ml
Clean steam condensate	Trimester	Not required	0.25 EU/ml

[a] Testing frequency of the whole water system. Individual sampling points may be tested based on a rotational basis.

[b] Additional alert levels to be defined based on the water system performance and historical data.

organisms and other water-borne opportunistic pathogens are among the most serious contaminants that can be found in pharmaceutical water systems. Moreover, *B. cepacia* complex organisms can survive or multiply in several types of non-sterile and water-based products. This is fostered by these organisms displaying resistance to some common types of preservatives and antimicrobial agents.

The test methods for the objectionable microorganisms should be generally based on the membrane filtration method using selective media or other validated methods.

It is recommended to include the absence of objectionable microorganisms during the validation phase for the water distribution system to detect them early on and introduce additional sanitization measures or system design change if necessary. In routine monitoring, the objectionable microorganism test may be omitted if the water system is qualified, if adequate sanitization practices that would kill objectionable microorganisms are implemented, and if during the water system qualification, no objectionable microorganism was detected. In some cases, objectionable microorganism tests are reintroduced, e.g. in case an objectionable microorganism is identified when the alert or action level is exceeded.

It is important to design sampling points in the processes that can be monitored for validation and for routine assessment. As an example, use-point samples for hose connections must be collected from actual production hoses using the same flush cycle used in production to prove proper water quality.

In order to avoid any subsequent changes in the count of microorganisms, the water samples must be transported and tested as rapidly as possible. The general requirement is whenever possible, to test the samples within two hours of collection when stored at room temperature.

If it is not possible to test the sample within about 2 hours after collection, the samples are held at refrigerated temperatures (2–8 °C) for a maximum of 24 hours. Longer sample hold times than those described above are implemented when they are supported with validation studies demonstrating that counts remain close to when tested immediately after collection (e.g. within $\pm 0.5 \log_{10}$). For such studies, if possible, it is preferable to use directly water samples from the water systems to cover the typical microflora of the water system used in the facilities.

Looking at other forms of microbial testing, bacterial endotoxin testing is not commonplace, unless the manufacturer of a non-sterile active pharmaceutical ingredient (API) either intends or claims that it is suitable for use in further processing to produce a sterile drug (medicinal) product, water used in the final isolation and purification steps should be monitored and controlled for total microbial counts, objectionable organisms, and endotoxins (ICH 2000).

Deviations related to microbiological water testing are written in Chapter 12.

7.3.4 RMMs in Monitoring

The EMA Q&A (2017) guideline states that *use of rapid microbiological methods should be considered as part of the control strategy to aid with rapid responses to deterioration of the system.*

Rapid microbiological methods (RMMs) use new technologies that may actually provide a better understanding of microbiological quality at a faster rate and with a higher level of sensitivity than the traditional methods described above. RMMs can be applied in replacement of traditional methods as long as they have been appropriately validated according to Ph. Eur. chapter 5.1.6, USP chapter <1223>, and PDA technical report No. 33. If alternative methods are applied, the user must have a clear understanding on how to use the method and how to interpret the results. Indeed for nongrowth-based RMMs, microorganisms in a viable but not culturable state might also be detected which might result in higher counts than with the traditional method even if there is actually no shift in the microbiological quality. This would mean that the action levels might need to be adapted or new ones to be defined based on the output of the new technology.

In water systems use of RMMs using viable particle measurement systems or flow cytometry may be an advantage in order to better evaluate the efficacy and frequency of water sanitization regimes. In such applications RMMs results do not necessarily need to be correlated with the CFU action levels. For more details on RMMs, please refer to Chapter 13.

7.3.5 Clean Steam

Clean steam is an important consideration where the facility has autoclaves in place. Clean steam (or pure steam) is the term applied to the required steam quality as applied for the sterilization or sanitization of equipment parts in contact with the product. The steam, as a condensate, must meet the same standards as per Water for Injections. With non-steriles the chemical purity may be the only concern. If microbial control is required then the testing conducted is commonly designed to show that the condensate is free from bacterial endotoxin (Annalaura *et al.* 2013).

In terms of microbial risk, with clean steam systems if condensate is allowed to collect in the system, and it cools, then stagnant water can provide an environment for bacterial growth. While the bacteria will probably be killed when the condensate is discharged into equipment, endotoxin will be unaffected at this temperature. Further with clean steam there needs to be (Sandle 2013c):

- Adequate insulation will reduce condensate formation.
- Pipes should be sloped (gradient 1 : 100) to direct condensate to low points. Here, steam traps can be used to remove condensate from the system.
- Pipes should be supported to avoid sagging.

- Steam traps should be located at the base of vertical risers and at branches to user points. Traps should remove condensate close to steam temperature. Steam traps are designed so that there is a physical difference between steam and water.
- Designing dead-legs out of the system, such as having instrument branches orientated vertically upwards.

7.3.6 Clean Equipment, Sanitization, and Cleaning Validation

In relation to water many items of equipment are required to be cleaned and disinfected in-between batches. The efficacy of the cleaning needs to be demonstrated and while an important requirement is the removal of product residues, the cleaning process should not add to the microbial bioburden. As an example, the Code of Federal Regulations for the cGMPs of Finished Pharmaceuticals, 21 CFR Part 211, contains a section under Subpart D – Equipment that specifies: Sec. 211.67 Equipment cleaning and maintenance Subpart (a) *Equipment and utensils shall be cleaned, maintained, and, as appropriate for the nature of the drug, sanitized and/or sterilized at appropriate intervals to prevent malfunctions or contamination that would alter the safety, identity, strength, quality, or purity of the drug product beyond the official or other established requirements.* In addition, where cleaning in itself is ineffective in reducing microbial bioburden down to an acceptable level, sanitization (or "disinfection") using chemical with disinfection capability is required.

Cleaning follows the following five standard steps:

- Prerinse step – the step to physically remove the bulk of remaining product.
- Wash step – here a cleaning solution is used (see below) to remove product residues. The cleaning solution may simply be a detergent or it may contain chemicals assessed to be antimicrobial, such as acid-based or alkali-based compounds.
- Rinse step – to remove the cleaning agent.
- Final rinse step – using a higher grade of water to ensure that no residues remain.
- Drying step – such as air drying or drying by heat.

There are different ways to clean equipment; broadly these are:

- Manual cleaning methods
- Agitation (vessels with agitators)
- Clean-In-Place
- Clean-Out-of-Place

Of these, manual cleaning is the most variable and the one most open to inconsistent practice. This form of cleaning may come under greater scrutiny from regulators; however, it is often required for small items of equipment or equipment that cannot be readily subjected to automated cleaning. Manual cleaning may include an additional "soak" step to the five steps of cleaning described earlier.

Clean-in-Place, if validated correctly (see below) is the most robust since the equipment is cleaned in situ. Clean-in-Place systems typically come with some degree of automatic control and are capable of performing all of the five standard cleaning steps described above automatically or manually step by step. Clean-out-of-Place is often as effective as Clean-in-Place but there can be some additional risks associated with the transfer of equipment into and out of the process area, and hence risk of recontamination depending on the storage location and safeguards in place. The advantages of these two approaches come with automation and also because cleaning solutions can be used at higher temperatures and higher concentrations than is often possible, due to health and safety concerns, with manual processes.

In terms of cleaning agents, there are several different types (Rohsner and Serve 1995), many of these have disinfection capabilities:

- Alkali based, e.g. NaOH and KOH.
- Acid based, e.g. HCl and HNO_3.
- Surfactants, e.g. Sodium lauryl sulfate.
- Complexing agents, e.g. EDTA.
- Oxidizers, e.g. Sodium hypochlorite.

Of these different agents, alkalis and oxidizers are theoretically the most destructive to microorganisms. Some surfactants have a degree of bactericidal or bacteriostatic activity, especially the cationic surfactants.

Once cleaning and disinfection has been conducted, the item of equipment should be maintained clean and dry. These are important points since provided an item is kept clean and dry, process equipment is unlikely to represent a significant source of microbiological contamination to medicines. However, poor design of equipment can result in the presence of "reservoirs" of potential contamination (Docherty 1999). Where equipment is inadequately cleaned between uses, or if it is allowed to collect moisture in ports and recesses which directly come into contact with the product, then they can represent significant sources of contamination. Contamination levels on the manufacturing environment may, however, be minimized by observing GMPs. For example, equipment and pipelines should be regularly cleaned and stored in a dry state, heating traps could be installed in sink U-bends thereby destroying the main reservoir of contaminants, and cleaning of production units by contractors should be carried out to pharmaceutical specification.

Status labels should be placed on equipment and equipment which is clean should be segregated from equipment which is dirty. In the event of an audit, equipment cleaning validation studies, using chemical (total organic carbon) and microbial (either contact plates/slides, swabs, or water rinse tests) should be available to present at the inspection. With cleaned equipment, it is also important to have established the time interval between the washing, drying, and the disinfection of components, containers, and equipment.

A factor for ease of cleaning and being maintained in a dry state relates to the design of the equipment. Equipment that has been improperly designed, especially of poor sanitary design, will be prone to microbial contamination (Clontz 2009). Where microbial contamination occurs this can lead to biofilm formation. As with water systems, the presence of biofilms in relation to equipment poses a significant contamination risk. Where bacteria are present in a biopharmaceutical process, such as equipment, there is a possibility of these organisms growing rapidly and forming a biofilm. The presence of contamination poses a risk to upstream processing and could render the final product unsuitable for use (Nims *et al.* 2012).

The ability of a process to clean and to disinfect, whether this is via a Clean-in-Place system or a manual process, is assessed through cleaning validation. This refers to the methodology applied that assures that a cleaning process removes residues of the API's of the product manufactured in a piece of equipment. The validation also assesses whether the cleaning agents themselves have been removed (such as detergents) and that microbial levels are reduced to an acceptable level. The aim is to ensure that each of these "residues" is removed to predetermined levels, sufficient to ensure that the quality of the next product manufactured is not compromised by "residues" from the previous product (Agalloco 1992).

While microorganisms are not a major regulatory concern in cleaning validation for non-sterile pharmaceuticals, there are several areas during cleaning where microorganisms pose a potential concern for product quality and therefore have importance from a contamination control perspective.

To approach cleaning validation, a protocol is required. The essential elements of the protocol are (LeBlanc 1998):

- Equipment description
- Description of the cleaning process
- Swab locations (both chemical and microbial)
  - Rationale for swab locations
- Equipment SOP description (how the process will be run)
- Equipment bracketing strategies, if applicable where there are several items of similar equipment
- Description of any loads pertinent to the equipment and additional items to be cleaned, such as utensils

The prerequisites for a successful cleaning validation study are ensuring that the scope of the study includes the following:

- Dirty equipment hold time studies
- Clean hold time studies
- Minimum cleaning agent contact time
- Rinse time
- Final rinse quality

There are different approaches that can be taken for setting appropriate microbial bioburden limits at the end of cleaning and disinfection of equipment. One approach is to base the limit on the quality of the final rinse water; for example, if the limit is 10 CFU/100 ml or 100 CFU/100 ml, then the equivalent limit is applied to the surface, taking into account the area of surface sampled.

An alternative approach is to base the limit for the surface on what is permissible within the finished product. An example is provided by LeBlanc (2002), based on the formula:

$$\frac{\text{Limit set for subsequent product} \times \text{minimum batch size}}{\text{Product contact surface area}}$$

Suppose there is a product bioburden limit of 70 CFU/g, a batch size of 2'000 kg, and a product contact surface area of 260'000 cm^2. Then,

$$\frac{70\,\text{CFU/g} \times 200'000\,\text{g}}{260'000\,\text{cm}^2}$$

equals a surface limit of 54 CFU/cm^2.

As discussed above, the dirty hold time is the greatest determinant of the ease or difficulty of removing product residues and minimizing bioburden buildup (since product residues may adhere to the equipment over time and be more difficult to remove; this can also “mask” microorganisms). Once the hold times have been established, they should not be exceeded in practice. A determinant of the clean hold time is the location where clean equipment will be stored. This should not be close to dirty equipment or in an area where equipment is cleaned. Water aerosols, in particular, pose a particular microbial recontamination risk.

With Clean-in-Place or Clean-out-of-Place, the control of cleaning can be assessed through the measurement of certain parameters. These parameters can be assessed as part of the cleaning validation study. Such parameters include:

- Prerinse temperature
- Prerinse time
- Cleaning agent type and quality
- Cleaning agent concentration
- Cleaning agent contact time

- Cleaning agent wash temperature
- Wash flow rate
- Post rinse time
- Post rinse temperature
- Final rinse water quality (conductivity and bioburden)

For all forms of cleaning, the cleaning validation study will need to use preset criteria to verify the success of the cleaning process. These are commonly:

- Organoleptic (or "visual") inspection
  Acceptable residuals limits, assessing:
  - API
  - Cleaning agent
  - Both of these are typically assessed as below 10 ppm, relating to a given product contact surface area
- Microbial bioburden (either in the rinse water or on surfaces or both)
  - A typical measure is either less than 25 or 10 CFU per unit of measurement

Typically, a minimum of three runs are performed for each bracketed item. To make such assessments it is important that the methods used have been qualified and that the following assay criteria are established (Jenkins *et al.* 1996):

- Limit of quantification
- Limit of detection
- Swab or contact plate/slide recovery range
- Surface recovery
- Swab stability

Establishing these parameters is not straightforward and there are differing studies and variables relating to devices (such as swab tip) and with the type of surface, condition of the surface, microbial population, and technique. A strong paper for swab recovery studies is by Goverde *et al.* (2017) and Pinto and colleagues have addressed the technicalities of using contact plates (Pinto *et al.* 2009).

This needs to be known for the microbiological methods involved as well as the chemical (degradents testing assessment, such as total organic carbon tests).

At the end of the cleaning validation all results should be captured onto results sheets and the final exercise written up as a report.

7.4 Conclusion

This chapter has looked at critical utilities for non-sterile pharmaceutical production, from the perspectives of good design and operational practices, and in relation to microbial risk. Good control of utilities is essential for the effective microbiological control of non-sterile products. As to the extent of this "control," this needs to be based upon an objective assessment of risk. To assess risk it is essential for the user to understand their products, processes, sources of contamination and mitigating factors; since these will vary, the required level of utility control will also vary.

Once the appropriate level of utility control has been established, then a monitoring regime needs to be constructed. It is important to note, and as emphasized elsewhere in this book, microbiological monitoring is not the same as microbiological control: monitoring exists to support controls and to verify that they remain operational. Appropriate control strategies in relation to utilities include having a sound knowledge of the risk areas; by consistently following good hygiene practices; ensuring that microbiological "hot spots" have been eliminated (such as dead-legs in water systems or exceeding validated hold-times with dirty equipment awaiting processing); and by following the basis of GMP, with a focus on what is actually a risk to patients rather than simply seeking to follow perceived regulatory compliance.

With testing, microbiologists need to sample and conduct regular tests of utilities in order to demonstrate that controls are working. This paper has assessed some of the appropriate requirements in relation to microbiological sampling. In doing so, the important features of sampling have been raised together with the factors that can lead to microbial contamination occurring. Each of these aspects should be built into a biocontamination control program for non-sterile manufacturing.

Bibliography

Agalloco, J, 1992, 'Points to consider' in the validation of equipment cleaning procedures', *PDA Journal of Pharmaceutical Science and Technology*, vol. **46**(5), pp. 163–168.

Aleem, H, Zhao, Y, Lord, S, McCarthy, T, Sharratt, P, 2003, 'Pharmaceutical process validation: an overview', *Journal of Process Mechanical Engineering*, vol. **217**(2), pp. 141–151.

Annalaura, C, Giulia, D, Stefano, C, 2013, 'The application of quality risk management to the bacterial endotoxins test: use of hazard analysis and critical control points', *PDA Journal of Pharmaceutical Science and Technology*, vol. **67**(5), pp. 553–567.

Baseman, H, 2010, 'Sterile product manufacture using form fill seal technologies', in J. Agalloco, J. Akers (eds.), *Advanced Aseptic Processing Technology*, Informa Healthcare, New York, pp. 164–165.

Bloomfield, S, Baird, R, 1996, *Microbial Quality Assurance in the Pharmaceutical, Cosmetics, and Toiletries Industries*, CRC Press, Boca Raton, pp. 108–113.

Chao, SB, 2005, 'Risk-based approach for biological process validation: integrating process validation with product development', *Journal of Validation Technology*, vol. **11**(1), pp. 54.

Charnock, C, 2004, 'The microbial content of non-sterile pharmaceuticals distributed in Norway', *The Journal of Hospital Infection*, vol. **57**, pp. 233–240.

Chu, JC, Shnayder, L, Maida, J, 2017, Facility utility systems, in T. Jacobs, A.A. Signore (eds.), *Good Design Practices for GMP Pharmaceutical Facilities*, 2nd ed., CRC Press, Boca Raton, pp. 125–158.

Clontz, L, 2009, *Microbial Limit and Bioburden Tests*, 2nd ed., CRC Press, Boca Raton, pp. 270–271.

Collentro, WV, 2011, *Pharmaceutical Water: System Design, Operation and Validation*, 2nd ed., Informa Healthcare, London, pp. 120–124.

Denyer, S, Baird, R, 1990, *Guide to Microbiological Control in the Pharmaceuticals and Medical Devices*, CRC Press, Boca Raton, pp. 280–285.

Docherty, SE, 1999, 'Establishing microbial cleaning limits for non-sterile manufacturing equipment', *Pharmaceutical Engineering*, vol. **19**(3), pp. 2–5.

EMA, 2017, 'Questions and Answers on Production of Water for Injections by Non-distillation Methods Reverse Osmosis and Biofilms and Control Strategies', *EMA/INS/GMP/443117/2017*, EMA, London.

EMA and PIC/S, 2017, 'Revision of Annex 1 of the Guidelines on Good Manufacturing Practice – Manufacture of Sterile Medicinal Products', https://ec.europa.eu/health/sites/health/files/files/gmp/2017_12_pc_annex1_consultation_document.pdf

EudraLex Volume 4, 2015, 'EU Guidelines for Good Manufacturing Practice for Medicinal Products for Human and Veterinary Use – Annex 15: Qualification and Validation'.

FDA, 2004, '*Guidance for Industry Sterile Drug Products Produced by Aseptic Processing – Current Good Manufacturing Practice*', U.S. Department of Health and Human Services Food and Drug Administration, Rockville, MD.

Goverde, M, Willrodt, J, Staerk, A, 2017, 'Evaluation of the recovery rate of different swabs for microbial environmental monitoring', *PDA Journal of Pharmaceutical Science and Technology*, vol. **71**(1), pp. 33–42.

Hart, KJ, 2005, 'Focus on facilities: compliance by design', *Bioprocess International*, vol. **3**(5), pp. 18–20.

Hinds, WC, 1982, *Aerosol Technology*, Wiley, New York, pp. 104–126.

ICH, 2000, 'Topic Q7 Good Manufacturing Practice for Active Pharmaceutical Ingredients', Step 5, Note for Guidance on Good Manufacturing Practice for Active Pharmaceutical Ingredients', *CPMP/ICH/4106/00.*

ICH, 2009, 'The International Conference on Harmonization of Technical Requirements for Registration of Pharmaceuticals for Human Use, Quality Guideline Q8(R2) Pharmaceutical Development', www.ich.org

IPSE, 2011, '*Good Practice Guide: Process Gases*', International Society for Pharmaceutical Engineering, Florida, pp. 81–86.

ISO 8573-1:2010, 2010, '*Compressed Air Contaminants and Purity Classes*', International Standards Organization, Geneva.

ISO 8573-7:2003, 2003, '*Test Method for Viable Microbiological Contaminant Content*', International Standards Organization, Geneva.

Jenkins, KM, Vanderwielen, AJ, Armstrong, JA, Leonard, LM, Murphy, GP, Piros, NA, 1996, 'Application of total organic carbon analysis to cleaning validation', *PDA Journal of Pharmaceutical Science and Technology*, vol. **50**(1), pp. 6–15.

LeBlanc, DA, 1998, 'Establishing scientifically justified acceptance criteria for cleaning validation of finished drug products', *Pharmaceutical Technology*, vol. **22**(10), pp. 136–148.

LeBlanc, D, 2002, 'Equipment cleaning validation', *Journal of Validation Technology*, vol. **8**(4), pp. 40–46.

MacGregor, JF, Bruwer, MJ, 2008, 'A framework for the development of design and control spaces', *Journal of Pharmaceutical Innovation*, vol. **3**, pp. 15–22.

Moissl-Eichinger, C, Rettberg, P, Pukall, R, 2012, 'The first collection of spacecraft-associated microorganisms: a public source for extremotolerant microorganisms from spacecraft assembly clean rooms', *Astrobiology*, vol. **12**(11), pp. 1024–1034.

Morring, KL, Worenson, WG, Attafield, MD, 1983, 'Sampling for airborne fungi: a statistical comparison of media', *American Industrial Hygiene Association Journal*, vol. **44**, pp. 662–664.

Nims, RW, Presente, E, Sofer, G, Phillips, C, Chang, A, 2012, 'Adventious agents: concerns and testing of biopharmaceuticals', in A.S. Rathore, G. Sofer (eds.), *Process Validation in Manufacturing Biopharmaceuticals*, CRC Press, Boca Raton, pp. 141–158.

Ocampo, A, Lum, S, Chow, F, Vogel, P, 2007, 'Current challenges for FDA-regulated bioanalytical laboratories for human (BA/BE) studies. Part I: defining the appropriate compliance standards – application of the principles of FDA GLP and FDA GMP to bioanalytical laboratories', *The Quality Assurance Journal*, vol. **11**(1), pp. 3–15.

Pinto, F, Hiom, S, Girdlestone, S, Maillard, JY, 2009, 'Evaluation of the effectiveness of commercially available contact plates for monitoring microbial environments', *Letters in Applied Microbiology*, vol. **48**(3), pp. 379–382.

Rohsner, D, Serve, W, 1995, 'The composition of cleaning agents for the pharmaceutical industries', *Pharmaceutical Engineering*, vol. **15**(3), pp. 20–25.

Sandle, T, 2010, 'Selection of active air samplers', *European Journal of Parenteral and Pharmaceutical Sciences*, vol. **15**(4), pp. 119–124.

Sandle, T, 2011, 'Key points for active air samplers', *Clean Air and Containment Review*, (5), pp. 8–10.

Sandle, T, 2013a, 'Bacterial adhesion: an introduction', *Journal of Validation Technology*, vol. **19**(2), pp. 1–10.

Sandle, T, 2013b, '*Contamination control risk assessment*', in R.E. Masden, J. Moldenhauer (eds.), Contamination Control in Healthcare Product Manufacturing, vol. **1**, DHI Publishing, River Grove, pp. 423–474.

Sandle, T, 2013c, *Sterility, Sterilisation and Sterility Assurance for Pharmaceuticals: Technology, Validation and Current Regulations*, Woodhead Publishing, Cambridge, pp. 316–317.

Sandle, T, 2014a, 'Examination of the order of incubation for the recovery of bacteria and fungi from pharmaceutical cleanrooms', *International Journal of Pharmaceutical Compounding*, vol. **18**(3), pp. 242–247.

Sandle, T, 2014b, 'Microbiology of pharmaceutical grade water', in G. Handlon, T. Sandle (eds.), *Industrial Pharmaceutical Microbiology: Standards & Controls*, Euromed Communications, Passfield, pp. 10.1–10.19.

Sandle, T, 2016, 'Control of WFI and clean steam systems for bacterial endotoxins', *Journal of GxP Compliance*, vol. **20**(4), pp. 1–15.

Somma, R, 2007, 'Development knowledge can increase manufacturing capability and facilitate quality by design', *Journal of Pharmaceutical Innovation*, vol. **2**, pp. 87–92.

Song, W, Gao, B, Wang, H, Xu X., Xue, M, Zha, M, Gong, B, 2017, "The rapid adsorption-microbial reduction of perchlorate from aqueous solution by novel amine-crosslinked magnetic biopolymer resin", *Bioresource Technology*, vol. **240**, pp. 68–76.

Stewart, SL, Grinshpun, SA, Willeke, K, Terzieva, S, Ulevicius, V, Donnelly, J, 1995, 'Effect of impact stress on microbial recovery on an agar surface', *Applied and Environmental Microbiology*, vol. **61**, pp. 1232–1239.

Sutton, S, 2006, 'Microbial limits tests: the difference between "absence of objectionable microorganisms" and "absence of specified microorganisms"', *PMF Newsletter*, vol. **12**(6), pp. 3–9.

Tyski, S, 2011, 'How to manufacture a pure and safe drug', *La Pharmacie Industrielle*, vol. **4**, pp. 57–61.

WHO, 2009, 'Application of Hazard Analysis and Critical Control Point (HACCP) Methodology to Pharmaceuticals'. *WHO Technical Report Series*, No. 908, Annex 7. 2003, http://www.apps.who.int/medicinedocs/en/d/Js5517e/23.html

WHO, 2016, 'Supplementary Guidelines on Good Manufacturing Practices for Heating, Ventilation and Air Conditioning Systems for Non-sterile Pharmaceutical Dosage Forms', *WHO Technical Report Series*, No. 961, 2016, Geneva, Switzerland.

Wijeysundera, NE, 2016, *Principles of Heating, Ventilation and Air Conditioning: With Worked Examples*, World Scientific, New Jersey, pp. 1–15.

Yang, X, Cui, L, 2004, 'Microbial process development for biopharmaceuticals: from science and engineering to regulatory management', *Bioprocess International*, vol. **2**(6), pp. 32–39.

Zoccolante, GV, 2005, 'High purity water', in A. Signore, T. Jacobs (eds.), *Good Design Practices for GMP Pharmaceutical Facilities*, Taylor & Francis, Boca Raton, pp. 119–162.

8

Microbiological Environmental Monitoring

Alexandra Stärk

MS&T Technical Steward Sterility Assurance, Novartis Pharma Stein AG, Stein, Switzerland

CONTENTS

Pharmaceutical Microbiological Quality Assurance and Control: Practical Guide for Non-Sterile Manufacturing, First Edition. Edited by David Roesti and Marcel Goverde.

8.1 Introduction

Microbiological environmental monitoring is employed to confirm the effectiveness of the operational controls in reducing microbial populations to acceptable levels. Microbiological environmental monitoring allows to get information about microbiological contamination levels on the sampling locations of interest and is performed by using different sampling techniques on a microbiological nutrient medium followed by subsequent incubation.

The microbiological environmental monitoring program of cleanrooms and production equipment is one important part of the overall microbial control and monitoring strategy in order to produce products with an appropriate microbiological quality. This microbiological environmental monitoring program must be defined together with the relevant supportive units as Quality Assurance or Quality Control.

By defining a state-of-the-art microbiological environmental monitoring program, the following questions need to be addressed:

- What is the overall microbial control strategy for the production of non-sterile products and how are microbiological environmental controls embedded in this strategy?
- Which methods have been used to perform the microbiological environmental monitoring and what needs to be monitored?
- How can these methods be validated and what is the recovery rate of these methods?
- What are the adequate control levels (e.g. action and alert levels) for the microbiological environmental monitoring?
- Which sampling frequencies should be applied?
- What kind of investigation is required and expected when a microbiological environmental monitoring result exceeds the defined control levels?

The following sections will guide through the relevant perquisites, methods, validations, and definitions for a sustainable and expressive microbiological environmental monitoring program.

8.2 Microbiological Control Strategy

The microbiological contamination for non-sterile products is controlled and monitored not only with microbiological environmental controls but also based on multiple parameters which are part of the microbial control strategy. An extensive overview of these controls is described in Chapter 1.

8.3 Cleanliness Zoning Concept for Non-sterile Products

The field of regulatory guidelines related to cleanliness zones for the production of non-sterile products and the associated control levels for microbiological environmental monitoring in these areas is rare compared to the high number of guidelines that can be found for the production of sterile products (e.g. EU GMP Guide – Annex 1 (EudraLex 2008), FDA Guidance on aseptic processing (FDA 2004)). For non-sterile product manufacturing, the rare guidance recommend an ISO 14644-1 class 8 or grade D "at rest" classification (Chinese GMP 2010; WHO 2016; USP <1115> 2018) or grade E/F (ZLG 2010).

Non-sterile pharmaceutical manufacturers generally apply a controlled not classified (CNC), ISO 8 or D, E, F grade for their cleanrooms. Actually, the definition of the cleanliness zoning concept should be based on the respective activities performed in these areas and the related contamination risks. Hygiene procedures, physical and microbiological environmental requirements are associated to the different cleanliness zones.

Examples of classification of production activities related to different cleanliness zones:

- Cleanliness zone with activities of higher contamination risk to the non-sterile product and therefore of higher criticality:
  - Manufacturing and primary packaging areas where the non-sterile product is permanently or intermittently exposed to the surrounding area (open product handling).
  - Washing rooms for product contacting equipment used for non-sterile products.
- Cleanliness zone with activities of a very low or no contamination risk to the non-sterile product and therefore of lower criticality:
  - Manufacturing areas for closed product operations.
  - Secondary and final packaging of the non-sterile product.
  - Visual inspection of primary packed products.
  - Ancillary production rooms in which product is not exposed to the surrounding area.
  - Warehouses.

- The cleanliness zones should also take into consideration the product characteristics like their route of administration and their composition:
  - Nonaqueous preparations for oral use (e.g. tablets) have a low risk for microbial proliferation due to their low water activity value and lower patient contamination risk due to the route of administration. The cleanliness zone for such products can be defined as less critical compared to other non-sterile products.
  - Aqueous or semi-aqueous preparations for oral use have an increased risk for microbial proliferation due to their higher water activity value and a lower patient contamination risk due to the route of administration. The cleanliness zone for such products must be defined with a higher criticality compared to nonaqueous preparation for oral use.
  - Dry powders or capsules for inhalation should be classified with a higher criticality due to the route of administration.

8.4 Microbiological Environmental Monitoring Strategy

Microbiological environmental monitoring is performed to get important information about the contamination level in the environment where the product is manufactured. The environment is mainly monitored on cleanroom and working surfaces, on the outside of production equipment, but also on the production equipment parts that are in direct contact with the product. In addition, the air in the production cleanrooms, the detergents and disinfectants, which are used to clean and disinfect the surfaces in the production-environment, are part of the microbiological environmental monitoring program.

For the production of non-sterile products, microbiological monitoring of personnel is not mandatory (as it is for aseptic processing) but might be required for investigation purposes in case of contaminations in the production environment above the defined microbiological monitoring levels.

The definition of an adequate and meaningful microbiological environmental monitoring program is key and should consider the following:

- Choice of nutrient medium (e.g. nutrient medium for bacteria, for yeast and molds)
- Choice of sampling method (e.g. surface sampling by swabbing or contact plates, active or passive air sampling)
- Choice of sampling location (e.g. critical surfaces)
- Choice of incubation temperature (e.g. lower temperature for the detection of yeast and molds)
- Choice of monitoring frequency (e.g. higher frequency for more critical areas)
- Choice of microbiological control levels (e.g. lower control levels for more critical areas as for noncritical areas)

Even by defining and executing a well-designed microbiological environmental monitoring program, the results of this program can never provide a full microbiological overview of the respective area or equipment. These are just considered as monitoring results which give some restricted microbiological information about this area. Each environmental monitoring result gives a so-called point-in-time information of the respective sampling location. By combing all obtained results from different sampling locations and by comparing these results to relevant historical data, a quite good evaluation of the level of control in the respective cleanroom can be made.

Dependent on the product characteristics, this evaluation and other parameters which are part of the microbial control strategy might be considered for the release or rejection decision on products which were produced in the respective cleanroom.

8.5 Microbiological Environmental Monitoring Methods

8.5.1 Microbiological Monitoring of Air

Microbiological monitoring of air can be performed either actively to get a quantitative result or passively to get a qualitative or a semiquantitative result of the airborne microorganisms.

8.5.1.1 Active Air Monitoring

The volume of air, against which the microbial environmental levels are defined, is generally $1\,m^3$. When a monitoring level of $200\,CFU/m^3$ or less applies, then a volume of $1\,m^3$ air is sampled. For higher monitoring levels a smaller volume of air may be sampled to avoid counts that are too numerous to count on the nutrient media plate. Very popular is the use of the impaction method. Other methods like impingement or gelatin membrane filtration are also available on the market.

For the active air monitoring with the impaction method the predefined volume of air is drawn by vacuum through a perforated plate and then accelerated and directed toward the surface of a Petri dish containing nutrient media. The Petri dish is located below the perforated plate and in the path of airflow. The microorganisms, because of their higher mass, become impacted on the nutrient media surface, while the rest of the air mass flows around the plate and exits the air sampler.

Any active air sampler should be evaluated for its suitability based on collection efficiency (as an example, see the study of Meier and Zingre (2000)), ability to be cleaned/disinfected (see, e.g. Sandle and Satyada 2015), and the possibility to be located on the defined sampling location.

The final air monitoring result, which is evaluated against the microbiological control level, is stated as the number colony forming units (CFU) of total aerobic microbial counts or molds per volume sampled or per acceptance criterion volume. Depending on the sampling device used, a conversion of the colony counts obtained may be necessary, i.e. at higher CFU densities, this conversion compensates for the probability that more than one CFU was impacted through the same hole or slit and is not recognizable as a separate CFU.

The active air sampler should be regularly calibrated (e.g. at least once per year) in order to obtain robust air monitoring results.

8.5.1.2 Settle Plate

In the settle plate test, which is regarded as a passive air monitoring, the microorganism-bearing particles are collected on a horizontally open exposed Petri dish containing a nutrient medium for a defined time period; mostly for four hours. In a passive air monitoring, only the relatively large, rapidly settled microorganism-bearing particles are recovered. The result depends very much on the respective turbulences in the area tested.

The final air monitoring result, which is evaluated against the microbiological control level, is stated as the number CFU of total aerobic microbial counts or molds per plate diameter (e.g. per 90 mm settle plate) per exposure time.

8.5.2 Microbiological Monitoring of Surfaces

Three methods can be applied for the determination of microorganism on surfaces:

- Contact (= imprint) technique for microbiological examination of even or slightly curved surfaces.
- Swabbing technique for sampling locations which are difficult to access or uneven surfaces.
- Rinsing technique for closed filling systems and production units

Contact samples are to be preferred for routine monitoring since the enumeration per surface is less dependent on the sampling technique compared to the swabbing method. The swabbing method should only be taken into consideration in cases where the surfaces are difficult to access, or the surfaces are uneven.

Rinse samples are primarily suited for surfaces which cannot be reached by contact plates or by the swabbing technique and are uncommonly used in routine.

8.5.2.1 Contact Samples

To collect the samples, the lid from the contact nutrient medium plate (normally 25 cm^2) is removed and the nutrient medium is placed on the sampling site and pressed together evenly. According to the ISO norms, in general 5–10 seconds with a pressure of approximately 500 g should be applied (ISO 18593, ISO 14698-1 Annex 3, DIN 10113-3). The sampling location must be wiped off the nutrient medium residual traces left by the plate by using a disposable disinfectant wipe.

The final surface monitoring result, which is evaluated against the microbiological control levels, is stated as the number CFU of total aerobic microbial counts or molds per 25 cm^2 or per plate.

8.5.2.2 Swab Samples

Sterile swabs for microbiological surface monitoring consist in general of a wad of absorbent material wound around one end of a small stick. The swab must be moist (moisten by a buffer solution) to ensure a better recovery of microorganisms on the surfaces. The swab is gently rubbed in a twisting motion against the sampling location to be examined (normally a surface equivalent to 25 cm^2).

After sampling, the swab can be subsequently managed in two ways, either by directly spreading onto the surface of a nutrient medium plate or by placing the swab into a tube of buffered solution and transferring the sample to the lab for processing. This sample can either be processed using the membrane filtration method or by using the pour/spread plate method and cultivation on a nutrient media (for comparison of these two methods, see Goverde *et al.* 2017).

In addition, the media residual trace left on the surface by the moistened swab should be cleaned and disinfected after sampling, e.g. by using a disposable disinfectant wipe.

The final surface monitoring result, which is evaluated against the microbiological control level, is stated as the number CFU of total aerobic microbial counts or molds per equivalent 25 cm^2.

8.5.2.3 Rinse Samples

An adequate volume of a sterile rinsing fluid (e.g. buffer solution) is prepared in a suitable container. A defined surface is rinsed or the fluid is passed via the intake, hose, etc., into the production equipment with the surface to be examined. The fluid is afterwards collected in a sterile container. The total volume of sample removed is normally not less than 100 ml. Depending on the microbial count to be expected, the whole volume of rinse sample, only part of it, or suitable dilutions is processed by using the membrane filtration method and cultivation on a nutrient media.

The final surface monitoring result which is evaluated against the microbiological control level is stated as the number CFU of total aerobic microbial counts or molds per 25 cm^2.

8.5.3 Microbiological Monitoring of Detergents and Disinfectants

The microbial count of disinfectants and detergents (tested either as concentrate or as cleaning solution) is commonly determined by membrane filtration and cultivation on a suitable nutrient media. As a matter of principle, samples are collected in sterilized bottles. If necessary, suitable devices, such as pipettes, to collect the samples are used.

The final monitoring result, which is evaluated against the microbiological control level, is stated as the number CFU of total aerobic microbial counts of molds per volume tested.

8.6 Method Validations and Suitability Tests for Microbiological Environmental Monitoring

As a prerequisite to get valid results out of the microbiological environmental monitoring program, the following points should be considered and supported by adequate studies:

- Choice and validation of nutrient media
- Incubation temperature and incubation period
- Evaluation or validation studies of the applied testing method
- Recovery rate for the applied testing method
- Sample hold time prior to incubation

Dependencies from one point to the other should also be evaluated, e.g. choice of nutrient media together with the incubation scheme (temperature and period) will lead to the recovery of the expected microorganisms.

8.6.1 Choice and Validation of Nutrient Media

For the microbiological environmental monitoring of surfaces and air, a nonselective nutrient media suitable for the determination of total aerobic microbial counts (e.g. TSA, Tryptic Soy Agar or CSA, Casein Soybean Digest Agar) is commonly used. TSA medium is generally sufficient to recover mesophilic bacteria, yeasts, and molds for microbiological environmental monitoring purposes.

If, based on a risk assessment, a particular microbial risk for a specific product is acknowledged, a specialized media should be used. For instance, in some conditions, detection of molds may be slightly improved by the use of the specialized medium Sabouraud Dextrose Agar (SDA). Other cases may concern specialized media for specific species or objectionable microorganisms that should be absent from the environment.

If, due to previously applied disinfection measures, the ambient air or surfaces contain a residual concentration of a medium with inhibitory effect (residuals of the disinfectant), a nutrient media with an antagonist should be used in order to ensure adequate microbial growth on the nutrient media plates. The inactivation of the disinfectant agent should be validated (EudraLex 2015, Annex 15).

For disinfectant monitoring, a nutrient medium which contains disinfectant antagonists must be used definitely. Commonly applied antagonists are lecithin, Tween 80, histidine, and sodium thiosulfate. These antagonists are able to inactivate a wide range of disinfectants. There could be a combination of at least three to four antagonists incorporated into the nutrient media in order to effectively neutralize specific disinfectants.

Each nutrient media must be validated regarding the expected growth promotion properties. These growth promotion properties must be demonstrated prior to the first use of the nutrient media for monitoring purposes. Adequate growth promotion properties should be confirmed at the end of the expiry date.

It is highly recommended not only to use the well-known pharmacopoeial-test strains (e.g. *Bacillus subtilis, Candida albicans, Aspergillus niger, Pseudomonas aeruginosa, Staphylococcus aureus, Escherichia coli,* and *Salmonella*) for nutrient media validation purposes but also to include one to three in-house isolates. The in-house isolates are mostly derived from environmental monitoring results – these are the most frequently identified microorganisms in the production-area of interest from a past time period. Re-evaluation and potential new definition of in-house isolates should be considered as the typical microbial flora might be changed over a certain time period.

The validation of the nutrient media should be performed with a low number of microorganisms (10–100 CFU). Adequate growth promotion properties of the nutrient media are confirmed with a recovery of at least 50% compared to the inoculum of the microbial suspension. The nutrient media should be incubated at the same incubation temperature as it is foreseen for routine microbiological environmental monitoring.

The growth promotion studies should be repeated on three different nutrient media batches to include batch variations in this study.

If the nutrient media contains antagonists, these should be evaluated for their absence of toxicity to microbial growth and their suitability to neutralize commonly used disinfectants. This study can be performed by confirming adequate growth-promoting properties in the presence and absence of residuals of disinfectants (see, e.g. Müller *et al.* 2017).

It is expected that the nutrient media validation is regularly assessed. In case of any change to the nutrient media itself, to the incubation parameters, or to the spectrum of the in-house isolates, revalidation of the nutrient media might be appropriate.

Although the general suitability of the nutrient media has been proven, each batch of nutrient media should be subjected to an incoming release control, performed as growth promotion test with comparable test strains as used for suitability tests (see, e.g. USP <61> and USP <62>). Only released nutrient media should be used for environmental monitoring purposes.

8.6.2 Incubation Temperature and Incubation Period

The incubation scheme defined by the incubation temperature and the incubation period highly affects the number of recovered microorganisms. Therefore, the incubation scheme must be well defined and supportive data should be available.

Microbiological environmental monitoring incubation schemes (see Table 8.1) required in guidelines for the manufacturing of sterile products can also be applied for the manufacturing of non-sterile products. Incubation schemes are defined within temperature ranges from 20 to 35 °C and an incubation period for a minimum of 72 hours.

Based on the different publications (see, e.g. Sandle *et al.* 2013; Gordon *et al.* 2014; Sandle 2014; Symonds *et al.* 2016; Goverde and Herzog 2017; Guinet *et al.* 2017) it can be concluded that, in general, recovery of bacteria is slightly increased by incubating at 30–35 °C and recovery of molds is slightly increased by incubating at 20–25 °C. Comparison of different incubation temperatures and incubation schemes (20–25 °C, 30–35 °C, 20–25 °C followed by 30–35 °C, 30–35 °C followed by 20–25 °C) did not, in general, result in a significant different recovery.

Therefore, total combined mold samples may typically be incubated at 20–25 °C for 5–7 days and total aerobic bacterial count samples at 30–35 °C for 3–5 days. As an alternative the same sample may be incubated serially to cover both optimum temperature ranges or at an intermediate temperature range (e.g. 25–30 °C).

The optimal incubation scheme strategy of the respective cleanrooms or production equipment can be supported by either in-vitro studies (spiking different test microorganisms including in-house isolates on media) or by in-situ studies (counts of microorganisms directly from environmental test samples are evaluated).

Table 8.1 Incubation schemes defined in different guidelines/regulations.

Guideline/Regulation	Incubation scheme
USP, chapter <1116>	20–25 °C, 3 d
FDA-Guidance for Industry – Sterile Drug Product produced by Aseptic Processing (2004)	30–35 °C, 2–3 d 20–25 °C, 5–7 d
FDA Pharmaceutical Microbiology Manual (2014) – chapter 9. Environmental monitoring section F Analytical Procedure – RODAC plates	20–35 °C, 5–7 d
JP, chapter G4	30–35 °C, >5 d 20–25 °C, >5 d 25–30 °C, >5 d
ZLG Aide Mémoire 07120605 (Germany)	20–25 °C, 5–7 d followed by 30–35 °C, 2–3 d
ANVISA – Quality Guide for Air Treatment and Environmental Monitoring Systems in the Pharmaceutical Industry	20–25 °C, 3–5 d followed by 30–35 °C, 2–3 d

The defined incubation scheme might be adapted for special purposes, e.g. environmental monitoring performed as root-cause investigation for mold contaminations.

If environmental monitoring is carried out in areas which have been shown to have a relatively high microbial level, it may be necessary – depending on the microbial count expected – to count the plates for the first time at an earlier incubation time (e.g. two to three days). After, e.g. five to seven days, the final count is undertaken; in each case, the highest result obtained during the evaluation is considered.

It is expected that the validation of the incubation scheme is regularly assessed. In case of any change of the nutrient media, the method applied, the spectrum of the in-house isolates, a revalidation of the incubation parameter might be appropriate.

8.6.3 Experimental Studies to Evaluate the Impact of the Applied Testing Method Toward Microbial Recovery

Each environmental monitoring sample – except of the surface contact samples – is somehow handled during the monitoring activity, e.g. exposure as settle plate, exposure in an active air sampling device, and filtration through a membrane.

Experimental studies would demonstrate that microbial counts are not negatively affected by the test method itself as required, e.g. in the Guide for Quality of Air Treatment Systems and Environmental Monitoring in the Pharmaceutical Industry (ANVISA 2013). Significant impact of the method on the recovery of microorganisms may be expected, for instance, due to loss of humidity in the nutrient media plate, e.g. during exposure as a settling plate (e.g. Sandle 2015) or the force of impaction (Stewart *et al.* 1995).

It is expected that the experimental studies are regularly assessed. In case of any change of the nutrient media, the method applied, and the incubation parameters, the spectrum of the in-house isolates method revalidation might be appropriate.

Example 8.1 Example of Experimental Study to Evaluate the Impact of the Air Monitoring Method on Microbial Recovery

Comparison studies should be performed with nutrient media plates which were exposed to the air monitoring activity (exposure for, e.g. four hours as settle plate or exposure for 1 m^3 in the active air monitoring device) and with nutrient media plates which are not exposed to such activities. Inoculation should be performed with a low number of microorganisms (10–100 CFU) and should – beside the well-known pharmacopoeial tests strains – also consider in-house isolates (see additional information in Section 8.6.1). Adequate growth promotion properties of the nutrient media after applying the air monitoring method or reduced influence of the sampling method to the microorganism recovery confirmed with a recovery of at least 50% or statistically non-inferior as compared to the inoculum of the microbial suspension. The nutrient media should be incubated at the same incubation temperature as it is foreseen for routine microbiological environmental monitoring. This experimental study should preferably be performed with nutrient media batches at the end of the expiry.

If a lower recovery rate is obtained, the monitoring activity must be adapted (e.g. reduction of exposure time or change of sample collecting instrument) or the nutrient media must be replaced by one which achieves a higher recovery rate (e.g. through increasing the nutrient media volume per plate).

Example 8.2 Method Suitability Test for Disinfectant or Detergent Monitoring Samples

Disinfectant or detergents are most frequently examined for microbial contamination by the membrane filtration method. To recover the present number of microorganisms in the solution to be tested, the membrane filter should be rinsed with an appropriate type and volume of rinsing fluid in order to remove residuals of the solution on the membrane filter. Residuals might have a negative influence on the ability for microorganisms to grow on the membrane filter which is put on a nutrient media plate. The method suitability test is based on USP <61>. Inoculation should be performed with a low number of microorganisms (10–100 CFU) and should – beside the well-known pharmacopoeial test-strains – also consider in-house isolates (see additional information in Section 8.6.1). The nutrient media should be incubated at the same incubation temperature as it is foreseen for routine microbiological environmental monitoring. Adequate growth promotion properties of the nutrient media after filtration of the respective agent are confirmed with a recovery of at least 50–200% compared to the inoculum of the microbial suspension. This method suitability test should also preferably be performed with nutrient media batches at the end of the expiry.

In case of a lower recovery rate, alternative procedures such as an increase of the volume of rinsing fluid or the use of another type of rinsing fluid may be tried.

8.6.4 Recovery Rate

It is known that the above-mentioned environmental monitoring methods (see Section 8.5) do not fully recover all viable and culturable microorganisms present on the respective sampling location or in the sample. The microorganisms might not get recovered since they might adhere strongly on a surface and would not get picked up by the contact plate, by the swab or by rinsing;

microorganisms might not settle directly on the plate or might not get impacted on the plate by using the impaction method.

The Brazilian health authority writes in a guideline on environmental monitoring (ANVISA 2013) that *The recovery of microorganisms from swabs must be validated, including the sampling method chosen, suitability of the swab wetting liquid, the suitability of the swab wetting liquid and the transference of microorganisms to the growth medium. The validation studies must prove a recovery greater than 50% of each of the microorganism strains used.* Furthermore, there is a similar expectation for the contact plates. Here, the just say "should be validated."

In reality, recovery rates from contact plates or swabs on different materials are reported to be not higher than 40–60% (Berchtold and Staerk, unpublished data; Goverde *et al.* 2017) and actually range from 5 to 90% depending on the type and status of the test strain and material used (e.g. Maunz and Kanz 1969; Obee *et al.* 2007; Van Horn *et al.* 2008). Therefore, a "validation" of the recovery with a strict acceptance criterion of 50% has a high probability to fail if the experimental design is close to real environmental conditions.

8.6.5 Sample Hold Time

The sample hold time is the time period between sampling till testing (e.g. membrane filtration of detergents) or start of incubation (e.g. surface contact samples). This time period must be kept as short as possible in order to

Example 8.3 Evaluation of Recovery Rate for Contact Plates

The recovery rate should be evaluated by using the relevant surface contact plate with the relevant nutrient media on the different materials representing the different sampling locations.

To determine the recovery rate, different materials are inoculated with different low-level number of microorganisms (10–100 CFU). In-house isolates should also be considered (see additional information in Section 8.6.1). After the microorganism's suspension becomes dry on the surface (dying-off effect during drying should also be taken into consideration), this surface is sampled several times with a new contact plate every time. The recovery rate can be calculated by comparing the results from the first contact plate with the results from all contact plates together which represents the total number of inoculated microorganism on this surface.

The recovery rates should therefore not be validated in terms of an analytic test with a strict acceptance criterion as described above but the method used can be evaluated for its suitability to recover microorganisms for the intended purpose taking into account the capability of the test method.

avoid die-off of microorganisms present in the sample or on the nutrient media plate and to get a representative result of the respective sampling location.

The sample hold time can be either defined by a scientific rationale or based on validation studies.

8.7 Initial Validation of Cleanrooms and Production Equipment

8.7.1 Initial Validation of Cleanrooms

Validation for new or existing cleanrooms whose modification could impact the already validated cleanroom status has to be performed for all classified areas.

Example 8.4 Validation of a Sample Hold Time for Surface Contact Samples or Air Monitoring Samples

Comparison studies should be performed with nutrient media plates (contact plates or plates for air monitoring) which are directly incubated, and which are incubated after the desired sample hold time.

Inoculation of these samples is performed with a low-level number of microorganisms (10–100 CFU). In-house isolates should also be considered (see additional information in Section 8.6.1).

The comparison study should yield at least 50% or statistically non-inferior recovery after the sample hold time.

If a lower recovery rate is obtained, either the sample hold time should be reduced or the factors during the sample hold time should be modified (e.g. storage conditions [storage at 2–8 °C versus storage at room temperature]).

It is expected that the sample hold time validation is regularly assessed. In case of any change of the nutrient media, the method applied, and the incubation parameters, the spectrum of the in-house isolates revalidation of the sample hold time might be appropriate.

An example of the validation of the sample hold time is given by Goverde and Herzog (2017). This study validates different incubation conditions and additionally the hold time of the used contact plates for three days at room temperature. In general, the hold time did not show any negative effect; in contrary, for the detection of molds compared to the control groups, higher mold numbers were found probably due to the longer incubation period than the control group (which was inoculated for five or three days at 20–25 °C or 30–35 °C, respectively).

In principle, the microbiological initial cleanroom validation is performed after the gowning, cleaning, and disinfection procedures have been defined and the technical qualification phase including HVAC qualification has been successfully completed. The cleanroom validation must be completed with compliant results prior to release of the cleanroom for use in manufacturing.

The microbiological initial cleanroom validation normally consists of two steps. First, the "at rest" validation followed by the "in operation" validation. One validation approach may consist of at least four consecutive validation runs: one run "at rest" and three runs "in operation."

The "at rest" validation has to be executed in the respective cleanroom which is fully equipped, cleaned, and disinfected, with HVAC operational and therefore ready for manufacturing but without production personnel. For "at rest" validation the sampling should generally be taken after the validated disinfectant contact time. The "in operation" validation must be performed for the respective cleanroom with production areas operating and the standard number of working personnel present per planned routine process.

In the course of the microbiological initial clean room validation also the "cleanroom clean hold time" may be considered and might be integrated in the validation activities. The "cleanroom clean hold time" is the idle time after cleaning and disinfection of the cleanroom till start of operations. This idle time should be supported with experimental data, e.g. results of microbiological environmental monitoring performed after this idle time.

8.7.2 Initial Validation of Production Equipment

The cleanliness of product-contacting surfaces of production equipment plays a key role in the microbial contamination control of products as these surfaces can directly impact the microbiological quality of the product. Hence, microbiological initial validation of such production equipment should be performed prior to the production of the first batch.

A commonly used approach for such validation activities is the monitoring of the product-contacting surfaces after all applied cleaning and disinfection steps to confirm the adequateness of these applied procedures. The focus should be set on the monitoring of worst-case sampling locations which are either based on risk rationales or based on a risk analysis.

The following rationales might be applied:

- Uneven equipment surfaces.
- Porous equipment surfaces.
- Equipment surfaces which are less accessible for cleaning/disinfection.
- Equipment surfaces with residual humidity after the last cleaning/disinfection steps.

- Equipment surfaces with cumbersome cleaning/disinfection, e.g. due to high number of small parts which needs to be disassembled and reassembled.
- Equipment surfaces which are hard to handle during cleaning/disinfection, e.g. due to a big size of the equipment.

Equipment hold times should be taken into consideration and might be included in the validation of production equipment, e.g. dirty hold time (time after manufacturing and before cleaning) or clean hold time (time after final disinfection till restart of production). Such equipment hold times should be supported with experimental data, e.g. results of the microbiological environmental monitoring performed after this hold time.

8.8 Definition of a Microbiological Environmental Routine Monitoring Program

To establish a program for microbiological routine environmental monitoring, the following parameters should be considered:

- Definition of microbiological control levels
- Definition of a monitoring frequency
- Definition of sampling locations
- Definition of measures if the microbiological control levels are exceeded
- Definition of trending of results

8.8.1 Definition of Microbiological Control Levels

Control levels for microbiological environmental monitoring are commonly defined as alert levels and as action levels.

An action level, when exceeded, signals an apparent drift from normal operating conditions and requires immediate actions by previously defined measures. The alert level, when exceeded, signals only a potential drift from normal operating conditions and triggers appropriate scrutiny and follow-up to address the potential problem.

Alert levels are always lower than action levels and generally based on the historical performance of the area monitored, whereas action levels are based on the acceptance criteria related to the cleanliness zone.

The microbiological environmental monitoring program is assigned to cleanliness zones. Stricter microbiological control levels are applied for higher cleanliness zones and more relaxed control levels are applied for lower cleanliness zones.

Due to already mentioned little regulatory guidance for the production of non-sterile products, the definition of adequate control level for the microbiological environmental monitoring is in the responsibility of the respective company or site.

The action levels for a cleanliness zone with higher criticality are derived in the examples (Tables 8.2–8.4) on hand from the action levels which are defined for cleanliness zone grade D for the production of sterile product (see EudraLex 2008, Annex 1). The action levels for a cleanliness zones with lower criticality are set around 50% higher. For further examples on action and alert levels, see Goverde and Roesti (2018) or ZLG (2010).

The action levels are fixed values, whereas the alert levels should be based on historical data analysis and trending as an indication of unusual high counts or potential hygiene problems (for calculation approaches for alert levels based on historical data, see Gordon *et al.* 2015) or Chapter 10. The alert level counts provided in Tables 8.2–8.4 can be used as initial levels if sufficient data for historical data analysis are not yet available.

Beside the microbiological control levels for "in operation," lower control levels may be defined for "at rest" (e.g. 20–50% of the "in operation" levels).

Two or three times consecutively exceeding the alert level is considered commonly as exceeding the action level.

Table 8.2 Examples for microbiological control levels for active air monitoring[a], "in operation."

	Total microbial count (CFU) per m^3	
Cleanliness zone for non-sterile products	**Alert level**	**Action level**
Higher criticality[b]	100	200
Lower criticality[b]	250	500

[a] In the production of non-sterile products the exposure of settle plates is not an established procedure for routine monitoring; only for investigation purposes or special conditions settle plates might be used.
[b] For examples of products with higher and lower criticality, see Section 8.3.

Table 8.3 Examples for microbiological control levels for surface monitoring on non-product-contacting surfaces, "in operation."

	Total microbial count (CFU) per 25 cm^2 contact plate	
Cleanliness zone for non-sterile products	**Alert level**	**Action level**
Higher criticality[a]	25	50
Lower criticality[a]	50	100

[a] For examples of products with higher and lower criticality, see Section 8.3.

Table 8.4 Examples for microbiological control levels for surface monitoring on product-contacting surfaces, "at rest"[a].

	Total microbial count (CFU) per 25 cm² contact plate	
Cleanliness zone for non-sterile products	**Alert level**	**Action level**
Higher criticality[b]	6	12
Lower criticality[b]	12	25

[a] Product-contacting surfaces are generally not monitored during "in operation" due to potential product residuals on the equipment surfaces which might influence the growth-promoting properties of the nutrient media. Therefore, monitoring of such surfaces can only be performed "at rest" or "before production." Intention of this monitoring is the confirmation of the correctly applied cleaning/disinfection procedures prior to production.
[b] For examples of products with higher and lower criticality, see Section 8.3.

If there is a relevant product contamination risk regarding molds (e.g. for inhalation products) or if molds should be separately detected and counted as a hygiene value, a second lower control level for molds can be set (e.g. 20% of the alert and of the action level for total aerobic microbial count).

8.8.2 Definition of Monitoring Frequencies

It is also the responsibility of the respective company or site to define – as part of the microbiological environmental monitoring program – an adequate monitoring sampling frequency which gives on one hand an overview of the current hygiene situation in the area of interest and on the other hand allows short-, mid-, and long-term trending in order to detect adverse shifts from the normal hygiene situation.

In general, higher cleanliness zones require a more frequent monitoring in order to get alarmed in time about higher contamination incidents as expected. This allows the initiation of adequate immediate actions in order to prevent a product contamination or even further product contaminations.

Areas with lower criticality and no direct product impact of a microbiological environmental contamination can be monitored with a reduced frequency. An example of such frequencies is given in Table 8.5 or for further references and examples, see Goverde and Roesti (2018).

Besides the monitoring frequency which is initially defined per cleanliness zone, an increased frequency of sampling may be temporally required for areas due to increased contamination levels during routine monitoring.

Table 8.5 Examples for microbiological environmental monitoring frequencies.

Microbiological environmental monitoring	Cleanliness zone for non-sterile products	
	Higher criticality[a]	Lower criticality[a]
Active air monitoring[b]	Monthly	Quarterly
Surface monitoring on non-product-contacting surfaces	Monthly	Quarterly
Surface monitoring on product-contacting surfaces[c]	Monthly	Quarterly

[a] For examples of products with higher and lower criticality, see Section 8.3.
[b] In the production of non-sterile products the exposure of settle plates is not an established procedure for routine monitoring; only for investigation purposes or special conditions settle plates might be used.
[c] Product-contacting surfaces cannot be microbiologically monitoring "in operation" due to potential product residuals on the equipment surfaces which might influence the growth-promoting properties of the nutrient media. Therefore, monitoring of such surfaces can only be performed "at rest" or "before production." Intention of this monitoring is the confirmation of the correctly applied cleaning/disinfection procedures prior production.

8.8.3 Definition of Sampling Locations

A result obtained from the microbiological environmental monitoring program should give meaningful information about the microbiological contamination status of the respective cleanroom or production equipment. Hence, the selection of sampling locations providing the excepted information is a key factor in defining the microbiological environmental monitoring program.

The sampling locations should be defined either by using risk rationales or by performing cleanroom-specific process risk analysis (e.g. HACCP or FMEA) to identify the most critical areas in the cleanroom. The risk rationale or the risk analysis should be available in a written form.

By using the risk rationale approach, the following rationales might be applied:

- Proximity to the product exposed to the surrounding environment.
- Proximity to the primary packaging material exposed to the surrounding environment.
- Highly frequented areas due to personnel activities or personnel flow.
- Highly frequented areas due to material flow.
- Areas of production manipulations, moving equipment parts, or critical connection steps.

- Areas of certain relevance due to critical operations (e.g. opening of a closed vessel for adding substances).
- Areas of certain relevance due to interfaces with lower cleanliness zone areas.
- Areas which are less accessible for cleaning/disinfection.
- Areas which are known to have a higher bioload based on initially performed cleanroom validation activities.
- General monitoring (hygiene mapping).

The sampling locations should be defined prior to start of the initial microbiological validation of the cleanroom or the production equipment. In general, the number of sampling locations depends on the layout and the cleanliness zone of the cleanroom or production equipment concerned. The number should be chosen to get a reliable “picture” of the microbiological contamination.

Review of the defined sampling locations must be done after significant changes which might have an influence on the microbiological contamination, e.g. change of the size and the layout of the cleanroom or production equipment, change of activities, change in proximity to open product, change in material and personnel flow, and change in microbiological trends. The sampling locations must be modified according to current needs as necessary.

8.8.4 Measures if Microbiological Control Levels Are Exceeded

Once the alarm and/or action level is exceeded, a formal deviation process should be triggered. Immediate actions (e.g. stop of a production line) should be followed by investigation of the root cause. Once the root cause is identified, adequate CAPAs and their required effectiveness check should be defined in order to correct and to prevent similar contamination (see also Chapter 12). Based on the outcome of the investigation, further decision on the potentially affected product, area, or equipment needs to be taken.

If the alert level is exceeded, the following elements can be verified:

- Review if the result is not a secondary contamination that occurred during sampling/transport or analytical testing.
- Identification of the isolate(s) and, if necessary, evaluation if the microorganism is potentially objectionable or from an unexpected origin.
- Review of major incidents occurred during production that would have explained a sudden increase in microbial counts.
- Trending of results over a relevant time period to ensure that no adverse trend is occurring.
- Additional sampling of the affected area if it has not been already resampled.

Based on the identified root cause (if the root cause could be identified), adequate CAPAs can be defined.

If the action level is exceeded, a deeper root cause analysis is expected. It is also expected that in this case a clear root cause or at least that a potential or most probable root cause is identified. If no root cause can be found, then all the investigated elements must be listed in the deviation.

The following elements can be verified to identify the root cause:

- Review if the result is not a secondary contamination that occurred during sampling/transport or analytical testing.
- Identification of the isolate(s) and, if necessary, evaluation if the microorganism is potentially objectionable or from an unexpected origin.
- Review of historical data trends.
- Review of area or equipment maintenance documentation.
- Review of cleaning and disinfection documentation and effectiveness.
- Review of the inherent physical or operational parameters such as changes in environmental temperature and relative humidity.
- Review of training status of the personnel involved in the process, in disinfection procedures and in environmental monitoring sampling.
- Review of incidents occurred during production that would have explained a sudden increase in microbial counts.
- Trending of results over a relevant time period to ensure that no adverse trend is occurring.
- Additional sampling of the affected area if it has not been already resampled.

All investigation steps and results must be summarized in an investigation report where the root cause, batch release decision (if relevant). and CAPAs are documented.

Following the investigation and depending on the root cause assessment, CAPAs taken may include: reinforcement of training of personnel, additional sampling at increased frequency, additional disinfection, additional product testing, or QA-Oversight.

In addition, depending on the outcome of the investigation and root cause assessment, production may be stopped, evaluation of a contamination risk on products which have already been released and might be impacted by the contamination (mainly relevant, if the contamination was found on equipment with product-contact and which is not sampled on a batchwise basis), equipment may not be used, and access of personnel may be forbidden until adequate CAPAs are implemented.

Further details on deviation handling for environmental monitoring in non-sterile manufacturing are given by Goverde and Roesti (2018).

8.8.5 Trending of Results

The microbiological environmental monitoring results have to be reviewed periodically and trended to assess the capability of the measures to control microbial contamination and verify that no adverse trend is happening. The results of the trend analyses should be summarized regularly in trend reports.

Trending must be performed according to predefined requirements and frequencies and has to take into account all appropriate environmental aspects including, e.g. cleanliness zone and, in some cases, specific type of isolated microorganisms. Trending is of essential relevance if a recurring isolated microorganism has been identified as potentially objectionable, it must be evaluated if these microorganisms would contaminate the final product.

Criteria defining an adverse trend must be in place. If an adverse trend occurs, the event should be handled as exceeding an action level and should trigger a deviation process.

8.8.5.1 Trending Areas and Periods

Trend areas must be composed of microbiological results that are grouped according to a rationale. For instance, a trend area can group sampling sites of the same production areas and cleanliness zones or from similar production lines. Results of these sampling sites are grouped within a given period. If there are only a few samples or high variability of sampling size per time interval, a longer trend period can be chosen to avoid strong dependence on single events or on sample size.

8.8.5.2 Trending of Identified Isolates

The types of isolates may also be trended in order to verify if there is a reoccurrence of microorganisms in a particular area. In microbiological environmental monitoring, if the same microorganism is isolated in high counts or high frequency, this may indicate potential adaptation to disinfectants or may be inherent to the process such as increased presence of Gram-negative bacteria by processes using water.

8.8.5.3 Adverse Microbiological Trend

An adverse trend is an early warning of a potential degradation or loss of control within the environment. Only a single excursion above a defined microbiological alert or action level is not considered as an adverse trend. An adverse trend can be defined as repeating higher than usual counts or increasing number of microorganisms or contamination occurrences over a certain time period. The adverse trend corresponds to the trend method used.

Depending on the evaluation method used, adverse microbiological trend may, for instance, be:

- Three times exceeding the alert level in a row.
- Two times exceeding the action level in a row.
- Increased counts or frequency of occurrence in the graphical interpretation of data.
- Higher proportion of samples exceeding microbiological alert and action levels from one time period to another or from more than one site from the same room (regardless of source: personnel, surface, or air) on the same day.

- Regular but isolated exceeding microbiological alert and action levels may have a common cause, for example, single excursions that always follow planned preventative maintenance.
- Repeated occurrence of a specific microorganism.

Further guidance is provided in Chapter 10.

An investigation must be performed to determine the root cause explaining the adverse trend as well as to assess the impact on the product's microbiological quality.

Corrective actions may include but are not limited to the following:

- Revision of the cleaning and disinfection program, including selection of other disinfectants, application methods, and frequencies.
- Increased oversight of personnel practices.
- Review of microbiological sampling methods and techniques.

If CAPAs are put in place to resolve the adverse trend, their effectiveness should be demonstrated with an improvement in the microbiological trend. If an improvement has been demonstrated, the effectiveness check of the adverse trend deviation may be closed.

8.9 Microbiological Environmental Monitoring: Examples for Users

8.9.1 Example for the Definition of Sampling Locations for a Production Area

In the course of the reconstruction of a production area where inhalation products are primary packed, an initial microbiological cleanroom validation including the redefinition of sampling locations had to be performed (Figure 8.1). The affected area includes the production room, the material entry, and the personnel entry.

As a prerequisite to start the initial microbiological cleanroom validation, the HVAC system was re-qualified, and the gowning, cleaning, and disinfection procedures which were defined for the respective cleanroom prior to reconstruction were found to be still valid. Cleaning and disinfection of the cleanroom was performed.

Sampling locations for the "at rest" and the "in operation" microbiological cleanroom validation were defined by applying the following risk rationales (Table 8.6 and Figure 8.2):

- Rationale 1: Area where open product is handled, area in proximity of handling of open product or product-contacting equipment, and areas where cleaned/disinfected product-contacting equipment is stored.
- Rationale 2: Area which is frequently used, e.g. due to higher personnel or material flow, and areas with connection to lower/higher cleanroom zones.

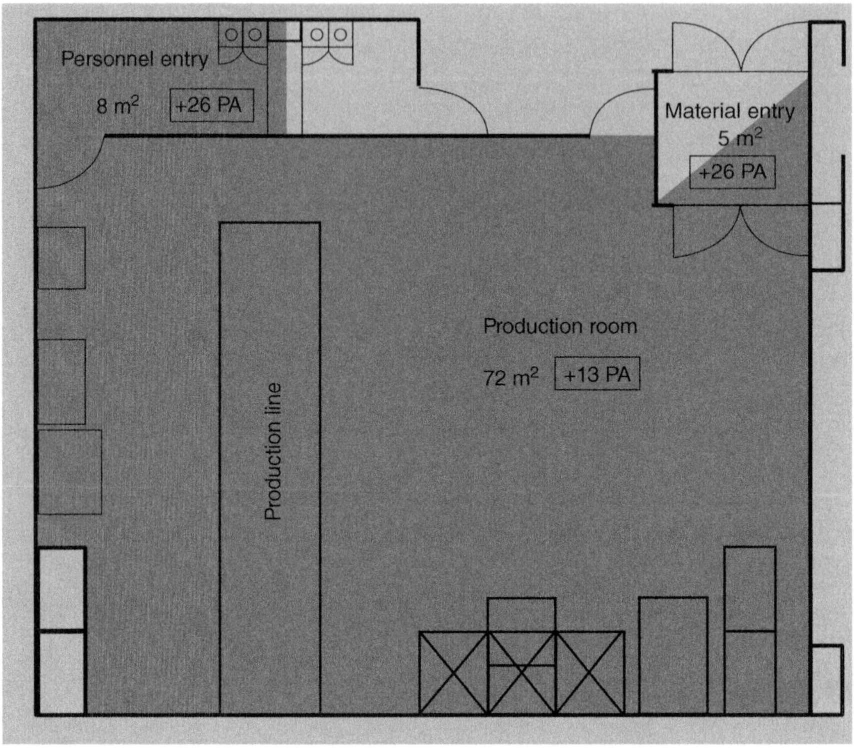

Figure 8.1 Production area with personnel and material entry room.

Table 8.6 Summary of the sampling locations and methods based on the risk rationale mentioned in the text.

Sample no	Description of sampling location	Method	Risk rationale
Material entry			
01	Middle of the room	Active air	2, 3
02	Door to room	Surface	3
03	Floor between mat and the wall	Surface	3, 4
Personnel entry			
04	Middle of the room	Active air	2, 3
05	Wall at the sink, behind the bench	Surface	3
06	Floor in the middle of the room	Surface	2, 3
07	Floor at the locker	Surface	3

Table 8.6 (Continued)

Sample no	Description of sampling location	Method	Risk rationale
Production area			
08	Air where product is put on the packaging line	Active air	1, 2
09	Air in front of RABS	Active air	1, 2
10	Working surface for cleaned/disinfected equipment	Surface	1, 2
11	Outside of the machine, funnel	Surface	1, 2
12	Outside of the machine, RABS	Surface	1, 2
13	Wall in the left edge at the equipment storing area	Surface	3
14	Floor nearby the RABS	Surface	2, 3
15	Floor where the product is put on the packaging line	Surface	2, 3
16	Floor nearby the personnel entry	Surface	2, 3
17	Floor nearby the material entry	Surface	2, 3
18	Floor nearby the funnel	Surface	3

- Rationale 3: Area which is not regarded as critical due to potential higher bioload, but which is monitored in order to check the general hygiene status of the cleanroom and the respective cleaning and disinfection efficacy within the cleanroom.
- Rationale 4: Area which is difficult to reach for cleaning and disinfection and area where, in general, a higher bioburden could be expected due to several reasons.

It was decided that also the cleaning and disinfectant agent has to be monitored during "at rest" and "in operation" cleanroom activities in order to have this supportive information in case of any deviation.

8.9.2 Examples for Deviation Handling

Once an exceeded action level of a microbiological environmental control has been noticed in the microbiological control lab, the company-specific formal deviation process should be followed. The main elements after the initiation of the deviation in the respective deviation system are first the definition of immediate action and the thorough root cause investigations followed by the definition of the respective CAPAs and effectiveness checks, and finally the decision on production release or rejection.

The following two examples will guide through the above-mentioned steps.

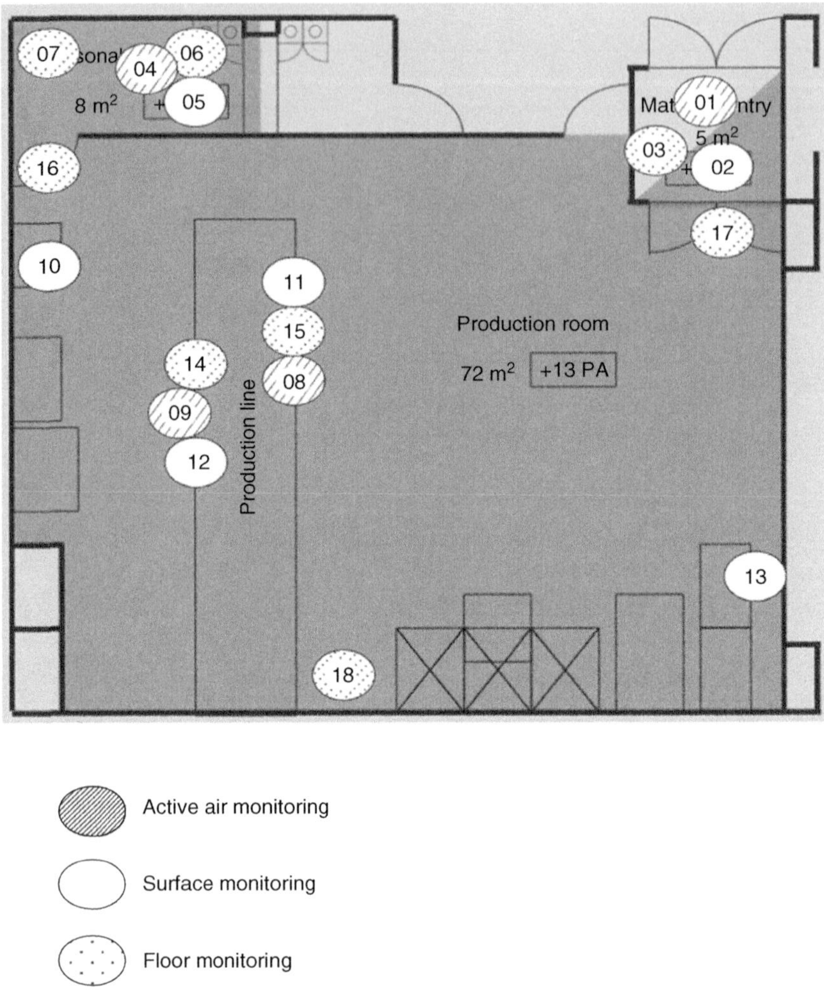

Figure 8.2 Production area from Figure 8.1 with the sampling locations according to the rationale from Table 8.6.

8.9.2.1 Example No. 1: Equipment-Surface Monitoring with Exceeded Action Level

8.9.2.1.1 Initial Situation In the course of the semiannually microbiological environmental surface monitoring of the product-contacting equipment of a vacuum conveying machine to produce oral tablets, the action level was exceeded.

- Sampling location: Container funnel inside
- Number of colonies per plate: 26 CFU/25 cm^2

- Action level defined for surface monitoring of product-contacting equipment in this area: 20 CFU/25 cm^2

8.9.2.1.2 *Immediate Actions* The following immediate actions were initiated:

- Opening of a deviation in the deviation system
- Information to the respective production head and responsible QA
- Identification of the detected microorganisms
- Cleaning/disinfection of the affected equipment and resampling

8.9.2.1.3 *Investigation* To find the root cause and to assess the contamination risk, the following points were investigated.

- *Identification:*
  The most frequently found isolate was identified as *Corynebacterium tuberculostearium*. This is a Gram-positive rod from the environment or humans.
- *Secondary contamination:*
  A secondary contamination could be excluded as the sampling and the transport to the microbiology lab was performed correctly by trained and qualified analysts.
- *Associated microbiological environmental monitoring results:*
  The results of the microbiological surface monitoring that was performed on other sampling locations (16 further sampling locations) on the same equipment on the same day according to the defined program did not show any exceeded action or alert levels. The results were within the requirements and not the same isolate was found.
- *Cleaning, disinfection, and resampling:*
  Cleaning, disinfection, and resampling were performed immediately after detection of the exceeded action level. This sample complies with the requirement. During resampling of the affected sampling location, surface monitoring was also performed on the other 16 sampling locations on this equipment. These samples also showed no contamination. The procedure and the agent for a correct cleaning and disinfection of the relevant equipment prior and after the concerned microbiological environmental monitoring was checked and confirmed by the respective equipment log book.
- *Historical data:*
  The historical data (two years back due to the sampling frequency of semiannually) of the relevant equipment were evaluated; no further exceedance of action levels were recorded within this time period. Therefore, a negative trend could be excluded.

- *Products already on the market:*
  Since the last complying sampling of the affected sampling point, all product batches manufactured on the concerned equipment were listed and the microbiological risk for "products on the market" was assessed. Due to the good historical data of the equipment, the good historical data of the products produced (confirmed by microbial enumeration testing), and the low water activity of these products, no risk was found to be present.
- *Directly affected product batches:*
  As directly after the microbiological surface monitoring the concerned equipment was used for production purposes, the first three batches produced were tested with the Microbial Enumeration Test. All results did comply with the specifications without microbial growth.
- *Investigations within production area:*
  The container funnel is cleaned and disinfected in an equipment washing machine according to SOP. An automatic cleaning and disinfection program is used. The equipment parts are transported back into the production cleanroom on a trolley; the equipment parts are covered with plastic foil. As typical environmental bacteria were identified, which get inactivated by the normal cleaning and disinfection procedure, it can be assumed that the contamination of the affected sampling point occurred after leaving the equipment washing machine. A contamination could have occurred, e.g. during transport to the production cleanroom by contaminated working clothes or during preparation for sampling, or during the sampling process itself.

8.9.2.1.4 Root Cause The root cause for the exceeded action level could not be found. It was supposed that a recontamination after the cleaning and disinfection, during transport, or during sampling occurred as a single event. Due to the good historical data, a systematic problem could be excluded.

8.9.2.1.5 Risk Assessment and Decision A risk for product or patient could be excluded as

- the resampling did comply with the requirements
- the historical data of the concerned equipment showed no negative trend; therefore, a single incident was present
- the most critical batches were tested for microbial contamination and all results did comply with the specification without microbial growth.

In conclusion, no microbial risk is seen. The equipment and the products can be released from a microbiological point of view. Since no clear root cause could be found and it was not a recurring event, no further CAPAs can be defined.

8.9.2.2 Example No 2: Cleanroom Surface Monitoring with Exceeded Action Level

8.9.2.2.1 Initial Situation In the course of the initial cleanroom validation of a cleanroom corridor and elevator, the action level of a surface sample on the floor of the elevator was exceeded during the second "in operation" run.

The elevator is connecting a nonclassified and controlled area in the basement to a corridor which leads to rooms for solid oral production. Due to a reconstruction of the corridor and the elevator, a revalidation for these cleanroom areas was required by one "at rest" and three "in operation" runs.

- Sampling location: middle of the floor of the elevator
- Number of colonies per plate: not countable as the microorganisms were swarming over the plate; regarded – as worst case – as an exceedance of the action level
- Action level defined for surface floor monitoring in this area: 100 CFU/25 cm^2

8.9.2.2.2 Immediate Actions The following immediate actions were initiated:

- Opening of a deviation in the deviation system
- Information to the respective production head and responsible QA
- Identification of the detected microorganisms
- Performance of an additional surface sampling on the floor in the affected elevator

8.9.2.2.3 Investigations To find the root cause and to assess the contamination risk, the following points were investigated.

- *Identification:*
  Paenibacillus lautus. It is a Gram-positive rod and can mainly be found in the environment (air, water, and soil) and is a swarming bacterium. Therefore, it might be that only a few (below the action level) bacteria were sampled, but due to the swarming of this bacterium the plate was overgrown and could not be quantitatively evaluated. Thus, the worst-case approach was chosen and the result was notified as exceeding the action level.
- *Secondary contamination:*
  A secondary contamination could be excluded as the sampling and the transport to the microbiology lab was performed correctly by trained and qualified analysts.
- *Associated microbiological environmental monitoring results:*
  The results of the microbiological environmental monitoring that was performed on other sampling locations in the corridor and in the elevator (in total: five further surface samples on the floor and five active air samples) on the same day (same "in operation" validation run) according to the defined validation protocol did not show any exceeded action or alert levels. The

results were within the requirements and no other swarming bacterium on plates was detected.

- *Additional active air sampling:*
  The additional surface sampling on the floor of the elevator was performed immediately after detection of the exceeded action level. This sample fully complies with the requirement. Surface sampling was also performed on the other floor sampling location of the elevator and on four further floor sampling locations of the corridor. These samples showed no contamination.
- *Last cleaning/disinfection:*
  The procedure and the agent for correct cleaning/disinfection of the relevant corridor and elevator prior and also after the occurrence of the exceeded action level was checked and confirmed by the respective cleanroom log book. However, since the elevator connected two different cleanliness zones (unclassified and uncontrolled basement to a defined cleanliness zone for solid oral dosage forms in the corridor), the cleaning frequency of daily might not be intensive enough for the elevator and the corridor.
- *Historical data:*
  The historical data of the room was checked. The room is new and only the "at rest" and the first "in operation" measurement were available. Both were within the requirements. Since it was a new room, no trending according the local SOP was possible.
- *Investigations within production area:*
  The elevator was used for the transportation of material from the basement (unclassified and uncontrolled area) into the production area for solid oral dosage forms. The material is placed in the elevator using a lift truck. The floor in the elevator was disinfected once a day according to an internal procedure by using a non-sporicidal disinfectant.

  The disinfection of the floor in the elevator was carried out by an external cleaning agency. The disinfection was performed correctly before and after the sampling. However, it was realized that disinfection once a day might not be enough due to the high frequency of loadings of the elevator.
- *Further validation runs:*
  It was decided to perform three further "in operation" validation runs in the corridor and the elevator. All three runs were within the requirements.

8.9.2.2.4 Root Cause The root cause for the detection of swarming bacteria on a surface sample on the floor was identified. A contaminated lift truck in the basement (unclassified and uncontrolled area) which was used for loading materials into the elevator contaminated the floor in the elevator. The disinfection frequency of the floor in the elevator which was performed once a day was evaluated as insufficient.

8.9.2.2.5 ***Risk Assessment and Decision*** The root cause could be identified and eliminated based on the defined CAPAs. The three further "in operation" validation runs were within the requirements; therefore, the area of the corridor and the elevator are regarded as microbiologically validated.

As during these validation runs no productions occur, a product-risk needs not to be evaluated.

8.9.2.2.6 CAPAs

- Additional sporicidal cleaning was performed since a spore-forming bacterium was found.
- The disinfection frequency for the room/elevator was increased to twice per day and a sticky entrance mat will be added to the area.
- Three additional validation runs "in operation" were performed.

8.10 Conclusion

Microbiological environmental monitoring can be considered as a very good tool within the overall microbial control strategy to evaluate a microbiological status of a cleanroom or production equipment.

In order to obtain significant results from the microbiological environmental monitoring program, the following needs to be considered concretely: choice of the method for the microbiological environmental monitoring, adequate performance of the respective method, definition of a sound environmental monitoring program considering sampling locations, monitoring frequencies, and deviation handling.

Furthermore, one should be aware that microbiological environmental controls are not

- a release test: contamination levels detected with environmental controls do not necessarily result in a rejection of the affected batch as it is the case of release tests. Detected contaminations by environmental controls lead – if the alert or action level is exceeded – to root cause investigations, definition of CAPAs, and a decision on batch release or batch rejection.
- highly recoverable: the number of microorganisms which will be recovered with the respective environmental control is dependent on the nutrient medium, the testing type, and other non-standardized factors. E.g. sampling of surfaces will not result in a 100% recovery rate; the recovery rate varies from one material to the other, from the contact pressure, the contact time, and the kind of microorganisms present on the surface.
- always repeatable: contamination level on a surface can differ from one moment to the other, e.g. a working surface gets contaminated by manual

manipulations on this surface or microorganisms might die-off or get eliminated by disinfection. Environmental monitoring results are just a snapshot in time.

Bibliography

ANVISA, 2013, *Guide for Quality of Air Treatment Systems and Environmental Monitoring in the Pharmaceutical Industry*, N.A.o.H. Surveillance, Editor. Brasilia.

Chinese GMP, 2010, *Good Manufacturing Practice for Drugs (2010 Revision)*, MOH Decree No. 79, State Food & Drug Administration P.R. China.

DIN 10113-3, 1997, *Determination of Surface Colony Count on Fitment and Utensils in Food Areas – Part 3: Semiquantitative Method with Culture Media Laminated Taking up Equipment (Squeeze Method)*, Deutsche Institut für Normung.

EudraLex, 2008, *The Rules Governing Medicinal Products in the European Union, Volume 4, EU Guidelines to Good Manufacturing Practice Medicinal Products for Human and Veterinary Use; Annex 1 – Manufacture of Sterile Medicinal Products (corrected version).*

EudraLex, 2015, *The Rules Governing Medicinal Products in the European Union, Volume 4, EU Guidelines for Good Manufacturing Practices for Medicinal Products for Human and Veterinary Use, Annex 15: Qualification and validation.*

FDA, 2004, *Guidance for Industry – Sterile Drug Products Produced by Aseptic Processing – Current Good Manufacturing Practice*, U.S. Department of Health and Human Services, Food and Drug Administration.

FDA, 2014, *Pharmaceutical Microbiology Manual*, U.S. FDA, Office of Regulatory Affairs, Office of Regulatory Science, Medical Products and Tobacco Scientific Staff, pp. 1–86.

Gordon, O, Berchtold, M, Staerk, A, Roesti, D, 2014, 'Comparison of different incubation conditions for microbiological environmental monitoring', PDA Journal of Pharmaceutical Science and Technology, vol. 68(5), pp. 394–406.

Gordon, O, Goverde, M, Pazdan, J, Staerk, A, Roesti, D, 2015, 'Comparison of different calculation approaches for defining microbiological control levels based on historical data', PDA Journal of Pharmaceutical Science and Technology, vol. 69, pp. 383–398.

Goverde, M, Herzog, S, 2017, 'In Situ Studie zur Validierung der Inkubationsbedingungen für Abklatschplatten aus dem mikrobiologischen Umgebungsmonitoring', Pharmazeutische Industrie, vol. 80(3), pp. 434–439.

Goverde, M, Herzog, S, 2018, 'Validierung der Inkubationsbedingungen. In-situ-Studie zu Abklatschplatten aus dem mikrobiologischen Umgebungsmonitoring', Pharmazeutische Industrie, vol. 80(3), pp. 434–439.

Goverde, M, Roesti, D, 2018, *Guidance for deviation handling of microbiological environmental monitoring excursions in non-sterile pharmaceutical manufacturing*, ECA Pharmaceutical Microbiological Working Group.

Goverde, M, Willrodt, J, Staerk, A, 2017, 'Evaluation of the recovery rate of different swabs for microbial environmental monitoring', PDA Journal of Pharmaceutical Science and Technology, vol. 71, pp. 33–42.

Guinet, R., *et al.*, 2017, 'Multicenter study on incubation conditions for environmental monitoring and aseptic process simulation', PDA Journal of Pharmaceutical Science and Technology, vol. 71, pp. 43–49.

ISO 14698-1, 2003, *Cleanrooms and Associated Controlled Environments – Biocontamination Control, Part 1: General Principles and Methods*, International Organization for Standardization.

ISO 18593, 2004, *Microbiology of Food and Animal Feeding Stuffs – Horizontal Methods for Sampling Techniques from Surfaces Using Contact Plates and Swabs*, International Organization for Standardization.

JP, 2017, Chapter G4, *Microorganisms – Microbiological Evaluation of Processing Areas for Sterile Pharmaceutical Products*, Japanese Pharmacopeia, 17th ed.

Maunz, U, Kanz, E, 1969, 'Beiträge zur quantitativen Auswertung von Abklatschkulturen', Gesundsheitswesen und Desinfektion, vol. 9, pp. 129–143.

Meier, R, Zingre, H, 2000, 'Qualification of air sampler system: the MAS-100', Swiss Pharma, vol. 22(1–2), pp. 15–21.

Müller, R, Eikmanns, U, Schindler, H, 2017, 'Neutralisation von Desinfektionsmittelrückständen', Pharmazeutische Industrie, vol. 79(9), pp. 1298–1302.

Obee, P, Griffith, CJ, Cooper, RA, Bennion, NE, 2007, 'An evaluation of different methods for the recovery of meticillin-resistant *Staphylococcus aureus* from environmental surfaces', Journal of Hospital Infection, vol. 65, pp. 35–41.

Sandle, T, 2014, 'Examination of the order of incubation for the recovery of bacteria and fungi from pharmaceutical-grade cleanrooms', International Journal of Pharmaceutical Compounding, vol. 18(3), pp. 242ff.

Sandle, T, 2015, 'Settle plate exposure under unidirectional airflow and the effect of weight loss upon microbial growth', European Journal of Parenteral and Pharmaceutical Sciences, vol. 20(2), pp. 45–50.

Sandle, T, Satyada, R, 2015, 'Assessment of the disinfection of impaction air sampler heads using 70% IPA, as part of cleanroom environmental monitoring', European Journal of Parenteral and Pharmaceutical Sciences, vol. 20(3), pp. 94–99.

Sandle, T, Skinner, K, Yeandle, E, 2013, 'Optimal conditions for the recovery of bioburden from pharmaceutical processes: a case study', European Journal of Parenteral and Pharmaceutical Sciences, vol. 18(3), pp. 84–91.

Stewart, SL, Grinshpun, SA, Willeke, K, Terzieva, S, Ulevicius, V, Donnelly, J, 1995, 'Effect of impact stress on microbial recovery on an agar surface', Applied and Environmental Microbiology, vol. 61(4), pp. 1232–1239.

Symonds, ID, Martin, DL, Davie, MC, 2016, 'Facility-based case study: a comparison of the recovery of naturally occurring species of bacteria and fungi on semi-solid media when incubated under standard and dual temperature conditions and its impact on microbial environmental monitoring approach', European Journal of Parenteral and Pharmaceutical Sciences, vol. 21(1), pp. 7–15.

USP 41-NF 36 chapter <1115>, 2018, 'Bioburden control of nonsterile drug substances and products', United States Pharmacopeia.

USP 41-NF 36 chapter <1116>, 2018, 'Microbiological control and monitoring of aseptic processing environments', United States Pharmacopeia.

Van Horn, KG, Audette, CD, Tucker, KA, Sebeck, D, 2008, 'Comparison of 3 swab transport systems for direct release and recovery of aerobic and anaerobic bacteria', Diagnostic Microbiology and Infection Disease, vol. 62, pp. 471–473.

WHO, 2016, *Supplementary Guidelines on Good Manufacturing Practices for Heating, Ventilation and Air Conditioning Systems for Non-Sterile Pharmaceutical Dosage Forms*, Revision Draft. World Health Organization, Geneva.

ZLG, 2010, *Aide Mémoire 07121105 – Inspektion von Qualifizierung und Validierung in pharmazeutischer Herstellung und Qualitätskontrolle.* Zentralstelle der Länder für Gesundheitsschutz bei Arzneimitteln und Medizinprodukten, Germany.

ZLG, 2016, *Aide Mémoire 07120605 – Überwachung von Sterilherstellern.* Zentralstelle der Länder für Gesundheitsschutz bei Arzneimitteln und Medizinprodukten, Germany.

9

Identification of Microorganisms

Christine E. Farrance

Senior Director of R&D and Scientific Affairs, Charles River Laboratories, Inc., Newark, DE, USA

CONTENTS

Pharmaceutical Microbiological Quality Assurance and Control: Practical Guide for Non-Sterile Manufacturing, First Edition. Edited by David Roesti and Marcel Goverde.

9.1 Introduction

A program to identify microorganisms is integral to the pharmaceutical microbiology laboratory. Pharmaceutical companies are expected to determine which microorganisms may be in their process, components, or products that are not required to be sterile, and develop appropriate testing to detect these organisms (21 CFR Part 211). They are also expected to determine what organisms would be objectionable in those products and have adequate detection methods for them. The United States Pharmacopoeia (USP) outlines how the testing can be done and states that the significance of the recovered microorganisms be evaluated by a risk-based assessment of relevant factors (USP 2012a, 2012c). For non-sterile products, identification of recovered microorganisms is also needed to rule out specific pathogens as defined by the use and nature of the product.

Microbial identification is dependent on established taxonomy, where taxonomy is the study of the organization and prioritization of microbial diversity based on actual relationships. Without taxonomy, microbial identification is not possible. Taxonomy, or systematics, is divided into three parts. First is classification which is the orderly arrangement of organisms into taxonomic groups on the basis of similarity. Second is nomenclature which labels those groups. Lastly, identification determines if an unknown belongs to one of those groups or taxonomic ranks. The group boundaries are made by taxonomists. They are flexible, dynamic, and evolving just like the organisms themselves. The classification scheme should facilitate the identification, and should be robust and logical to recognize organisms previously encountered and categorize ones newly found. A phenetic classification is based on relationships as they exist now without reference to evolution or ancestry. A phylogenetic classification is based on relationships described by ancestry, and not according to present properties (Vandamme *et al.* 1996). Polyphasic taxonomy is a general classification methodology combining the strengths of both phenetic and phylogenetic studies (Vandamme *et al.* 1996). It describes the integration of all available genotypic and phenotypic information into a consensus type of classification. The polyphasic approach is necessary for taxonomic classification, and relevant at times for biotechnology research, ecology, and bioremediation studies (Das *et al.* 2014). In 1987, it was stated that species taxonomy should be determined phylogenetically (Wayne *et al.* 1987). The term "species" is used to express membership of organisms in a taxonomic rank (Stackebrandt and Goebel 1994) and was defined as a group of strains, including the type strain, that share 70% or greater DNA relatedness (Wayne *et al.* 1987). This species definition was based on a large amount of empirical data. While the designated type strain of a species serves as the name bearer of the species and as a reference specimen, it is essential that the boundaries of species demarcation be flexible in order to achieve a classification scheme that facilitates identification.

Identification and characterization of recovered microorganisms are some of the most important tasks of a QC microbiology laboratory. Correct identifications are critical for in-process controls, environmental monitoring (EM), finished product release, and investigations. Routine environmental and utility monitoring, as described in 21 CFR Part 211, are done to demonstrate the state of control of a facility and facilitate in-process monitoring. Routine identifications of recovered organisms for trending provide useful information on the baseline picture of the normal microbiota and categorize the resident bioburden. Then, if there is an excursion or negative trend, there would be baseline information to know if this organism is normal or novel. Significant change in the species diversity could indicate the potential routes of contamination allowing implementation of corrections before product contamination occurs. Routine identifications can also determine the efficacy of sanitization or cleaning protocols, and the suitability of rotational frequency of disinfectants and sporocides. Additionally, excursions, out of specifications, action level overages, and product failures require identification of organisms to the species level and maybe to the strain level. Important strains or isolates should be banked to use later for comparisons and for use in microbial challenge testing, validation of microbial content test methods, and growth promotion tests, for example. Additional instances of when an identification is needed include the confirmation of the species identity of QC stock cultures, biological indicators, and cell banks. The identification should be confirmed using a method that is suitable for each species (USP 2012c).

Many times, species-level identifications are required, however, this is not always necessary. The amount of effort devoted to the analysis and description of a microbe should be proportional with the contamination risk to the product. The monitoring and reaction is driven by risk and the product. There should be a hierarchy of response from characterization to identification to strain typing (Sutton and Cundell 2004). The level of characterization depends on the criticality of the area involved and whether an investigation warrants further identification.

All characterizations and identifications must begin with core microbiological techniques as pure cultures are essential for all further analyses. Characterization comprises a morphological description consisting of the Gram-staining reaction, colony morphology, cell morphology, key biochemical parameters (e.g. the results of coagulase, catalase, and oxidase tests), presence of spores, motility, etc. Many times this is a sufficient description for an alert or no alert response. The broad characterization can give an indication of crude or large-scale issues such as people, soil, or water sources of an organism (Sutton and Cundell 2004). Accurate genus and species identification will give a high-resolution map of the microbiota of a facility. However, characterization or primary screening is sufficient for risk assessment in non-sterile pharmaceutical manufacturing to either recommend a corrective action or dictate the need for testing to obtain a species-level identification or strain differentiation of isolates within a species. Basic characterization will not be discussed further. Instead, emphasis will be on the technologies available for the identification of unknown microorganisms and strain-level differentiation. First, it is necessary to provide a background on the history of microbial taxonomy and classification as it relates to current technologies utilized for microbial identification and strain differentiation in pharmaceutical manufacturing. Understanding the science and history of taxonomy sets the stage for a description of the methodologies that can be generally applied for the identification of microorganisms today. Identification consists of providing a taxonomic description with a genus and species designation. The identification methods need to be relatively inexpensive, fast, yet accurate with the most relevant and appropriate breadth of organisms in the reference database. The data must also be available for easy analysis during tracking and trending responsibilities. In order for an organism to be identified, there must already be an established nomenclature and classification scheme.

9.2 History and Challenges of Bacterial Taxonomy and Classification

Microbes have been around for 3.8 billion years; yet, they have only been known to humans for 300 years. Estimates suggest that the described bacterial species in 2014 ranged from 11 000 to 13 000 and new species are published at a rate of 650–1 000 per year (Kyrpides *et al.* 2014; Federhen 2015). However, this may represent significantly less than 0.1% of the existing microbial diversity (Tamames and Rossello-Mora 2012). Classification of bacterial species with the Linnaean approach to naming life forms began in the late 1700s. The descriptions were based on cell shape, behavior, and habitat since the primary analysis tool was the microscope. Simple Linnaean descriptions continued for bacteria through the late 1800s. With the development of the capacity to grow organisms on agar plates and pure culture techniques, the capacity to recognize and describe species expanded to include growth characteristics and the results of biochemical tests. The early 1900s brought studies and delineation of genera and species based on characteristics such as metabolic byproducts, fermentation of sugars, temperature ranges for growth and morphology, and started to introduce the concept of natural relationships (Orla-Jensen 1921). However, there was still very little consistent classification or organization to the descriptions. Originally, bacteria were thought of as plants, and discussions of their taxonomy were part of Botany Societies and Conferences and type specimens were deposited in herbariums. These early studies would lead to the grouping of all known bacteria, and were arranged by properties which were documented to facilitate determination of hierarchical arrangements. In the early 1900s, Robert Buchanan published 10 papers on the nomenclature and classification of bacteria. These papers were revolutionary and included all described bacteria and included higher groupings thereby launching the systematic efforts (e.g. Buchanan 1916, 1918). Bergey's Manual for Determinative Bacteriology was published in 1923 and represented a major step toward a practical system of nomenclature. By the 1950s, the *International Bulletin of Bacteriological Nomenclature and Taxonomy* was established which was later renamed as the *International Journal of Systematic Bacteriology*, now known as the *International Journal of Systematic and Evolutionary Microbiology* (IJSEM). IJSEM is the "official journal of record for novel prokaryotic taxa. It is the official publication of the International Committee on Systematics of Prokaryotes" (ICSP) (http://ijs.microbiologyresearch.org/content/journal/ijsem). The ICSP is the body that oversees the nomenclature of prokaryotes, determines the rules by which prokaryotes are named and whose Judicial Commission issues Opinions concerning taxonomic matters, and revisions to the Bacteriological Code (www.the-icsp.org).

9.2.1 Definition of Strains

Species are classified as distinct groups of strains that have certain distinguishing features and strains within the species generally share close resemblance to one another in those essential features. A strain is a descendent of a single isolation in pure culture. It represents a succession of cultures ultimately derived from an initial single colony. The type strain is the original reference specimen for the name, it is the permanent example of the species. The species is composed of the type strain plus all other strains considered to be sufficiently similar to the type to warrant inclusion with it in the species. When new species are published, the very first strain is isolated and characterized and is defined as the type, but it does not mean that the type is the typical strain in the species. The type strain should be deposited into at least two global culture collections such as the American Type Culture Collection, the Deutsche Sammlung von Mikroorganismen und Zellkulturen, the English National Collection of Type Cultures, and the Korean Collection for Type Cultures, for example. Along with the type strain, many other non-type strains are deposited in the collections to be used as reference specimens (Table 9.1).

9.2.2 Evolution of a Phylogenetic Marker

In the late 1800s classification was based on colony morphology, the results of Gram's differential staining method which provides information about the cell wall structure (Gram 1884), and on basic biochemical traits passed to the next generation of cells. As a result, in the early 1900s, different identities were assigned to organisms with only minor phenotypic differences. In the 1950s and 1960s, the discovery of DNA and the field of genetics allowed the comparison of bacterial genomes with the calculation of overall base composition (mol% G + C). Organisms with different values were shown to be different species. If they had the same overall base composition, they may or may not be the same species. This technology was used to differentiate taxa, but no insight into the genetic or phylogenetic relationship was generated. An alternative approach and a phylogenetically informative molecule was needed. Zuckerkandl and Pauling proposed the study of primary structures of macromolecules to deduce phylogenetic histories (Zuckerkandl and Pauling 1965). Thus, there was a shift from gross organismal properties to a molecular basis for

Table 9.1 Example of the diversity of type and non-type strains for *Micrococcus luteus* and the different strain designations in the global culture collections.

Micrococcus luteus	**Strains deposited in global culture collections**
Type strain designations (not a complete list)	ATCC 15307^T, ATCC 4698^T, CIP A270^T, DSM 20030^T, JCM 1464^T, KCTC 1056^T, KCTC 3063^T, LMG 4050^T, NBRC 3333^T, NCIMB 10474^T, NCIMB 9278^T, NRRL B-287^T
Non-type strains (not a complete list)	ATCC 49732, CIP 103664, ATCC 9341, DSM 3448, JCM 20050, KCTC 1071, LMG 8194, NBRC 12708, NRRL B-1018, ATCC 10786, CIP 830160, DSM 1790, KCTC 2177, NBRC 13867, NCIMB 8166

Superscript "T" indicates that the strain is the type strain for the species.

phylogenetic interpretation. Investigations into molecular markers of evolution were initiated in the 1960s and 1970s. Scientists began to focus on evaluating different genes or gene regions and showed that the rRNA operon (*rrn*) was more conserved than the rest of the genome (Doi and Igarashi 1965; Dubnau *et al.* 1965). The genes in this operon encode for the structural RNA moieties that are the backbone of the ribosome. The *rrn* operon produces a single-precursor RNA molecule that undergoes maturation to generate the three different-sized bacterial rRNA molecules (16S, 23S, and 5S, see Figure 9.1). Additionally, unlike most operons in bacterial cells which only have one copy, the *rrn* operon can have one or many copies, up to 15 in some cases (Andersson *et al.* 1995; Rainey *et al.* 1996). The copy number for the bacterial rRNA operon varies by species and has been linked to different ecological strategies (Klappenbach, Dunbar, and Schmidt 2000). Studies using rRNA:DNA hybridization were able to demonstrate the discriminatory power of the region when it was determined that there were actually five genera that comprised the "Pseudomonads" (Palleroni *et al.* 1973).

Fox and colleagues proposed the usefulness of the 16S small subunit of the rRNA operon and a sequence-based analysis for the phylogenetic marker. They stated that primary structures could be aligned and changes at positions with common ancestry could be seen. The number and character of the positional differences could be a basis for inference of relationships (Fox, Pechman, and Woese 1977). The pioneering work by Woese and Fox using the conserved small subunit rRNA gene demonstrated that there were three domains among living things, not just the tradition two, the Archaea, Bacteria, and Eukarya (1977). Further work 20 years later elucidated that these three domains evolved by different pathways from a common ancestor (Woese, Kandler, and Wheelis 1990). Woese *et al.* promoted the use of the 16S rRNA gene as a molecular chronometer (1985). Research in the following years built on that concept, concluding that the absolute rate of change is not known, but evolutionary distances and relatedness of organisms can be tracked (Kimura 1980; Pace 1997; Thorne, Kishino, and Painter 1998).

In the 1970s, molecular and genetic technologies started influencing the taxonomy of bacteria. Molecular traits were used to separate phenotypically

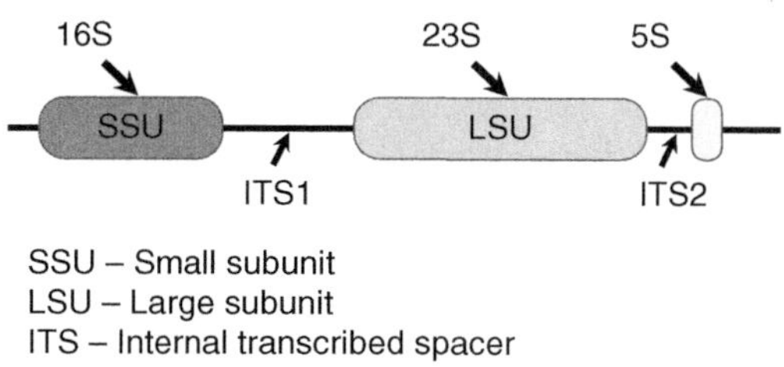

Figure 9.1 The rRNA region for bacteria.

similar, but genetically distinct groups of organisms. With the development of more complex molecular methods, new standards were beginning to emerge such as DNA–DNA hybridization, the concept of a genomospecies, and sequencing of the 1500 base pair (bp) 16S rDNA, and it became clear that the molecular data were more discriminatory than biochemical. Still, into the 1980s, taxa were being divided based on their Gram reaction and their medical significance.

9.2.3 Setting a New Starting Point

In 1980, IJSEM published a list of approved bacterial names which contained all the bacterial names having standing in nomenclature and set a new starting point (Skerman, McGowan, and Sneath 1980). From the 1980s until today, the availability of 16S rDNA sequence data has demonstrated the utility of this method for the phylogenetic placement of bacteria. The 16S rRNA gene is a strong phylogenetic marker since it is functionally conserved, thus the sequence changes are accurate measures of time and evolution. It is present in almost all bacteria often as multigene operons, and the 1500 bp is large enough in size with sufficient polymorphisms to have sufficient informational content for bioinformatics and to give distinguishing and statistically valid measurements. The gene has evolutionarily conserved regions and variable elements; the combination of conserved and variable domains allows the gene to classify organisms from the domain to the species level. The base positions 60–110 (based on *Escherichia coli* numbering) are some of the most discriminatory for closely related organisms as they contain a high level of variability (Ludwig and Klenk 2005). Since there are functional constraints in the structural molecule, changes are not free and linear. Therefore, there is no correlation to time (Olsen 1987; Woese 1987; Olsen and Woese 1989, 1993; Ludwig and Schleifer 1994; Olsen, Woese, and Overbeek 1994; Ludwig *et al.* 1998; Patel 2001; Clarridge 2004).

Analysis of the 16S rRNA gene sequence was revolutionary. The data are unambiguous and reproducible both within and between laboratories which allowed for comparisons between datasets. In fact, sequencing techniques have changed the definition of microorganisms. With the introduction of a sequence-based phylogeny, the inter- and intra-relatedness of species could now be determined objectively, but guidelines were needed in order to provide stability, reproducibility, and transparency with molecular techniques for taxonomy (Stackebrandt and Goebel 1994). Analyses of the 16S rDNA sequence-based phylogenies have the robust capacity to show taxonomic relationships between organisms not seen with any other method. Studies of sequence diversity within a “species” continues to reveal multiple sequence clusters that are ecologically distinct (e.g. Cohan 2002) and phylogenetic evaluation suggests that the taxonomy for certain groups of organisms is not appropriate, for example, organisms in the *Erwinia* and *Pantoea* genera (Zhang and Qiu 2015; Palmer *et al.* 2017). Many times, the phylogeny has little correlation to the phenotypic or phenetic grouping that was core to the initial characterization.

9.2.4 The Contemporary View of Microbial Taxonomy

The contemporary view of microbial taxonomy is determined mainly by the availability, applicability, and resolving powers of the methods used to elucidate their phylogenetic relationship. Bacterial taxonomy now reflects a phylogenetic classification based on the 16S rRNA gene sequence as visualized by dendrograms or trees whose structure is based on different models of evolution (Figure 9.2). Fungal taxonomy also reflects a phylogenetic classification as visualized by dendrograms, but uses the ITS rRNA region (see Section 9.3). The primary sequence data is processed using algorithms and the branching pattern in the resulting tree indicates the path of evolution. There are different models that reflect different aspects of evolutionary processes such as base frequencies and substitution types. These dendrograms can reflect evolutionary distances, maximum parsimony, or maximum likelihood.

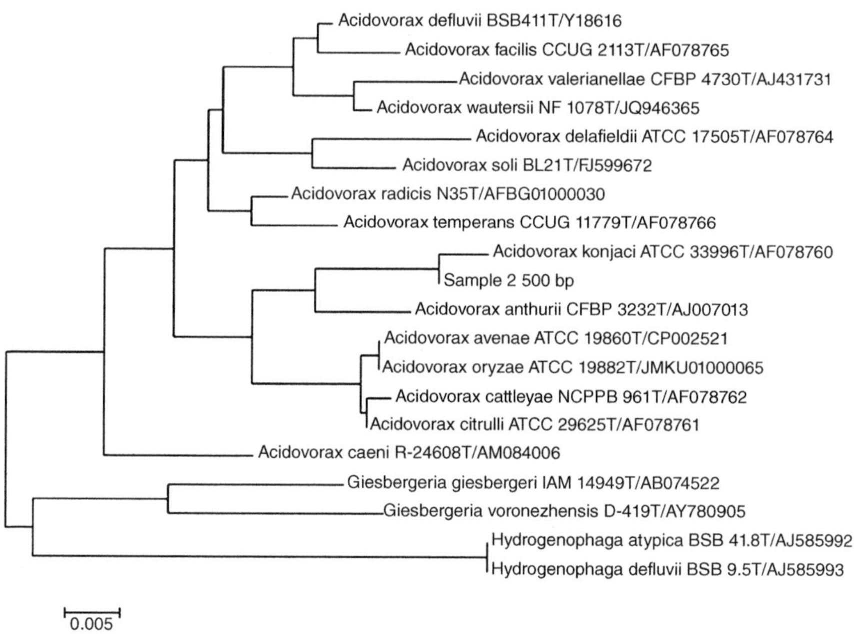

Figure 9.2 Example dendrogram or tree showing phylogenetic relationships. Neighbor-joining phylogenetic tree based on the first 500 bp of the 16S rRNA gene sequences. Bar, 0.005 substitutions per nucleotide position.

Bootstrapping values can be displayed on the dendrograms. The process of bootstrapping entails randomly sampling or deleting columns in an alignment or distance matrix until 100–1 000 datasets are created. These datasets are used to compile a consensus tree and bootstrap values are assigned to the branches. The value provides an evaluation of confidence for each branch of an observed tree (Efron, Halloran, and Holmes 1996).

Distance Matrix models are based on what is expected to give rise to differences between present-day sequences by comparing them, counting differences, and applying corrections for superimposed mutations (Olsen and Woese 1993). The evolutionary distance values of the Neighbor Joining tree are calculated by evaluating and transforming the measured base differences. The real number of changes can be underestimated by just looking at the difference seen in the present sequences. Thus, the differences are transformed to distances using models such as Jukes and Cantor (1969). In this model, the distance is considered, but not the character of the change.

Maximum Parsimony models assume that preservation is more likely than change and these models strive to minimize the total tree length and do not calculate branch lengths.

Maximum Likelihood models are the most sophisticated and the topology of the tree is optimal if it reflects a path of evolution that most likely resulted in the current sequences. The analysis is more complex and needs more computing time than either Neighbor Joining Distance or Maximum Parsimony.

There are errors in every inferred phylogenetic tree (Olsen and Woese 1993). The best tree for a given molecule is not necessarily correct. It is the result of a finite body of data used for the inference. Researchers should expect some level of disagreement when comparing different molecular phylogenies. Only when there are significant disagreements is it time to look at alternative explanations. Are there bad data, flawed analyses, lateral gene transfer, recombination, or gene divergence within a species? All interesting avenues for the academics to pursue.

9.2.5 Limits of Resolution

There is no phylogenetic standard for family, genus, or species demarcation. Stackebrandt and Goebel (1994) stated that ≤97% similarity in the full-length 16S rRNA genes between two bacterial isolates indicates that they are different species, but if there was a ≥97% similarity in the 16S rRNA genes then the alternative method of DNA relatedness must be used to determine if the isolates are different species. Many scientists criticize the use of DNA relatedness from a scientific and theoretical standpoint, to a methodological one (Istock *et al.* 1996; Vandamme *et al.* 1996). While other researchers state that although it has been demonstrated that the 16S rRNA gene sequence relatedness of <97% represents a new species, the meaning of a similarity score of >97% is unclear (Fox, Wisotzkey, and Jurtshuk 1992; Palys, Nakamura, and Cohan 1997; Petti 2007). Strains with >97% similarity could represent a new species or clustering of strains within a previously defined species (Janda and Abbott 2007). There is no universal agreement on a threshold value that constitutes a definitive species or species delineation (Stackebrandt and Goebel 1994; Harrington and On 1999; Vandamme *et al.* 2000). In fact, more recently ranges of 98.7–99.0% and 98.2–99.0% have been proposed as the species threshold (Stackebrandt and Ebers 2006; Meier-Kolthoff *et al.* 2013). Because bacterial genera do not evolve at the same speed, and many organisms have been classified as different species prior to the evaluation of the rRNA gene sequence, it is necessary to use different threshold values depending on the genus in question. The threshold value will vary by the taxonomic group as some species are too closely related, or there is little information of normal strain to strain variation within a species and inter-operon variation within a strain.

The 16S rRNA gene sequence may not be completely sufficient or reliable for understanding evolutionary phylogenies (Fox, Wisotzkey, and Jurtshuk 1992; Stackebrandt and Goebel 1994), but it is the best available scheme especially considering horizontal gene transfer (Janda and Abbott 2007) and has revolutionized bacterial taxonomy. Still, in 2002, there were new recommendations to the species definition as improved molecular methods, such as multi-locus sequence alignments (MLSA) and more recently core genome identity (CGI) comparisons, became available to systematists (Stackebrandt *et al.* 2002). MLSA uses phylogenetic analyses of different alleles, generally protein-coding gene targets, to evaluate different species and genera, and has been scientifically validated through the use of well-characterized reference strains (Gevers *et al.* 2005; Fraser *et al.* 2009; Tindall *et al.* 2010; Vandamme *et al.* 2013). Appropriate protein-coding gene targets are universally distributed, not transmitted horizontally, and have molecular evolution rates that are somewhat higher than the 16S rRNA gene, but comparable to it. For example, species having identical 16S rDNA sequences can be differentiated with other protein-coding gene sequences. The beta subunit of RNA polymerase, *rpoB*, is single copy, universal, and hypervariable allowing it to be more discriminatory than the 16S rRNA gene (Gundi *et al.* 2009). The DNA gyrase enzyme, which is a type II topoisomerase that negatively supercoils closed circular double-stranded DNA, comprises two subunits, GyrA and GyrB. The *gyrA* and *gyrB* genes can be used to differentiate closely related taxa as they have a much higher mutation rate than the 16S rRNA gene (Onodera and Sato 1999). The next generation of molecular methods for taxonomy are being built on the next generation of sequencing technologies and bioinformatics. Whole genome sequences of type strains are being generated and species analysis based on average nucleotide identity (ANI) and CGI are being performed. ANI and CGI are strengthening molecular taxonomy since much more information is used during the analysis.

9.2.6 Summary: The Whole Picture

Taxonomy, or classification, requires a description of the morphology, biochemistry, physiology, and genetics. Taxonomic classification is dynamic. New findings can lead to a change and reclassification of an organism. A change in the classification can lead to a new name and new criteria to identify an existing species or publication of a previously undescribed, new species. Not all "no identification" results are bad. Many times these results indicate the isolation of a new species or even a new genus. Taxonomic classification that is poorly defined hampers identification. The identification of a microorganism can be made on one or more characteristics that uniquely define a taxon. These characteristics can be based on phenotype (cell shape, Gram reaction, and fermentation characteristics) or nucleic acid sequence. The caveat is that the characteristic being used to determine the identity must be **unique**, that is, not found in any other group, and that is nearly impossible to do when basic cellular descriptors are used for the characterization.

9.3 History and Challenges of Fungal Taxonomy and Classification

Fungi are complex members of the domain Eukarya. They can be innocuous, toxic, or flavorful, and can be of concern in manufacturing facilities. Fungi cause damage to infrastructure by growing on fibrous materials such as drywall and pallets, and can cause spoilage in materials, ingredients, and products. Most critically, fungi can be pathogenic. Speciation of this diverse group of organisms is important, and in cases of product contaminants and investigations, strain-level differentiation can also be necessary. There are very few mycologists in industrial pharmaceutical manufacturing with identification expertise using classical mycology or morphological methods for environmental fungi, but there are many reference texts that focus on the descriptions of medically important fungi. Fungi can reproduce by asexual spore formation (anamorph stage) and by sexual reproduction (teleomorph stage), and produce a different type of spore. Spore formation is essential for morphological identification, thus if spores are not produced the organisms cannot be identified. The identification of filamentous fungi using morphological characters is very labor and time intensive and many times only results in a genus-level, or higher, confidence. Better methods are needed.

There are >100 000 described species of fungi and there are estimates that the undiscovered and unclassified species range from 1 to 10 million (Guarro, Gené, and Stchigel 1999). They can be unicellular, or form hyphal and mycelium masses, and produce fruiting bodies (Figure 9.3). Yeast are single-celled fungi that usually form a loose arrangement of budding cells. This growth form spans a wide range of unrelated fungi. Molds are filamentous fungi that have multicellular mycelium with a mass of branching hyphae which are a chain of filament-like cells. Mushrooms are filamentous fungi that form spore-bearing fruiting bodies. The classification and nomenclature of fungi is governed by the International Code of Botanical Nomenclature which was renamed to the International Code of Nomenclature (ICN) for algae, fungi, and plants. Historic descriptions based on morphology have confounded current taxonomy. Many fungi have more than one name, and many times this is due to the anamorph and the teleomorph being described or classified at different times, or due to the same organism looking different when grown on different media, and the connection between the different forms or morphologies had not been made (Hawksworth 2011). As fungal taxonomy and classification moves to a

(a) (b)

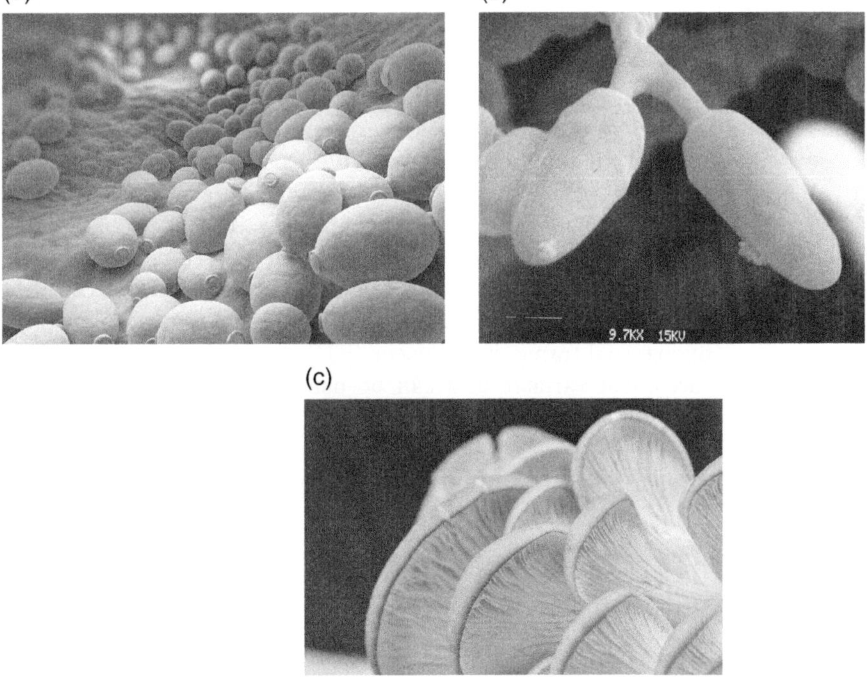

(c)

Figure 9.3 Fungi can be unicellular like (a) the yeast *Candida albicans*, form conidia at the tips of hyphae or mycelium masses like (b) *Culicinomyces clavosporus*, or produce complex fruiting bodies like the oyster mushroom (c), *Pleurotus ostreatus*.

phylogenetic approach, the issue is becoming more apparent (Taylor *et al.* 2000; Hawksworth 2006). The practice of assigning separate names for anamorphs and teleomorphs was discontinued on 1 January 2013 as dictated in the ICN Congress Melbourne Code of 2011.

9.3.1 Identification of Fungi Using the rRNA Region

Sequence data derived from the ribosomal RNA region and the analysis of phylogenetic trees are also used for taxonomic classification of fungi. The different regions of the rRNA operon have different levels of conservation and variability which allow them to be used for different levels of classification. The rRNA region for fungi consists of the 18S small subunit rRNA gene, the ITS region comprising the first and second internal transcribed spacer regions (ITS1 and ITS2) flanking the 5.8S rRNA gene, and the large subunit rRNA gene which ranges in size from 25 to 28S (Hibbett *et al.* 2007, see Figure 9.4). Generally, the 18S rDNA sequence is appropriate for the taxonomic domain and phylogenetic aspects and relationships at the class level, while the 28S rRNA gene is more variable and can be used to classify fungi from the phylum to the genus level. The evaluation of sequence information arising from the large subunit of the rRNA that is described as 26S in size is most often associated with the phylogeny of yeast (Kurtzman and Robnett 1998). The 5.8S gene does not contain much usable information, but the regions flanking this gene, the ITS1 and ITS2 regions, are less conserved. These regions are considered the optimal barcodes for fungal taxonomy (Petti 2007; Porter and Golding 2012), and are used to differentiate fungi to the species level (Mitchell and Zuccaro 2006). Although some organisms are better resolved by evaluating the sequence information in the D1 and D2 region of the 28S rRNA gene (Hall, Wohlfiel, and Roberts 2003, 2004), there are a lack of sequences from this region for effective analysis (Das *et al.* 2014). The regions outside the rRNA genes, called the external transcribed spacer (ETS) and the intergenic spacer (IGS) regions, are most variable and can be used for strain or population characterization.

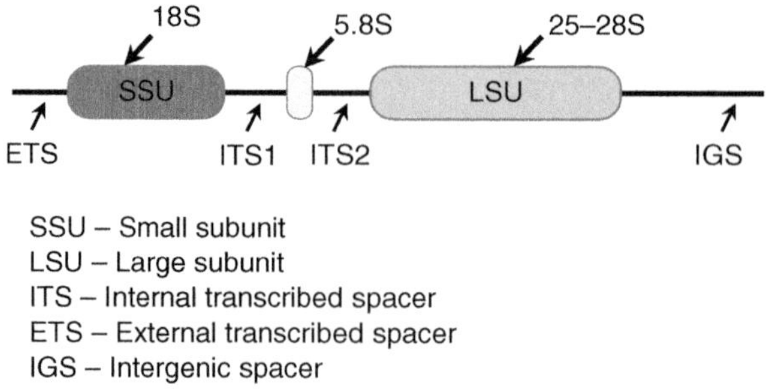

Figure 9.4 The rRNA region for fungi.

In addition to its use in phylogenetics and systematics, the ITS region is the barcode chosen for the identification of fungi (Vralstad 2011; Schoch *et al.* 2012). This applies to yeast, molds, and mushrooms. Using sequence-based techniques implies that the organisms do not have to be viable and more importantly, the classification is not dependent on the sporulation of the fungal isolate (Petti 2007). Like bacteria, sequence conservation in the rRNA region of fungi can impact species-level resolution and protein-coding gene targets, such as the β-tubulin gene, must be used for adequate discrimination for closely related groups (Huang, Lee, and Tai 2009). Also similar to bacteria, the phylogenetic analysis of sequence data from the rRNA region is the most accurate method for fungal identification.

9.4 Current Identification Technologies

Microorganisms from the environment that can be isolated, can be characterized by various culture-dependent classical techniques such as plating, staining, and biochemical tests. These tests can be time consuming and can be variable, but should be able to determine the identification to the genus level. The new protein-based and sequence-based methods decrease time to result, increase speciation, and decrease errors for microbial identification. This is especially true with fungi which are slow growing, and conventional methods are often ambiguous (Siqueira and Rocas 2005). However, the new protein-based and sequence-based methods increase instrumentation costs or assay complexity, or both.

Identification systems are based on different analytical methods. They can be manual or automated, or are a combination of both with simple or extensive sample preparation prior to analysis. Manual systems, or methods with extensive sample manipulations, depend of the skills of the technicians performing the assay. Automated systems allow for less direct sample contact, and thereby permit the technicians more time for other duties and allow for more standardization. Different systems have different strengths and weaknesses. The limitations can be inherent to the test method, technology, or the science. In all systems, the accuracy and performance are dependent on the database being used as a reference. If an organism is not in the database, it cannot be identified. The user must be aware of the version of the database on the instrument and be aware that the database may be outdated. They should fully understand the limitations of the identification platform and the reference database. Microbial identification systems that are purchased and used in a facility that adheres to cGMPs will need to plan and execute a validation of the system. This internal validation is generally planned and executed by the end user, but can have support from the vendor. The company's quality system and standard operating procedures should govern the validation plans and required documentation for both the specific system itself and its reference database. The systems can be categorized as phenotypic, proteotypic, or genotypic as described below and in Figure 9.5.

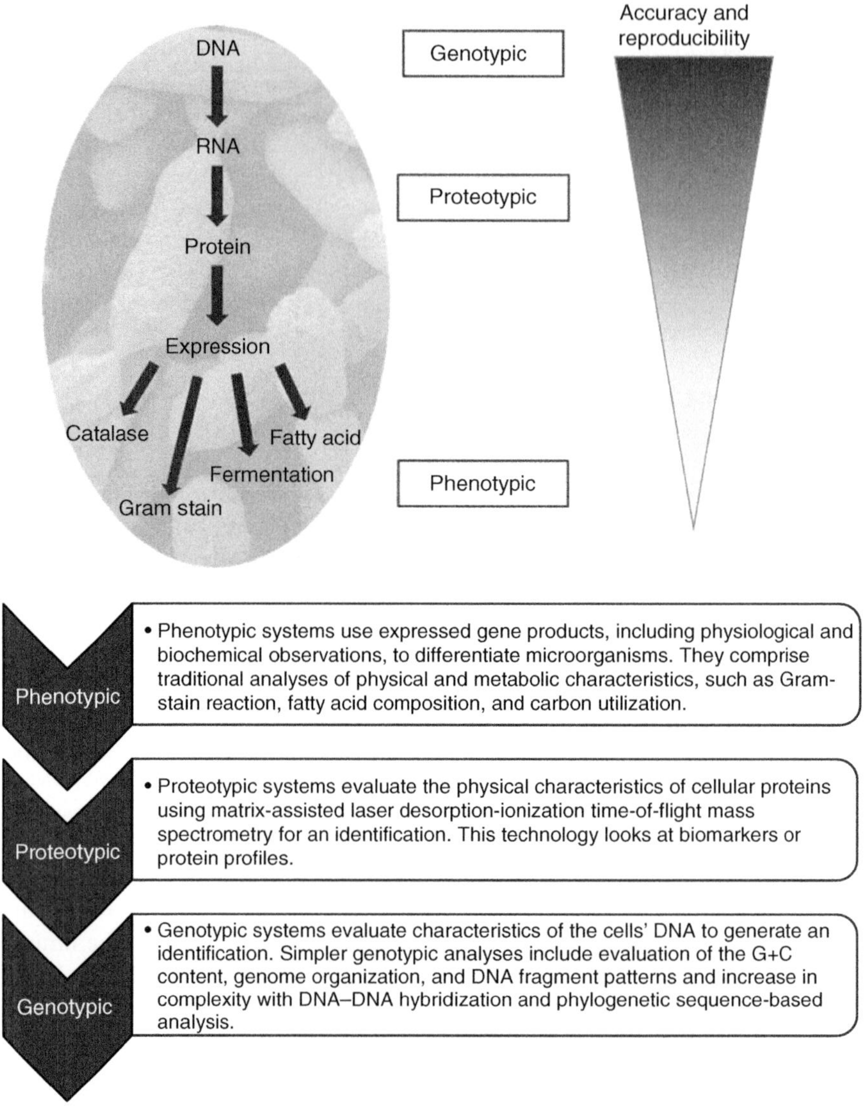

Figure 9.5 Summary of the three different systems to differentiate microorganisms: relative accuracy and reproducibility.

9.4.1 Considerations of the Different Systems

Genotypic identifications have been shown with convincing scientific data to be the most accurate; they supersede phenotypic systems (Drancourt *et al.* 2000; Fontana *et al.* 2005; Petti, Polage, and Schreckenberger 2005; Zbinden *et al.* 2007). Genotypic identifications are more discriminatory, robust, and reproducible as the classification is based on nucleic acid sequences which are highly conserved, the data less subjective than conventional practices, and the technology is not dependent on culture conditions (Clarridge 2004). For example, Drancourt *et al.* presented an analysis of the 16S rRNA gene sequences of a collection of environmental and clinical isolates that were previously unidentifiable. As shown in Figure 9.6, the sequence information indicated that 78.5% of the isolates had a close match to a described species, 11.4% represented a potential new species within a genus, and the remaining 10.2% may represent novel genera. These results could only have been determined through sequence-based identification, and not through conventional methodologies (Drancourt *et al.* 2000).

It was also shown that for isolates identified to the species level using sequence information (the 78.5% described above), the conventional identification produced an inaccurate result due to inappropriate biochemical profiles for 58.7%, inaccurate Gram-stain results for 11.6%, and incorrect catalase and oxidase activity for 3.6% of the samples (Drancourt *et al.* 2000). Genotypic identifications are the more informative, accurate, and reproducible.

That being said, there are varied levels of operational needs and these needs dictate different approaches to microbial identification and strain differentiation.

One technology will not be sufficient, as a technology must be suitable for the intended purpose, and the system chosen must be appropriate to the desired outcome.

The limitations and strengths of the different technologies must be understood in order to choose the most appropriate method for the desired level of resolution. Closely related species may be difficult or impossible to distinguish using certain systems. In some cases, the most appropriate methods for speciation, for example, can vary based on the species under examination.

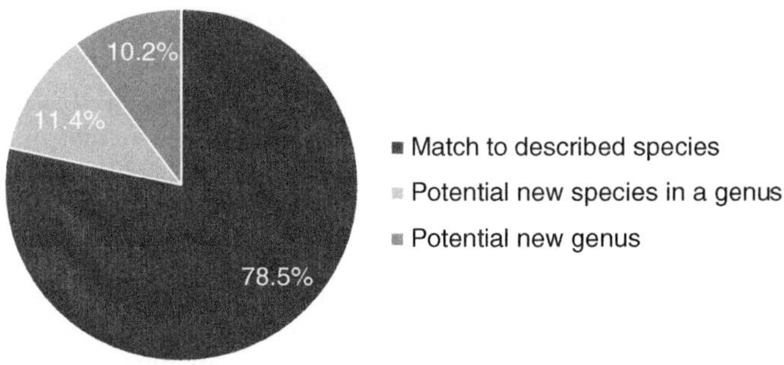

Figure 9.6 The identification rate of previously unidentified isolates using 16S rDNA sequence analysis. Source: Data from Drancourt *et al.* (2000).

Many references propose a polyphasic approach, especially for clinical samples, where the methods are run simultaneously. All the genotypic, phenotypic, and phylogenetic information is generated for a microbial taxon at one time, but this may not be feasible in resource-limited environments. Polyphasic taxonomy also does not make sense for routine identifications, yet in many cases conventional techniques describing physical and biochemical traits are not sufficient (Vandamme *et al.* 1996; Gillis *et al.* 2005; Prakash *et al.* 2007; Cleenwerck and De Vos 2008). It is difficult for a single technology to meet the needs of all situations. Thus, having access to methods with overlapping resolution allows for greater accuracy when there are clear expectations for each technology and the understanding that every technology is dependent on the robustness of its database for the correct identification. Whether using a phenotypic, proteotypic, or genotypic system, accuracy is linked to the quality of the underlying data and the associated taxonomic descriptor for those data. Phenotypic systems must have accurate and complete morphological and biochemical descriptors of the type or typical reference strains. Proteotypic systems must have high-quality reference spectra while genotypic systems are dependent on the quality and accuracy of the reference sequences. Additionally, the appropriate names must be associated with the reference biochemical, spectral, or sequence data.

9.4.2 Challenges with Nomenclature and Reference Data

There are many reasons why the names associated with reference data are not accurate. Sequences in genotypic databases could have been taken from noncurated public databases with incorrect information, or the type strain for a species in the culture collections was named by biochemical means and is incorrect (Coyle *et al.* 1993). There are many challenges with nomenclature that were assigned to organisms before taxonomy was determined by rDNA sequencing. There are many species that are "misplaced"; they are found within various taxonomic groups. There is an initiative to sequence these "orphan" species to help determine where they belong based on phylogenetic analyses. Researchers are identifying type strains with no associated 16S rDNA information, or type strains with poor-quality sequence information, and generating new sequence data for those strains. Upon analysis, they are finding multiple species with unexpected affiliations. That is, they are misclassified (Yarza *et al.* 2013). Curation of the reference databases must also occur to reflect current taxonomic classifications to produce the most accurate identifications. It should be noted that identification rates and accuracy in published studies, and internal qualifications, are snapshots in time and are highly dependent on the library being used as the reference, the taxonomic distribution of the samples themselves, clinical or environmental, and in some cases with the sample preparation method itself. This is particularly true with the publication of the performance of the matrix-assisted laser desorption-ionization time-of-flight (MALDI-TOF) systems over time. Consistent methods for sample preparation have since been developed, so results are easier to compare from lab to lab, but the impact of differential depth and breadth of library coverage remains.

9.4.3 Considerations for Using an In-house System or a Contract Testing Lab

In order to choose an identification platform for each isolate, the limitations, breadth of library coverage, and ability to identify different organisms – bacteria, yeast, and mold – must be understood for each system. The required level of resolution must also be understood since the system chosen will vary if a taxonomic descriptor (genus and species name) is needed or a strain-level comparison is required. An additional consideration is whether these different technologies and expertise will be obtained and run in-house or if the work is outsourced to a contract testing laboratory to take advantage of a wider breadth of experience and systems. Consideration for obtaining and running a system in-house includes the cost and effort to validate a system and the reference database. The validation of a new identification tool involves determining its taxonomic resolution using reference strains followed by testing with new strains or isolates. The evaluation should also consider the costs to run and maintain the system, cost of the reagents and consumables, the system throughput, the technical ability required to perform the assay, time to result, system resolution, accuracy, and ease of accessing, interpreting, integrating, and tracking the data (Stager and Davis 1992; Janda and Abbott 2007). Selection of a system to use will depend on particular needs such as number of tests per week, time needed to result, cost of consumables, experience and education required of the technician to perform the test, amount of labor for the test, the database depth and breadth in support of the test, manufacturing niche, impact of the organisms on the product or people, and criticality of the sample to the process. The best system for the type of sample or the identification needed must be chosen. Procedures should be in place to clearly dictate the path to be followed and systems used for an identification during the different processes. The appropriateness of each test method should be specified through qualifications, internal data, or published literature, and when the test method should be used needs to be defined (USP 2012b). USP Chapter <1113> requires verification of the accuracy of the method and indicates that there is a hierarchy to the impact of the identification errors with a misidentification to the genus level, the greatest impact, followed by a misidentification to the species level and then no identification having the lowest impact (USP 2012b). Performing appropriate testing and obtaining an accurate identification is challenging. Understanding how the methods work can be achieved, and that can help ensure that work is being performed within specified parameters or constraints. The method must be appropriate for the desired results. Many new technologies require an in-depth understanding of chemistry and molecular biology, but they can be assimilated.

9.4.4 Phenotypic Systems

9.4.4.1 Biochemical Approaches

Phenotypic systems use physical attributes of the cells to generate an identification, specifically the growth of the cells in differential and selective media. These systems are based on simple growth-based assays looking at colorimetric changes and results of carbon utilization. Traditional phenotypic identifications can be difficult, time consuming, and subjective (Stager and Davis 1992). Strains within species can show phenotypic variability and can generate atypical results leading to poor identifications.

The manual systems are very basic with a limited number of tests which results in a limited resolution, but do provide additional information over the basic characterization. These tests were first introduced in the 1970s as strips of differential media for identification of broad types of bacteria and yeast, and represented the first step in automation and standardization of the growth assays which were previously done in traditional culture tubes. The identification is determined by interpreting the results of color changes and reactions in the differential media. The manual strips are generally not used as a primary identification system since they have low throughput and are labor intensive to read the results and determine the identification code. The subjectivity of the manual interpretation and the limited reference library effect the accuracy and reproducibility of the result.

Slightly improved performance is found with the automated versions of the phenotypic identification systems first introduced in the late 1970s. One system evaluates the growth of the bacteria or yeast on a specific card, selected after determination of the Gram reaction and the results of additional tests, by the automated incubation and reading of its reaction with various media. The method is simple to perform. Cells are resuspended in saline to the correct optical density as dictated by the card. The card is loaded with the sample with the aid of a vacuum and sealed. After placement into the system carousel, the cards are incubated and read at short intervals of time using different wavelengths of light to determine the percent transmittance. This proceeds until threshold values are reached. The results are compared to reference data to determine the percent probability and confidence of the match, which ranges from excellent to no match. As the results' interpretation is automated, the test is less subjective than manual review, but the library in support of this system is very limited.

Many of the systems require the determination of the Gram-staining reaction of the cell and additional ancillary tests such as the results of catalase and oxidase tests to proceed with the correct suite of phenotypic tests. If these tests are performed or interpreted incorrectly, this erroneous result, which is most often not detected, affects all downstream analyses and will result in an incorrect identification. The Gram stain is an essential first step, and is itself a subjective technique. The result is affected by the physiological state of the cell and timing nuances of the method. The Gram-staining method often results in false positives and false negatives and caution needs to be had in the consideration of species that are known to be Gram variable in their staining reaction. Also, if bacteria lack a cell wall, like mycoplasmas, they will still produce a Gram-negative reaction. Results will vary based on changes in cell wall thickness, and on the presence of capsules and envelopes. Most frequently, the errors made in phenotypic identifications can be linked back to the determination of cell shape and the Gram reaction.

An additional commercial system has taken a similar phenotypic approach to identifications of bacteria, yeast, and filamentous fungi. However, in this case metabolism is indicated through a redox indicator dye (Miller and Rhoden 1991). This system was dependent on the Gram-staining reaction initially, but later versions which are also more automated, have become independent of this requirement. The wells in this platform contain tetrazolium violet, buffered nutrient medium, and a different carbon source. The system determines the ability of an organism to metabolize a specific carbon source. The tetrazolium violet salt is reduced by an electron donation from NADH during metabolism (Tachon *et al.* 2009). The salts are colorless, however, the reduced formazans are purple. When a carbon source is not utilized, the well remains colorless. Thus, a metabolic fingerprint is generated and probability and similarity indices are calculated (Stager and Davis 1992). However, not all samples produce identification results due to database limitations.

9.4.4.2 Fatty Acid Approach

In the 1980s, researchers determined that differences in fatty acid chain length were a good taxonomic marker. Fatty acid composition shows conservation and that significant change occurs over a long period of time. More than 300 fatty acids can be detected with double bond positions, functional group bindings, and chain length differences characterized. Fatty acids could be extracted from bacteria and yeast, and methylated for a qualitative analysis on a gas chromatograph (Dawyndt *et al.* 2006). This identification system was not dependent on a Gram-stain result. Sample preparation is more extensive for this method than the other phenotypic methods. The cells must be grown under specific conditions and harvested in the log phase of growth. The cells are resuspended, saponified by heating in the presence of sodium hydroxide and methanol, then methylated, and extracted in an organic phase prior to placement in the gas chromatograph (Stager and Davis 1992). The fatty acid profile is reproducible if the growth conditions are controlled carefully. Identifications are made based on a similarity index which indicates how close an unknown's fatty acid profile compares with an average profile of strains in the reference library. The similarity index is a measure of the relative distance of the unknown to the population mean.

9.4.4.3 Summary of Phenotypic Identification Systems

Phenotypic identification systems are hampered as cells being tested do not react in a typical manner for their species. They are affected by their origin and growth conditions. Organisms recovered from manufacturing facilities are stressed as they have been exposed to pharmaceutical products, nutrient-poor environments, and potentially to chemicals used for disinfection and cleaning. They are then being analyzed on growth-based systems and the results are being compared to reference profiles from primarily clinical isolates. Stressed cells will not express the proteins needed to get an accurate and reproducible result. In fact, it has been shown that the similarity or probability numbers produced by some phenotypic systems do not correlate to the accuracy of the result. One study of a commonly used system states that *The most important aspect of this study is that even excellent identification by the ... colorimetric card assay allows no prediction of the correctness of the results* (Zbinden *et al.* 2007) (Table 9.2).

Table 9.2 Different identification systems have different strengths and weaknesses that are related to the technology and/or the reference database.

Pros and cons of phenotypic systems
Measuring parameters that cannot reflect the complexity of an organism
Measuring parameters that are not consistently expressed
Outdated and limited reference databases lead to incorrect or no identifications
Must ensure sample analyzed is pure or a blended phenotypic description will result
Low accuracy and reproducibility, especially when dependent on the Gram stain determination
Simple to perform, but time consuming
Subjective analysis
Genus-level confidence

9.4.5 Proteotypic Systems

9.4.5.1 History of MALDI-TOF

MALDI-TOF mass spectrometry (MS) generates identifications based on physical characteristics or profiles that are based on stable cellular proteins. It is now shown to be a reliable method of protein-based identification of intact bacteria and yeast (Sauer *et al.* 2008; Marklein *et al.* 2009; Bizzini and Greub 2010; Carbonnelle *et al.* 2011; Rosenvinge *et al.* 2013). The use of MALDI-TOF for the identification of mold or filamentous fungi is complex and will be discussed in more detail in Section 9.4.5.4.

In order to use MALDI-TOF for the identification of intact cells, significant technological and methodological developments were needed in the field of MS. Early attempts at using MS for the analysis of microbes was limited by data comprising very small mass ranges. The development of MALDI-TOF MS was advanced in the late 1980s with profound expansions in the ionization of higher molecular weight molecules. These discoveries included the use of a UV laser and an organic matrix to control energy and produce a soft desorption resulting in ions over 10 000 Da (Karas *et al.* 1987; Karas and Hillenkamp 1988). Further work expanded on these concepts and described the development of a laser ionization TOF mass spectrometer, and the use of an ultrafine metal powder to enhance heating for molecular ion formation. The novel sample preparation method and laser ionization system resulted in the production of molecular ions up to 100 000 mass to charge (m/z) (Tanaka *et al.* 1988). This work lead to a Nobel Prize in Chemistry in 2002, and enabled rapid analysis of protein macromolecules with the application of a crystalizing, light-absorbing matrix compound to maintain the integrity of the molecules. The matrix crystals absorb energy from a nitrogen laser, resulting in desorption and soft ionization of large intact biomolecules (Tanaka *et al.* 1988). The soft ionization is critical in ion formation. There is the addition or loss of one proton, but no loss in sample integrity (Everley *et al.* 2008).

In 1996, three publications described the analysis of intact or whole cells. Claydon *et al.* used MALDI-TOF to identify intact Gram-negative and Gram-positive microorganisms directly from single colonies in culture. They demonstrated that the procedure provided a unique mass spectral fingerprint from the components of the cell (Claydon *et al.* 1996). Additional work demonstrated successful, and rapid, species-specific identification of bacteria by comparison to reference spectra or with the co-analysis of known strains (Holland *et al.* 1996; Krishnamurthy and Ross 1996). Whole-cell MS could produce mass spectra of total cellular components by analyzing cells without laborious extraction procedures (Claydon *et al.* 1996; Holland *et al.* 1996; Krishnamurthy and Ross 1996).

After demonstration of the ability of MALDI-TOF to produce mass spectra with very little sample preparation, studies addressed the concerns over stability and consistency of the mass spectral patterns. Valentine *et al.* demonstrated that the identification was independent from culture conditions using three species grown on four different media. This is significantly different from phenotypic methods where the culture conditions can affect the physiology and protein expression profiles of a cell (Valentine *et al.* 2005). Small variations in growth conditions has little to no effect on identifications indicating an increased reliability in the results (Welker and Moore 2011). However, the growth phase does affect the outcome and data need to be collected during the active or log growth phase of the organisms (Valentine *et al.* 2005; Wunschel *et al.* 2005a, 2005b). With the advancements in the field of genomics, the large percentage of mass spectral peaks in whole-cell analysis were determined to comprise ribosomal proteins (Teramoto *et al.* 2007; Welker and Moore 2011). This observation explained the stability of the spectra under varying culture conditions as ribosomal proteins are synthesized under all growth conditions, and are the most abundant cellular proteins representing approximately 60–70% of the dry weight of a microbial cell (Ryzhov and Fenselau 2001). Approximately 50 individual ribosomal proteins were identified in the mass range of 3 000–20 000 Da while other proteins in this range were shown to be housekeeping or structural proteins which are also present in the cell under all conditions. These proteins are constitutively expressed as part of cellular structure and function. They include carbon regulators and DNA-binding proteins (Dieckmann *et al.* 2008). The proteins detected in the MALDI-TOF spectra are conserved, ubiquitous, and abundant. These characteristics lead to stable mass spectral patterns, and thus more reproducible identifications. The more similar the patterns, the more likely the identification.

9.4.5.2 MALDI-TOF Used for Microbial Identification Today

Over time, the technology has become commercialized as databases of clinically relevant taxa were built to allow for reliable identification. An essential prerequisite for accurate identification is the inclusion of a reliable, scientifically validated, and comprehensive reference database (Rahi, Prakash, and Shouche 2016). Species show intraspecific variation with respect to mass spectral patterns, like they do with both sequence and biochemical properties. Thus, species need to be represented by reference spectra from multiple strains to cover the natural diversity within the species (Lartigue *et al.* 2009). There are two primary commercially available instruments. The identification systems differ in their databases, identification algorithms, and the numerical rankings are reported differently. However, they function similarly.

The identification method in its simplest form is to have cultures on solid media, place a small amount of cells (10^5–10^6) on a target plate, and extract the cells on the target with a matrix solution. Some genera require an extra, yet still simple, extraction step prior to placing the supernatant on the target plate (Alatoom *et al.* 2011). The matrix solution contains an organic solvent and strong acids. The solvent lyses the cells and makes the proteins accessible for analysis. The proteins and matrix co-crystalize and once in the MALDI instrument, the laser bombards the sample, proteins undergo desorption and soft ionization. The matrix's acidic components provide the positive charge to the proteins and positively charged molecules are accelerated up a flight tube as MALDI-TOF systems are run in positive ion mode for protein-based microbial identification (Horneffer *et al.* 2001). The ratio of mass to charge is measured by a time-of-flight analyzer and proteins in the mass range of 2 000–20 000 Da are measured. Spectral data are collected and software automatically performs peak assignments and collects the necessary data for analysis and evaluation of quality parameters. These data can include the presence and absence of peaks, the peak height, and the signal-to-noise ratio. Although the highly abundant proteins comprising the mass spectrum are not individually characterized, they produce a profile unique to types of microorganisms. The mass fingerprint is compared by the software to reference entries for an identification based on a list of closely related organisms with numerical scores or confidence values. The highest score, above predefined acceptance thresholds, is the identification match. The species-level threshold is an empirical value, determined by comparisons of known strains with genotypic reference classifications. Different algorithms for analyzing mass data generally lead to similar results which underlies the robustness of the MALDI-TOF method (De Bruyne *et al.* 2011). There is spectral variation that can be seen after running duplicate samples of individual strains, but even with that, whole-cell MS with the MALDI-TOF is very accurate and reproducible for species-level differentiation.

9.4.5.3 Performance

Comparison of identifications made by the MALDI-TOF to phenotypic systems have demonstrated that the MALDI is very fast and more accurate for the identification of clinical organisms (Seng *et al.* 2009; Cherkaoui *et al.* 2010; Van Veen, Claas, and Kuijper 2010; Bessede *et al.* 2011; Bizzini *et al.* 2011; Carbonnelle *et al.* 2012). When there were discordant results between the MS result and the phenotypic system, the 16S rRNA gene sequence most often confirmed the MALDI result (Benagli *et al.* 2011). Benagli *et al.* also note that a reliable identification is only possible if the reference database contains the relevant strains and that these strains have been characterized by sequencing the informative gene regions. Organisms are either correctly identified or yield a low score or match, resulting in no identification. This no identification can be due to no match in the database or because of poor sample preparation. Incorrect identifications are rarer than with phenotypic systems and usually occur with closely related organisms (Bessede *et al.* 2011; De Bel *et al.* 2011; Neville *et al.* 2011; Dubois *et al.* 2012). Resolution of certain taxonomic groups of closely related species complexes is still a challenge for MALDI, and most of the time that is also the case with the genotypic resolution for that group, examples include the *Bacillus cereus* group, the *Burkholderia cepacia* complex, *E. coli*/*Shigella*, and the *Enterobacter cloacae* complex (ECC) (Pavlovic *et al.* 2012; Khot and Fisher 2013; Almuzara *et al.* 2015). New or alternative analyses of the data can provide increased resolution in some groups of organisms (Sato *et al.* 2011; Shah *et al.* 2011; Khot and Fisher 2013) (Table 9.3).

9.4.5.4 The Use of MALDI-TOF for the Identification of Filamentous Fungi

As discussed earlier, the identification of filamentous fungi can be done by macro and microscopic assessment by trained mycologists. Microscopy evaluates the sexual and asexual spores' color, shape, and surface and also evaluates the branching of the spore carrying structures, their color, and growth rates. These methods are time consuming, due to extended incubation periods, and often require extensive experience for accurate interpretation. Molecular methods are highly accurate and reproducible, but also require well-equipped laboratory and highly trained personnel to perform the assay (see Section 9.4.6). With the adoption of MALDI-TOF MS methods in the clinical, environmental, and QC microbiology laboratories for the rapid and accurate identification of bacteria and yeast, evaluation turns to the identification of filamentous fungi by this method. The option to identify filamentous fungi with MALDI-TOF is available, although the emphasis in the literature, and in the reference databases, for the systems is on clinically relevant organisms. Thus, the identification of filamentous fungi with this technique is less advanced (Huang *et al.* 2017), and few filamentous fungi are represented in the commercial systems'

Table 9.3 Different identification systems have different strengths and weaknesses that are related to the technology and/or the reference database.

Pros and cons of proteotypic systems
Simple to use
Extremely fast time to result
Large capital expenditure
Very low reagent cost
Highly accurate and reproducible for species-level resolution
Data can indicate if culture is mixed
No need for a Gram stain determination
Resolution of certain closely related species complexes is a challenge
Commercial systems have limited databases
Performance can be improved by the creation and addition of reference spectra, but this can be difficult

databases. However, MALDI-TOF could allow for the fast identification of spore-producing and sterile molds, and also those with similar morphology.

Even though the identification process for filamentous fungi with the MALDI-TOF is similar to that for yeast and bacteria, the sample preparation methods are not as simple or straightforward which has hindered its widespread implementation as a routine identification method for molds. Filamentous fungi have a higher level of biological complexity than yeast and bacteria with the coexistence of different growth forms, and a more extensive and time-consuming sample preparation protocol is necessary due to the rigidity of their cell wall. Many sample preparation methods for fungal hyphae and spores have been investigated, and in general, the methods give good results and obtain a high level of accuracy (Hettick *et al.* 2008; Lau *et al.* 2013). Filamentous fungi do exhibit variable phenotypes under different growth conditions which result in variable protein spectra when analyzed. The media on which the sample is grown and the temperature of incubation have no effect, but age of culture, source of the mycelium, and pigment production do have an effect (Chalupova *et al.* 2014). Thus, there is a need to standardize culture conditions. Importantly, for successful identification, it is critical that the same method used to generate the reference spectra be used with the test samples. Attempts to simplify and standardize the sample growth and preparation methods, as well as expand on the appropriate reference spectral, have been published and have produced data supporting an 87–90% accuracy (Cassagne *et al.* 2011; Lau *et al.* 2013). However, consistent generation of spectra remains problematic. Studies on the identification of filamentous fungi have used different instruments, different commercial databases, custom-developed databases, and different extraction protocols, making it difficult to compare results. One recent study evaluated clinically relevant filamentous fungi and found that 25% of the tested isolates were not in the reference database (McMullen *et al.* 2016), and that number increases dramatically when environmental fungi are evaluated. The technology will fail to produce an identification if there is no reference spectrum. However, when species in the sample cohort are chosen based on their presence in the reference databases, the reportable rates and accuracy of the identifications are very high (100 and 99%, respectively) (Huang *et al.* 2017). To use MALDI-TOF for the identification of mold recovered from environmental sources, intensive library development will be needed to have a high reportable rate.

9.4.6 Genotypic Systems

9.4.6.1 Sequence-Based Identification Technologies

The history of the discovery of DNA and the determination of its structure can be found in any biology, microbiology, biochemistry, genetics, or molecular biology text book. DNA sequencing (1968) was first performed 15 years after the discovery of the double helix (1953) (Hutchison 2007). However, the targeted DNA cleavage chemical method of Maxam and Gilbert (1977) and the more well-known dideoxy chain-terminator method of Sanger, Nicklen, and Coulson (1977) were not published until the mid-1970s. Incremental improvements in Sanger sequencing have allowed for a significant increase in the amount and quality of the data, simplification of the chemistry, and a decrease in the time needed to generate the results. These improvements have impacted nearly all fields of biological and medical science. The acquisition of sequence information has led to an overwhelming amount of data that also required improvements in the computer systems used for analysis which in turn has created the field of bioinformatics.

Sanger sequencing is based on a dideoxy chain-termination during DNA synthesis. Synthesis is carried out in the presence of all four DNA nucleotides, with a proportion of each base modified and labeled. Originally, this was a radiolabel, ^{32}P, and four separate reactions were completed, but now, with the most often used systems, the modified bases carry one of four fluorescent tags. By using labeled nucleotides lacking a critical 3′ hydroxyl group, DNA polymerase adds a labeled base to a nascent chain, but cannot extend the strand further (Sanger, Nicklen, and Coulson 1977). A series of chain-terminated products are produced, each ending with a labeled base. The different-sized molecules are separated by electrophoresis and visualized. When radionuclides were used, this was accomplished by the use of slab gels and exposure to X-ray film, but with the application of fluorescent tags, this is now done in real time as the samples run through a capillary loaded with a polymer matrix. Fluorescently tagged DNA chains are observed as they pass a detector and the four dyes are distinguished by their wavelength. The sequence can be deduced from the order in which the four different dyes pass the detector.

Sequence-based identification technologies can be applied to bacteria and fungi and are the most robust as the methods are rapid, accurate, reproducible; not constrained by the organisms' growth characteristics (e.g. slow and anaerobic); nor affected by environmental factors. Further advantages for sequence-based approaches are that the raw data will indicate if the sample culture is pure, there is no requirement to perform a Gram stain, and the sample can be alive or dead, young or old. With the appropriate expertise, it can also be determined if a novel organism, that is perhaps a new species or even a new genus, has been recovered.

9.4.6.2 Sequence-Based Identification Technologies and Phylogenetic Analyses

There is one primary commercial system available for in-house use, but its libraries are limited. However, contract testing laboratories can also generate sequence information from the conserved ribosomal DNA regions of both bacteria (16S) and fungi (ITS and 28S) to determine an identification. The methods are more labor intensive than both phenotypic and proteotypic methods and require additional equipment in the laboratory to generate the data, such as PCR machines. The process uses standard molecular biology techniques to isolate total DNA from the bacteria or fungi, PCR amplify the target sequences, and to perform Sanger sequencing (Sanger, Nicklen, and Coulson 1977). The resulting DNA sequence is analyzed and an identification should be determined by a phylogenetic analysis. Combining DNA sequencing with phylogenetic analysis, and not just evaluating the sequence alignment and similarity results, permits the most accurate identification and also the determination of novel species. That is because the relatedness between organisms is better shown in a dendrogram than by percent similarity or difference alone (Clarridge 2004). As discussed earlier, different DNA targets document different phylogenetic lineages. The 16S rRNA gene is recognized as the most suitable target for bacterial classification and identification. The ITS1 and ITS2 regions in fungi have been shown to be the most useful markers for fungal classification, and thus should be used for identification (Leaw *et al.* 2006). The commercial sequence-based identification system uses the fungal 28S rRNA-based gene target (D2), which does not provide sufficient species resolution in many cases, and the fungal database is limited (Hall, Wohlfiel, and Roberts 2003, 2004). However, its availability in the market meant that fungal identifications could be generated in-house in less than one day using this sequence-based approach.

9.4.6.3 Considerations When Using a Sequence-Based Approach

The 16S rRNA gene and the ITS region are universal, so the relationships among all bacteria and fungi can be seen to the species and subspecies level. There is the occasional exception where more than one species has the same or similar sequence and thus cannot be resolved (Clarridge 2004). Sometimes sequencing the entire 1500 bp 16S rRNA gene is needed to distinguish particular bacterial taxa, but usually the first 500 bp is adequate to differentiate isolates. Clarridge compared 100 type strains from clinical isolates using the first 500 bp and full 1500 bp and found that the dendrograms were similar, but not identical. She concluded that evaluating the first 500 bp is sufficient for bacterial identification and that the first 500 bp may even show increased resolution between certain species since this region can show slightly more diversity than the remaining sequence (Clarridge 2004). An additional study of more than 200 isolates of interest to the pharmaceutical industry compared the phylogenetic trees which resulted from the sequences of the first 500 bp and the nearly full-length 1500 bp 16S rRNA gene (Farrance and Hong 2015). Again, the relationships were similar, but not identical. In this study, 93.7% of the samples resulted in the same identification regardless of which sample sequence, 500 or 1500 bp, was used for the initial comparison query. Of these, 78.8% resulted in the same exact match to a single species, 13.5% of the queries had the same outcome but neither the 500 bp nor the 1500 bp 16S rDNA sequence was able to result in a definitive species designation, that is the unknown matched two or more closely related species. The remaining 1.4% of the unknowns also had the same outcome, but were only able to be identified to the genus or order level. One percent of the samples resulted in a different species-level identification when evaluating the 500 vs. 1500 sequence, but it was thought to have been due to poor-quality reference sequences in the curated public database. Finally in this study, 5.3% of the samples had better resolution with the 1500 bp 16S rRNA gene sequences than with the first 500 bp alone. The nearly full-length gene was able to provide a definitive species designation, while the first 500 bp of the 16S rRNA gene sequence matched two or more closely related species (Farrance and Hong 2015). Considering these data, and the discussion of the limitations of using the 16S rRNA gene sequence in general for certain closely related species that have a high degree of similarity, targeting alternate regions with increased variability, such as a protein-coding gene, is appropriate.

As stated above, it is well understood that the phylogenetic analysis of the 16S or ITS gene regions is not always sufficient for species resolution and that sequencing information from other genetic markers such as housekeeping genes can be used to increase discrimination. Housekeeping genes are used as they evolve more slowly than other protein-coding genes, but variations do occur that are usually selectively neutral and occur at a higher frequency than in the more conserved ribosomal gene regions. These include housekeeping or protein-coding genes such as *gyrB*, *rpoB*, *recA*, *ppk1*, and *dnaJ* (Rowland, Aboshkiwa, and Coleman 1993; Karlin, Weinstock, and Brendel 1995; Mollet, Drancourt, and Raoult 1997; Onodera and Sato 1999; Shi, Rao, and Kornberg 2004; Alexandre *et al.* 2008). For example, the average base substitution in the 16S rRNA gene is 1% per every 50 million years, but is 0.7–0.8% per every 1 million years for *gyrB* (Chun and Bae 2000). If the 16S rRNA gene or ITS region are identical, more changes can be seen in protein-coding genes and the species can usually be differentiated. The phylogenetic analysis and identification of any bacteria or fungi are dependent on the selection of the suitable gene targets specific for each taxon. The resulting phylogenetic trees from the alternative gene targets reveal relationships at a level where 16S rDNA and ITS sequences are not discriminatory. This method can have superior resolution to the traditional ribosomal region sequence analysis for closely related species (Figure 9.7 and Table 9.4).

9.4.6.4 Cautions Using Public Databases for Sequence-Based Identifications

If a commercial sequence-based identification system is being utilized, sometimes an organism cannot be identified due to database discrepancies or limitations. Often, the unknown has a poor or no match in the commercial reference database. A standard practice is to compare the unknown's sequence

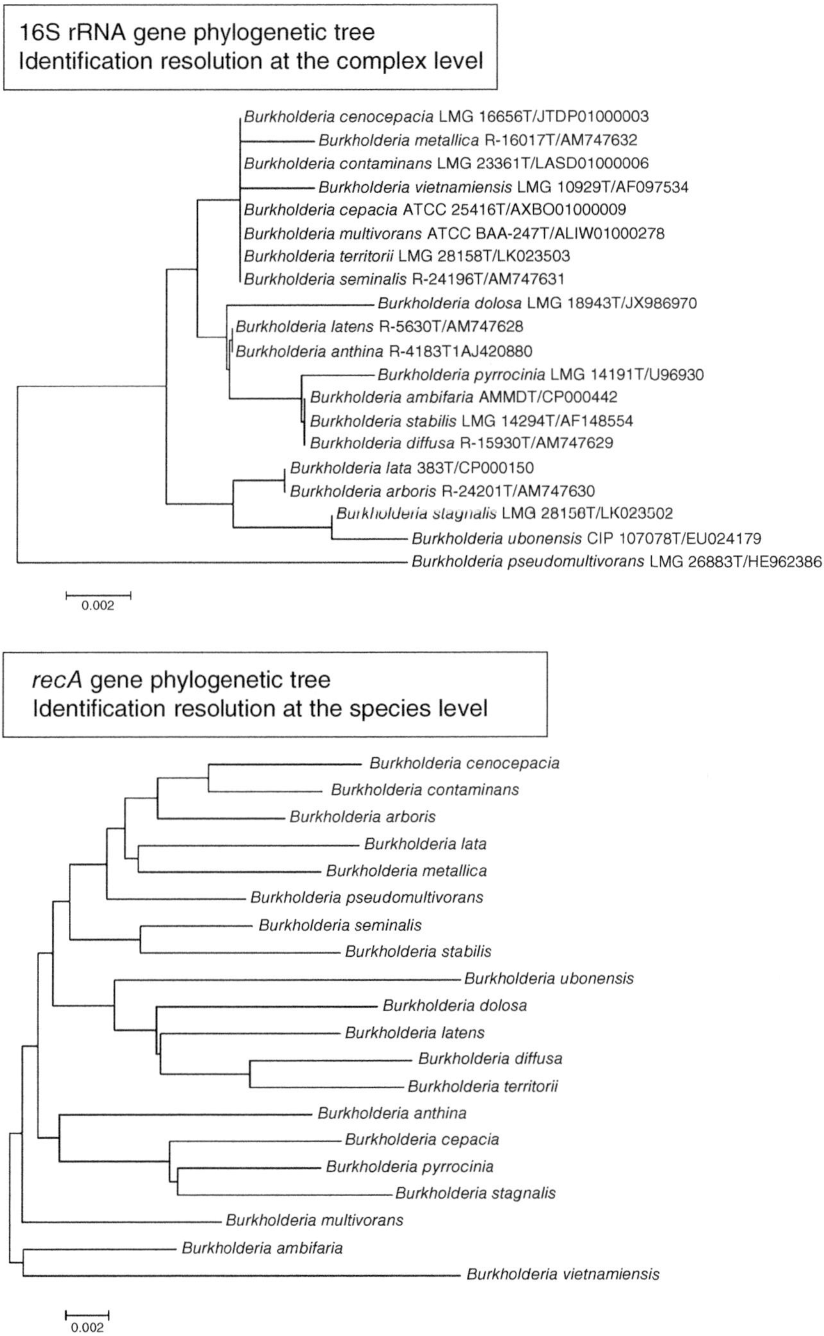

Figure 9.7 Resolving the *Burkholderia cepacia* complex (Bcc).

Table 9.4 Different identification systems have different strengths and weaknesses that are related to the technology and/or the reference database.

Pros and cons of sequence-based genotypic systems
High level of skill needed to perform
Short time to result
Large capital expenditure
High reagent cost
Produce data of highest specificity, reliability, and reproducibility for species and strain-level resolution
Data indicate if culture is mixed
No need for a Gram stain determination
Sequence-based methods independent of the growth stage of the organism and the media
Resolution of certain closely related species is a challenge when using conserved rRNA genes
Data interpretation requires a high level of knowledge
Commercial system has limited databases
Phylogenetic analyses indicate when species-level confidence is not attainable with the database
Reference libraries can be expanded with type strain sequences

to public databases or curated databases whose sequences were ultimately derived from a public database to generate a closer match. Caution should be taken with this approach from a scientific and compliance perspective. Sequence data generated today with the available equipment, reagents, and data analysis tools have the potential to be of high accuracy, but there is no guarantee (Clayton *et al.* 1995). Public databases contain many poor-quality sequences, thus the sequence quality needs to be evaluated before using the information. Conservative estimates from one study showed that at least 5% of 1399 sequences searched had substantial errors associated with them ranging from chimeras (64%) to sequencing errors or anomalies (35%) (Ashelford *et al.* 2005). In another study, at least 26% of 16S rRNA gene sequence pairs (two sequences deposited for the same species) in GenBank had 1% random sequencing errors and, of these, almost half had 2% random sequencing errors (Clayton *et al.* 1995). The presence of several classes of compromised sequences, such as chimeras and reverse complement sequences, seem to be increasing over time (Abarenkov *et al.* 2010). There are also misidentified sequences and other annotation issues in the public databases (Federhen 2015). Sequences of compromised technical quality or inaccurate taxonomic annotations are major contributors to incorrect sequence-based identifications when using sequences from public databases (Nilsson *et al.* 2012). Additionally, there are multiple copies of the rRNA gene regions (16S, ITS, 28S) within the bacterial and fungal genomes. Therefore, polymorphic bases are expected when generating data with Sanger sequencing methods. These polymorphisms can cause difficulties in generating and interpreting sequence data, especially when there are insertions or deletions in different copies of the rRNA operon. Yet, these polymorphisms are important and should be considered (Pace 1997; Sacchi *et al.* 2002). The entries in public databases are owned by the submitter, the quality of the data, and the assignment of the correct original taxonomy is dependent on them. Public databases are not curated, thus information derived from them must be validated. All identification technologies are dependent on a database of organisms for the correct identification. As stated above, there should be validation documents for the reference database, and there should be established procedures for controlling updates to the database, and documentation of such activities.

9.4.7 Next-Generation Sequencing Systems

The new "massively parallel" sequencing methods are greatly increasing sequencing capacity. These advances will continue to allow for new approaches to be taken for a variety of problems in biology, evolution, and the environment. These next-generation methods take a very different technological approach to generating sequence data versus the traditional Sanger dideoxy method. The common feature of these methods is that they are massively parallel, which means that the number of simultaneous sequence reads from a single experiment is tremendous as compared to capillary electrophoresis-based Sanger sequencers. At present, this very high throughput is achieved at the great expense of length of read and accuracy of the individual reads. However, because of the vast amounts of data and the assembly of the data, high overall accuracy can be achieved because of the high degree of sequence coverage.

Next-Generation Sequencing

- Massively parallel sequencing that creates megabases of information
- Shorter reads than the traditional Sanger method, in general
- Multiple platforms with same basic method, but different chemistries
  - Prepare sample "library" by creating fragments and adding adaptors
  - Amplify each library fragment
  - Sequence
  - Analyze the tremendous amount of data

The first commercially available massively parallel method was based on pyrosequencing (Nyren, Pettersson, and Uhlen 1993). An array of randomly sheared DNA is attached to linker sequences and placed in emulsion droplets. These templates are sequentially exposed to each of the four nucleotides, and the amount of incorporation is monitored by the detection of the luminescence released via pyrophosphate. The next generation of sequencing technologies have various strategies for sample and template preparation, sequencing, imaging, genome alignment, data analysis, and assembly methods. The unique combination of protocols is what separates one method from the others and determines what type of data are produced by each platform. Template preparation can occur by randomly fragmenting the target DNA, immobilizing the fragments to a solid surface, and clonally amplifying the templates using emulsion PCR or solid-phase amplification. The sequencing and imaging strategies can be done by cyclic reversible termination, single nucleotide addition, real-time sequencing, and sequencing by ligation. Imaging methods vary from measuring bioluminescence signal, proton-induced voltage changes, to four-color fluorescent imaging of single-molecule events. These technologies can also target specific regions for analysis using multiplex PCR, hybridization capturing, and microarrays to generate the templates (Metzker 2010; Liu *et al.* 2012). An alternative next-generation sequencing technology that does not require any labels on the DNA or the nucleotides is the nanopore approach. Nanopore technologies rely on the electronic or chemical signature of the different nucleotides for discrimination. The pores can be constructed from carbon nanotubes or even based on biological pore proteins. The pores are engineered to optimize the translocation rate and the detection of specific bases as the DNA passes through the pore.

Each generation of sequencing technologies is addressing the costs per run, sample throughput, accuracy, and read lengths. Each system has its more appropriate application based on its inherent strengths and weaknesses and each should be used for the biological problems to which they are most applicable. With the continued advancement of the next-generation sequencing platforms and the genome analysis software, the accumulation of data on a large scale has become faster and less expensive. Current applications for the next-generation sequencing platforms include genomics, epigenomics, and metagenomics. These applications are becoming well suited for different stages of pharmaceutical manufacturing including viral detection, cell line authentication, EM, microbiome analysis, and identifications using shotgun whole genome sequencing or targeted amplicon sequencing, or metagenomics approaches.

9.4.7.1 Whole Genome Sequencing

Whole genome sequencing and genomics applications can be used for large-scale alignment and comparative analysis with both bacteria and fungi. Shotgun sequencing permits the comprehensive analysis of all genes present in an organism. The analysis can be done with different software tools and statistical packages. The researcher still needs to select the phylogenetic marker genes that are relevant for particular groups of organisms while still using the maximum information available (Capella-Gutierrez, Kauff, and Gabaldon 2014). With the abundance of genome sequence data, target genes or allele sequences can simply be pulled out of the genome data and used for the analysis. Consideration of the presence or absence of gene families, conserved insertion or deletion on the large scale, differences in gene content, conservation of gene order, and biases of nucleotide composition of the genomes can also be done. With the increase in whole genome sequences, researchers have demonstrated that a comparison of ANI of the shared genes between two strains can be a robust way to compare genetic relatedness (Goris *et al.* 2007). There is now a shift from DNA–DNA hybridization to ANI in the publication of new species (Richter and Rossello-Mora 2009; Kim *et al.* 2014).

9.4.7.2 Metagenomics

Simply defined, metagenomics is the study of DNA or RNA directly obtained from samples from an environment – cultured or uncultured. Targeted amplicon metagenomics permits the analysis of all the 16S rRNA genes or the ITS gene regions present in a sample and the identification of species in a microbial population. All current, commonly used methods for microbial identification utilize culturable organisms. Metagenomics can utilize next-generation sequencing methods to generate a taxonomic profile of a microbial community by investigating microbial genomes obtained directly from environmental samples without cultivation and without prior knowledge of the constituent communities (Riesenfeld, Schloss, and Handelsman 2004). Next-generation sequencing of multiple hypervariable regions of the 16S rRNA gene or the ITS regions is ideal for a thorough characterization of low complexity metagenomes or for a broad genus-level overview of populations in highly complex communities or microbiomes.

As of 2018, these technologies are not in routine use for microbial identification in most industry laboratories and still lie more in the realm of research science, but commercial kits for the 16S and ITS rRNA amplicons are available for biome and other metagenomics applications. However, these tools have not been optimized for routine microbial identification or characterization for industrial microbiology laboratories. They remain significantly more resource and time intensive to perform as compared with the more traditional Sanger sequencing applications.

9.4.8 Other Spectroscopy or Spectrometry Methods for Identification and/or Strain Typing

9.4.8.1 FT-IR Spectroscopy

Traditionally used for identity confirmation of materials or chemicals, Fourier transform-infrared (FT-IR) spectroscopy is establishing itself as a method for the rapid differentiation and identification of microorganisms. Spectroscopy involves the observation of molecular vibrations, in this case, excited by an IR beam. Molecules absorb energy and start rotational movement. The absorption is correlated to a concentration of specific components and the spectrum reflects the composition of the sample. The IR spectrum is a biochemical fingerprint used to characterize a substance. IR spectroscopy has been used since the 1950 to analyze biomolecules such as proteins, nucleic acids, carbohydrates, and lipids (Beekes, Lasch, and Naumann 2007). An IR spectrum is produced by measuring the intensity of IR radiation, emitted from a heat source, both before and after it interacts with a sample. Absorption bands are seen due to the interaction of the light and absorption of the IR radiation by the biological sample. Fourier transformation is a mathematical means of extracting individual frequencies from the "interferogram" for final representation in an IR spectrum (Beekes, Lasch, and Naumann 2007). Spectra of complex biological samples, such as intact cells, represent the superposition of all vibrational modes of all the molecules in the sample and give a spectral fingerprint. This fingerprint is dependent on the physiological state of the cells. The method is simple; cells are suspended in solution then dried on a carrier, and the spectra generated (Figure 9.8). However, the analysis of the data is complex (Grunert *et al.* 2013). The fingerprint can be used as a comparative tool to look at strain-level differentiation. Additionally, like many other fingerprint-based systems,

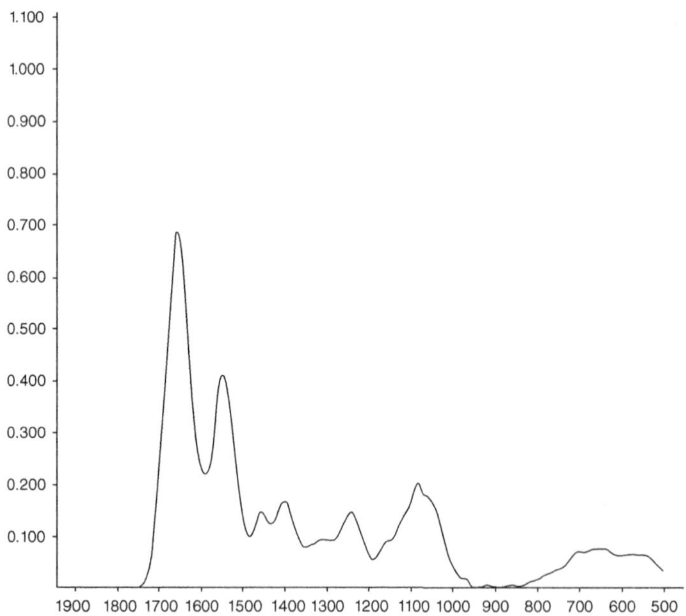

Figure 9.8 Example portion of a spectrum of *Cronobacter sakazakii* using IR spectroscopy.

an identification is achieved by comparing an unknown microorganism to a reference database (Nie *et al.* 2007). Thus, the reference spectral database plays a key role in the identification process (Jaureguiberry *et al.* 2016). This method is not frequently used in regulated environments.

9.4.8.2 Raman Spectroscopy

Another method that is being used for whole organism characterization with a spectroscopic fingerprint is based on Raman spectroscopy. When light interacts with matter it can be transmitted, absorbed, reflected, or scattered. Scattered photons have either the same energy or frequency as the incident light (Rayleigh or elastic scattering) or, a small fraction, have different energy or frequency and that is the Raman or inelastic scattering. The shift occurs because photons exchange part of their energy with molecular vibrations in the material. The amount of energy corresponds to specific molecular vibrations. Indian physicist C.V. Raman was awarded the 1930 Nobel Prize in Physics for this discovery. A commercial system can sample directly from the air, via a nebulizer or through filtration, and deposits the cells directly on a surface for scanning. Once introduced in to the system, the Raman spectra are collected and compared with reference spectra in a database. The technology is nondestructive. Combining Raman spectroscopy with optical microscopy increases the sensitivity and resolution, resulting in the ability to record spectra from a single cell. Interpretation of the spectra is still complex, and robust analysis and extensive databases are needed for reliable interpretation (Ashton *et al.* 2011). Like phenotypic methods, sample treatment, physiological state of the cells, and measurement parameters effect the identification process. Raman is also dependent on the quality of the underlying database and the statistical algorithms used in the analysis (Pahlow *et al.* 2015). Adoption of this technology has been hindered by inherent issues with the technology such as weak signal, long data acquisition times, high background signals, high instrument costs, and lack of automation in sample processing (Ashton *et al.* 2011). This method is not frequently used in regulated environments.

9.4.8.3 Other Technologies

There are additional MS technologies used for microbial identification and the targeted detection of specific microorganisms, but these are also not utilized frequently. They include the combination of PCR and MALDI-TOF where multiple protein-coding genes and the conserved 16S rRNA region are amplified by PCR followed by resolution of the amplicons using MALDI. Another PCR-based approach uses universal primers for bacteria, yeast, or filamentous fungi followed by electrospray ionization (ESI) time of flight MS to resolve the amplicons (Baldwin *et al.* 2009; Simner *et al.* 2013). Advantage of PCR/ESI-MS over sequencing is that it is fast and has high throughput, but it is costly. Surface-enhanced laser desorption/ionization uses a chip with a chemical or biochemical molecule present to capture specific proteins of interest. This protein profiling technique is used in diagnostics for targeted detection and identification. Finally, flow cytometry can also be used for the targeted identification of microorganisms. Antibodies can be used to bind to specific targets and be detected via flow cytometry using light, electrical conductance, or fluorescence.

9.4.9 Strain-Level Differentiation

When working under cGMPs, it is extremely important to be able to accurately identify organisms to the species level. In addition, characterization of microorganisms to the strain level is critical to permit tracking potential origins of a contamination, to avoid delays in product release, and complete investigations. Increased discrimination or resolution at the strain level is often required during manufacturing process investigations of out of specification situations to help find the root cause of a contamination, or during the investigation into a sampling, operator, or laboratory error. There are different methods available to the microbiologist for strain-level resolution.

9.4.9.1 Fragment-Based Genotypic Technologies for Strain-Level Differentiation

9.4.9.1.1 Restriction Fragment Strain-level characterization of bacterial cultures can be provided by a commercial system which uses restriction fragment length polymorphisms and Southern blot (Southern 1975) technologies. This system can also generate an identification. It is based on the premise that the DNA restriction sites in the bacterial rRNA operon are conserved in the 16S, 5S, and 23S genes, and the sites are more variable in the internal transcribed spacer (ITS), intergenic, and flanking sequences. Restriction sites are specific palindromic DNA sequences that are recognized by specific restriction endonucleases. These enzymes cut the DNA within those sequences, and a specific enzyme only cuts within that sequence. For example, the *Eco*RI enzyme only recognizes the GAATTC palindromic sequence and specifically cuts between the G and the A on both DNA strands. A strain comparison, or an identification, is determined based on the DNA banding patterns. The method has been partially automated and consists of diluting a test culture, and extracting the DNA by heating to rupture the cell walls with the aid of a proteolytic enzyme to also help lyse cells, especially Gram-positive cells. The released genomic DNA is digested with one of two commercially available restriction enzymes, the DNA fragments are separated by size on an agarose gel, and transferred to a nylon membrane. The membrane is then probed with the rRNA region and flanking DNA from the *rrnB* operon from *E. coli*. This region is approximately 1440 bp and includes the 16S, 23S, 5S, ITS regions and flanking glutamic acid tRNA gene. After the probe is hybridized, it is detected with an alkaline phosphatase conjugated antibody and a chemiluminescent reaction. The signal is captured by a camera and the banding pattern visualized (Bruce *et al.* 1995; Bruce 1996). For comparison of isolates within the same species, the banding patterns are grouped together if they are matching or placed into different groups if banding patterns are distinguishable (Figure 9.9). Increased discrimination is achieved with additional restriction enzyme reactions (De Cesare *et al.* 2001). Algorithms can also compare the test pattern bands to references and if the pattern is in the reference database, an identification is generated. However, this method is more appropriate for use as a strain-level comparative tool within a species (Brisse *et al.* 2000). Even with the automated system, there can be issues with the data quality that can interfere with the accuracy and repeatability of this method. The DNA can be uncut, partially cut or sheared, the gel may have bubbles or be too dry, and the DNA may not migrate consistently.

9.4.9.1.2 PCR Amplification Additional fragment-based genotypic analyses for strain-level differentiation involve the amplification of different DNA sequences using PCR to generate DNA barcodes visualized through gel-based technologies. Some of the PCR primers recognize repetitive genome-wide sequences or random sequences and generate specific patterns that are used for comparison. The comparison of DNA fingerprinting banding patterns between taxa is based on the assumption that same-sized bands are homologous, but this is only correct for closely related taxa. All PCR-based technologies are affected by DNA quality and PCR temperatures that can be inconsistent between laboratories and instruments. This effects amplification and the

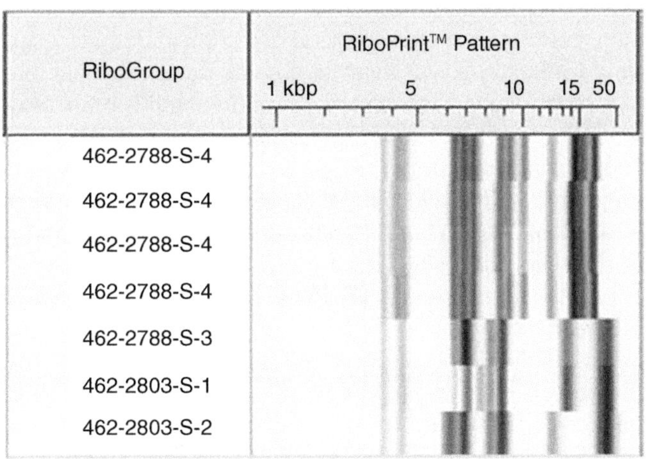

Figure 9.9 Example of the restriction patterns for seven isolates of *Bacillus cereus* after cutting with *PvuII*.

reproducibility of the data and as a result comparison from lab to lab is difficult. Despite these concerns, PCR-based fingerprinting for microbial typing can be reliable, rapid, and highly discriminatory.

Short interspersed repetitive DNA sequences have been identified at distinct locations in the genomes of all organisms, including bacteria and fungi. The use of defined primers for PCR amplification of these repetitive elements is referred to as repetitive element sequence-based polymerase chain reaction (rep-PCR) (Versalovic, Koeuth, and Lupski 1991; Hiett and Seal 2009). The initial discovery of repetitive extragenic palindromic (REP) elements occurred in the genomes of *E. coli* and *Salmonella* (Lupski and Weinstock 1992). Other examples of well-characterized repetitive DNA sequences in bacteria include the enterobacterial repetitive intergenic consensus (ERIC) sequence (Hulton, Higgins, and Sharp 1991) and the interspersed repetitive BOX sequence from *Streptococcus pneumoniae* (Koeuth, Versalovic, and Lupski 1995). These repetitive DNA sequences can be utilized for DNA fingerprinting or strain differentiation of microbes by amplifying the DNA via different, but specific, primer annealing sites. The primer sets anneal to the conservative repetitive DNA sequences that are present throughout the genome. The amplified DNA fragments, when separated by electrophoresis, constitute a genomic fingerprint that has shown to be a reproducible tool for strain-level discrimination of bacteria and fungi.

Alternatively, the PCR primers can short arbitrary primers that amplify random DNA fragments (RAPD). The benefit to this PCR technology is that no prior knowledge of the targeted DNA is needed as the primers are arbitrary and will bind somewhere in the targeted genome to create a PCR profile for comparison (Espinasa and Borowsky 1998).

Amplified fragment length polymorphism (AFLP) PCR is a highly sensitive method for detecting polymorphisms in different strains based on selective amplification of fragments obtained after restriction digestion and is also used as a DNA fingerprinting tool. The AFLP methodology is more reproducible than other amplification technologies, has higher resolution, and is more sensitive (Vos *et al.* 1995). The method entails digesting genomic DNA with restriction enzymes and ligating adaptors to the ends of the fragments. A subset of the restriction fragments is then selected to be amplified by using primers complementary to the adaptor sequence, the restriction site sequence, and a few nucleotides inside the restriction site. The amplified fragments are separated and visualized usually through fluorescence methodologies via automated capillary sequencing instruments.

PCR-based nucleic acid amplification tests (NAAT/NAT) also allow for the detection of specific organisms of concern, and by the selective amplification with target-specific primers, the identity is deduced or presumed upon successful amplification.

Pulsed-field gel electrophoresis (PFGE) was first developed in the mid-1980s to separate large yeast chromosome-sized DNA fragments with a changing polarity of an electric field (Schwartz and Cantor 1984). Since then, it has been used for strain-level characterization especially in the food and probiotic industries. Genomic DNA is digested with specific, infrequently cutting, restriction enzymes that have been determined to be appropriate for resolution of specific species. The high molecular weight restriction fragments are separated by PFGE and produce a DNA fingerprint with a specific pattern. Organisms with the same PFGE banding pattern are viewed as the same strain (Tenover *et al.* 1995). However, the technology is time consuming, requires delicate sample handling to avoid sheering the larger DNA molecules, and has not achieved broad acceptance in the pharmaceutical industry. It was, however, the gold standard for typing during investigations into foodborne illness outbreaks, but is being replaced by next-generation whole genome sequencing technologies.

Multiple locus variable number of tandem repeats analysis (MLVA) is another technique used to generate DNA fingerprints for bacterial isolates. PCR is used along with species-specific primers that will amplify the variable-number tandem repeat (VNTR) regions. The VNTR PCR products are separated via capillary gel electrophoresis and detected with fluorescence. The sizes of the PCR products are converted into allele types to determine how closely related the bacterial strains are to each other. MLVA may be able to differentiate bacterial strains that look the same using other methods of DNA fingerprinting, such as PFGE. Thus, MLVA is a complementary technique to PFGE, allowing more detailed differences between bacteria that have similar PFGE patterns to be visualized (Table 9.5).

9.4.9.2 Sequence-Based Genotypic Technologies for Strain-Level Differentiation

Multi-locus sequence typing (MLST) is a well-established, highly accurate sequence-based method for strain-level differentiation for microorganisms (Enright *et al.* 2000). Strain-level differentiation is made by analyzing

Table 9.5 Different identification systems have different strengths and weaknesses that are related to the technology and/or the reference database.

Pros and cons of fragment-based genotypic systems
Good to very good resolution
Fingerprints generated by the different methods are not comparable to each other
Test methods are complex to perform
Data quality can interfere with accuracy and repeatability
Require a well-equipped molecular biology laboratory and expensive instruments
Different methods have had successful applications in the food industry and in clinical investigations

sequence variations in multiple protein-coding genes or housekeeping genes that encode for proteins necessary for the normal cellular functions and which contain a high degree of sequence variability. Depending on the degree of discrimination needed, strains can be evaluated by a different number of genes. Single locus strain typing (SLST) methods involve sequencing one gene that is known to harbor highly variable DNA sequences. The challenge is the selection of suitable housekeeping genes for acceptable resolution for each species, as these will differ from species to species (Achtman 2008). Confirmation of species identity for all isolates can be first determined by standard rDNA sequencing (16S or ITS). Accurate identification is essential to choose the correct primer sets for strain-appropriate target gene amplification for the species under investigation. The MLST target gene sequences from each isolate can be aligned and compared in a phylogenetic tree to show the amount of conservation and divergence in the DNA of that gene region. Allele numbers can also be assigned to all unique sequences from each locus to determine the sequence type. Strain comparison is made by looking for a unique combination of alleles across all loci, the allelic profile. The effectiveness of MLST as a portable, universal, and conclusive method for characterizing bacteria was first demonstrated by sequencing internal fragments of genes in *Neisseria meningitidis* (Maiden *et al.* 1998; Maiden 2006). Since then, numerous methods have been published. The advantage of this technology is that it is appropriate for both bacteria and fungi (Maiden 2006), and is highly reproducible and unambiguous. The disadvantages are that it requires a fully functioning sequencing facility, knowledge of the appropriate primers to use for each species of interest, and trained personnel to perform the assay and analyze the data appropriately. By combining standard genotypic identification methods with MLST, it is possible to resolve some of the most difficult organisms commonly observed in manufacturing facilities. Given the reproducibility of MLST over time, and between laboratories, these methods can more reliably determine the relatedness of strains and is superior to the fragment-based and PCR-based methods discussed above.

9.5 Strengths and Weaknesses with Each Categorical Method

No system is 100% accurate, but each system has its strengths and limitations. Some highlights are described below and must be balanced by the need of the manufacturer. There are varied levels of operational needs and these needs dictate different approaches to microbial identification and strain differentiation. The limitations and strengths of the different technologies must be understood in order to choose the most appropriate method for the desired application. Closely related species may be difficult or impossible to distinguish using certain systems. In some cases, the most appropriate methods can vary based on the species under examination. A system that generates a genus and species designation is providing a microbial identification. A different system or technology will be needed for a strain-level comparison.

Common platforms for microbial identification include phenotypic commercial systems and MALDI-TOF proteotypic systems. These systems should only be used to generate an identification. If using phenotypic systems, confidence to the genus level should be considered as accuracy to the species level is not high, while MALDI-TOF provides a very confident species-level identification. The most highly confident and accurate resolution at the species or strain level can be achieved through sequence-based approaches. The target gene will dictate the level of resolution – speciation or strain typing. Strain-level differentiation can also be achieved through different fragment-based approaches. However, given the reproducibility and increased resolution of the sequence-based approach, it is superior to fragment-based and PCR-based methods for determination of strain relatedness.

- **Phenotypic systems** used for identification are measuring parameters that cannot reflect the complexity of an organism and these measured parameters are also not consistently expressed. The commercial systems available have outdated and very limited reference databases, leading to incorrect identifications or no identifications. The user must be vigilant about ensuring the purity of the sample being applied to the phenotypic systems as it is not clear from the data if a sample is mixed and a blended phenotypic description will result. The accuracy and reproducibility of these systems is low. However, for genus-level confidence, they could be sufficient.
- **Proteotypic systems** used for identification come with a large capital expenditure, but are simple to use, have an extremely fast time to result, and are highly accurate and reproducible. The turnaround time for a result can be minutes as compared to days with conventional biochemical methods. Testing can be performed on very little sample and has a very low reagent cost (Tan *et al.* 2012). The commercial systems have traditionally had clinical applications and this is reflected in the relevancy of their databases, but there is an extensive publication history in support of the accuracy of the results obtained for many, but not all, groups of organisms that have representation in the reference libraries. Performance can be improved by the creation and addition of reference spectra by the user for underrepresented species, but this can be beyond the capability for some laboratories.
- **Fragment-based genotypic technologies** for identification and strain delineation have good resolution, but the fingerprints generated by the

different methods are not comparable to each other. These test methods are complex to perform, can have issues with repeatability, and require a well-equipped molecular biology laboratory and expensive instruments, in some cases. Different methods have had successful applications in the food industry and in clinical investigations.

- **Sequence-based genotypic technologies** for identification and strain differentiation also come with a large capital expense, have a relatively short time to result, and produce data of highest specificity, reliability, and reproducibility. Sequence-based methods will generate consistent results independent of the growth stage of the organism and the media upon which it was grown which is not the case for the phenotypic and proteotypic systems. Using the conserved rRNA genes for taxonomic classification has limitations within certain groups of bacteria and fungi, and due to the lack of a universal rule for species cutoff, interpretation of the data takes a high level of knowledge. Bacterial identification using the first 500 bp of the 16S rRNA gene sequence can provide a species-level identification up to approximately 87% of the time, genus-level identifications another 11%, with the remainder at family or above (unpublished data). Inability to generate species-level identifications is due to species with identical or nearly identical reference sequences, poor-quality reference sequences, issues with published taxonomic classification, or novel taxa (Janda and Abbott 2007). Using alternative gene regions, such as protein-coding genes, can increase resolution in closely related groups. Nonetheless, there are some groups of organisms that cannot yield clear or confident species-level results due to their inaccurate or muddled taxonomy. Additionally, new species are being published each day based on a type strain descriptor and sequence analysis of additional protein-coding genes. These new species are not in the commercial system's databases. However, phylogenetic analyses can indicate when a species-level confidence is not attainable with the current database. With the appropriate quality procedures in place for experimentally confirming type strain sequences, reference libraries can be expanded and the assignment of a novel taxon can be made to novel isolates.
- **Reference libraries** are used with all commercial systems, while contract laboratories can also use proprietary validated reference databases. In each case, consideration must be made to the reclassification of microorganisms in the scientific literature. Many commercial databases are not updated frequently, nor are they curated well to reflect the current taxonomy. All identification methods rely on the quality, breadth, and depth of the underlying reference database. Commercial systems perform very well on bacteria studied from clinical sources, but differences are seen when analyzing environmental isolates (Urwyler and Glaubitz 2016).
- **Summary attributes of the methods** (see Table 9.6)

Table 9.6 Categories to consider prior to adopting an identification or strain characterization technology.

	Phenotypic	Proteotypic	Genotypic
Application	Routine	Routine	Routine and critical
Resolution	Low (genus)	Medium (species)	High (species and strain)
Accuracy	Low	Med–high	High–very high
Organism range (bacteria, fungi)	Low	Med–broad	Broad
Assay throughput	Medium	High	High
Assay time	Medium	Fast	Medium
Capital costs	Medium	High	High
Consumables cost	High	Low	High
Operational skill and knowledge	Low	Low	High

9.6 Case Studies from a Contract Testing Lab

The following examples of evaluating the impact of choosing the most appropriate identification system for the required application, of poor-quality reference data, and of the impact of taxonomic name changes, and the curation of reference material are taken from blog posts originally on the Charles River Eureka site. Available at eureka.criver.com.

9.6.1 A Caveat to Identifying QC Strains

Within the realm of EM, clinical microbiology, or industrial microbiology, QC strains (also called reference organisms) are a vital part of a company's compliance umbrella and their accreditation process. QC analyses are required for consistent and reliable results. QC strains are also used during process validations, proficiency testing, and instrument qualifications to evaluate performance so that the desired analytical and diagnostic standards can be achieved.

If, as part of the Quality system, the identity of the QC strain must be confirmed, it is imperative that the identification system being used for confirmation of identity has the correct species-level resolution for that QC strain. This is very important because species-resolution capabilities of identification technologies are not the same for different closely related species. It is critical, therefore, to understand this before choosing an identification method, so that deviations and investigations can be avoided.

First, consider the example of *Enterobacter hormaechei*, ATCC 700323, which is a QC strain for a phenotypic identification system. This organism could also make its way into proficiency testing or process validation cohorts. *Enterobacter hormaechei* (type strain, ATCC 49162T) was named after Estenio Hormaeche, a Uruguayan microbiologist who, along with P.R. Edwards, proposed the genus *Enterobacter* (O'Hara *et al.* 1989). *Enterobacter hormaechei* strains have been isolated from human sources such as blood, respiratory tract, and wounds (Mezzatesta, Gona, and Stefani 2012). It belongs to the *Enterobacter cloacae complex (ECC)* which consists of the additional species *E. asburiae, E. cloacae, E. kobei, E. ludwigii,* and *Lelliottia nimipressuralis* (*Enterobacter nimipressuralis*). In general, species within the ECC are well-recognized nosocomial pathogens.

Based on our experience with repeatedly testing *E. hormaechei* ATCC 700323 by MALDI-TOF MS, it incorrectly identifies as *E. cloacae*. On the other hand, *E. hormaechei* ATCC 700323 correctly identifies as *E. hormaechei* by 500 bp 16S rDNA sequencing and phylogenetic analysis. Our MALDI reference library contains a total of 23 entries representing six ECC species. Evaluating the accuracy of these ECC MALDI entries by reviewing the taxonomic placement of the strains and the quality of the spectra used for each library entry showed there were no inconsistencies. Thus, erroneous MALDI library entries were not a factor in the incorrect identification we observe using the MALDI-TOF. Therefore, we surmised that MALDI cannot reliably discriminate between species of the ECC most likely due to the lack of diversity in their protein profiles. A recent study that used several culture collection strains of species from the ECC observed a similar pattern with MALDI (Pavlovic *et al.* 2012). Therefore, if a customer submits *E. hormaechei* ATCC 700323 for MALDI-based identification, the result is correct only to the species-complex level due to lack of species-level resolution for these organisms using this identification platform.

An incorrect identification of a QC isolate can cause a deviation, an out of specification, or a run failure leading to delayed release of product or test results. However, if the species resolution of the identification system is known in advance, then the appropriate choice of a system can be made so that the expected identification matches the system output. In this case, for the ECC, it would be 16S rDNA sequencing and phylogenetic analysis.

9.6.2 *Odoribacter* Are Not *Bacillus*: The Importance of Accurate Reference Data

Multiple identification technologies applied under the right circumstances can find and fix errors that may not be obvious when only a single technology is used. While attempting to identify an isolate with our MALDI-TOF service, the top library match was *Odoribacter splanchnicus*, usually an inhabitant of the human intestine. Upon quality review of the result, it was determined that the colony morphology did not resemble *O. splanchnicus*. In such cases, rDNA sequencing is used for resolution. The identification by 16S rDNA sequencing was *Bacillus pumilus/safensis* which is found in diverse environments and which matched the colony morphology.

There are typically three main reasons that may lead to discrepant results from presumably the same isolate by different technologies. The first occurs when the two discrepant species in question are closely related as in the previous ECC example. Different identifications may also occur if the culture plate contains mixed colonies, or if the library entries of one of the two technologies is erroneous. The first reason was ruled out because the two identifications in question represent species that are phylogenetically distant. *Odoribacter* species are Gram-negative, anaerobic, nonspore forming, and fusiform, whereas *Bacillus* species are Gram-positive, aerobic, spore forming, and rod shaped. The original culture plate was a pure culture, therefore the second reason was also ruled out. This left us with the third reason, which was the possibility of a problematic library entry.

The accuracy of the DNA sequence data in our library was confirmed by the correct phylogenetic placement of the type strains of *B. pumilus* and *B. safensis*. The MALDI entry for *O. splanchnicus* was created by the instrument manufacturer using the type strain (DSM 20712T). Since this entry was not created by us, we could not analyze the exact same isolate. However, we could purchase the type strain from the American Type Culture Collection (ATCC 29572T). We verified its identity with 16S rDNA sequencing and created a new *O. splanchnicus* MALDI entry. None of the previous MALDI spectra from the isolate in question matched the new *O. splanchnicus* entry that was created. When the previous *O. splanchnicus* MALDI entry was excluded, the isolate in question now had a best MALDI match of *Bacillus*. Based on the results of this investigation, we inactivated the original *O. splanchnicus* MALDI entry in 2015, and the manufacturer also deleted this entry during their 2016 library release.

This investigation shows that using identification technologies with the highest accuracy can resolve discrepancies quickly. Furthermore, it enables MALDI library curation that is continually verified with genotypic and phylogenetic approaches.

9.6.3 Synonyms of Fungal Strains: Taxonomic Investigations

In Industrial Microbiology, the identity of production strains must be confirmed. Taxonomic classification can be a problem with many groups of organisms, especially fungi! A fungal strain which can be used to synthesize the antibiotic penicillin was identified by us as *Penicillium rubens* using the ITS2 rRNA region. The expectation was for an identification of *Penicillium notatum* (ATCC 9478) and *Penicillium chrysogenum* (ATCC 11709). How could this happen? It is not just an academic question. Mixing up strains can lead to QC failures and result in issues with maintaining compliance under regulatory guidance documents.

At the time we generated these data, the samples were indeed cataloged by the ATCC as *P. chrysogenum*. A commonly occurring mold in indoor environments, *P. chrysogenum* has gained much attention in the pharmaceutical industry for production of penicillin. Yet, the taxonomy of fungi is a complicated affair. For instance, Alexander Fleming's strain, the original penicillin-producing strain, was initially classified as *Penicillium rubrum*, but was later reclassified as *P. notatum* and finally as *P. chrysogenum*. All these classifications were based on morphology.

This classification issue explains why different culture collections have cataloged these strains with different names. Nonetheless, recent studies using rDNA sequencing analysis have shown that penicillin-producing strains, originally described by Fleming, are not *P. rubrum, P. notatum,* or *P. chrysogenum,* but *P. rubens* (Houbraken, Frisvad, and Samson 2011). Fleming's strain is the strain that we received (ATCC 9478, CBS 205.57, NRRL 824, and IMI 015378). Many other culture collections have since changed their species identification for this strain, and the user must be alert to these changes, especially if they have obtained the stock cultures prior to any taxonomic change. The naming issue can be further complicated when confirmatory identifications are generated by certain commercial identification systems that continue to misidentify this species as *P. chrysogenum*. Additionally, many systems lack the relevant species entries in their reference libraries, and thus they will not be able to produce the name, *P. rubens*.

This is the first of two examples of why it is critical to curate the names of the species in your reference library in order to ensure you have the most recent classification and contain enough species diversity to generate an accurate identification.

9.6.4 Synonyms of Bacterial Strains: Taxonomic Investigations

An accrediting body informed us that although we had passed, we had misnamed the species of our *Pseudomonas* sample as part of a routine quality assurance test intended to measure proficiency and competence at identifying microbes correctly. As part of the accreditation process, we received three blinded samples to process and identify in-house. One of the samples appeared to be *Pseudomonas oleovorans*, a Gram-negative bacteria found in oil–water emulsions used as lubricants and cooling agents in the cutting and grinding of materials. The company providing the sample said it was actually *Pseudomonas pseudoalcaligenes*, a strain first isolated in swimming pools.

As scientists confident in our results, we naturally investigated this discrepancy. An article from 2010 stated that *P. pseudoalcaligenes* was a later heterotypic synonym of *P. oleovorans* (Saha *et al.* 2010). In other words, it had undergone a name change, but the National Collection of Type Culture (NCTC), the repository that supplied the reference strains for our quality assurance test, had not updated their nomenclature. *Pseudomonas pseudoalcaligenes* was reclassified to *P. oleovorans* and this name change has already been applied to our reference library as well as many other culture collections including the ATCC and Deutsche Sammlung von Mikroorganismen und Zellkulturen GmbH (DSMZ). We provided an update to the company that administered the compliance test, and they amended their report to reflect a higher score.

Communicating with the appropriate organizations to share instances of inaccurate classification is very important. Many times, these institutions express to us that they are grateful for our efforts and will review the data and make any necessary amendments to their catalog entry. Keeping taxonomic names current for the libraries and databases used for a microbial identification system is very important for providing accurate identifications. Misidentification could be detrimental for the end user and also to a company's finances and reputation.

9.7 Conclusion

A program to identify microorganisms is integral to the pharmaceutical microbiology laboratory. Identification is dependent on established taxonomy; classification must facilitate the identification and be robust, yet flexible, to allow for the expansion and rearrangement of taxonomic ranks. The identification and characterization of microorganisms are some of the most important tasks of a QC microbiology laboratory, but they are not trivial. The amount of effort devoted to the analysis and description of a microbe should be proportional with the contamination risk to the product. Many times, species-level identifications are required, however, this is not always necessary. There should be a hierarchy of response from characterization to identification to strain typing. The level of characterization depends on the criticality of the area involved and whether an investigation warrants further identification. The primary screening is sufficient for risk assessment in non-sterile pharmaceutical manufacturing to either recommend a corrective action or dictate the need for testing to obtain a species-level identification or strain differentiation of isolates within a species. The identification of microorganisms can be done through different processes, each with its own level of accuracy and reproducibility. Accuracy of identification, and strain-level characterization, is dependent on the method used to generate and interpret the data as well as the library database used as a reference. The methods chosen will impact the confidence in the data and decisions made based on those data. The focus should be on microbial safety and risk assessment. Accurate and consistent identification methods, such as MALDI-TOF, are critical in managing risk by yielding data that allow for comprehensive and reliable tracking and trending during routine monitoring and more confidence in any resulting decisions. During an investigation into excursions, it can be extremely important to be able to accurately identify organisms to the species level and distinguish between strains, using a sequence and phylogenetic approach, to determine the potential origin of the contamination and complete the investigation. The priority of the samples will correlate with their criticality, and will dictate which method should be used for the most accurate identification, and strain-level differentiation, of the microorganisms.

Bibliography

Abarenkov, K, Tedersoo, L, Nilsson, RH, Vellak, K, Saar, I, Veldre, V, Parmasto, E, Prous, M, Aan, A, Ots, M, Kurina, O, Ostonen, I, Jõgeva, J, Halapuu, S, Põldmaa, K, Toots, M, Truu, J, Larsson, K-H, Kõljalg, U, 2010, 'PlutoF: a web based workbench for ecological and taxonomic research, with an online implementation for fungal ITS sequences', Evol. Bioinformatics Online, 6, pp. 189–196.

Achtman, M, 2008, 'Evolution, population structure, and phylogeography of genetically monomorphic bacterial pathogens', Annu Rev Microbiol, vol. 62, pp. 53–70.

Alatoom, AA, Cunningham, SA, Ihde, SM, Mandrekar, J, Patel, R, 2011, 'Comparison of direct colony method versus extraction method for identification of Gram-positive cocci by use of Bruker Biotyper matrix-assisted laser desorption ionization-time of flight mass spectrometry', J Clin Microbiol, vol. 49, pp. 2868–2873.

Alexandre, A, Laranjo, M, Young, JPW, Oliveira, S, 2008, 'dnaJ is a useful phylogenetic marker for alphaproteobacteria', Int J Syst Evol Microbiol, vol. 58, pp. 2839–2849.

Almuzara, M, Barberis, C, Traglia, G, Famiglietti, A, Ramirez, MS, Vay, C, 2015, 'Evaluation of matrix-assisted laser desorption ionization-time-of-flight mass spectrometry for species identification of nonfermenting Gram-negative bacilli', J Microbiol Methods, vol. 112, pp. 24–27.

Andersson, SG, Zomorodipour, A, Winkler, HH, Kurland, CG, 1995, 'Unusual organization of the rRNA genes in *Rickettsia prowazekii*', J Bacteriol, vol. 177, pp. 4171–4175.

Ashelford, KE, Chuzhanova, NA, Fry, JC, Jones, AJ, Weightman, AJ, 2005, 'At least 1 in 20 16S rRNA sequence records currently held in public repositories is estimated to contain substantial anomalies', Appl Environ Microbiol, vol. 71, pp. 7724–7736.

Ashton, L, Lau, K, Winder, CL, Goodacre, R, 2011, 'Raman spectroscopy: lighting up the future of microbial identification', Future Microbiol, vol. 6, pp. 991–997.

Baldwin, CD, Howe, GB, Sampath, R, Blyn, LB, Matthews, H, Harpin, V, Hall, TA, Drader, JJ, Hofstadler, SA, Eshoo, MW, Rudnick, K, Studarus, K, Moore, D, Abbott, S, Janda, JM, Whitehouse, CA, 2009, 'Usefulness of multilocus polymerase chain reaction followed by electrospray ionization mass spectrometry to identify a diverse panel of bacterial isolates', Diagn Microbiol Infect Dis, vol. 63, pp. 403–408.

Beekes, M, Lasch, P, Naumann, D, 2007, 'Analytical applications of Fourier transform-infrared (FT-IR) spectroscopy in microbiology and prion research', Vet Microbiol, vol. 123, pp. 305–319.

Benagli, C, Rossi, V, Dolina, M, Tonolla, M, Petrini, O, 2011, 'Matrix-assisted laser desorption ionization-time of flight mass spectrometry for the identification of clinically relevant bacteria', PLoS One, vol. 6, pp. e16424.

Bessede, E, Angla-Gre, M, Delagarde, Y, Sep Hieng, S, Menard, A, Megraud, F, 2011, 'Matrix-assisted laser-desorption/ionization biotyper: experience in the routine of a university hospital', Clin Microbiol Infect, vol. 17, pp. 533–538.

Bizzini, A, Greub, G, 2010, 'Matrix-assisted laser desorption ionization time-of-flight mass spectrometry a revolution in clinical microbial identification', Clin Microbiol Infect, vol. 16, pp. 1614–1619.

Bizzini, A, Jaton, K, Romo, D, Bille, J, Prod'hom, G, Greub, G, 2011, 'Matrix-assisted laser desorption ionization-time of flight mass spectrometry as an alternative to 16S rRNA gene sequencing for identification of difficult-to-identify bacterial strains', J Clin Microbiol, vol. 49, pp. 693–696.

Bradley, F, Halloran, E, Holmes, S, 1996, 'Bootstrap confidence levels for phylogenetic trees', Proceedings of the National Academy of Sciences, vol. 93, pp. 7085–7090.

Brisse, S, Verduin, CM, Milatovic, D, Fluit, A, Verhoef, J, Laevens, S, Vandamme, P, Tummler, B, Verbrugh, HA, van Belkum, A, 2000, 'Distinguishing species of the *Burkholderia cepacia* complex and *Burkholderia gladioli* by automated ribotyping', J Clin Microbiol, vol. 38, pp. 1876–1884.

Bruce, JL, 1996, 'Automated system rapidly identifies and characterizes microorganisms in food', Food Technol, vol. 77, 77–81.

Bruce, JL, Hubner, RJ, Cole, EM, McDowell, CI, Webster, JA, 1995, 'Sets of EcoRI fragments containing ribosomal RNA sequences are conserved among different strains of *Listeria monocytogenes*', Proc Natl Acad Sci USA, vol. 92, pp. 5229–5233.

Buchanan, RE, 1916, 'Studies in the nomenclature and classification of bacteria: the problem of bacterial nomenclature', J Bacteriol, vol. 1, pp. 591–596.

Buchanan, RE, 1918. 'Studies in the nomenclature and classification of the bacteria: X. Subgroups and genera of the Myxobacteriales and Spirochaetales', J Bacteriol, vol. 3, pp. 541–545.

Capella-Gutierrez, S, Kauff, F, Gabaldon, T, 2014, 'A phylogenomics approach for selecting robust sets of phylogenetic markers', Nucleic Acids Res, vol. 42, pp. 54.

Carbonnelle, E, Mesquita, C, Bille, E, Day, N, Dauphin, B, Beretti, JL, Ferroni, A, Gutmann, L, Nassif, X, 2011, 'MALDI-TOF mass spectrometry tools for bacterial identification in clinical microbiology laboratory', Clin Biochem, vol. 44, pp. 104–109.

Carbonnelle, E, Grohs, P, Jacquier, H, Day, N, Tenza, S, Dewailly, A, Vissouarn, O, Rottman, M, Herrmann, JL, Podglajen, I, Raskine, L, 2012, 'Robustness of two MALDI-TOF mass spectrometry systems for bacterial identification', J Microbiol Methods, vol. 89, pp. 133–136.

Cassagne, C, Ranque, S, Normand, AC, Fourquet, P, Thiebault, S, Planard, C, Hendrickx, M, Piarroux, R, 2011, 'Mould routine identification in the clinical laboratory by matrix-assisted laser desorption ionization time-of-flight mass spectrometry', PLoS One, vol. 6, pp. e28425.

Chalupova, J, Raus, M, Sedlarova, M, Sebela, M, 2014, 'Identification of fungal microorganisms by MALDI-TOF mass spectrometry', Biotechnol Adv, vol. 32, pp. 230–241.

Cherkaoui, A, Hibbs, J, Emonet, S, Tangomo, M, Girard, M, Francois, P, Schrenzel, J, 2010, 'Comparison of two matrix-assisted laser desorption ionization-time of flight mass spectrometry methods with conventional phenotypic identification for routine identification of bacteria to the species level', J Clin Microbiol, vol. 48, pp. 1169–1175.

Chun, J, Bae, KS, 2000, 'Phylogenetic analysis of *Bacillus subtilis* and related taxa based on partial gyrA gene sequences', Antonie Van Leeuwenhoek, vol. 78, pp. 123–127.

Clarridge, JE, 3rd, 2004, 'Impact of 16S rRNA gene sequence analysis for identification of bacteria on clinical microbiology and infectious diseases', Clin Microbiol Rev, vol. 17, pp. 840–862.

Claydon, MA, Davey, SN, Edwards-Jones, V, Gordon, DB, 1996, 'The rapid identification of intact microorganisms using mass spectrometry', Nat Biotech, vol. 14, pp. 1584–1586.

Clayton, RA, Sutton, G, Hinkle, PS Jr., Bult, C, Fields, C, 1995, 'Intraspecific variation in small-subunit rRNA sequences in GenBank: why single sequences may not adequately represent prokaryotic taxa', Int J Syst Bacteriol, vol. 45, pp. 595–599.

Cleenwerck, I, De Vos, P, 2008, 'Polyphasic taxonomy of acetic acid bacteria: an overview of the currently applied methodology', Int J Food Microbiol, vol. 125, pp. 2–14.

Cohan, FM, 2002, 'What are bacterial species?', Annu Rev Microbiol, vol. 56, pp. 457–487.

Coyle, MB, Leonard, RB, Nowowiejski, DJ, Malekniazi, A, Finn, DJ, 1993, 'Evidence of multiple taxa within commercially available reference strains of *Corynebacterium xerosis*', J Clin Microbiol, vol. 31, pp. 1788–1793.

Das, S, Dash, HR, Mangwani, N, Chakraborty, J, Kumari, S, 2014, 'Understanding molecular identification and polyphasic taxonomic approaches for genetic relatedness and phylogenetic relationships of microorganisms', J Microbiol Methods, vol. 103, pp. 80–100.

Dawyndt, P, Vancanneyt, M, Snauwaert, C, De Baets, B, De Meyer, H, Swings, J, 2006, 'Mining fatty acid databases for detection of novel compounds in aerobic bacteria', J Microbiol Methods, vol. 66, pp. 410–433.

De Bel, A, Wybo, I, Vandoorslaer, K, Rosseel, P, Lauwers, S, Pierard, D, 2011, 'Acceptance criteria for identification results of Gram-negative rods by mass spectrometry', J Med Microbiol, vol. 60, pp. 684–686.

De Bruyne, K, Slabbinck, B, Waegeman, W, Vauterin, P, De Baets, B, Vandamme, P, 2011, 'Bacterial species identification from MALDI-TOF mass spectra through data analysis and machine learning', Syst Appl Microbiol, vol. 34, pp. 20–29.

De Cesare, A, Bruce, JL, Dambaugh, TR, Guerzoni, ME, Wiedmann, M, 2001, 'Automated ribotyping using different enzymes to improve discrimination of *Listeria monocytogenes* isolates, with a particular focus on serotype 4b strains', J Clin Microbiol, vol. 39, pp. 3002–3005.

Dieckmann, R, Helmuth, R, Erhard, M, Malorny, B, 2008, 'Rapid classification and identification of salmonellae at the species and subspecies levels by whole-cell matrix-assisted laser desorption ionization-time of flight mass spectrometry', Appl Environ Microbiol, vol. 74, pp. 7767–7778.

Doi, RH, Igarashi, RT, 1965, 'Conservation of ribosomal and messenger ribonucleic acid cistrons in *Bacillus* species', J Bacteriol, vol. 90, pp. 384–390.

Drancourt, M, Bollet, C, Carlioz, A, Martelin, R, Gayral, JP, Raoult, D, 2000, '16S ribosomal DNA sequence analysis of a large collection of environmental and clinical unidentifiable bacterial isolates', J Clin Microbiol, vol. 38, pp. 3623–3630.

Dubnau, D, Smith, I, Morell, P, Marmur, J, 1965, 'Gene conservation in *Bacillus* species. I. Conserved genetic and nucleic acid base sequence homologies', Proc Natl Acad Sci USA, vol. 54, pp. 491–498.

Dubois, D, Grare, M, Prere, MF, Segonds, C, Marty, N, Oswald, E, 2012, 'Performances of the Vitek MS matrix-assisted laser desorption ionization-time of flight mass spectrometry system for rapid identification of bacteria in routine clinical microbiology', J Clin Microbiol, vol. 50, pp. 2568–2576.

Efron, B, Halloran, E, Holmes, S, 1996, 'Bootstrap confidence levels for phylogenetic trees', Proc Natl Acad Sci USA, vol. 93, pp. 13429–13429.

Enright, MC, Day, NP, Davies, CE, Peacock, SJ, Spratt, BG, 2000, 'Multilocus sequence typing for characterization of methicillin-resistant and methicillin-susceptible clones of *Staphylococcus aureus*', J Clin Microbiol, vol. 38, pp. 1008–1015.

Espinasa, L, Borowsky, R, 1998, 'Evolutionary divergence of AP-PCR (RAPD) patterns', Mol Biol Evol, vol. 15, pp. 408–414.

Everley, RA, Mott, TM, Wyatt, SA, Toney, DM, Croley, TR, 2008, 'Liquid chromatography/mass spectrometry characterization of *Escherichia coli* and *Shigella* species', J Am Soc Mass Spectrom, vol. 19, pp. 1621–1628.

Farrance, CE, Hong, S, 2015, 'Is it essential to sequence the entire 16S rRNA gene for bacterial identification?', Am Pharm Rev, vol. 18(7), pp. 1–7.

Federhen, S, 2015, 'Type material in the NCBI taxonomy database', Nucleic Acids Res, vol. 43, pp. 1086–1098.

Fontana, C, Favaro, M, Pelliccioni, M, Pistoia, ES, Favalli, C, 2005, 'Use of the MicroSeq 500 16S rRNA gene-based sequencing for identification of bacterial isolates that commercial automated systems failed to identify correctly', J Clin Microbiol, vol. 43, pp. 615–619.

Fox, GE, Pechman, KR, Woese, CR, 1977, 'Comparative cataloging of 16S ribosomal ribonucleic acid: molecular approach to procaryotic systematics', Int J Syst Evol Microbiol, vol. 27, pp. 44–57.

Fox, GE, Wisotzkey, JD, Jurtshuk, P, Jr., 1992, 'How close is close: 16S rRNA sequence identity may not be sufficient to guarantee species identity', Int J Syst Bacteriol, vol. 42, pp. 166–170.

Fraser, C, Alm, EJ, Polz, MF, Spratt, BG, Hanage, WP, 2009, 'The bacterial species challenge: making sense of genetic and ecological diversity', Science, vol. 323, pp. 741–746.

Gevers, D, Cohan, FM, Lawrence, JG, Spratt, BG, Coenye, T, Feil, EJ, Stackebrandt, E, Van de Peer, Y, Vandamme, P, Thompson, FL, Swings, J, 2005, 'Opinion: re-evaluating prokaryotic species', Nat Rev Microbiol, vol. 3, pp. 733–739.

Gillis, M, Vandamme, P, De Vos, P, Swings, J, Kersters, K, 2005, 'Polyphasic taxonomy', in GM Garrity, DJ Brenner, NR Krieg, JT Staley (eds.), Bergey's Manual of Systematic Bacteriology, Springer-Verlag, New York.

Goris, J, Konstantinidis, KT, Klappenbach, JA, Coenye, T, Vandamme, P, Tiedje, JM, 2007, 'DNA–DNA hybridization values and their relationship to whole-genome sequence similarities', Int J Syst Evol Microbiol, vol. 57, pp. 81–91.

Gram, C, 1884, 'The differential staining of Schizomycetes in tissue sections and in dried preparations', Fortschr Med, vol. 2, pp. 185–189.

Grunert, T, Wenning, M, Barbagelata, MS, Fricker, M, Sordelli, DO, Buzzola, FR, Ehling-Schulz, M, 2013, 'Rapid and reliable identification of *Staphylococcus aureus* capsular serotypes by means of artificial neural network-assisted Fourier transform infrared spectroscopy', J Clin Microbiol, vol. 51, pp. 2261–2266.

Guarro, J, Gené, J, Stchigel, AM, 1999, 'Developments in fungal taxonomy', Clin Microbiol Rev, vol. 12, pp. 454–500.

Gundi, VA, Dijkshoorn, L, Burignat, S, Raoult, D, La Scola, B, 2009, 'Validation of partial rpoB gene sequence analysis for the identification of clinically important and emerging *Acinetobacter* species', Microbiology, vol. 155, pp. 2333–2341.

Hall, L, Wohlfiel, S, Roberts, GD, 2003, 'Experience with the MicroSeq D2 large-subunit ribosomal DNA sequencing kit for identification of commonly encountered, clinically important yeast species', J Clin Microbiol, vol. 41, pp. 5099–5102.

Hall, L, Wohlfiel, S, Roberts, GD, 2004, 'Experience with the MicroSeq D2 large-subunit ribosomal DNA sequencing kit for identification of filamentous fungi encountered in the clinical laboratory', J Clin Microbiol, vol. 42, pp. 622–626.

Harrington, CS, On, SL, 1999, 'Extensive 16S rRNA gene sequence diversity in Campylobacter hyointestinalis strains: taxonomic and applied implications', Int J Syst Bacteriol, vol. 49, pp. 1171–1175.
Hawksworth, DL, 2006, 'Pandora's mycological box: molecular sequences vs. morphology in understanding fungal relationships and biodiversity', Rev Iberoam Micol, vol. 23, pp. 127–133.
Hawksworth, DL, 2011, 'A new dawn for the naming of fungi: impacts of decisions made in Melbourne in July 2011 on the future publication and regulation of fungal names', IMA Fungus, vol. 2, pp. 155–162.
Hettick, JM, Green, BJ, Buskirk, AD, Kashon, ML, Slaven, JE, Janotka, E, Blachere, FM, Schmechel, D, Beezhold, DH, 2008, 'Discrimination of *Penicillium* isolates by matrix-assisted laser desorption/ionization time-of-flight mass spectrometry fingerprinting', Rapid Commun Mass Spectrom, vol. 22, pp. 2555–2560.
Hibbett, DS, Binder, M, Bischoff, JF, Blackwell, M, Cannon, PF, Eriksson, OE, Huhndorf, S, James, T, Kirk, PM, Lucking, R, Thorsten Lumbsch, H, Lutzoni, F, Matheny, PB, McLaughlin, DJ, Powell, MJ, Redhead, S, Schoch, CL, Spatafora, JW, Stalpers, JA, Vilgalys, R, Aime, MC, Aptroot, A, Bauer, R, Begerow, D, Benny, GL, Castlebury, LA, Crous, PW, Dai, YC, Gams, W, Geiser, DM, Griffith, GW, Gueidan, C, Hawksworth, DL, Hestmark, G, Hosaka, K, Humber, RA, Hyde, KD, Ironside, JD, Koljalg, U, Kurtzman, CP, Larsson, KH, Lichtwardt, R, Longcore, J, Miadlikowska, J, Miller, A, Moncalvo, JM, Mozley-Standridge, S, Oberwinkler, F, Parmasto, E, Reeb, V, Rogers, JD, Roux, C, Ryvarden, L, Sampaio, JP, Schussler, A, Sugiyama, J, Thorn, RG, Tibell, L, Untereiner, WA, Walker, C, Wang, Z, Weir, A, Weiss, M, White, MM, Winka, K, Yao, YJ, Zhang, N, 2007, 'A higher-level phylogenetic classification of the fungi', Mycol Res, vol. 111, pp. 509–547.
Hiett, KL, Seal, BS, 2009, 'Use of repetitive element palindromic PCR (rep-PCR) for the epidemiologic discrimination of foodborne pathogens', Methods Mol Biol, vol. 551, pp. 49–58.
Holland, RD, Wilkes, JG, Rafii, F, Sutherland, JB, Persons, CC, Voorhees, KJ, Lay, JO, 1996, 'Rapid identification of intact whole bacteria based on spectral patterns using matrix-assisted laser desorption/ionization with time-of-flight mass spectrometry', Rapid Commun Mass Spectrom, vol. 10, pp. 1227–1232.
Horneffer, V, Forsmann, A, Strupat, K, Hillenkamp, F, Kubitscheck, U, 2001, 'Localization of analyte molecules in MALDI preparations by confocal laser scanning microscopy', Anal Chem, vol. 73, pp. 1016–1022.
Houbraken, J, Frisvad, JC, Samson, RA, 2011, 'Fleming's penicillin producing strain is not *Penicillium chrysogenum* but *P. rubens*', IMA Fungus, vol. 2, pp. 87–95.
Huang, CH, Lee, FL, Tai, CJ, 2009, 'The beta-tubulin gene as a molecular phylogenetic marker for classification and discrimination of the *Saccharomyces* sensu stricto complex', Antonie Van Leeuwenhoek, vol. 95, pp. 135–142.
Huang, Y, Zhang, M, Zhu, M, Wang, M, Sun, Y, Gu, H, Cao, J, Li, X, Zhang, S, Wang, J, Lu, X, 2017, 'Comparison of two matrix-assisted laser desorption

ionization-time of flight mass spectrometry systems for the identification of clinical filamentous fungi', World J Microbiol Biotechnol, vol. 33, pp. 142.
Hulton, CS, Higgins, CF, Sharp, PM, 1991, 'ERIC sequences: a novel family of repetitive elements in the genomes of *Escherichia coli, Salmonella typhimurium* and other enterobacteria', Mol Microbiol, vol. 5, pp. 825–834.
Hutchison, CA, 3rd, 2007, 'DNA sequencing: bench to bedside and beyond', Nucleic Acids Res, vol. 35, pp. 6227–6237.
Istock, CA, Bell, JA, Ferguson, N, Istock, NL, 1996, 'Bacterial species and evolution: theoretical and practical perspectives', J Ind Microbiol, vol. 17, pp. 137–150.
Janda, JM, Abbott, SL, 2007, '16S rRNA gene sequencing for bacterial identification in the diagnostic laboratory: pluses, perils, and pitfalls', J Clin Microbiol, vol. 45, pp. 2761–2764.
Jaureguiberry, M, Madoz, LV, Giuliodori, MJ, Wagener, K, Prunner, I, Grunert, T, Ehling-Schulz, M, Drillich, M, De la Sota, RL, 2016, 'Identification of *Escherichia coli* and *Trueperella pyogenes* isolated from the uterus of dairy cows using routine bacteriological testing and Fourier transform infrared spectroscopy', Acta Vet Scand, vol. 58, pp. 81.
Jukes, TH, Cantor, CR, 1969, 'Evolution of protein molecules', in HN Munro (ed.), Mammalian Protein Metabolism, Academic Press, New York.
Karas, M, Hillenkamp, F, 1988, 'Laser desorption ionization of proteins with molecular masses exceeding 10,000 daltons', Anal Chem, vol. 60, pp. 2299–2301.
Karas, M, Bachmann, D, Bahr, U, Hillenkamp, F, 1987, 'Matrix-assisted ultraviolet laser desorption of non-volatile compounds', Int J Mass Spectrom Ion Process, vol. 78, pp. 53–68.
Karlin, S, Weinstock, GM, Brendel, V, 1995, 'Bacterial classifications derived from recA protein sequence comparisons', J Bacteriol, vol. 177, pp. 6881–6893.
Khot, PD, Fisher, MA, 2013, 'Novel approach for differentiating *Shigella* species and *Escherichia coli* by matrix-assisted laser desorption ionization-time of flight mass spectrometry', J Clin Microbiol, vol. 51, pp. 3711–3716.
Kim, M, Oh, HS, Park, SC, Chun, J, 2014, 'Towards a taxonomic coherence between average nucleotide identity and 16S rRNA gene sequence similarity for species demarcation of prokaryotes', Int J Syst Evol Microbiol, vol. 64, pp. 346–351.
Kimura, M, 1980, 'A simple method for estimating evolutionary rates of base substitutions through comparative studies of nucleotide sequences', J Mol Evol, vol. 16, pp. 111–120.
Klappenbach, JA, Dunbar, JM, Schmidt, TM, 2000, 'rRNA operon copy number reflects ecological strategies of bacteria', Appl Environ Microbiol, vol. 66, pp. 1328–1333.
Koeuth, T, Versalovic, J, Lupski, JR, 1995, 'Differential subsequence conservation of interspersed repetitive *Streptococcus pneumoniae* BOX elements in diverse bacteria', Genome Res, vol. 5, pp. 408–418.

Krishnamurthy, T, Ross, PL, 1996, 'Rapid identification of bacteria by direct matrix-assisted laser desorption/ionization mass spectrometric analysis of whole cells', Rapid Commun Mass Spectrom, vol. 10, pp. 1992–1996.

Kurtzman, CP, Robnett, CJ, 1998, 'Identification and phylogeny of ascomycetous yeasts from analysis of nuclear large subunit (26S) ribosomal DNA partial sequences', Antonie Van Leeuwenhoek, vol. 73, pp. 331–371.

Kyrpides, NC, Hugenholtz, P, Eisen, JA, Woyke, T, Goker, M, Parker, CT, Amann, R, Beck, BJ, Chain, PS, Chun, J, Colwell, RR, Danchin, A, Dawyndt, P, Dedeurwaerdere, T, DeLong, EF, Detter, JC, De Vos, P, Donohue, TJ, Dong, XZ, Ehrlich, DS, Fraser, C, Gibbs, R, Gilbert, J, Gilna, P, Glockner, FO, Jansson, JK, Keasling, JD, Knight, R, Labeda, D, Lapidus, A, Lee, JS, Li, WJ, Ma, J, Markowitz, V, Moore, ER, Morrison, M, Meyer, F, Nelson, KE, Ohkuma, M, Ouzounis, CA, Pace, N, Parkhill, J, Qin, N, Rossello-Mora, R, Sikorski, J, Smith, D, Sogin, M, Stevens, R, Stingl, U, Suzuki, K, Taylor, D, Tiedje, JM, Tindall, B, Wagner, M, Weinstock, G, Weissenbach, J, White, O, Wang, J, Zhang, L, Zhou, YG, Field, D, Whitman, WB, Garrity, GM, Klenk, HP, 2014, 'Genomic encyclopedia of bacteria and archaea: sequencing a myriad of type strains', PLoS Biol, vol. 12, pp. e1001920.

Lartigue, MF, Hery-Arnaud, G, Haguenoer, E, Domelier, AS, Schmit, PO, Van der Mee-Marquet, N, Lanotte, P, Mereghetti, L, Kostrzewa, M, Quentin, R, 2009, 'Identification of *Streptococcus agalactiae* isolates from various phylogenetic lineages by matrix-assisted laser desorption ionization-time of flight mass spectrometry', J Clin Microbiol, vol. 47, pp. 2284–2287.

Lau, AF, Drake, SK, Calhoun, LB, Henderson, CM, Zelazny, AM, 2013, 'Development of a clinically comprehensive database and a simple procedure for identification of molds from solid media by matrix-assisted laser desorption ionization-time of flight mass spectrometry', J Clin Microbiol, vol. 51, pp. 828–834.

Leaw, SN, Chang, HC, Sun, HF, Barton, R, Bouchara, JP, Chang, TC, 2006, 'Identification of medically important yeast species by sequence analysis of the internal transcribed spacer regions', J Clin Microbiol, vol. 44, pp. 693–699.

Liu, L, Li, Y, Li, S, Hu, N, He, Y, Pong, R, Lin, D, Lu, L, Law, M, 2012, 'Comparison of next-generation sequencing systems', J Biomed Biotechnol, vol. 2012, pp. 251364.

Ludwig, W, Klenk, HP, 2005, 'Overview: a phylogenetic backbone and taxonomic framework for prokaryotic systematics', in GM Garrity, DJ Brenner, NR Krieg, JT Staley (eds.), Bergey's Manual of Systematic Bacteriology, Springer-Verlag, New York.

Ludwig, W, Schleifer, KH, 1994, 'Bacterial phylogeny based on 16S and 23S rRNA sequence analysis', FEMS Microbiol Rev, vol. 15, pp. 155–173.

Ludwig, W, Strunk, O, Klugbauer, S, Klugbauer, N, Weizenegger, M, Neumaier, J, Bachleitner, M, Schleifer, KH, 1998, 'Bacterial phylogeny based on comparative sequence analysis', Electrophoresis, vol. 19, pp. 554–568.

Lupski, JR, Weinstock, GM, 1992, 'Short, interspersed repetitive DNA sequences in prokaryotic genomes', J Bacteriol, vol. 174, pp. 4525–4529.

Maiden, MC, 2006, 'Multilocus sequence typing of bacteria', Annu Rev Microbiol, vol. 60, pp. 561–588.

Maiden, MC, Bygraves, JA, Feil, E, Morelli, G, Russell, JE, Urwin, R, Zhang, Q, Zhou, J, Zurth, K, Caugant, DA, Feavers, IM, Achtman, M, Spratt, BG, 1998, 'Multilocus sequence typing: a portable approach to the identification of clones within populations of pathogenic microorganisms', Proc Natl Acad Sci USA, vol. 95, pp. 3140–3145.

Marklein, G, Josten, M, Klanke, U, Muller, E, Horre, R, Maier, T, Wenzel, T, Kostrzewa, M, Bierbaum, G, Hoerauf, A, Sahl, HG, 2009, 'Matrix-assisted laser desorption ionization-time of flight mass spectrometry for fast and reliable identification of clinical yeast isolates', J Clin Microbiol, vol. 47, pp. 2912–2917.

Maxam, AM, Gilbert, W, 1977, 'A new method for sequencing DNA', Proc Natl Acad Sci USA, vol. 74, pp. 560–564.

McMullen, AR, Wallace, MA, Pincus, DH, Wilkey, K, Burnham, CA, 2016, 'Evaluation of the Vitek MS matrix-assisted laser desorption ionization-time of flight mass spectrometry system for identification of clinically relevant filamentous fungi', J Clin Microbiol, vol. 54, pp. 2068–2073.

Meier-Kolthoff, JP, Auch, AF, Klenk, HP, Goker, M, 2013, 'Genome sequence-based species delimitation with confidence intervals and improved distance functions', BMC Bioinformatics, vol. 14, pp. 60.

Metzker, ML, 2010, 'Sequencing technologies: the next generation', Nat Rev Genet, vol. 11, pp. 31–46.

Mezzatesta, ML, Gona, F, Stefani, S, 2012, '*Enterobacter cloacae* complex: clinical impact and emerging antibiotic resistance', Future Microbiol, vol. 7, pp. 887–902.

Miller, JM, Rhoden, DL, 1991, 'Preliminary evaluation of Biolog, a carbon source utilization method for bacterial identification', J Clin Microbiol, vol. 29, pp. 1143–1147.

Mitchell, JI, Zuccaro, A, 2006, "Sequences the environment and fungi', Mycologist, vol. 20, pp. 62–74.

Mollet, C, Drancourt, M, Raoult, D, 1997, 'rpoB sequence analysis as a novel basis for bacterial identification', Mol Microbiol 26, pp. 1005–1011.

Neville, SA, Lecordier, A, Ziochos, H, Chater, MJ, Gosbell, IB, Maley, MW, Van Hal, SJ, 2011, 'Utility of matrix-assisted laser desorption ionization-time of flight mass spectrometry following introduction for routine laboratory bacterial identification', J Clin Microbiol, vol. 49, pp. 2980–2984.

Nie, M, Zhang, WQ, Xiao, M, Luo, JL, Bao, K, Chen, JK, Li, B, 2007, 'FT-IR spectroscopy and artificial neural network identification of *Fusarium* species', J Phytopathol, vol. 155, pp. 364–367.

Nilsson, RH, Tedersoo, L, Abarenkov, K, Ryberg, M, Kristiansson, E, Hartmann, M, Schoch, CL, Nylander, JAA, Bergsten, J, Porter, TM, Jumpponen, A, Vaishampayan, P, Ovaskainen, O, Hallenberg, N, Bengtsson-Palme, J, Eriksson,

KM, Larsson, KH, Larsson, E, Kõljalg, U, 2012, 'Five simple guidelines for establishing basic authenticity and reliability of newly generated fungal ITS sequences', MycoKeys, vol. 4, pp. 37–63.

Nyren, P, Pettersson, B, Uhlen, M, 1993, 'Solid phase DNA minisequencing by an enzymatic luminometric inorganic pyrophosphate detection assay', Anal Biochem, vol. 208, pp. 171–175.

O'Hara, CM, Steigerwalt, AG, Hill, BC, Farmer, JJ, 3rd, Fanning, GR, Brenner, DJ, 1989, '*Enterobacter hormaechei*, a new species of the family Enterobacteriaceae formerly known as enteric group 75', J Clin Microbiol, vol. 27, pp. 2046–2049.

Olsen, GJ, 1987, 'Earliest phylogenetic branchings: comparing rRNA-based evolutionary trees inferred with various techniques', Cold Spring Harb Symp Quant Biol, vol. 52, pp. 825–837.

Olsen, GJ, Woese, CR, 1989, 'A brief note concerning archaebacterial phylogeny', Can J Microbiol, vol. 35, pp. 113–123.

Olsen, GJ, Woese, CR, 1993, 'Ribosomal RNA: a key to phylogeny', Faseb J, vol. 7, pp. 113–123.

Olsen, GJ, Woese, CR, Overbeek, R, 1994, 'The winds of (evolutionary) change: breathing new life into microbiology', J Bacteriol, vol. 176, pp. 1–6.

Onodera, Y, Sato, K, 1999, 'Molecular cloning of the gyrA and gyrB genes of *Bacteroides fragilis* encoding DNA gyrase', Antimicrob Agents Chemother, vol. 43, pp. 2423–2429.

Orla-Jensen, S, 1921, 'The main lines of the natural bacterial system', J Bacteriol, vol. 6, pp. 263–273.

Pace, NR, 1997, 'A molecular view of microbial diversity and the biosphere', Science, vol. 276, pp. 734–740.

Pahlow, S, Meisel, S, Cialla-May, D, Weber, K, Rösch, P, Popp, J, 2015, 'Isolation and identification of bacteria by means of Raman spectroscopy', Adv Drug Deliv Rev, vol. 89, pp. 105–120.

Palleroni, NJ, Kunisawa, R, Contopoulou, R, Doudoroff, M, 1973, 'Nucleic acid homologies in the genus *Pseudomonas*', Int J Syst Evol Microbiol, vol. 23, pp. 333–339.

Palmer, M, Steenkamp, ET, Coetzee, MPA, Chan, WY, Van Zyl, E, De Maayer, P, Coutinho, TA, Blom, J, Smits, THM, Duffy, B, Venter, SN, 2017, 'Phylogenomic resolution of the bacterial genus *Pantoea* and its relationship with *Erwinia* and *Tatumella*', Antonie Van Leeuwenhoek, vol. 110, pp. 1287–1309.

Palys, T, Nakamura, LK, Cohan, FM, 1997, 'Discovery and classification of ecological diversity in the bacterial world: the role of DNA sequence data', Int J Syst Bacteriol, vol. 47, pp. 1145–1156.

Patel, JB, 2001, '16S rRNA gene sequencing for bacterial pathogen identification in the clinical laboratory', Mol Diagn, vol. 6, pp. 313–321.

Pavlovic, M, Konrad, R, Iwobi, AN, Sing, A, Busch, U, Huber, I, 2012, 'A dual approach employing MALDI-TOF MS and real-time PCR for fast species identification within the *Enterobacter cloacae* complex', FEMS Microbiol Lett, vol. 328, pp. 46–53.

Petti, CA, 2007, 'Detection and identification of microorganisms by gene amplification and sequencing', Clin Infect Dis, vol. 44, pp. 1108–1114.
Petti, CA, Polage, CR, Schreckenberger, P, 2005, 'The role of 16S rRNA gene sequencing in identification of microorganisms misidentified by conventional methods', J Clin Microbiol, vol. 43, pp. 6123–6125.
Porter, TM, Golding, GB, 2012, 'Factors that affect large subunit ribosomal DNA amplicon sequencing studies of fungal communities: classification method, primer choice, and error', PLoS One, vol. 7, pp. e35749.
Prakash, O, Verma, M, Sharma, P, Kumar, M, Kumari, K, Singh, A, Kumari, H, Jit, S, Gupta, SK, Khanna, M, Lal, R, 2007, 'Polyphasic approach of bacterial classification: an overview of recent advances', Indian J Microbiol, vol. 47, pp. 98–108.
Rahi, P, Prakash, O, Shouche, YS, 2016, 'Matrix-assisted laser desorption/ionization time-of-flight mass-spectrometry (MALDI-TOF MS) based microbial identifications: challenges and scopes for microbial ecologists', Front Microbiol, vol. 7, pp. 1359.
Rainey, FA, Ward-Rainey, NL, Janssen, PH, Hippe, H, Stackebrandt, E, 1996, '*Clostridium paradoxum* DSM 7308T contains multiple 16S rRNA genes with heterogeneous intervening sequences', Microbiology, vol. 142, pp. 2087–2095.
Richter, M, Rossello-Mora, R, 2009, 'Shifting the genomic gold standard for the prokaryotic species definition', Proc Natl Acad Sci USA, vol. 106, pp. 19126–19131.
Riesenfeld, CS, Schloss, PD, Handelsman, J, 2004, 'Metagenomics: genomic analysis of microbial communities', Annu Rev Genet, vol. 38, pp. 525–552.
Rosenvinge, FS, Dzajic, E, Knudsen, E, Malig, S, Andersen, LB, Lovig, A, Arendrup, MC, Jensen, TG, Gahrn-Hansen, B, Kemp, M, 2013, 'Performance of matrix-assisted laser desorption-time of flight mass spectrometry for identification of clinical yeast isolates', Mycoses, vol. 56, pp. 229–235.
Rowland, GC, Aboshkiwa, M, Coleman, G, 1993, 'Comparative sequence analysis and predicted phylogeny of the DNA-dependent RNA polymerase beta subunits of *Staphylococcus aureus* and other eubacteria', Biochem Soc Trans, vol. 21, pp. 40S.
Ryzhov, V, Fenselau, C, 2001, 'Characterization of the protein subset desorbed by MALDI from whole bacterial cells', Anal Chem, vol. 73, pp. 746–750.
Sacchi, CT, Whitney, AM, Reeves, MW, Mayer, LW, Popovic, T, 2002, 'Sequence diversity of *Neisseria meningitidis* 16S rRNA genes and use of 16S rRNA gene sequencing as a molecular subtyping tool', J Clin Microbiol, vol. 40, pp. 4520–4527.
Saha, R, Sproer, C, Beck, B, Bagley, S, 2010, '*Pseudomonas oleovorans* subsp. lubricantis subsp. nov., and reclassification of *Pseudomonas pseudoalcaligenes* ATCC 17440T as later synonym of *Pseudomonas oleovorans* ATCC 8062 T', Curr Microbiol, vol. 60, pp. 294–300.
Sanger, F, Nicklen, S, Coulson, AR, 1977, 'DNA sequencing with chain-terminating inhibitors', Proc Natl Acad Sci USA, vol. 74, pp. 5463–5467.

Sato, H, Teramoto, K, Ishii, Y, Watanabe, K, Benno, Y, 2011, 'Ribosomal protein profiling by matrix-assisted laser desorption/ionization time-of-flight mass spectrometry for phylogenety-based subspecies resolution of *Bifidobacterium longum*', Syst Appl Microbiol, vol. 34, pp. 76–80.

Sauer, S, Freiwald, A, Maier, T, Kube, M, Reinhardt, R, Kostrzewa, M, Geider, K, 2008, 'Classification and identification of bacteria by mass spectrometry and computational analysis', PLoS One, vol. 3, pp. e2843.

Schoch, CL, Seifert, KA, Huhndorf, S, Robert, V, Spouge, JL, Levesque, CA, Chen, W, 2012, 'Nuclear ribosomal internal transcribed spacer (ITS) region as a universal DNA barcode marker for fungi', Proc Natl Acad Sci USA, vol. 109, pp. 6241–6246.

Schwartz, DC, Cantor, CR, 1984, 'Separation of yeast chromosome-sized DNAs by pulsed field gradient gel electrophoresis', Cell, vol. 37, pp. 67–75.

Seng, P, Drancourt, M, Gouriet, F, La Scola, B, Fournier, PE, Rolain, JM, Raoult, D, 2009, 'Ongoing revolution in bacteriology: routine identification of bacteria by matrix-assisted laser desorption ionization time-of-flight mass spectrometry', Clin Infect Dis, vol. 49, pp. 543–551.

Shah, HN, Rajakaruna, L, Ball, G, Misra, R, Al-Shahib, A, Fang, M, Gharbia, SE, 2011, 'Tracing the transition of methicillin resistance in sub-populations of *Staphylococcus aureus*, using SELDI-TOF mass spectrometry and artificial neural network analysis', Syst Appl Microbiol, vol. 34, pp. 81–86.

Shi, X, Rao, NN, Kornberg, A, 2004, 'Inorganic polyphosphate in *Bacillus cereus*: motility, biofilm formation, and sporulation', Proc Nat Acad Sc USA, vol. 101, pp. 17061–17065.

Simner, PJ, Uhl, JR, Hall, L, Weber, MM, Walchak, RC, Buckwalter, S, Wengenack, NL, 2013, 'Broad-range direct detection and identification of fungi by use of the PLEX-ID PCR-electrospray ionization mass spectrometry (ESI-MS) system', J Clin Microbiol, vol. 51, pp. 1699–1706.

Siqueira, JF, Jr., Rocas, IN, 2005, 'Exploiting molecular methods to explore endodontic infections: part 1 Current molecular technologies for microbiological diagnosis', J Endod, vol. 31, pp. 411–423.

Skerman, VBD, McGowan, V, Sneath, PHA, 1980, 'Approved lists of bacterial names', Int J Syst Evol Microbiol, vol. 30, pp. 225–420.

Southern, EM, 1975, 'Detection of specific sequences among DNA fragments separated by gel electrophoresis', J Mol Biol, vol. 98, pp. 503–517.

Stackebrandt, E, Ebers, J, 2006, 'Taxonomic parameters revisited: tarnished gold standards', Microbiol Today, vol. 33, pp. 152–155.

Stackebrandt, E, Goebel, BM, 1994, 'Taxonomic note: a place for DNA-DNA reassociation and 16S rRNA sequence analysis in the present species definition in bacteriology', Int J Syst Evol Microbiol, vol. 44, pp. 846–849.

Stackebrandt, E, Frederiksen, W, Garrity, GM, Grimont, PAD, Kämpfer, P, Maiden, MCJ, Nesme, X, Rosselló-Mora, R, Swings, J, Trüper, HG, Vauterin, L, Ward, AC, Whitman, WB, 2002, 'Report of the ad hoc committee for the

re-evaluation of the species definition in bacteriology', Int J Syst Evol Microbiol, vol. 52, pp. 1043–1047.
Stager, CE, Davis, JR, 1992, 'Automated systems for identification of microorganisms', Clin Microbiol Rev, vol. 5, pp. 302–327.
Sutton, SVW, Cundell, AM, 2004, 'Microbial identification in the pharmaceutical industry', Pharmacopeial Forum, vol. 30, pp. 1884–1894.
Tachon, S, Michelon, D, Chambellon, E, Cantonnet, M, Mezange, C, Henno, L, Cachon, R, Yvon, M, 2009, 'Experimental conditions affect the site of tetrazolium violet reduction in the electron transport chain of *Lactococcus lactis*', Microbiology, vol. 155, pp. 2941–2948.
Tamames, J, Rossello-Mora, R, 2012, 'On the fitness of microbial taxonomy', Trends Microbiol, vol. 20, pp. 514–516.
Tan, KE, Ellis, BC, Lee, R, Stamper, PD, Zhang, SX, Carroll, KC, 2012, 'Prospective evaluation of a matrix-assisted laser desorption ionization-time of flight mass spectrometry system in a hospital clinical microbiology laboratory for identification of bacteria and yeasts: a bench-by-bench study for assessing the impact on time to identification and cost-effectiveness', J Clin Microbiol, vol. 50, pp. 3301–3308.
Tanaka, K, Waki, H, Ido, Y, Akita, S, Yoshida, Y, Yoshida, T, Matsuo, T, 1988, 'Protein and polymer analyses up to m/z 100 000 by laser ionization time-of-flight mass spectrometry', Rapid Commun Mass Spectrom, vol. 2, pp. 151–153.
Taylor, JW, Jacobson, DJ, Kroken, S, Kasuga, T, Geiser, DM, Hibbett, DS, Fisher, MC, 2000, 'Phylogenetic species recognition and species concepts in fungi', Fungal Genet Biol, vol. 31, pp. 21–32.
Tenover, FC, Arbeit, RD, Goering, RV, Mickelsen, PA, Murray, BE, Persing, DH, Swaminathan, B, 1995, 'Interpreting chromosomal DNA restriction patterns produced by pulsed-field gel electrophoresis: criteria for bacterial strain typing', J Clin Microbiol, vol. 33, pp. 2233–2239.
Teramoto, K, Sato, H, Sun, L, Torimura, M, Tao, H, Yoshikawa, H, Hotta, Y, Hosoda, A, Tamura, H, 2007, 'Phylogenetic classification of *Pseudomonas putida* strains by MALDI-MS using ribosomal subunit proteins as biomarkers', Anal Chem, vol. 79, pp. 8712–8719.
Thorne, JL, Kishino, H, Painter, IS, 1998, 'Estimating the rate of evolution of the rate of molecular evolution', Mol Biol Evol, vol. 15, pp. 1647–1657.
Tindall, BJ, Rossello-Mora, R, Busse, HJ, Ludwig, W, Kampfer, P, 2010, 'Notes on the characterization of prokaryote strains for taxonomic purposes', Int J Syst Evol Microbiol, vol. 60, pp. 249–266.
Urwyler, SK, Glaubitz, J, 2016, 'Advantage of MALDI-TOF-MS over biochemical-based phenotyping for microbial identification illustrated on industrial applications', Lett Appl Microbiol, vol. 62, pp. 130–137.
USP, 2012a, 'Chapter <1111> Microbiological Examination of Non-Sterile Products: Acceptance Criteria for Pharmaceutical Preparations and Substances for Pharmaceutical Use.' In *USP 35*, 691. Rockville, MD: The United States Pharmacopeial Convention/National Formulary.

USP, 2012b, 'Chapter <1113> Microbial Characterization, Identification, and Strain Typing.' In *USP 35*, 694–97. Rockville, MD: The United States Pharmacopeial Convention/National Formulary.

USP, 2012c, 'Chapter <1117> Microbial Best Laboratory Practices.' In *USP 35*, 707–12. Rockville, MD: The United States Pharmacopeial Convention/National Formulary.

Valentine, N, Wunschel, S, Wunschel, D, Petersen, C, Wahl, K, 2005, 'Effect of culture conditions on microorganism identification by matrix-assisted laser desorption ionization mass spectrometry', Appl Environ Microbiol, vol. 71, pp. 58–64.

Van Veen, SQ, Claas, EC, Kuijper, EJ, 2010, 'High-throughput identification of bacteria and yeast by matrix-assisted laser desorption ionization-time of flight mass spectrometry in conventional medical microbiology laboratories', J Clin Microbiol, vol. 48, pp. 900–907.

Vandamme, P, Pot, B, Gillis, M, De Vos, P, Kersters, K, Swings, J, 1996, 'Polyphasic taxonomy, a consensus approach to bacterial systematics', Microbiol Rev, vol. 60, pp. 407–438.

Vandamme, P, Harrington, CS, Jalava, K, On, SL, 2000, 'Misidentifying helicobacters: the *Helicobacter cinaedi* example', J Clin Microbiol, vol. 38, pp. 2261–2266.

Vandamme, P, Moore, ER, Cnockaert, M, De Brandt, E, Svensson-Stadler, L, Houf, K, Spilker, T, Lipuma, JJ, 2013, '*Achromobacter animicus* sp. nov., *Achromobacter mucicolens* sp. nov., *Achromobacter pulmonis* sp. nov. and *Achromobacter spiritinus* sp. nov., from human clinical samples', Syst Appl Microbiol, vol. 36, pp. 1–10.

Versalovic, J, Koeuth, T, Lupski, JR, 1991, 'Distribution of repetitive DNA sequences in eubacteria and application to fingerprinting of bacterial genomes', Nucleic Acids Res, vol. 19, pp. 6823–6831.

Vos, P, Hogers, R, Bleeker, M, Reijans, M, Van de Lee, T, Hornes, M, Frijters, A, Pot, J, Peleman, J, Kuiper, M, Zabeau, M, 1995, 'AFLP: a new technique for DNA fingerprinting', Nucleic Acids Res, vol. 23, pp. 4407–4414.

Vralstad, T, 2011, 'ITS, OTUs and beyond-fungal hyperdiversity calls for supplementary solutions', Mol Ecol, vol. 20, pp. 2873–2875.

Wayne, LG, Brenner, DJ, Colwell, RR, Grimont, PAD, Kandler, O, Krichevsky, MI, Moore, LH, Moore, WEC, Murray, RGE, Stackebrandt, E, Starr, MP, Truper, HG, 1987, 'Report of the ad hoc committee on reconciliation of approaches to bacterial systematics', Int J Syst Evol Microbiol, vol. 37, pp. 463–464.

Welker, M, Moore, ER, 2011, 'Applications of whole-cell matrix-assisted laser-desorption/ionization time-of-flight mass spectrometry in systematic microbiology', Syst Appl Microbiol, vol. 34, pp. 2–11.

Woese, CR, 1987, 'Bacterial evolution', Microbiol Rev, vol. 51, pp. 221–271.

Woese, CR, Fox, GE, 1977, 'Phylogenetic structure of the prokaryotic domain: the primary kingdoms', Proc Natl Acad Sci USA, vol. 74, pp. 5088–5090.

Woese, CR, Stackebrandt, E, Macke, TJ, Fox, GE, 1985, 'A phylogenetic definition of the major eubacterial taxa', Syst Appl Microbiol, vol. 6, pp. 143–151.

Woese, CR, Kandler, O, Wheelis, ML, 1990, 'Towards a natural system of organisms: proposal for the domains Archaea, Bacteria, and Eucarya', Proc Natl Acad Sci USA, vol. 87, pp. 4576–4579.

Wunschel, DS, Hill, EA, McLean, JS, Jarman, K, Gorby, YA, Valentine, N, Wahl, K, 2005a, 'Effects of varied pH, growth rate and temperature using controlled fermentation and batch culture on matrix assisted laser desorption/ionization whole cell protein fingerprints', J Microbiol Methods, vol. 62, pp. 259–271.

Wunschel, SC, Jarman, KH, Petersen, CE, Valentine, NB, Wahl, KL, Schauki, D, Jackman, J, Nelson, CP, White, E, 2005b, 'Bacterial analysis by MALDI-TOF mass spectrometry: an inter-laboratory comparison', J Am Soc Mass Spectrom, vol. 16, pp. 456–462.

Yarza, P, Sproer, C, Swiderski, J, Mrotzek, N, Spring, S, Tindall, BJ, Gronow, S, Pukall, R, Klenk, HP, Lang, E, Verbarg, S, Crouch, A, Lilburn, T, Beck, B, Unosson, C, Cardew, S, Moore, ER, Gomila, M, Nakagawa, Y, Janssens, D, De Vos, P, Peiren, J, Suttels, T, Clermont, D, Bizet, C, Sakamoto, M, Iida, T, Kudo, T, Kosako, Y, Oshida, Y, Ohkuma, M, Arahal, R, Spieck, DE, Roeser, AP, Figge, M, Park, D, Buchanan, P, Cifuentes, A, Munoz, R, Euzeby, JP, Schleifer, KH, Ludwig, W, Amann, R, Glockner, FO, Rossello-Mora, R, 2013, 'Sequencing orphan species initiative (SOS): filling the gaps in the 16S rRNA gene sequence database for all species with validly published names', Syst Appl Microbiol, vol. 36, pp. 69–73.

Zbinden, A, Böttger, EC, Bosshard, PP, Zbinden, R, 2007, 'Evaluation of the colorimetric VITEK 2 card for identification of Gram-negative nonfermentative rods: comparison to 16S rRNA gene sequencing', J Clin Microbiol, vol. 45, pp. 2270–2273.

Zhang, Y, Qiu, S, 2015, 'Examining phylogenetic relationships of *Erwinia* and *Pantoea* species using whole genome sequence data', Antonie Van Leeuwenhoek, vol. 108, pp. 1037–1046.

Zuckerkandl, E, Pauling, L, 1965, 'Molecules as documents of evolutionary history', J Theor Biol, vol. 8, pp. 357–366.

10

Calculating Alert Levels and Trending of Microbiological Data

David Roesti

Novartis Pharma Stein AG, Stein, Switzerland

CONTENTS

Pharmaceutical Microbiological Quality Assurance and Control: Practical Guide for Non-Sterile Manufacturing, First Edition. Edited by David Roesti and Marcel Goverde.

10.1 Introduction

Trending of in-process control or release data is an integral part of current cGMP and quality risk management. In addition, the recommended monitoring levels or acceptance criteria in the guidelines or Pharmacopoeias do not take into account the uniqueness of each manufacturing line or product type and it is more and more expected that these levels take into account the expected performance data.

The ICH Q9 quotes that *It is important to understand that product quality should be maintained throughout the product lifecycle such that the attributes that are important to the quality of the drug (medicinal) product remain consistent with those used in the clinical studies. An effective quality risk management approach can further ensure the high quality of the drug (medicinal) product to the patient by providing a proactive means to identify and control potential quality issues during development and manufacturing.*

The ICH Q10 writes that firms should establish and maintain a state of control and facilitate continuous improvement. Trending analysis of data is needed *To develop and use effective monitoring and control systems for process performance and product quality, thereby providing assurance of continued suitability and capability of processes. Quality risk management can be useful in identifying the monitoring and control systems.*

Some major health authority guidelines explicitly require trending of microbiological data and setting monitoring levels based on the process performance.

For instance, the new EU annex 1 2017 revised draft writes *Regular ongoing chemical and microbial monitoring of water systems should be performed with alert limits based on the qualification that will identify an* ***adverse trend*** *in the performance of the systems. (...). A breach of an alert limit should trigger review and follow-up, which might include investigation and corrective action. Any breach of an action limit should lead to a root cause investigation and risk assessment. (...) Appropriate alert and action limits should be set for the results of particulate and microbiological monitoring. Alert levels should be established based on results of Performance Qualification (PQ) tests or* ***trend data*** *and should be subject to periodic review.* and defines alert levels as *An established microbial or airborne particle level giving early warning of potential drift from* *normal operating conditions and triggers appropriate scrutiny and follow up to address the potential problem. Alert levels are always lower than action levels and are established based on historical and qualification trend data and periodically reviewed.*

The 2004 FDA Guidance for industry writes *Microbiological monitoring levels (…) should be based on the need to maintain adequate microbiological control throughout the entire sterile manufacturing facility. One should also consider* ***environmental monitoring data from historical databases,*** *media fills, cleanroom qualification, and sanitization studies, in developing monitoring levels. Data from similar operations can also be helpful in setting action and alert levels, especially for a new operation. (…) Each individual sample result should be evaluated for its significance by comparison to the alert or action levels. (…) We recommend review of trends in product bioburden and consideration of whether adverse bioburden trends have occurred.*

FDA Compliance Program Guidance Manual, Sterile Drug Process Inspections 2016 writes: ***Are the microbial alert and action levels based on the historical EM data*** *derived from the manufacturing operations, support utilities and personnel practices performed at the manufacturing site? (…) Review and assess the EM trending data, which will provide a good indication if the viable and non-viable particles are maintained within the established levels or drifting out of control. What are the causes of the aberrant events? Were corrective actions and preventive measures taken to preclude the reoccurrence of the viable and non-viable particle anomalies?*

The above-mentioned guidelines are nevertheless written for the manufacturing of sterile products and few guidance is provided for trending of microbiological data for non-sterile product manufacturing.

Actually, the Chinese FDA guidance Good Manufacturing Practice for Drugs (2010 Revision) also valid for non-sterile product manufacturing does write that: *For some kinds of data (e.g. testing results, environment monitoring data, microorganism monitoring data of water for pharmaceutical use), it is recommended that records be kept in a manner permitting* ***trend*** *evaluation.*

Anvisa (2013) Guide for Quality of Air Treatment Systems and Environmental Monitoring in the Pharmaceutical Industry writes: *Alert limit values for particles or microorganisms are values lower than the regulatory maximum, but they must be sufficiently* ***above the normal variation*** *of historically found results of contaminants. The response to a value above the alert limit is many times only a note regarding the event that will serve as the basis for a potential* ***trend analysis,*** *that is, to verify if the event is not part of a set of abnormally high values. (…) In the absence of other aggravating circumstances, result* ***trends*** *are an important instrument to determine if an event is indicative of a serious problem or not. (…) The* ***trends*** *over time must be documented and presented in such a manner that normal and abnormal values are readily identified during the analysis.*

In addition, it is expected that non-sterile product manufacturing also meet the requirements of the ICH guidelines.

It is then no surprise that trending of microbiological data and setting of monitoring levels using historical data as part of the quality management system of non-sterile product manufacturing is now increasingly being requested by health authorities during inspections.

Definitions

Alert level
Quality level that, when exceeded, signals a potential drift from normal operating conditions and triggers appropriate scrutiny and follow-up to address the potential problem. Alert levels are always lower than action levels

Action level
Quality level that, when exceeded, signals an apparent drift from normal operating conditions and which requires an immediate action(s) by previously defined documented measures.

Exceeding the alert level or action level is usually a *single* event.

Trend
A statistical term referring to the direction or rate of change of variables.

Adverse trend
An adverse microbiological trend is an early warning of a potential degradation or loss of control within the environment, the utility, raw material, or product tested. An adverse trend can, for instance, be defined as repeating higher than usual counts or increasing amount of microorganisms or contamination occurrences over a certain time period.

Adverse trend is systematically related to *multiple* events (Figure 10.1).

10.2 Goal of the Chapter

The goal of this chapter is to give the reader an insight of different tools that may be used to calculate alert levels and trend data using highly variable microbiological results. The tools described below may then be used for multiple applications such as monitoring of the environment or water, or microbiological testing of excipients, drug substances, and drug products.

10.3 Alert Levels Based on Historical Data

10.3.1 Introduction

The alert level must be set as to differentiate single unusual high counts from the normal baseline of counts. It must be set not too high as exceeding

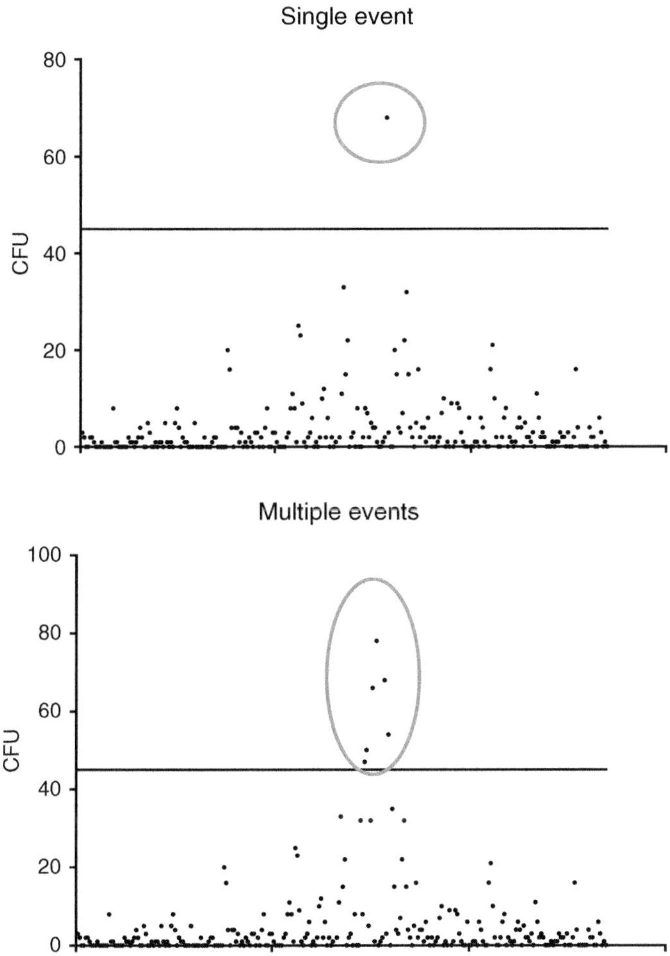

Figure 10.1 Plotting of microbiological monitoring data representing single and multiple events.

counts would not be captured and not too low as "false alarms" would be excessively raised.

A very pragmatic way to set alert levels based on historical data would be to plot the microbial counts on a X/Y graph with X being the timescale and Y the counts and then to set the level based on visual analysis of the graph. Whereas this approach is very simple to apply, it is totally subjective and would depend on the level of experience and personality (conservative or not) of the microbiologist. In the two data set examples shown in Figure 10.2 that are derived from environmental monitoring data of non-sterile manufacturing areas, setting of

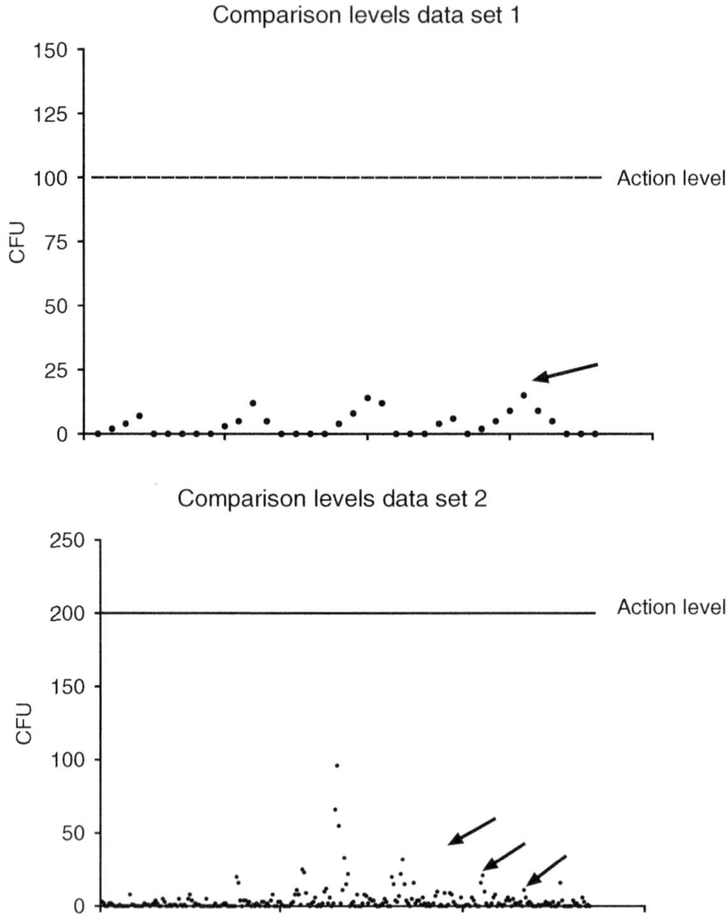

Figure 10.2 Plotting of microbiological monitoring data. Arrow represents potential areas to set the alert level.

an alert level is quite straightforward for the first set of data and less evident for the second data set.

To reduce the part of subjectivity, the alert levels could be calculated using statistical methods. In the pharmaceutical industry a large variety of statistical tools exist to monitor process performance and define microbiological monitoring levels that differentiate single deviating events from baseline variability. However, these tools are generally based on normal distributed data and might not fit to microbiological data. The following sections will discuss the different control chart tools that may be potentially used as well as the calculation of levels using the percentile method assuming or not distribution fits.

10.3.2 Distribution of Microbiological Data

The main distribution models that will be discussed in this section are described in Figures 10.3–10.5.

Microbiological data from bioburden of finished products, product intermediates, or from environmental monitoring are generally not normally distributed, contain many zero values, and are highly dispersed.

As suggested by Yang *et al.* (2013) and Gordon *et al.* (2015), microbiological data resulting from monitoring activities tend to follow a negative binomial distribution. The negative binomial model is capable of dealing with highly dispersed data as, for instance, the natural bioburden of finished products, product intermediates, or the environment. Because it is a discrete distribution, it may also be used for data obtained with microbial counts.

Figure 10.6 illustrates the relation of two microbiological data sets from environmental monitoring results with the different distribution models. As shown by Figure 10.6, the negative binomial distribution fits better data with many zero counts and high variability.

Yang *et al.* (2013) suggest that a zero inflated negative binomial model would fit better microbiological data than negative binomial. It is applied to explain extra zeroes in the data (it is parameterized with the same parameters as the negative binomial distribution with an additional parameter that explains the additional proportion of zeroes). Gordon *et al.* (2015) also suggest that

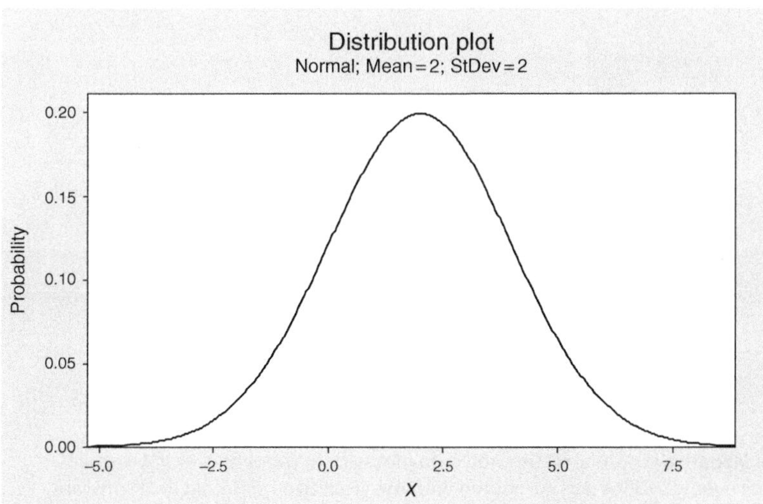

Figure 10.3 Normal distributed data plot with mean = 2 and standard deviation of 2. Plot was computed with the Minitab 17 (Minitab Inc.) software.

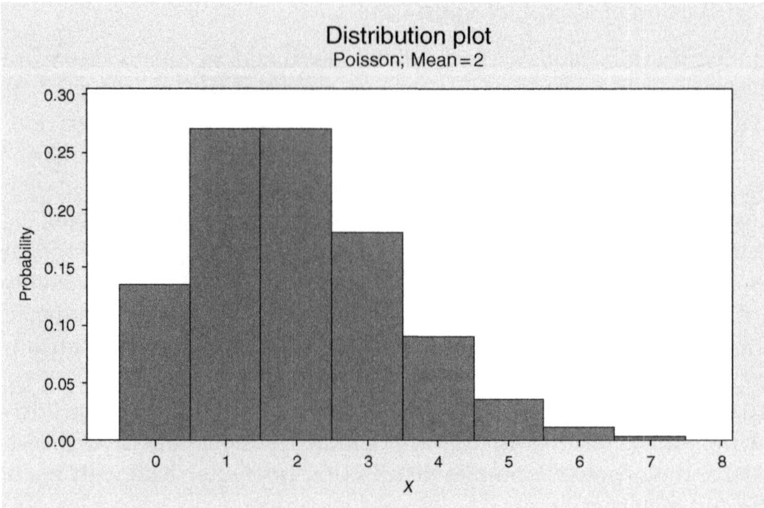

Figure 10.4 Poisson distributed data plot with average = variance = 2. Plot was computed with the Minitab 17 (Minitab Inc.) software.

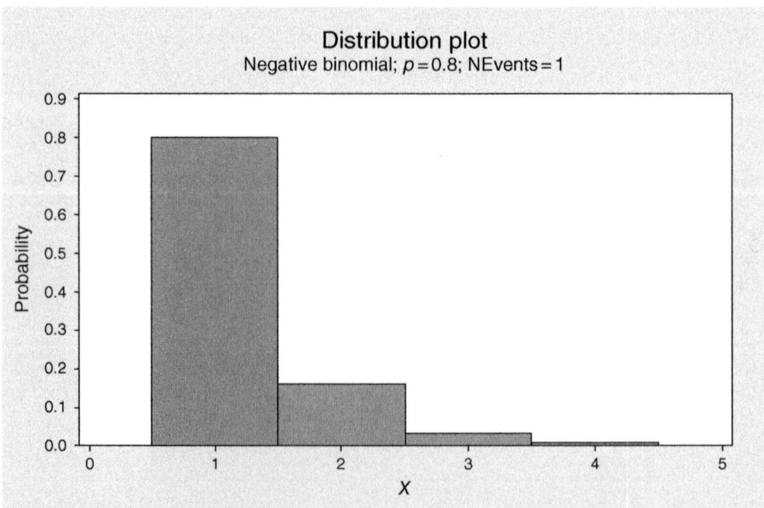

Figure 10.5 Negative binomial distributed data plot with number of events 1 and probability of event 0.8. Plot was computed with the Minitab 17 (Minitab Inc.) software.

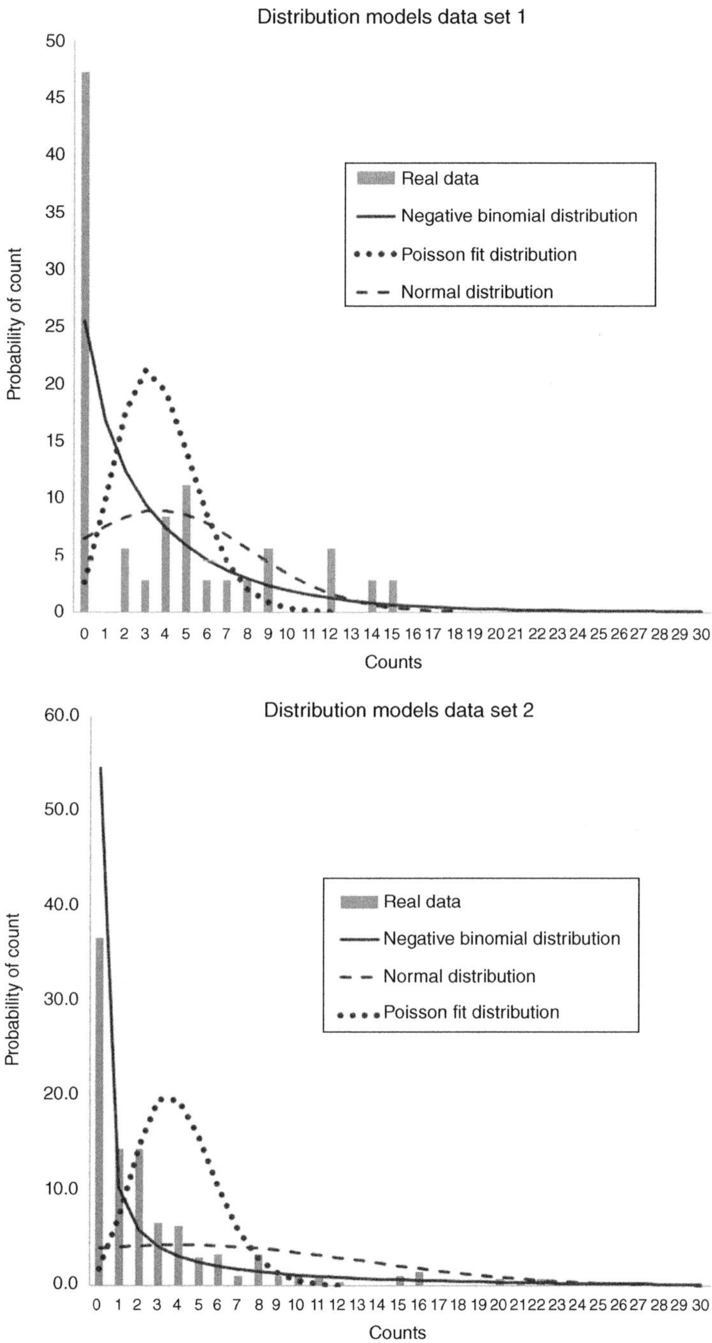

Figure 10.6 Plotting of two different data sets of microbiological data against different distribution models.

the Gamma fit model may be used as unique model to calculate alert levels. This method applies continuous data (however, microbiological data are discrete).

Between the negative binomial distribution, the zero-inflated negative binomial variant and the gamma fit model differences in the obtained calculated levels are generally marginal (e.g. 92 CFU instead of 91 CFU) using microbiological data. In addition, non-homogenous distribution of microorganisms in products, raw materials, utilities, or in the environment and due to the limitations of the microbiological sampling and test methods, expecting a perfect fit of data to a particular distribution model is unlikely.

10.3.3 Data to Be Included in the Calculations

Data may be retrieved from the LIMS systems and copied directly in statistical calculation tools or Excel sheets. This process should meet data integrity rules to avoid mixing up or modification of data. Automatic or semiautomatic calculation tools (e.g. Excel spreadsheets) may be used as long as they are appropriately validated.

Depending on the type of data analyzed, data may be taken from single source of testing (e.g. microbiological test results of different batches from a product) or multiple sources (e.g. different sampling locations in a cleanroom).

Especially for monitoring, data are recommended to cover more than a year's time, e.g. to capture the seasonal variation (if any) or modification in production rhythms.

10.3.3.1 Clustering of Data

Often not enough data points are obtained in non-sterile manufacturing due to low testing frequencies. When justified, clustering data together overcome the limitations of the sample size.

Data from different sampling locations may be clustered if:

- They are of the same test type (e.g. active air monitoring data cannot be combined with surface monitoring)
- They originate from areas of comparable design/process (e.g. compounding cleanrooms might not necessarily be combined with washing rooms; rooms with different grades)
- They originate from areas in which a similar microbiological burden is expected (e.g. packaging area not combined with filling area)

10.3.3.2 Data that Can Be Excluded from the Calculation

Outliers may be excluded from the calculations if there is a suitable justification. Utilization of statistical tools to remove outliers or manual exclusion to "smoothen" the pattern is not recommended since the remaining data points would not represent the overall variability of data and calculated levels might be underestimated.

If the typical counts of, e.g. a cleanroom lie at the level of the action level or even above it, the cleanroom hygiene performance may not be satisfactory for its grade. Corrective actions should then be implemented (e.g. increase room ventilation, change personnel flow, modifying the cleaning/disinfection procedure).

The following data may, for instance, be excluded from the calculation (list not exhaustive):

- Deviation counts that resulted in a clear root cause assessment and corrective and preventive actions.
- Data from an uncontrolled area following temporary shutdown.
- Older counts before a major change that affected the bioburden (e.g. upgrade/downgrade of a cleanroom).
- Former excipient microbial enumeration counts following a change in supplier/quality grade.

10.3.3.3 Periodic Reassessment of Alert Levels

Microbial alert levels may be reassessed periodically if sufficient additional results can be integrated in the calculation. Generally, a period of one to two years suffices to obtain sufficient data. Another reason for reassessing data is when significant changes occur that may significantly affect the level of microorganisms (e.g. change in clean room grade, significant increase in production activity, change in the product's formulation).

Some regulators are of the opinion that even bioburden levels in processes should be continuously improved implying that the recalculated levels must be systematically lower than the previous one.

Nonetheless, this would mean that levels would be tightened so much that they would approach process capability without any added value in terms of product safety and quality. It should be allowed not to tighten the levels systematically with a scientifically justified risk assessment. A justified example could be that production volumes had momentarily decreased resulting in a diminution of cleanroom activity hence temporarily improving the overall bioburden level. Taking into account the normal production activity, this level would be expected to be higher (as demonstrated in previous calculations with normal production activity data).

10.3.4 Calculating Alert Levels Using Control Charts

Control charts also known as Shewhart charts are a statistical process control tool used to determine if a manufacturing process is in a state of control (e.g. PDA TR-59). The basic idea of control charts is to differentiate common cause variation (= noise or acceptable count levels) from single causes (= atypical high count or deviation).

In process control charts the variation of data is assessed with the standard deviation or sigma (σ) value. The standard deviation is a measure that is used to quantify the amount of variation or dispersion of a set of data values. A low standard deviation indicates that the data points tend to be close to the mean (also called the expected value) of the set, while a high standard deviation indicates that the data points are spread out over a wider range of values.

With normal distributed data 99.73% of the data would fall within the mean $\pm 3 \times \sigma$. The rule of $3 \times \sigma$ limits from the center line is applied as a convention and is considered by the industry and regulators as an appropriate way to determine if a process is under control. Mean $+ 3 \times \sigma$ defines the upper control limit (UCL). The UCL may vary depending on multiple factors such as the type of control chart used, the type of sigma used, and utilization of unbiased constants or not.

It is to note that for common control charts (e.g. Xbar-R charts or I-charts) a prerequisite is that data are normally distributed which is generally not the case for microbiological counts. These charts require also data to be continuous which CFU counts are not (they are actually discrete data). There exists literature report (Bar 2015) that plotting data on control charts still may depict realistically the behavior of the microbial monitoring process even if the data are not normally distributed.

Nonetheless, it is better practice to first normalize data before plotting the data on control charts to make a more precise estimation of upper microbiological monitoring levels.

Data may be transformed to approximate a normal distribution using the following methods:

$$Y' = \mathrm{Log}_{10}(Y+1)$$

$$Y' = LN(Y+1)$$

$$Y' = \sqrt{Y}$$

$$Y' = \frac{1}{\sqrt{Y+1}}$$

$$Y' = \frac{1}{Y+1}$$

$$Y' = Y^2$$

To evaluate which value would result in the closest approximation to a normal distribution curve method, the Box Cox transformation may be used.

The Minitab 17 (Minitab Inc.) contains the Box Cox function which estimates an appropriate exponent lambda that varies from −5 to 5. The data are then transformed using the transformation method associated to the lambda value (Table 10.1).

Table 10.1 Common transformations where Y' is the transform of the data Y using the Box Cox transformation.

Lambda	Transformation method
2	$Y' = Y^2$
0.05	$Y' = \sqrt{Y}$
0	$Y' = LN(Y)$
−0.5	$Y' = \frac{1}{\sqrt{Y}}$
−1	$Y' = \frac{1}{Y}$

To use the Box Cox transformation the data must be greater than zero. In that case if data contain zero values, they can be transformed as count +1 before running the Box Cox transformation.

The lambda for the data sets 1 and 2 from Figure 10.2 was calculated using the Box Plot function from Minitab. Transformed data were then subjected to a normality test (Table 10.2).

It follows that even using the best transformation method, the data do not necessarily approximate a normal distribution. The more the data deviate from normality, the less precise the estimation. Wheeler (2018) has calculated that three-sigma limits yield values 97.5% quantiles with any probability distribution that provides a reasonable model for a predictable process. Control charts have per default an UCL that would correspond to a 99.86th percentile. When comparing the UCLs calculated with the control charts of untransformed data and the alert levels calculated using percentiles and distribution models (Table 10.4), the UCL is systematically lower whatever the sigma used even if theoretically one would expect a higher count (99.86% quantile is higher than a 99% quantile). This difference results from the fact that the 99.86% quantile applies for data that strictly follow a normal distribution.

Table 10.2 Probability of a normal distribution calculated with the Shapiro–Wilk test using untransformed and transformed data sets 1 and 2 from Figure 10.2.

	Data set 1	Data set 2
Lambda	−0.5	0.26
Shapiro–Wilk test	$p = 0.0001$	$p < 0.0001$

If $p > 0.05$, then a normal distribution is assumed. Data were first transformed $Y + 1$ to rule out 0 values and analyzed with the Box Cox function of Minitab 17.

Table 10.3 Percentage of observations falling either within mean ± k sigma ranges or the upper level of mean + k sigma assuming a normal distribution.

k	$\mu \pm k \times \sigma$	$\mu + k \times \sigma$
3	99.73	99.86
2.576	99.00	99.50
2.326	98.00	99.00
2	95.46	97.73
1.960	95.00	97.50
1.645	90.00	95.00
1	68.27	84.13

10.3.4.1 Examples of Control Charts

The simplest variable control chart is the I-chart where I stands for individuals. I-charts are used as process control tools for continuous data to detect the presence of special causes when the sample size is 1. By default, I-charts computed in the Mintab 17 (Minitab Inc.) estimates the process variation, σ, with the average of the moving range divided by an unbiasing constant based on Wheeler and Chambers (1992). Often, the choice to use them depends on company policy or industry standards. Unbiasing constants reduce the bias that can occur when a parameter is estimated from a small number of observations. As the number of observations increases, unbiasing constants have less effect on the calculated results.

The UCL is calculated as a specified number of standard deviations above the center line which is the average of all individual observations.

10.3.4.1.1 Upper Control Limit Calculation

$$\text{UCL} = \mu + 3 \times \sigma$$

The classical formulation of standard deviation of individual measurements from a population is shown in the following section.

10.3.4.1.2 Classical Standard Deviation

$$\sigma = \sqrt{\frac{\sum_{i=1}^{N}(x_i - \bar{x})^2}{N-1}}$$

$\bar{x}$ = average of measurements
x_i = individual measurement
N = number of measurements

For the I-chart calculation the Minitab 17 software uses different types of calculations for sigma. The most commonly used is the square root of the mean of the squared differences (MSSD) between consecutive points. This method is used when one cannot reasonably assume that at least two consecutive points were collected under similar conditions (which is generally the case for microbiological data).

10.3.4.1.3 MSSD Standard Deviation

$$\sigma = \frac{\sqrt{\frac{\sum_{i=2}^{N}(x_{i+1}-x_i)^2}{2(N-1)}}}{C_4'(N)}$$

x_i = individual measurement
N = number of measurements
$C_4'(N)$ = unbiasing constant

The I-chart calculation results in a plot of individual values versus the observation number with upper and lower control limits (Figure 10.7). Outlier results may be flagged based on defined criteria (e.g. all values exceeding the center line plus three times the standard deviation). In order to further approximate the normal distribution, the I-chart of the data sets were recalculated following $\log_{10}(\text{count}+1)$ and compared with untransformed data. The UCL was back transformed to provide comparison data. Note that the transformed data still failed to pass a normality test (data not shown).

Other control charts that are typically used in process control evaluation are:

- **Xbar-R chart** which is a variable control chart for continuous data with a constant subgroup size of less than eight. Xbar-S charts are used for higher subgroup sizes. The Xbar-chart plots the average of a subgroup as a data point. The R-chart plots the difference between the highest and lowest values within a subgroup as a data point. Subgroups may, for instance, be composed of all samples taken in a cleanroom or multiple bags tested from an excipient batch at a certain timepoint.
- **C-charts** are attribute control charts similar in structure to variables control charts, except that they plot statistics from count data rather than measurement data. In this case, data are discrete (as for microbiological CFUs) and not continuous. C-chart is used to track the number of defects and detects the presence of special causes. C-charts are used when the subgroup size is constant. Interestingly, the C-chart assumes a Poisson and not a normal distribution which has often been proclaimed to be the better assumption for microbiological data. Calculated UCLs with microbiological data using C-charts will be set quite low as compared to other control chart and percentile-based calculations. This is probably due to the fact that the assumed Poisson distribution underestimates the variability. This underestimation was also shown by Gordon *et al.* (2015).
- **Laney's U-chart**. Classical U-charts allow to monitor over time the defect rates per unit sampled in each subgroup. U-charts are used when the subgroup size varies (for C-chart the subgroup size is constant). Laney (2002) has adjusted the U-charts to adjust over- and underdispersion. The calculations

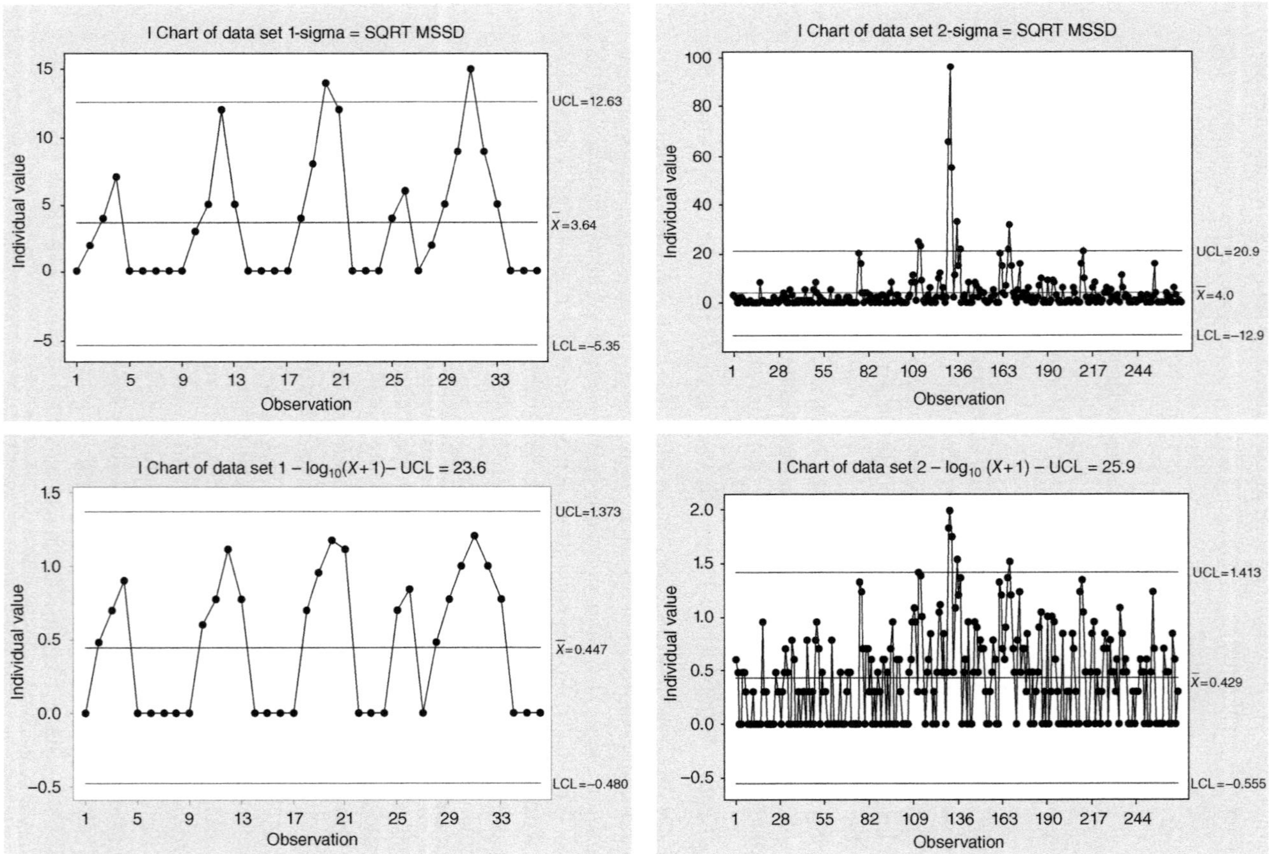

Figure 10.7 I-charts of data sets 1 and 2. Top row charts were computed with untransformed data and bottom row charts were plotted with the same data that were $\log_{10}$(count + 1) transformed and computed with the Minitab 17 (Minitab Inc.) software. Note that the UCL number given in the chart title was back-transformed (10UCL) from the figure.

for the Laney U′ attributes chart include Sigma Z, which is an adjustment for over- or underdispersion. A Sigma Z value of 1 indicates that no adjustment is necessary and that the Laney attributes chart is exactly the same as a traditional attributes chart. Since data are discrete and not continuous and overdispersion is adjusted, Laney's U-chart would seem in theory to be the best choice in terms of control charts for microbiological data. Nonetheless, the resulting UCLs are actually much lower than the ones calculated using other methods or the percentile ranking method (Table 10.4).

Table 10.4 Comparison of the calculated alert levels using control charts and percentiles assuming or not a distribution and estimates the delta with the optimal negative binomial method of moments function.

		Data set 1		Data set 2	
Calculation method	**Percentile**	**Alert level**	**Δ Negbin**	**Alert level**	**Δ Negbin**
Negative binomial method of moments	95.00	13	0	21	0
	99.00	22	0	45	0
	99.86	30	0	79	0
Percentile Excel	95.00	14	1	16	5
	99.00	NA[a]		58	13
	99.86	NA[a]		NA[a]	
Gamma Fit Method of Moments	95.00	13	0	21	0
	99.00	21	1	46	1
	99.86	30	0	79	0
Poisson	95.00	7	6	8	13
	99.00	9	13	9	36
	99.86	11	19	11	68
I-chart	95.00	9	5	13	8
	99.00	11	11	17	28
	99.86	13	17	21	58
I-chart $\log_{10}(\text{count} + 1)$	95.00	9	5	9	12
	99.00	15	15	16	29
	99.86	24	6	26	53
Xbar-chart	95.00	6	7	7	14
	99.00	7	15	9	36
	99.86	9	21	10	69

(Continued)

Table 10.4 (Continued)

Calculation method	Percentile	Data set 1		Data set 2	
		Alert level	Δ Negbin	Alert level	Δ Negbin
Xbar-chart $\log_{10}(\text{count}+1)$	95.00	5	8	5	16
	99.00	7	15	6	39
	99.86	8	22	8	71
R-chart	95.00	9	4	17	4
	99.00	11	11	19	26
	99.86	12	18	22	57
R-chart $\log_{10}(\text{count}+1)$	95.00	7	8	20	1
	99.00	11	11	32	13
	99.86	16	14	51	28
C-chart	95.00	7	8	7	14
	99.00	8	14	9	36
	99.86	9	21	10	69
Laney U-chart	95.00	8	5	10	11
	99.00	10	12	12	33
	99.86	12	18	14	65

Values in shade are the reference values.
[a] Not enough data points for this analysis.

10.3.5 Calculating Alert Levels Using Percentile Ranking

The percentile is a measure used in statistics that indicates the value below which a given percentage of observations fall. The percentile ranking calculation enables to determine alert levels defined at the *k*th percentile. For instance, if an alert level is defined as the 95th percentile, 95% of all data points fall within or below this level.

The formula is:

$$\text{Percentile rank} = \frac{\text{number of scores below } x}{n} \times 100$$

Percentile ranking may be executed parametrically either assuming a distribution model or non-parametrically.

10.3.5.1 Nonparametric Percentile Ranking Method

Nonparametric (distribution free) methods are used if no assumption for the distribution of the data is made.

For example, the Percentile.exc function from Excel may be used:

PERCENTILE.EXC (Array;k), where the Array is the data used and *k*th percentile.

The nonparametric percentile ranking may be used for large data sets. For low sample size data sets it is not appropriate. If the sample size is too small, the impact of the highest observed values is strong and alert levels might then be defined excessively high. For microbiology data due to their variability generally at least 60 values for a 95th percentile or 300 values for a 99th percentile are needed.

10.3.5.2 Parametric Percentile Ranking Method

In the parametric percentile ranking method the microbiological alert level is defined as the *k*th percentile assuming an appropriate distribution.

This is a more precise ranking method since it takes into account the distribution of data when determining the value corresponding to the *k*th percentile. Instead of ranking the untransformed counts, the parametric percentile ranks the distribution densities.

Based on the author's experience, applying the negative binomial model to microbiological data resulting from monitoring or material testing activities generally provides a satisfactory estimation of the alert levels. The Gamma fit model can also be considered as a suitable alternative.

As described in Section 10.3.2, microbiological data are generally not normally distributed; therefore, models based on normal distribution might not be the most appropriate models. The calculations shown in this chapter will be performed assuming a negative binomial distribution since this is the most appropriate distribution model for the majority of microbiological data obtained from monitoring or raw material/product bioburden testing.

The author recommends that enough data points are used in the calculations to capture the typical variation. Generally, it is assumed that at least 50 data points are necessary to calculate microbiology data with the distribution models cited below (Gordon *et al.* 2015). However, in cases where the data can be expected to be very stable, a smaller number (min. 20–30 data points) may be sufficient.

Different statistical software packages may be used to perform these calculations. For instance, the Excel add-on XLSTAT Pro Addinsoft can be used to calculate the percentiles using different distribution models as well as executing Chi-squared goodness-of-fit tests to determine the best fit.

Instructions shown in Figure 10.8 would apply.

Alternatively, the alert levels calculated in the SAS 9.4 are shown in Figure 10.9.

10.3.5.3 Comparing Calculation Methods Using Control Charts and Percentile Ranking

Assuming a normal distribution, a certain percentage of observations fall within a range of $\mu \pm k \times \sigma$ where μ is the average of data and σ the standard

XLSTAT 2015 (Addinsoft)	In the example below the XLSTAT 2015 version was used to fit negative binomial distribution. The method of moments was preferred over the maximum likelihood method to estimate the dispersion parameters since it works better with low samples sizes and lower means and microbiological data (Gordon *et al.* 2015)
Open XLSTAT 2015 Data set are placed in a column. Select the icon modelling data and "Distribution fitting" in the menu	
On the "General" sheet Select the data from the column under Data and select the Negative binomial (2) under distribution	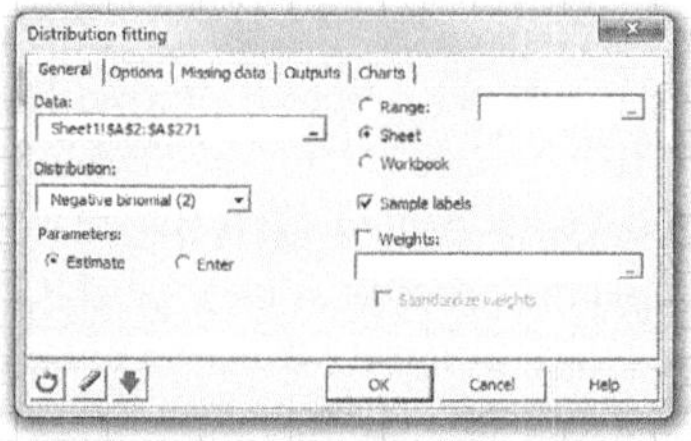
On the "Options" sheet For information, the Chi-square fit test may be performed but since microbiological data are highly variable and too few the Chi-square fit test often fails Estimation method of moments must be selected then press ok	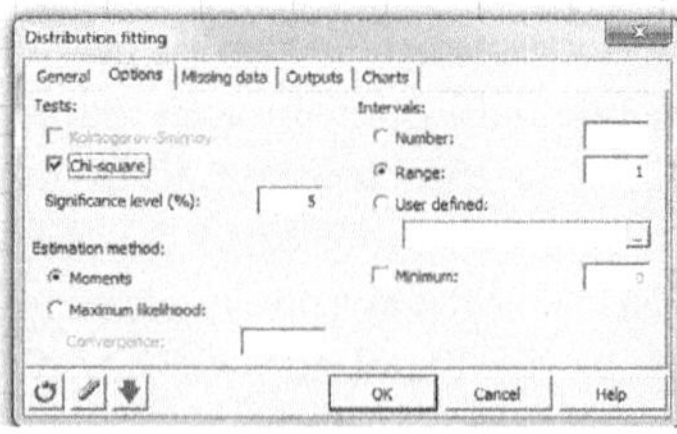

Figure 10.8 Example of alert level calculation using the percentile ranking method calculated with the XLSTAT 2015 software.

Under estimated statistics, the estimated parameters *k* and *p* will be used to fit the data into a negative binomial distribution. Data will be plotted automatically

Summary statistics:

Variable	bservation	ith missin	thout miss	Minimum	Maximum	Mean	d. deviation
3	269	0	269	0.000	96.000	3.974	9.151

Estimated parameters:

Parameter	Value
k	0.198
p	20.071

Statistic	Data	Parameters
Mean	3.974	3.974
Variance	83.734	83.734
Skewness	6.041	4.496
Kurtosis (I	47.665	30.315

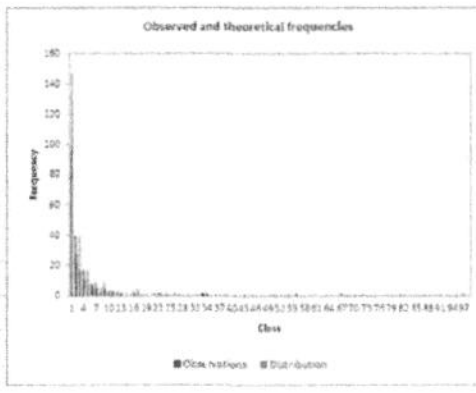

The density (distribution) provides the probability of the count assuming a negative binomial distribution

Descriptive statistics for the intervals :

ower boun	Upper bound	Frequency	Relative frequency	Density (Data)	Density (Distribution)
0	1	99	0.368	0.368	0.547
1	2	39	0.145	0.145	0.103
2	3	39	0.145	0.145	0.059
3	4	17	0.063	0.063	0.041
4	5	17	0.063	0.063	0.031
5	6	8	0.030	0.030	0.025
6	7	9	0.033	0.033	0.021
7	8	3	0.011	0.011	0.017
8	9	9	0.033	0.033	0.015
9	10	3	0.011	0.011	0.013
10	11	3	0.011	0.011	0.011
11	12	3	0.011	0.011	0.010
12	13	1	0.004	0.004	0.009
13	14	0	0.000	0.000	0.008
14	15	0	0.000	0.000	0.007
15	16	3	0.011	0.011	0.006
16	17	4	0.015	0.015	0.006
17	18	0	0.000	0.000	0.005
18	19	0	0.000	0.000	0.005

To determine the percentiles, the density (distribution) data must be multiplied by 100 and cumulated separately.

The alert level can then be defined based on the percentile selected. In this case with a 99th percentile (marked in gray) would correspond to a count of 45

Counts	Cumulative probability
0	54.6906
1	65.00538
2	70.89069
3	74.99802
4	78.12599
5	80.62758
6	82.69194
7	84.43304
8	85.92524
9	87.21997
10	88.35434
11	89.35609
12	90.24652
13	91.04237
14	91.75702
15	92.40136
16	92.98436
17	93.51348

36	98.28501
37	98.39253
38	98.49277
39	98.5863
40	98.6736
41	98.75514
42	98.83132
43	98.90253
44	98.96912
45	99.03142
46	99.08974
47	99.14433
48	99.19547

Figure 10.8 (Continued)

SAS 9.4 (SAS Institute Inc.)

Inthe SAS script box, write the script coding the Negbin calculation. The raw data should be copied under the term "datalines" (corresponds to 0, 2 ,4 ... in the snapshot) and then run the calculation.

The coding is

```
data a;
   input CFU;
datalines;
<INSERT DATA LINE>

proc countreg;
model CFU =/ dist=negbin method=qn;

   output out=prednb probcount(0 to 20);
   ods output Modelfit=fit;
run;

proc means data=prednb mean;
    var p_0-p_20;
 run;
```

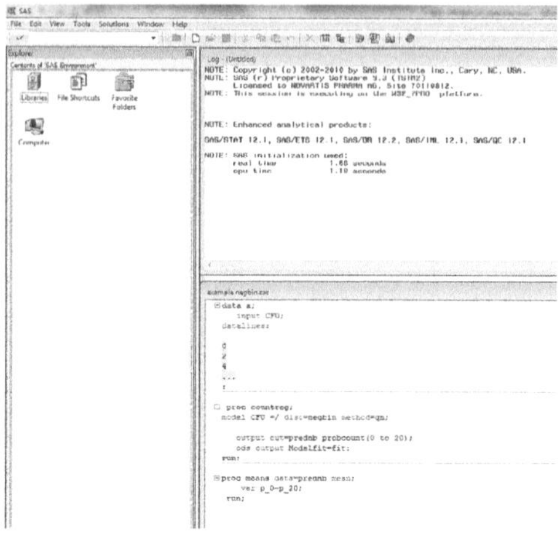

After the run, the probabilities of the counts according to a negative binomial distribution are reported

Copy the probability of counts in Excel and rank cumulatively to calculate the percentile level. In the example below, the 95th percentile would correspond to 16 CFU.

Variable	Label	Mean
P_0	Probability of CFU taking level=0	0.3720879
P_1	Probability of CFU taking level=1	0.1439822
P_2	Probability of CFU taking level=2	0.0926184
P_3	Probability of CFU taking level=3	0.0674909
P_4	Probability of CFU taking level=4	0.0520635
P_5	Probability of CFU taking level=5	0.0414970
P_6	Probability of CFU taking level=6	0.0337841
P_7	Probability of CFU taking level=7	0.0279171
P_8	Probability of CFU taking level=8	0.0233245
P_9	Probability of CFU taking level=9	0.0196535
P_10	Probability of CFU taking level=10	0.0166722
P_11	Probability of CFU taking level=11	0.0142208
P_12	Probability of CFU taking level=12	0.0121851
P_13	Probability of CFU taking level=13	0.0104809
P_14	Probability of CFU taking level=14	0.0090445
P_15	Probability of CFU taking level=15	0.0078271
P_16	Probability of CFU taking level=16	0.0067902
P_17	Probability of CFU taking level=17	0.0059035
P_18	Probability of CFU taking level=18	0.0051425
P_19	Probability of CFU taking level=19	0.0044873
P_20	Probability of CFU taking level=20	0.0039216

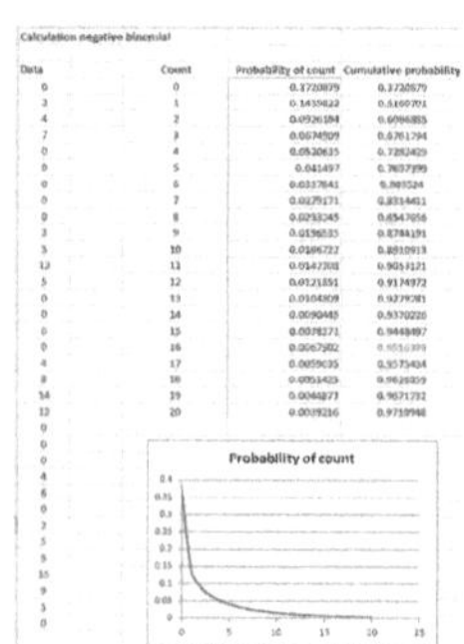

Figure 10.9 Example of alert level calculation using the percentile ranking method calculated with the SAS 9.4 software.

deviation. If we take into account only the upper level ($\mu + k \times \sigma$), then the percentage of observations falling within the level is higher as shown in Table 10.3.

To give the reader concrete examples, different calculation methods were compared with the parametric percentile ranking method using the negative binomial method of moments function (optimal method to calculate alert levels) and summarized in Table 10.4.

As shown in Table 10.4, the Gamma Fit Method of Moments has the shortest delta with the negative binomial. Some calculation methods seem obviously disqualified. Indeed, Poisson distribution model, C-chart, Laney's U-chart, XBar chart, and untransformed I-chart would set the levels far too low thereby generating an excessive amount of points exceeding the level. For attribute charts such as C-charts and Laney's U′ prime chart, data used in the calculation have to be discrete so transformations of data would not be possible in the calculation. Surprisingly, these attribute charts had even worse estimations even if the Laney U-chart is a control used for overdispersed and discrete data (such as microbiological data).

10.3.5.3.1 Selecting the Percentile The percentile directly impacts the threshold of acceptance below which a microbiological count result is considered part of the expected variation. Therefore, the choice of the percentile has consequences on the number of potential excursions which occur, even if the process or product of interest is actually under control. Table 10.5 shows when comparing different percentiles the expected values above the level.

The PDA technical report no. 13 recommends the 95th percentile to define alert levels for environmental monitoring based on historical data. Nonetheless, the percentile should be chosen so that the alert levels retain their function as an indication of abnormal counts or potential hygiene problems. As shown in Tables 10.4 and 10.5, whatever the method the 95th percentile may not be that appropriate as it might generate too many counts exceeding the alert level. Working with a too low percentile would waste resources by investigating nonissues or normal/controlled conditions. For sites in which many analyses are performed during a time period, a percentile higher than 95% would also be acceptable provided that unusual high counts remain detectable and the alert level remains below the action level.

Figure 10.10 illustrates examples of calculated alert levels using different methods and percentiles.

Experience has shown that a 99th percentile for nonparametric ranking percentile calculation is generally a good balance for microbiological environmental monitoring or testing of non-sterile products and the standard setting of 99.86th percentile for control charts corresponding to a $3 \times \sigma$ generally also suffices (Figure 10.11).

Table 10.5 Comparison of percentiles, number of analysis, and theoretical amount of excursions.

Number of samples	Percentile	Number of excursions
100	95	5
	99	1
	99.9	0.1
300	95	15
	99	3
	99.9	0.3
1 000	95	50
	99	10
	99.9	1
5 000	95	250
	99	50
	99.9	5
10 000	95	500
	99	100
	99.9	10

10.3.6 Conclusion Calculation of Alert Levels

There is no absolute standard that can be used to calculate levels using historical data as data distribution may differ between the different microbiological tests.

The author suggests careful evaluation of data before applying control charts since they might apply for microbiological data in very limited cases. Indeed, the calculated upper microbiological monitoring level might be systematically underestimated because of data overdispersion thus resulting in an increased amount of exceeding alert levels or "false alarms". It is therefore recommended to first transform the data using the most appropriate formula and even if the data transformation does not statistically pass the normality test, it would reduce the impact of the overdispersion and zero counts on the estimation of the alert level using the control chart method.

The most reliable method used to cover a large diversity of microbiological data is the percentile ranking method using the negative binomial or Gamma fit or, if enough values are available, a nonparametric fit. Nonetheless, not all

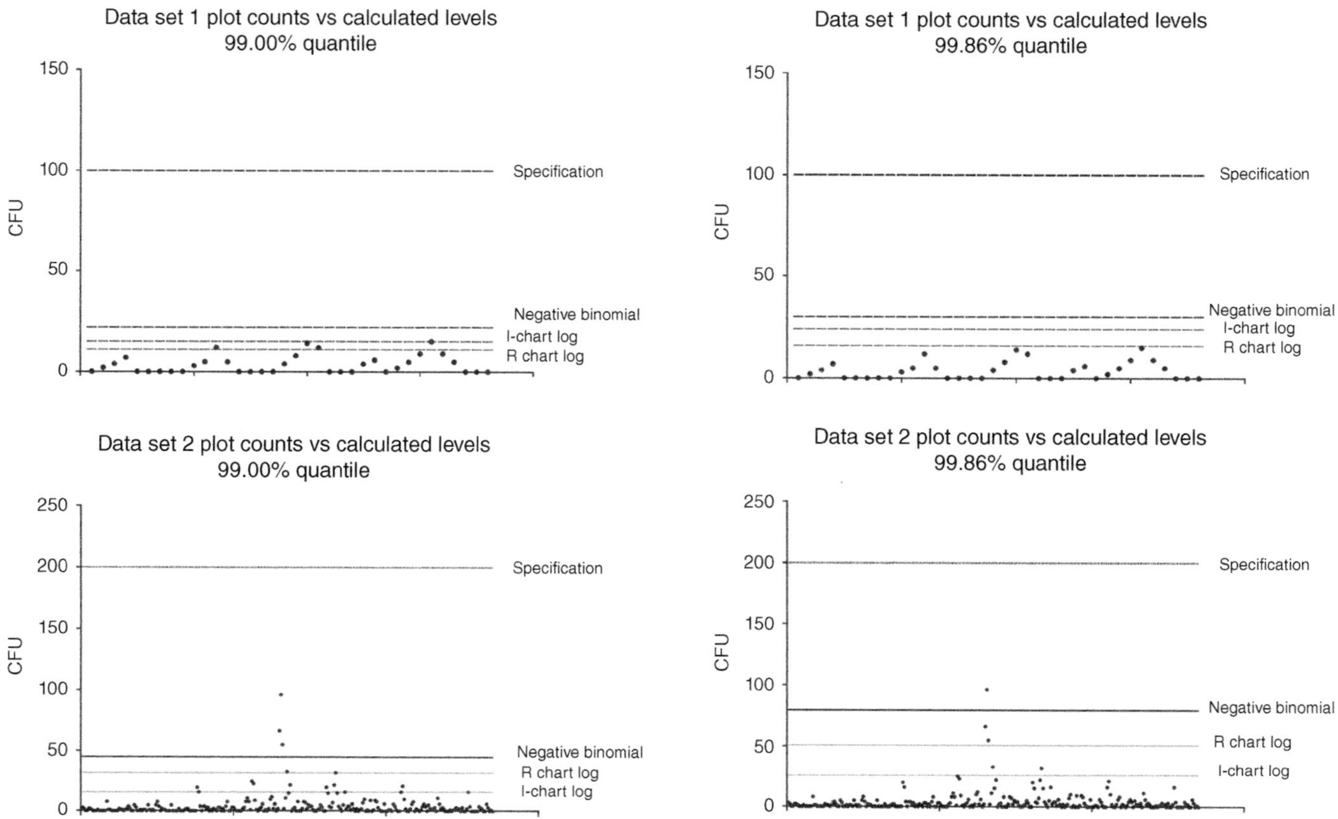

Figure 10.10 Plotting of data sets 1 and 2 as well as calculated alert levels using different methods. Data was plotted using the GraphPad Prism 7 (Graphpad Software).

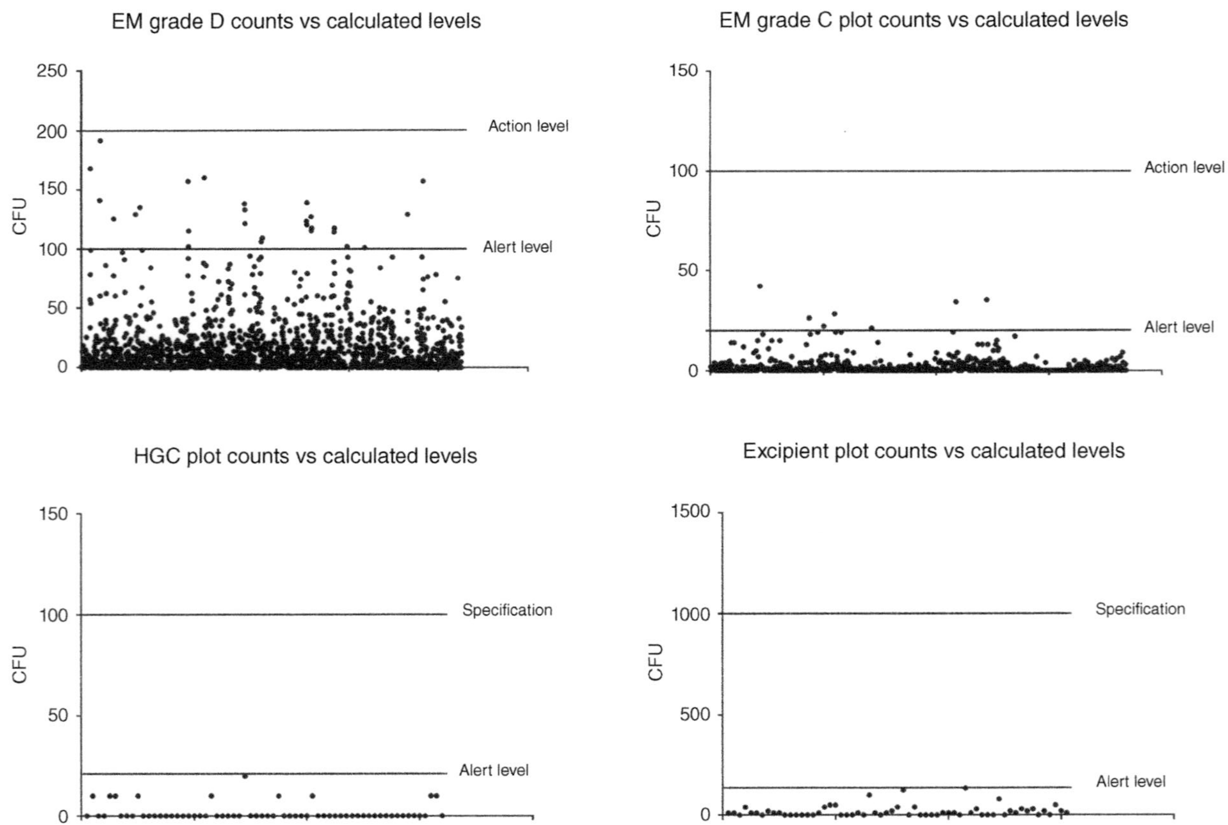

Figure 10.11 Alert levels of different microbiological data sets. The alert levels were calculated with negative binomial ranking method with a 99th percentile. Data sets were composed of viable air environmental monitoring data of a grade D (upper left) and grade C cleanroom (upper right) as well as batch product testing of a hard gelatin capsule (HGC) formulation (bottom left) and raw material testing of an excipient of natural origin (bottom right).

criteria can be automated or obtained from a statistical calculation and it remains under the microbiologist's responsibility to select the most appropriate percentile for the data set in question.

10.4 Trending

10.4.1 Introduction

Microbiological data are reviewed periodically and trended to assess the capability of the measures to control contamination and that the microbiological quality of products or raw materials would continuously remains within the standards defined and is not worsening.

Trending of microbiological data can be performed in several manners but the end goal is to determine if an adverse trend is arising. The results of the trend analyses are generally documented in trend reports that will contain also the corrective actions in case of adverse trend.

Microbiological data may be trended by plotting the data on a *X*/*Y* chart and an analyst visually assesses if there is an adverse trend or not. This remains one of the most common ways of trending data in pharmaceutical companies (Figure 10.12). However, as for setting alert levels, this subjective approach is highly variable among individuals and it is recommended to define clear adverse trend criteria and use statistical tools to support the decision-making.

In addition, due to the high variability of microbiological raw data it is sometimes not evident to visually detect an adverse trend (Data set 4 in Figure 10.13).

10.4.2 Defining and Investigating Adverse Trends

10.4.2.1 Grouping of Data

To perform trend analysis especially for monitoring results where low testing frequencies and multiple results from a monitored area occur, it would be relevant to group data of multiple sampling locations. Grouped data, which can also be referred to as a trend area, must be composed of microbiological results that are clustered according to a rationale. For instance, volumetric air sampling from the same production areas and cleanliness zones or from similar manufacturing lines. It might not be suitable to combine data from washing rooms and production rooms even if they are belonging to the same grade area.

10.4.2.2 Definitions of Adverse Trend

Multiple designations may be used to define adverse trends and it is the user to define which is the most relevant definition in the context of application.

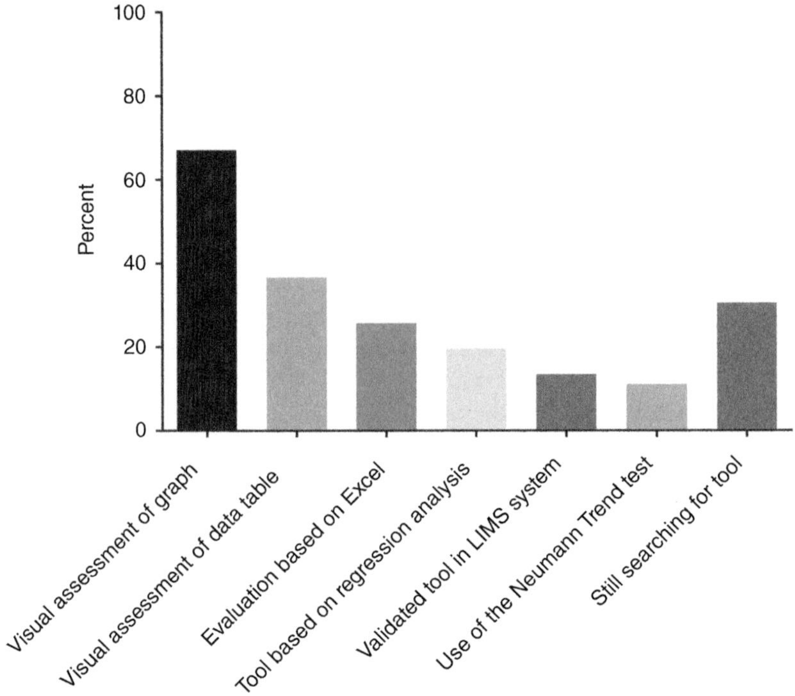

Figure 10.12 Survey trending tool used in pharmaceutical companies (107 responses) extracted and translated from GMP News. "Wann ist ein Trend ein Trend? Ergebnisse einer Umfrage" 11.07.2014, ECA.

Nonetheless, it should always be considered that an adverse trend takes into account multiple events.

Depending on the evaluation method used, adverse microbiological trend may, for instance, be:

- Increased counts or frequency of occurrence in the graphical interpretation of data.
- Two or three times exceeding the alert or action level in a row.
- Two or three times exceeding a trend limit in a row.
- Two to three times exceeding the contamination recovery rate limit in a row.
- Two to three times exceeding sigma limits in a row.
- Higher proportion of samples (e.g. 10%) exceeding microbiological alert and action levels from one time period to another or from more than one site from the same room (regardless of source: personnel, surface, or air).

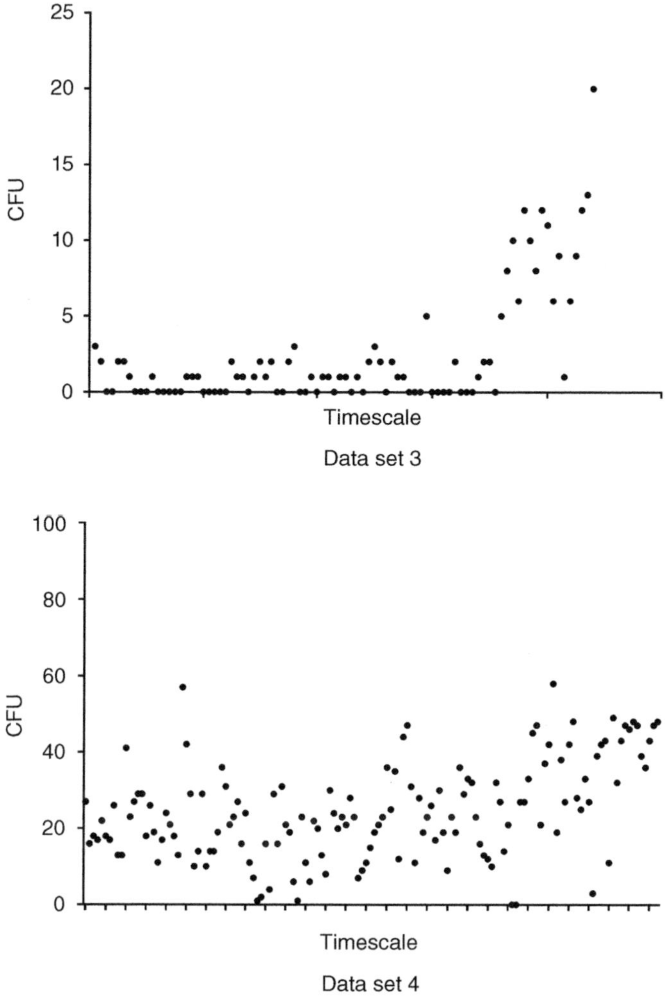

Figure 10.13 Examples of plotting microbiological data to evaluate the adverse trend.

- Concerned sampling point is exceeded more than 30% during the considered time period.
- Repeated occurrence of a specific microorganism.

10.4.2.3 Investigating Adverse Trends

When an adverse trend is detected, an investigation is performed to determine the root cause explaining the adverse trend as well as to assess the impact on the product's microbiological quality.

When CAPAs are put in place to resolve the adverse trend, their effectiveness should be demonstrated with an improvement in the microbiological trend.

More details can be found in Chapter 12.

10.4.3 Examples of Trending Methods

10.4.3.1 Graphical Interpretation Using Regression Analysis

The regression analysis may be used as a graphical support to determine if an adverse trend occurs or not. The slope indicates the steepness of a line and the graphical direction of the slope might indicate an adverse trend in time. It cannot be used as a true statistical test since the goal of the analysis is not to determine if a correlation exists between time and microbial counts and the R^2 and p-values are irrelevant for this graphical interpretation.

The formula used for the regression analysis is

$$y = \alpha x + \beta$$

y = intercept value
x = slope

In the example shown in Figure 10.14, the slope of both data sets is positive, meaning data are increasing in time. It should be noted in this case that significance testing depends strongly on the variance of the data. So, exactly the same slope may be significantly different from zero ($p < 0.05$) in the case of low variation, whereas the equal slope is nonsignificant in the case when the data are highly variable. Hence, a significance test is not valuable in this case. One possibility is to define a trend using a minimum slope limit. For instance, if the slope in a one-year period is less than 1, then there this is considered a trend.

10.4.3.2 Graphical Interpretation Using Rolling Averages

In general, control charts such as individual charts and moving range charts are not recommended for microbiological data since the latter are too variable and it is not evident to outline a trend. Rolling average charts compare cluster of means during a time period with one another which enables to better visualize main patterns in highly variable data. Rolling average charts are more appropriate to trend multiple events such as microbiological data over time in an easy to interpret graphical form. The observations can be either individual measurements or subgroup means.

The rolling average length should depend on the amount of data points to be trended per trend area and timepoints. The higher the rolling average length, the smoother the pattern.

An adverse trend may be defined using the graphical interpretation of the chart or by defining control limits which can be, for instance, set as three times sigma above the center limit.

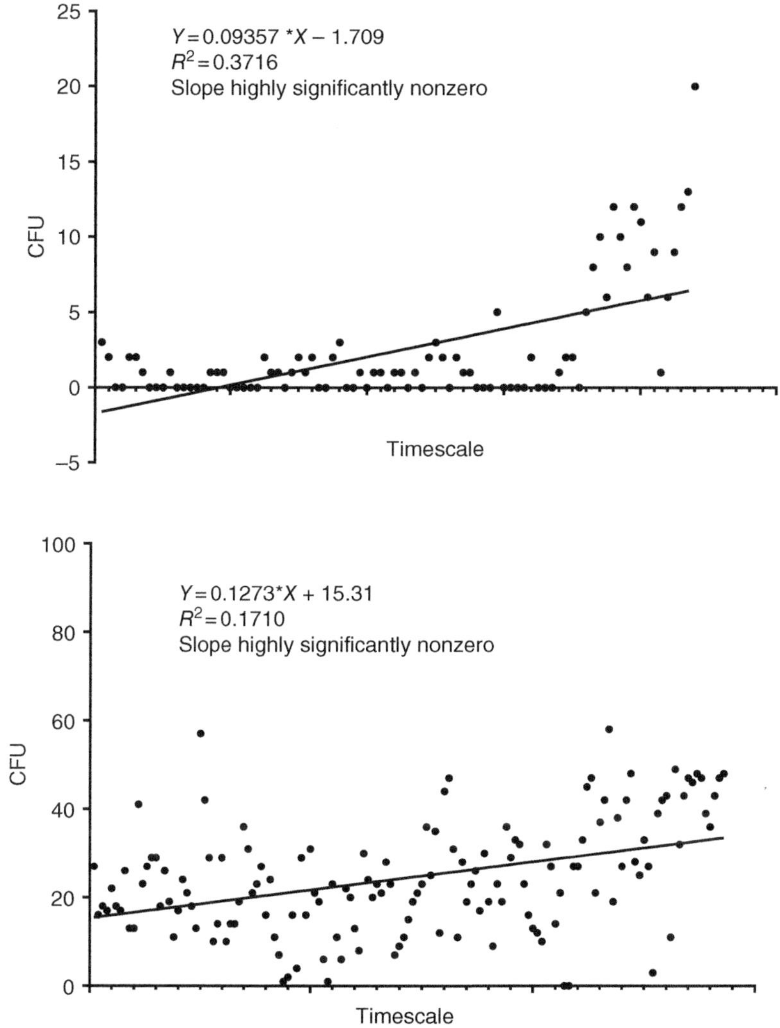

Figure 10.14 Regression analysis of data sets 3 and 4. The regression analysis was calculated and results plotted in Graphpad Prism 7.

Rolling average charts enable to measure effectiveness of measures taken in the long term. They do not enable to detect major changes immediately but provide a response after a certain time depending on the rolling average length (Figure 10.15).

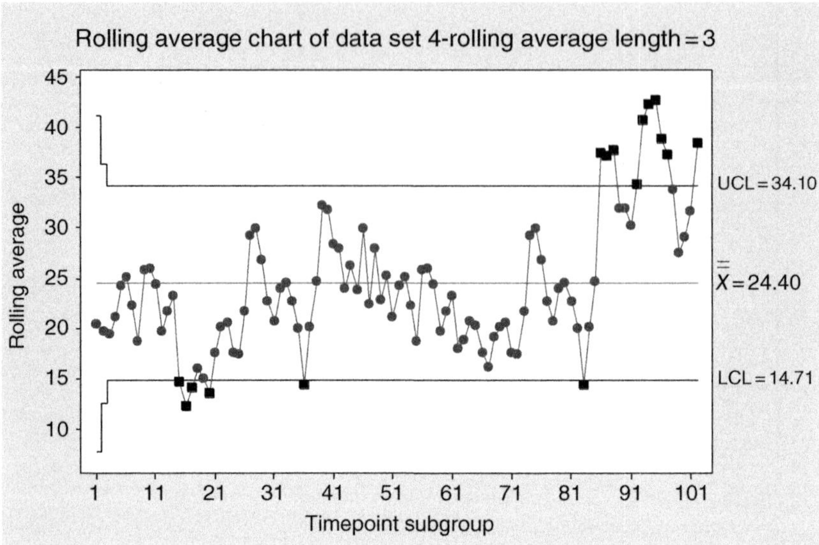

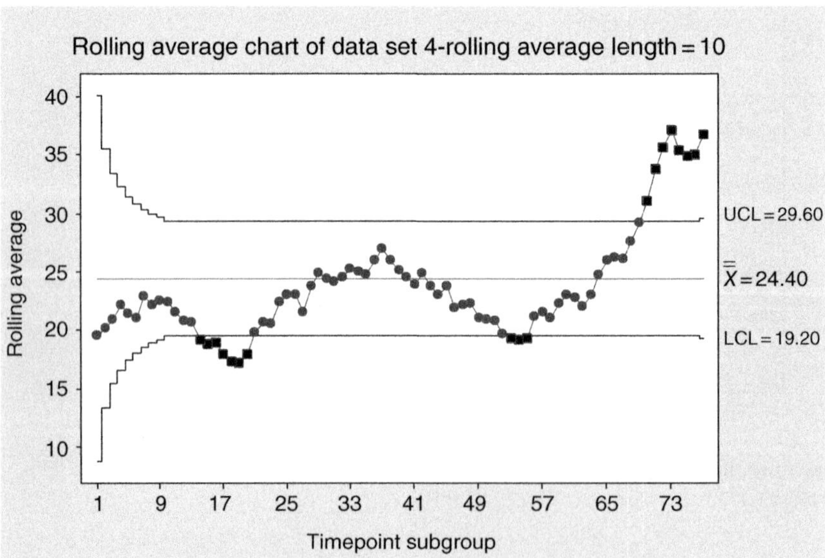

Figure 10.15 Rolling average calculation using data set 4 with a subgroup size of 3 and using moving averages of 3 (upper graph) and 10 (lower graph). Flagged values are above the center line +3 times sigma limit. The moving averages were calculated and results plotted in Minitab 17 Software.

10.4.3.3 Neumann Gradual Trend Test

The Neumann test enables to determine gradual trends in data points. It is used to determine the negative autocorrelation where the neighboring data are more different from each other than in average expected. The Neumann test assumes a normal distribution and might not be discriminant enough for highly variable data such as microbiological results.

The formula is as follows:

$$d_N = \frac{\sum_{i=2}^{N}\left(x_i - x_{i-1}\right)^2}{\sum_{i=1}^{N}\left(x_i - \overline{x}\right)^2}$$

d_N = sum of squares of differences of successfully formed pairs divided by the sum of squares of difference data point to the mean.

$d_{N,\,\mathrm{crit}}$ = expected average difference between data points. Comparison value that depends on the number of samples and the desired alpha significance. It is derived from Von Neumann *et al.* (1941) and can be found in http://www.faes.de/Basis/Basis-Statistik/Basis-Statistik-Tabelle-Neuman/basis-statistik-tabelle-neumann.html (accessed 9 March 2018).

A trend is when d_N is less than the compared value ($d_{N,\,\mathrm{crit}}$), based on the number of observations to be trended and the statistical significance (p) (Figure 10.16).

If $d_N < d_{N,\,\mathrm{crit}}$ gradual trend occurs
If $d_N > d_{N,\,\mathrm{crit}}$ no gradual trend occurs

10.4.3.4 Microbiological Contamination Recovery Rate and Number of Exceeding Alert Levels

The contamination recovery rate method for microbiological environmental monitoring data is described in USP Chapter <1116>. The recovery rate is compared with a contamination recovery rate limit that is predefined. The recovery rate of a trend area for a defined period is calculated as the percentage of samples where microbial growth occurred versus total amount of samples tested.

This method is suitable for data for which counts are not expected or in rare occurrences such as environmental monitoring for aseptic areas. For non-sterile production areas, a low level bioburden is generally expected so recovery rates based on microbial growth versus absence of microorganisms would not make sense.

In the case where a low level of bioburden is normally expected, the recovery rate should preferably be based on the amount of unexpected high counts as compared to the normal bioburden during a certain time period. Actually, calculated alert levels from Section 10.3 could be used as a threshold of the unexpected high count.

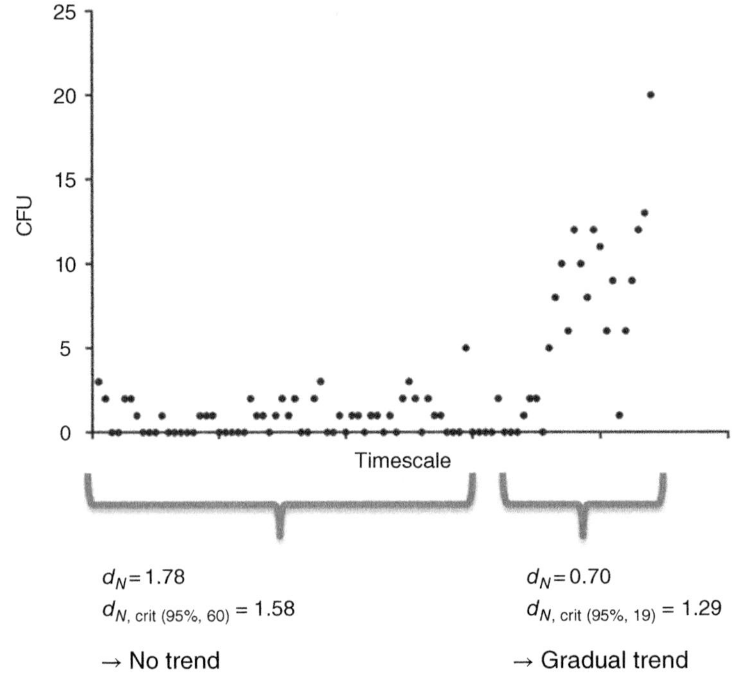

Figure 10.16 Calculation of data set 3 using the Neumann gradual trend method. Calculations were performed in Microsoft Excel 2013.

In this case trending may be performed either by

- Evaluating if the alert level is exceeded several times in a row.
- Evaluating if the ratio of unexpected high counts is higher than a prescribed recovery rate.
- Comparing the number of exceeded alert of one trending area from one time period to another (e.g. 6 months period).

Examples are shown in Figure 10.17.

10.4.3.5 Qualitative Value System

The target value concept is based on trending qualitative values of a trending area over time and evaluating if these qualitative values exceed a defined trend limit several times in a row and described in Pfohl *et al.* (2005a, 2005b).

These qualitative values may encompass several analytical results or different quality attributes. This would mean that the trending is less dependent on the data distribution or variability.

For instance, the distribution of qualitative values shown in Table 10.6 may be set for trending of environmental monitoring data.

Each result from each test would be transcribed into a qualitative value. In the example for the trending area of product manufacturing clean room, the sum qualitative values may be summed up as shown in Table 10.7.

The sum of the qualitative values point per trend area (e.g. product manufacturing cleanroom) per trend period is summed up and plotted (Figure 10.18).

The weekly sum of qualitative values is compared to a trend limit. If the qualitative values exceed, for instance, three times in a row the trend limit,

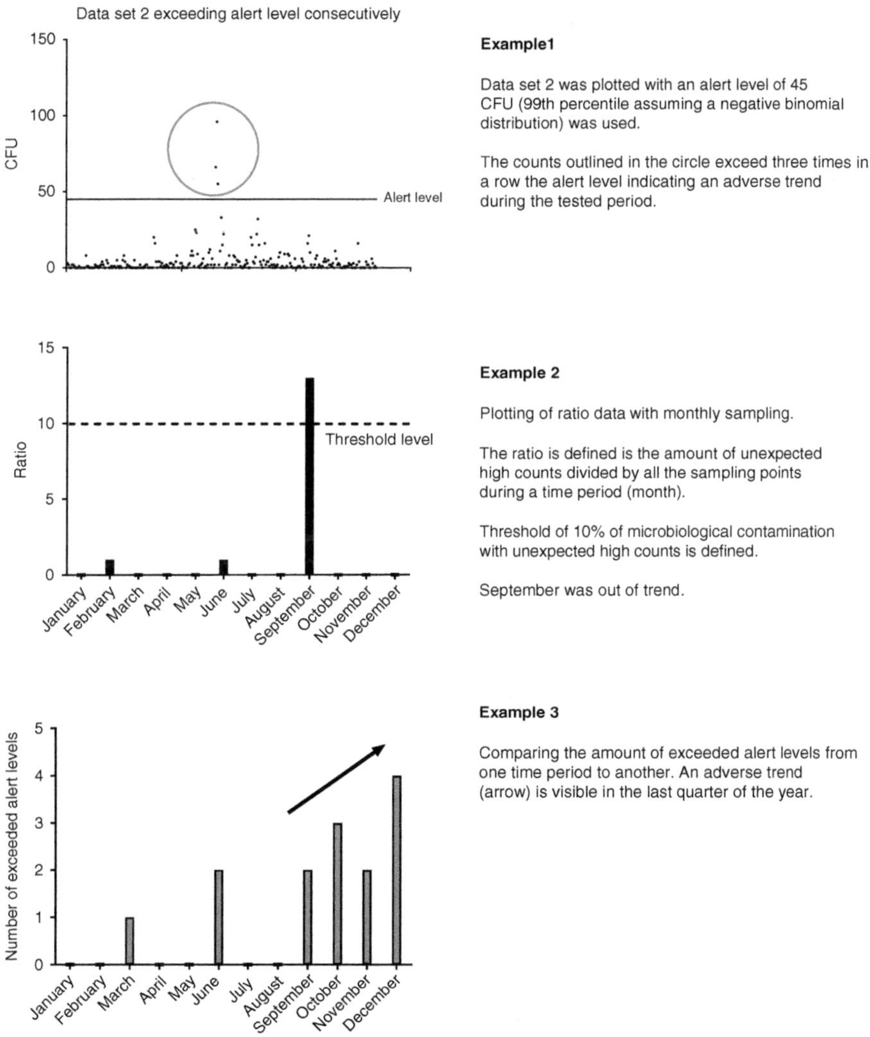

Figure 10.17 Examples of trending using alert levels and unexpected high counts.

Table 10.6 Example of evaluation of qualitative values for environmental monitoring results.

Trend area	Type of test	Range of results and corresponding qualitative values			
Product manufacturing clean room	Air sample 1	0 CFU/1 m^3	1–50 CFU/m^3	51–200 CFU/m^3	>200 CFU/m^3
	Air sample 2	0 CFU/1 m^3	1–50 CFU/m^3	51–200 CFU/m^3	>200 CFU/m^3
	Surface sample 1	0 CFU/25 cm^2	1–20 CFU/25 cm^2	1–10 CFU/25 cm^2	>50 CFU/m^3
	Surface sample 2	0 CFU/25 cm^2	1–20 CFU/25 cm^2	21–50 CFU/m^3	>50 CFU/m^3
	Surface sample 3	0 CFU/25 cm^2	1–20 CFU/25 cm^2	21–50 CFU/m^3	>50 CFU/m^3
	Surface sample 4	0 CFU/25 cm^2	1–20 CFU/25 cm^2	21–50 CFU/m^3	>50 CFU/m^3
Qualitative value		0	3	5	10

Table 10.7 Results of week 22 environmental monitoring testing product manufacturing clean room.

Trend area	Type of test	Result	Qualitative value
Product manufacturing clean room	Air sample 1	52 CFU/m^3	5
	Air sample 2	14 CFU/m^3	3
	Surface sample 1	3 CFU/25 cm^2	3
	Surface sample 2	0 CFU/25 cm^2	0
	Surface sample 3	22 CFU/25 cm^2	5
	Surface sample 4	0 CFU/25 cm^2	0
		Sum of qualitative value	**16**

then an adverse trend is considered (Figure 10.18). Depending on how the level is set, only one time exceeding the trend limit could also be considered.

The trend limits are set by the microbiologist taking into account the value of the criticality of the trend area evaluated and the historical performance of this trend area.

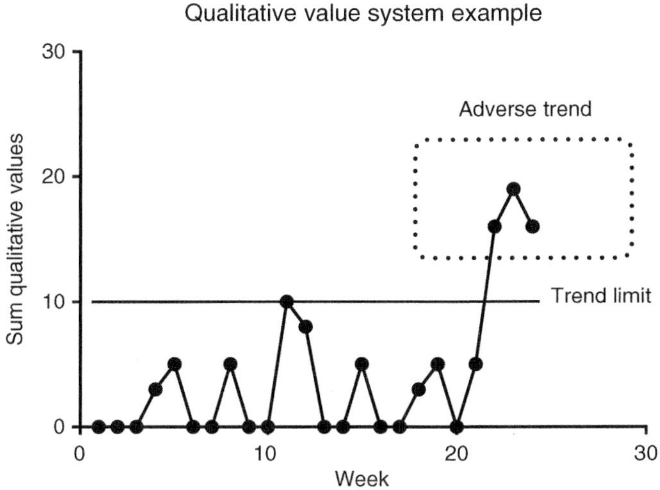

Figure 10.18 Example of plotting weekly sums of qualitative values per trend area against a trend limit using the GraphPad Prism 7 Software.

10.4.3.6 Trending of Identified Isolates

The types of isolates may also be trended from one time period to another in order to verify if there is a reoccurrence/increase of specific microorganisms that would indicate a deficiency.

One method would be to compare proportions of types of microorganisms from one time period to another. For instance, when by excursions the microorganisms are identified, they may be classified into different morphological types and compared from time period to another to evaluate if the proportion off one type of microorganism is increasing (Figure 10.19).

In Figure 10.19 the proportion of the different types of microorganisms is similar between 2016 and 2017. In 2018, however, a significant increase in the proportion of spore-forming bacteria has occurred.

Cases that may be considered worthwhile investigating would, for instance, be:

- In environmental monitoring, if spore-forming microorganisms are increasingly recovered, it could indicate that the cleaning/disinfectant regime is not sufficiently aggressive.
- If the proportion of molds is increasing in the cleanroom air results, this might indicate abnormal presence of humidity or water leakages in the cleanroom area.
- If objectionable microorganisms are increasing in the pharmaceutical-grade water, this might indicate the presence of a biofilm in the water system or that the sanitization regime is not adequate.
- Increasing presence of molds in hygroscopic raw materials might indicate that the storage conditions of this material are not adequate.

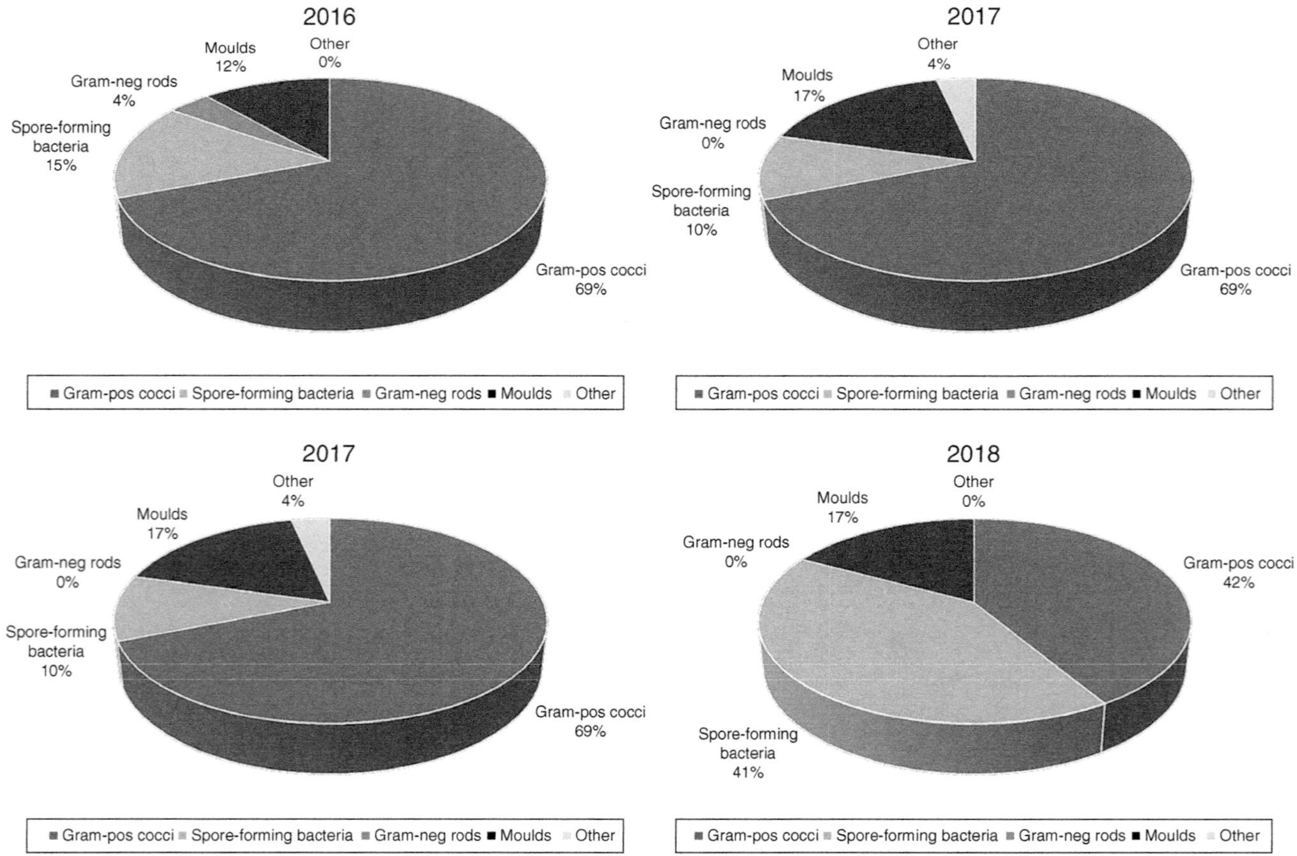

Figure 10.19 Example of comparison of proportions of different types of microorganisms identified in excursions from one year to another. pos, positive; neg, negative.

10.5 Conclusion

Setting of alert levels and trending of microbiological data using statistical tools may seem a complicated or irrelevant task for the microbiologist considering the high variability of data and lack of confidence or understanding using mathematical models. Nonetheless, with the support of statisticians and examples in the scientific literature, the task is not incommensurable.

There is not one preferred method of analysis that may be used and selection of analytical tools should fit the purpose of the task. However, mathematical methods do not completely replace the conceptual thinking of the microbiologist as the selection of the statistical tool, grouping of relevant data, and setting of appropriate acceptance criteria would still remain under his responsibility.

By including statistical models for the trending of microbiological data, the human variability and subjectivity can be lowered and the overall microbiological quality can better be assessed over time thus enabling the organizations to act in a proactive and not reactive manner before processes start getting out of control. In addition, the same methods of analysis would allow benchmarking of multiple production sites and would support standardization processes in multisite producing companies.

Acknowledgments

The editors would like to thank Dr. Urs Simmen (Simmen Statistical Consulting, Basel) for critical review of the manuscript.

Bibliography

Akers, J, Agallaco, J, 2001, 'Environmental monitoring: myths and misapplications', PDA J Pharm Sci Tech, vol. 55, pp. 176–184.

ANVISA, 2013, 'Guide for Quality of Air Treatment Systems and Environmental Monitoring in the Pharmaceutical Industry 1st Edition', March 2013.

Bar, R, 2015, 'Charting and evaluation of environmental microbial monitoring data', PDA J Pharm Sci Tech, vol. 69, pp. 743–761.

Borneman, J, 2018, 'A Study of Estimates of Sigma in Small Sample Sizes', viewed 16 May 2018, https://www.isixsigma.com/tools-templates/sampling-data/a-study-of-estimates-of-sigma-in-small-sample-sizes.

CFDA, 2010, Revision 'Good Manufacturing Practice for Drugs'.

EudraLex Volume 4, 2017 Consultation document, 'EU Guidelines to Good Manufacturing Practice Medicinal Products for Human and Veterinary Use: Annex 1 Manufacture of Sterile Medicinal Products'.

FDA, 2004, 'Guidance for Industry: Sterile Drug Products Produced by Aseptic Processing – Current Good Manufacturing Practice'.

FDA, 2015, 'Pharmaceutical Compliance Program Guidance Manual Chapter 56 – Drug Quality Assurance Sterile Drug Process Inspections'.

FDA, 2016, 'Compliance Program Compliance Manual, Sterile Drug Process Inspections'.

Gordon, O, Goverde, M, Pazdan, J, Staerk, A, Roesti, D, 2015, 'Comparison of different calculation approaches for defining microbiological control levels based on historical data', PDA J Pharm Sci Tech, vol. 69, pp. 383–398.

ICH, 2008, 'Harmonised Tripartite Guideline Pharmaceutical Quality System Q10 Current Step 4, Version 4', June 2008.

ICH, 2005, 'Harmonised Tripartite Guideline Quality Risk Management Q9, Current Step 4 Version 9', November 2005.

Laney, DB, 2002, 'Improved control charts for attributes', Qual Eng, vol. 14(4), pp. 531–537.

PDA Technical Report No. 13, revised 2014, Fundamentals of an Environmental Monitoring Program, Parenteral Drug Association.

PDA Technical Report No. 59, 2012, Utilization of Statistical Methods for Production Monitoring, Parenteral Drug Association.

Pfohl, M, Fricker, P, Kohler, D, Berchtold, M, 2005a, 'Trending von Umgebungskontroll-Daten für den Herstellbereich sterile Formen, Teil 1', Pharm Ind, vol. 67(5), pp. 598–602.

Pfohl, M, Fricker, P, Kohler, D, Berchtold, M, 2005b, 'Trending von Umgebungskontroll-Daten für den Herstellbereich sterile Formen, Teil 2', Pharm Ind, vol. 67(6), pp. 705–708.

Toft, N, Innocent, GT, Mellor, DJ, Reid, SWJ, 2006, 'The Gamma–Poisson model as a statistical method to determine if micro-organisms are randomly distributed in a food matrix' Food Microbiol, vol. 23(1), pp. 90–94.

USP 41-NF 36, Chapter <1116> 'Microbiological control and monitoring of aseptic processing environments'.

Von Neumann, BJ, Kent, RH, Bellinson, HR, Hary, BI, 1941, 'The mean square successive difference', Ann Math Stat, vol. 12(2), pp. 153–162.

Wheeler, DJ, 'Are you sure we don't need normally distributed data?', viewed 16 May 2018, https://www.qualitydigest.com/inside/twitter-ed/are-you-sure-we-don-t-need-normally-distributed-data.html (accessed 22 July 2019).

Wheeler, DJ, Chambers, DS, 1992, Understanding Statistical Process Control, (Second Edition). SPC Press.

Yang, H, Zhao, W, O'day, T, Fleming, W, 2013, 'Environmental monitoring: setting alert and action limits based on a zero-inflated model' PDA J Pharm Sci Technol, vol. 67(1), pp. 2–8.

Websites

https://support.minitab.com/en-us/minitab/18/help-and-how-to/quality-and-process-improvement/control-charts/how-to/variables-charts-for-individuals/i-mr-chart/methods-and-formulas/methods-and-formulas-for-individuals-chart (accessed 16 March 2018).

https://www.gmp-navigator.com/gmp-news/wann-ist-ein-trend-ein-trend-ergebnisse-einer-umfrage (accessed 18 May 2018).

11

Exclusion of Objectionable Microorganisms from Non-sterile Pharmaceutical Drug Products

Tony Cundell

Microbiological Consulting, LLC, Scarsdale, NY, USA

CONTENTS

Pharmaceutical Microbiological Quality Assurance and Control: Practical Guide for Non-Sterile Manufacturing, First Edition. Edited by David Roesti and Marcel Goverde.

11.1 Introduction

In non-sterile drug product manufacturing, low microbial counts are tolerated and the final drug product does not necessarily need to be free of microorganisms to be released to the market. Nonetheless, some microorganisms are considered as objectionable in that they can adversely affect the appearance, physicochemical attributes, or therapeutic effects of a non-sterile drug product or due to their numbers and/or pathogenicity, may cause infection, allergic response, or even toxemia in patients receiving the product. Recent surveys have found that the presence of objectionable microorganisms, and not microbiological numbers, represents the vast majority of microbiologically related FDA recalls of non-sterile drug products.

This chapter will focus on microbiological risk assessments, to evaluate if a recovered microorganism is objectionable in a specific dosage form or not, and will contain case studies related to objectionable microorganisms, that although having a US bias, may be useful to the readers from all parts of the globe.

Origin of the Term Objectionable Microorganisms

It is a current Good Manufacturing Practice (cGMP) requirement to exclude objectionable microorganisms from non-sterile pharmaceutical products. The pertinent sections of the US federal regulations are 21 CFR 211.113 Control of microbiological contamination (a) *Appropriate written procedures designed to prevent objectionable microorganisms in drug products not required to be sterile shall be established and followed.* Furthermore, 21 CFR 211.165 Testing and release for distribution (b) *There shall be appropriate laboratory testing, as necessary, of each batch of drug product required to be free of objectionable microorganisms that may cause infection when given by the route of administration of the drug product and/or cause physicochemical deterioration to the product.*

In addition, 21 CFR 211.84(d) (6) states *Each lot of a component, drug product container, or closure that is liable to microbiological contamination that is objectionable in view of its intended use shall be subjected to microbiological tests before use.*

Case-by-Case Risk Assessment or a List of Objectionable Microorganisms?

Many pharmaceutical professionals working in general management, manufacturing, quality control units, or even microbiology testing laboratories ask for a list of objectionable microorganisms against which they can evaluate the microorganisms isolated from their drug products while others, more thoughtfully, want rules for the evaluation whether a microorganism is objectionable in a specific dosage form or even individual drug products. A show of hands, when polling attendees at national and regional meetings in the United States, shows that opinion is evenly split on this issue.

A company-wide list may be attractive as it is unequivocal and leads to consistent decision-making across the company. This one-size-fits-all approach may be most suitable for pharmaceutical companies that manufacture only a limited number of drug products or specialize in a particular dosage form like compressed tablets, topical creams, or nasal sprays. However, the list may not reflect the route of administration of a drug product, pharmaceutical ingredients used in the formulation, manufacturing process, product attributes that may allow a microorganism to grow within the product during its shelf life, or the targeted patient population. The company must have access to microbiological expertise to assemble an objectionable microorganism list. This expertise is often lacking, even in large pharmaceutical companies. Furthermore, a list may not be updated when due to taxonomic advances a microorganism name is changed due to reclassification or when our clinical colleagues discover an emerging pathogen. Another disadvantage of referring to a list for release decision-making is that it does not encourage an understanding of your company's drug products and foster microbial contamination risk assessment that is encouraged by regulatory agencies and leading pharmaceutical microbiologists. The use of a list that will be reviewed by external auditors, who may not be microbiologists, can lead to disagreement as to the content of the list, resulting in frequent revisions in response to uninformed opinion from nonspecialists. Perhaps more important than the detection of an objectionable microorganism is whether it will survive and grow in the drug product. Only a risk assessment can determine this critical consideration. Lastly, no microorganisms should be added to an objectionable microorganism list unless it can be detected during routine microbiological testing. For example, adding a strict anaerobic pathogen or fastidious microorganism to the list when it cannot be detected using the methods described in USP <61> and <62> is not scientifically justified and intellectually dishonest.

As not unexpected, the author of this chapter, because of his role as the cochair of the PDA task force responsible for the 2014 Technical Report No. 67 Exclusion of Objectionable Microorganisms from Non-sterile Pharmaceutical and OTC Drug Products, Medical Devices and Cosmetics, advocates using risk assessment tools to determine if microorganisms isolated from non-sterile drug products using the methods described in USP <61> Microbiological Examination of Non-sterile Products: Microbial Enumeration Tests and <62> Microbiological Examination of Non-sterile Products: Tests for Specified Microorganisms is objectionable in that specific drug product. Perhaps, the best solution combines elements of both approaches with core list of objectionable microorganisms by dosage form and the option to evaluate any other microorganisms isolated from a non-sterile drug product to determine if they are objectionable.

11.2 What Is an Objectionable Microorganism?

It is notable that the 21 CFR 211.113 *Control of microbiological contamination* contains no actual definition of an objectionable microorganism and certainly does not provide a list of microorganisms to be excluded from our non-sterile drug products. The assignment of the responsibility is to the pharmaceutical manufacturer who must develop a written program to exclude objectionable microorganisms from their drug products was, in the author's opinion, the right decision, as only the manufacturer has the complete range of knowledge of the pharmaceutical ingredients, formulation, manufacturing processes, product attributes, and intended patient population to make these critical judgments. Furthermore, the FDA has not written a guidance document outlining how the regulations can be met, but CDER microbiologists at industry meetings have stated they intend to write Guidance for Industry on non-sterile drug product manufacturing in 2018. In the absence of regulatory guidance, the reader is directed to USP General Informational Chapter <1115> *Bioburden Control of Non-sterile Drug Substances and Products.*

In addition, 21 CFR 317 List of qualifying pathogens that have the potential to pose a serious threat to public health may be a useful reference for frank but not opportunistic pathogens.

Obviously, the concept of objectionable microorganisms does not apply to sterile drug products and medical devices as all viable microorganisms must be excluded from these products. Designating different species of microorganisms isolated during environmental monitoring in aseptic processing areas as objectionable is misguided as their identity indicates their origin not their level of objectionableness. In contrast, it is reasonable to react to objectionable microorganisms when found in pharmaceutical ingredients and water for pharmaceutical use, especially when those microorganisms are objectionable in the non-sterile drug products manufactured in your facility.

The industry challenge is that the absence of objectionable microorganisms requirement for a non-sterile drug product is a critical quality attribute, without a defined test method and acceptance criteria, making it a unique product specification. This is largely the source of much confusion. Currently, there is no consensus among manufacturers and regulators how to approach this issue.

Although cGMP regulations, i.e. CFR 211.113, do not define the term objectionable microorganisms, they can, for the purpose of this discussion, be broadly defined as:

1) Microorganisms that can proliferate in a product adversely affecting the chemical, physical, functional, and therapeutic attributes of that pharmaceutical product.
2) Microorganisms that due to their numbers in the product and their pathogenicity can cause infection in the patient via the route of administration when treated with that pharmaceutical product.

To identify potential objectionable microorganisms, three excellent sources of information are:

1) Microorganisms most frequently implicated with US non-sterile drug product recalls.
2) Major infection outbreaks related to non-sterile products investigated by the U.S. Federal Center for Disease Control and Prevention (CDC).
3) Nosocomial infections reported in the clinical literature, especially as related to non-sterile drug products and medical devices.

Based on the overall summation of the frequency of events from these three sources, the seven most common microorganisms found were *Staphylococcus aureus*, *Pseudomonas aeruginosa*, *Klebsiella pneumonia*, *Serratia marcescens*, *Burkholderia cepacia*, *Enterobacter cloacae*, and *Bacillus cereus*.

Recall histories are a useful source of information on microbial contamination of non-sterile drug products. Based on a survey of US product recalls of non-sterile products published by Sutton and Jimenez (2012), there are around 15–20 recalls annually. Perhaps surprisingly to some readers, the major reason for the recalls is the presence of objectionable microorganisms in these products (72%), not exceeding the microbial limit. During a 7-year period from 2004 to 2011, 144 non-sterile drug products were subject to voluntary recall. The recalls by product types and percentage of implication of different microorganisms are illustrated in Figure 11.1. Because of the strict adherence to cGMPs and greater financial and technical resources of large pharmaceutical companies, it may not be unexpected that pharmaceuticals have the least number of recalls among these product types.

The prominence of *B. cepacia* within the recalls is notable due to its well-known resistance of disinfectants and preservative systems.

The designation of the second highest category of objectionable microorganisms merely as unspecified fungi is a poor reflection of the job that the pharmaceutical industry does identifying fungi. The author has previously reviewed the issue of fungal contamination of pharmaceutical drug products (Cundell 2013).

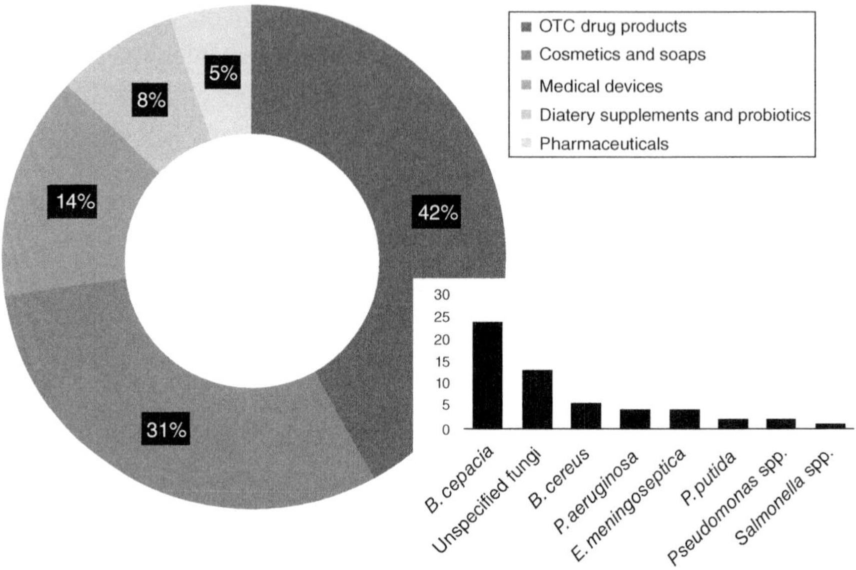

Figure 11.1 Percentage of recalls by product category and the percentage of implication of different microorganisms. *Source:* Data from Sutton and Jimenez (2012).

1022 nosocomial outbreaks were reported in the clinical literature for the period from 1966 to 2002, i.e. around 28 outbreaks annually. Outbreaks are clusters of infection associated with the source in multiple locations. The most frequent species implicated in clusters of hospital patient infection are illustrated in Figure 11.2. It is notable that in the vast majority of outbreaks in which drug products were implicated, the products were sterile, not non-sterile products, presumably due to the high-risk intravenous route of administration of many injectable drug products.

Another potential source of information limited to foodborne intestinal illness recommended by some subject matter experts is the online FDA Bad Bug Book published by the FDA Center for Food Safety and Applied Nutrition (Moldenhauer 2017). As stated in Section 11.1, the publication describes agents that range from live pathogenic organisms, such as bacteria, protozoa, worms, and fungi, to nonliving entities, such as viruses, prions, and natural toxins. Each section has a description of the organism, disease (mortality, infective dose, symptoms, duration of the illness, and route of entry), frequency of occurrence of the disease, sources of the organism, diagnosis, target populations, food analysis, and examples of outbreaks. Ninety percent of the outbreaks are attributable to five pathogens (Table 11.1).

However, the reader is cautioned that many foodborne pathogens listed in the FDA Bad Bug Book are not usually found in drugs and medical devices as

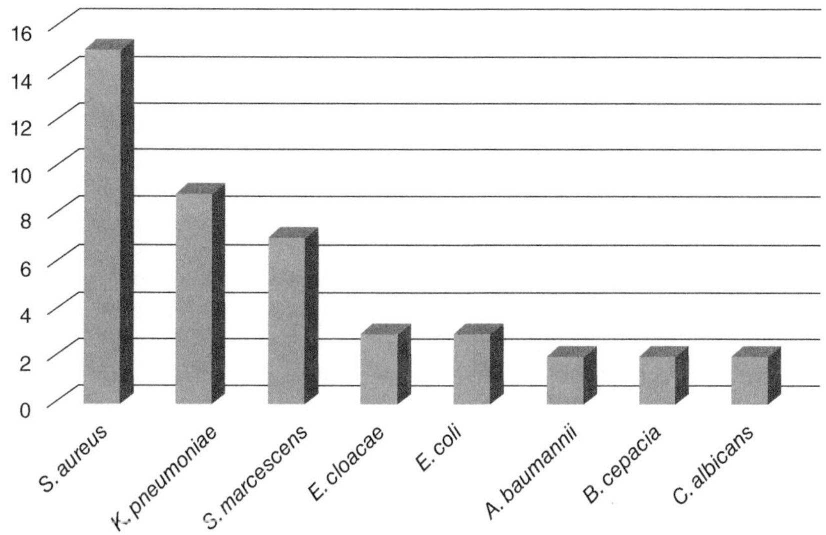

Figure 11.2 Most frequent species implicated in nosocomial outbreaks (expressed in percentage).

Table 11.1 Top five pathogens causing domestically acquired foodborne illnesses, United States (CDC report).

Pathogen	Estimated annual number of illnesses	90% Credible interval	% of all cases
Norovirus	5 461 731	3 227 078–8 309 480	58
Salmonella, non-typhoidal	1 027 561	644 786–1 679 667	11
Clostridium perfringens	965 958	192 316–2 483 309	10
Campylobacter spp.	845 024	337 031–1 611 083	9
Staphylococcus aureus	241 148	72 341–529 417	3
Subtotal		Cumulative	91

they are derived from fecal contamination from animals, humans, and untreated water that are not found in GMP facilities; the USP <61> and <62> microbial tests may not isolate them due to their physiological or fastidious nutritional requirements, and only *B. cereus*, *S. aureus*, and perhaps *Salmonella* spp. are common to both the Bad Bug Book list and recent US drug recalls.

Table 11.2 summarizes the information about these three microorganisms. It should be emphasized that large numbers of the bacteria *S. aureus* and *B. cereus* are needed to product levels of exotoxins to induce intestinal illness, so that low counts of these two bacteria may be present in, for example, a compressed tablet without concern. *Bacillus cereus* is a ubiquitous microorganism that can survive in many types of environments and totally eliminating it from non-aseptic areas is not realistic under actual cleanroom standards. Therefore, a particular attention should be paid in the manufacturing process at steps where microorganisms originating from the excipients or environment would proliferate. For these critical steps, a holding time should be defined from the start of a process step where microbial proliferation would occur (e.g. mixing in an aqueous diluent at ambient temperature) until a microbial reducing step (e.g. heating and drying). The holding time should support that the product or intermediate can be left to stand without a microbiological risk in the non-sterile product manufacturing process. Microbial controls are detailed in Chapter 2.

What is the experience outside of the United States? Recalls for microbial contamination of cosmetics from 2008 to 2014 in the European Union were analyzed from the Rapid Alert System (RAPEX) database (Neza and Centini 2016). Of a total of 527 alerts for microbial contaminated and over-preserved cosmetic products, 62 (12%) were contaminated with microorganisms. Thirty-five percent of the alerts were for contamination with *P. aeruginosa* followed in descending order unspecified aerobic microorganisms, *Klebsiella oxytoca*, *B. cepacia*, *S. aureus*, *Escherichia coli*, *Candida albicans*, *Enterobacter gergoviae*, and *S. marcescens*. Although there are differences in the frequency of the isolation of these different bacterial species, the same organisms appear in each recall database reinforcing the designation of these species as objectionable microorganisms.

The *Burkholderia cepacia* Case

Why has the bacterium *B. cepacia* become prominent among drug product recalls? *Burkholderia cepacia* is a Gram-negative, oxidase-positive, rod-shaped, opportunistic pathogen with a reputation of overcoming antimicrobial preservative systems and antiseptics and growing in multiple-use oral liquids and topical products. It is a member of a group of 20 closely related species in the *B. cepacia* complex (BCC) that share a high metabolic versatility and variable virulence due in part to their large genomic size, i.e. 8 million base pairs, and widespread distribution. BCC causes serious infections in individuals with cystitis fibrosis and chronic granulomatous disease. Furthermore, it is an opportunistic pathogen in mechanically ventilated patients, the immunosuppressed, surgical patients, and those with serious underlying disease.

In June 2017, the FDA in response to additional recalls of oral liquid products due mainly to BCC issued a warning to non-sterile drug product manufacturers to carry out the following:

- Establish procedures designed to prevent objectionable microorganism contamination of non-sterile drug products, such as procedures to assure adequate quality of incoming materials, sanitary design, maintenance and cleaning of equipment, production and storage time limitations, and monitoring of environmental conditions.
- Use scientifically sound and appropriate acceptance criteria and test procedures to assure that drug product components (including pharmaceutical water) and finished drug products conform to appropriate quality standards.
- Provide appropriate drug product specifications (tests, methods, and acceptance criteria) in applications submitted to the FDA for new drug applications (NDA) or for abbreviated new drug applications (ANDA). As appropriate, additional laboratory tests may be needed to determine whether products are suitable for release.
- Ensure that the methods used to test finished drug products prior to release for distribution are appropriately validated, accurate, sensitive, specific, and reproducible.
- Test in-process materials during the production process (e.g. at commencement or completion of significant phases, or after storage for long periods) using valid in-process specifications to assure – among other things – that the drug product will meet its final specification, including criteria for absence of microbial contamination, where appropriate.
- Investigate any failure to meet specifications, including other batches of the same drug product and other drug products that may have been associated with the specific failure or discrepancy, and implement appropriate corrective and follow-up actions to prevent recurrence.

Some countries actually cite objectionable microorganisms in their guidelines for non-sterile pharmaceutical drug products. This is the case for instance with the Australian Therapeutic Goods Authority (TGA) guidance in Section 17.3.2 Objectionable Microorganisms. The regulation states:

> In addition to being free from contamination with specified microorganisms, a non-sterile medicine should also be free from contamination with other microorganisms that might be objectionable in the dosage form.

Table 11.2 Descriptions of *Salmonella* spp., *Staphylococcus aureus*, and *Bacillus cereus* (FDA Bad Bug Book).

Organism	*Salmonella* spp.	*S. aureus*	*B. cereus*
Description	*Salmonella* is a motile, nonspore-forming, Gram-negative, rod-shaped bacterium in the family *Enterobacteriaceae* and the tribe *Salmonellae*	Staphylococcal species are Gram-positive, nonmotile, catalase-positive, small, spherical bacteria (cocci), in pairs, short chains, or bunched in grape-like clusters	*B. cereus* is a Gram-positive, facultatively anaerobic, endospore forming, large rod. *B. cereus* is motile and hemolytic
Disease	Non-typhoidal salmonellosis is generally self-limiting among healthy people with intact immune systems	*S. aureus* is a versatile human pathogen causing food poisoning, toxic shock syndrome, pneumonia, postoperative wound infection, and nosocomial bacteremia	*B. cereus* food poisoning is caused by two recognized types of illness that are caused by two distinct metabolites (toxins)
Mortality	Generally less than 1%; however, *Salmonella enteritidis* has a 3.6% mortality rate in outbreaks in nursing homes and hospitals	Death from staphylococcal food poisoning is uncommon, but has occurred among the elderly, infants, and severely debilitated people	In rare cases, the emetic enterotoxin of *B. cereus* foodborne illness has been implicated in liver failure and death in healthy individuals
Symptoms	Nausea, vomiting, abdominal cramps, diarrhea, fever, and headache	Ingested enterotoxin may rapidly produce nausea, abdominal cramping, vomiting, and diarrhea	*Diarrheal type*: Watery diarrhea, abdominal cramps, and pain. *Emetic type*: Nausea and vomiting
Duration	Symptoms generally last four to seven days	A few hours to one day	The symptoms usually subside after 24 hours of onset
Onset	6–72 hours after exposure	Symptoms usually in one to seven hours	*Diarrheal type:* 6–15 hours. *Emetic type:* 0.5–6 hours
Infective dose	As low as one cell, depending on age and health of host and strain differences	The intoxication dose of S.E. is less than 1 μg. This toxin level is reached when *S. aureus* populations exceed 10^5 organisms/g in food	The presence of large numbers of *B. cereus* ($>10^6$/g) is necessary to produce sufficient exotoxin to cause illness
Source	Oral (ingestion of contaminated food, fecal particles, or contaminated water)	Consumption of food contaminated with enterotoxigenic *S. aureus* or ingestion of the preformed enterotoxin	Consumption of food contaminated with enterotoxigenic *B. cereus* or with the emetic toxin

For example, pseudomonad-type bacteria are considered to be objectionable in aqueous dosage forms that are intended for inhalant, cutaneous, nasal, auricular, oromucosal, gingival or vaginal use and in transdermal patches. These dosage forms are expected to be free from contamination with these types of bacteria.

Drug product specifications for these dosage forms should include an absence of pseudomonads in 1 g or 1 mL, or per patch.

Evaluation of the significance of, and risk from, other objectionable microorganisms should consider:

- The formulation of the medicine,
- Its route of administration,
- Its method of application, and
- The population for which the medicine is intended, including: the possibility of
- Underlying illness in the user of the medicine and the possible concurrent use of immunosuppressive agents or corticosteroids.

Does Antibiotic Resistance Make a Microorganism Objectionable?

In a clinical setting, the antibiotic resistance pattern is critical information in the successful treatment of a microbial infection. The timely selection of the most suitable antibiotic will reduce the morbidity and mortality of patients. The question can be asked, if any weight should be given to antibiotic resistance in determining whether a microorganism found in a non-sterile drug product is objectionable or whether the speciation alone is sufficient. Currently, antibiotic resistance is not a consideration in terms of the objectionable microorganism risk assessment associated with a product release decision. Perhaps in the future both strain typing and antibiotic resistance will be a consideration in designating an isolate objectionable.

If antibiotic resistance were a consideration, what microorganisms would be objectionable? Based on an August 2017 press release, the WHO stated that the most critical group includes multidrug-resistant bacteria that pose a particular threat in hospitals, nursing homes, and among patients whose care requires devices such as ventilators and blood catheters (Table 11.3). They include *Acinetobacter*, *Pseudomonas*, and various Enterobacteriaceae (including *Klebsiella*, *E. coli*, *Serratia*, and *Proteus*). They can cause severe and often deadly infections such as bloodstream infections and pneumonia.

Table 11.3 WHO priority pathogens for the development of new antibiotics.

Pathogen	Antibiotic resistance	Priority level	Isolated on compendial media	Objectionable in non-sterile drug products
Acinetobacter baumannii	Carbapenem	Critical	Yes	Inhalation products
Pseudomonas aeruginosa	Carbapenem	Critical	Yes	Topical and inhalation products
Members of the family *Enterobacteriaceae*	Carbapenem and third-generation Cephalosporins	Critical	Yes	Oral products
Enterococcus faecium	Vancomycin	High	Yes	Oral products
Staphylococcus aureus	Methicillin and Vancomycin	High	Yes	Topical and inhalation products
Helicobacter pylori	Clarithromycin	High	No	N.A.
Campylobacter spp.	Fluoroquinolone	High	No	N.A.
Salmonella spp.	Fluoroquinolone	High	Yes	Oral products
Neisseria gonorrhoeae	Third generation Cephalosporins and Fluoroquinolone	High	No	N.A.
Streptococcus pneumoniae	Penicillin	Medium	Yes	Inhalation products
Haemophilus influenzae	Ampicillin	Medium	Yes	Inhalation products
Shigella spp.	Fluoroquinolone	Medium	Yes	Oral products

11.3 Screening for Objectionable Microorganisms

To demonstrate that objectionable microorganisms are excluded from non-sterile drug products, some type of screening would be necessary. With respect to product-release and shelf life microbial testing of non-sterile drug products, there are four general levels of testing that may be conducted:

1) Microbial enumeration tests
2) Testing for the absence of specified microorganisms
3) Screening for the absence of objectionable microorganisms if a specific risk for the product in question has been defined
4) Microbial identification

Microbial enumeration tests and testing for the absence of specified microorganisms is detailed in Chapter 5.

11.3.1 Microbial Characterization, Identification, and Strain Typing

USP General Informational Chapter <1113> *Microbial Characterization, Identification, and Strain Typing* emphasized the distinction between these three levels of microbial characterization, microbial identification, and strain typing and makes recommendations as to when they would be used in a pharmaceutical setting. Phenotypic methods based on Gram reaction and patterns of biochemical reactions continue to be reliable in terms of microbial identification, e.g. API 20E, Biolog, and Vitek Compact 2. MALDI-TOF mass spectrometric methods, with their rapidly expanding organism databases, are rapidly becoming the first-line identification method, especially in larger microbiology laboratories. Nucleic acid base sequencing may be considered the gold standard for microbial taxonomy and identification and should be used as a referee test when a critical product release decision must be made. Keep in mind that all microbial identification technologies and their databases have limitations and may not be able to differentiate between closely related species.

Experience in food, veterinary, and clinical microbiology shows us that many infectious agents are member of a species that contain both pathogenic and nonpathogenic strains due to differences in animal vectors, virulence factors, and antibiotic resistance patterns. This suggests that more attention should be given to strain typing when determining whether a microorganism isolated from a drug product would be objectionable. More information on microbiological identification can be read in Chapter 9.

11.3.2 Screening of Objectionable Microorganisms

Most US pharmaceutical companies include the compendial microbial test methods found in USP <61> *Microbiological Examination of Non-sterile Products: Microbial Enumeration Tests* and <62> *Microbiological Examination of Non-sterile Products: Tests for Specified Microorganisms* and the recommended microbiological requirements found in USP <1111> *Microbiological Examination of Non-Sterile Products: Acceptance Criteria for Pharmaceutical Preparations and Substances for Pharmaceutical Use* in their regulatory submissions for non-sterile drug products. In Europe and Japan, they would cite their harmonized counterparts from the European and Japanese Pharmacopoeia.

The screening for specified microorganisms for a particular dosage form, e.g. absence of *S. aureus* and *P. aeruginosa* for topical products, is included in the drug product specification. It can be argued that these requirements have been counterproductive in excluding objectionable microorganisms, as pharmaceutical microbiologist may mistakenly believe that meeting the absence of specified microorganism requirements is sufficient to manage risk.

The position of the tripartite pharmacopoeias in evaluating microorganisms recovered from non-sterile products is stated in USP <1111> as follows:

> In addition to the microorganisms listed in table 1, the significance of other microorganisms recovered should be evaluated in terms of the following:
>
> - The use of the product: hazard varies according to the route of administration (eye, nose, respiratory tract).
> - The nature of the product: does the product support growth? Does it have adequate antimicrobial preservation?
> - The method of application.
> - The intended recipient: risk may differ for neonates, infants, and the debilitated.
> - Use of immunosuppressive agents, corticosteroids.
> - The presence of disease, wounds, organ damage.
>
> Where warranted, personnel conduct a risk-based assessment of the relevant factors with specialized training in microbiology and in the interpretation of microbiological data. For raw materials, the assessment takes account of the processing to which the product is subjected, the current technology of testing, and the availability of materials of the desired quality.

Recent podium presentations by FDA microbiologists recommend having a BCC risk mitigation strategy especially for oral liquids and nasal sprays, provide test methods and acceptance criteria that demonstrate that drug products are free of BCC, and demonstrate the method suitability of those methods (Pfeiler 2017). In addition, pharmaceutical companies have received warning letters after inspection and deficiency letters in response to regulatory submission related to BCC screening. The author of this chapter supports the FDA recommendations.

11.3.3 How Should We Screen Non-sterile Drug Products for Objectionable Microorganisms?

Different schools of thoughts exist:

- Modifications to the existing compendial tests to increase their efficacy.
  - Identifying the representative colonies of microorganisms found on plates used for the microbial enumeration.
  - Streaking out from the general enrichment broth used in the tests for the absence of specified microorganisms onto a general microbiological growth medium like soybean–casein digest agar, MacConkey agar, or blood agar and identifying all bacterial isolates.
  - This approach would overcome the objection that solely screening for BCC by general enrichment and streaking out on selective/diagnostic media used in a clinical setting discounts the seriousness of other objectionable microorganisms like *K. pneumonia, S. marcescens, E. cloacae,* and *B. cereus.*
  - Adding tests for the absence of the most damaging objectionable microorganisms such as BCC.
- Another approach suggested by the predominance of Gram-negative bacteria among objectionable microorganisms is to conduct Gram strains and microscopic examination of all colonies isolated during microbial testing and identifying only the Gram-negative rods. This approach, seemingly attractive, would not identify Gram-positive cocci and rods that may be objectionable in a specific drug product.

11.3.4 Selectivity of the USP Absence of Specified Microorganism Tests with Regards to *B. cepacia* Complex

It is useful to compare the biochemical and physiological characteristics of the three most prominent important opportunistic pathogen from the BBC to two related pseudomonads *P. aeruginosa* and *Pseudomonas fluorescens* to understand their selective isolation (Table 11.4). The data from Bergey's Manual suggest tests that involve enrichment at 42 °C, and selective isolation on MacConkey or Cetrimide agar will have reduced recoveries of BCC members (Holt *et al.* 2000) so that the use of these screening strategies may be deficient.

Table 11.4 Key biochemical and physiological characteristics of *Burkholderia cepacia* complex.

Test	*Pseudomonas aeruginosa* (%)	*Pseudomonas fluorescens* (%)	*Burkholderia multivorans* (%)	*Burkholderia cenocepacia* (%)	*Burkholderia vietnamiensis* (%)
Oxidation of glucose	100	100	100	95	100
Ornithine decarboxylase	0	0	0	71	0
Growth at 42 °C	100	0	100	84	100
Oxidase	99	97	100	100	100
Nitrate reduction	98	19	94	31	47
Gelatin liquefaction	82	100	2	55	0
Growth on MacConkey agar	100	100	96	84	83
Growth on Cetrimide agar	94	89	54[a]	54[a]	54[a]
Growth of BCSA	0	0	100	100	100
Identification on API 20 NE	100	100	100	100	100

[a] Limited data are available (Lambe and Stewart 1971).

The limitation of the screening methods found in USP <62> in terms of potential BCC detection is discussed more fully in the following section.

- **Is the USP test for the absence of *P. aeruginosa* suitable for screening for BCC?** Not less than 1 g is used to inoculate soybean–casein digest broth, which is mixed and incubated at 30–35 °C for 18–24 hours. Subculture on a plate of Cetrimide agar, and incubate at 30–35 °C for 18–72 hours. Growth on the plate indicates the possible presence of *P. aeruginosa*. This is confirmed by an identification test. According to the description of Cetrimide agar published by a leading media manufacturer, the use of cetrimide (cetyltrimethylammonium bromide) was recommended by Lowbury (1951); this compound largely inhibits the growth of the accompanying microbial flora including other related Gram-negative bacteria and Gram-positive bacteria. According to Lowbury and Collins (1951), a concentration of 0.3 g/l inhibits the accompanying organisms satisfactorily and minimizes interference with the growth of *P. aeruginosa*. The pigment production of *P. aeruginosa* is not inhibited when grown on this medium. *P. aeruginosa*, as well as *Pseudomonas putida* and *P. fluorescens* are able to grow on Cetrimide agar at 30–35 °C, while *B. cepacia* and *Stenotrophomonas maltophilia* are inhibited. Reportedly *B. cepacia* is able to grow at an incubation temperature of approximately 25 °C.
- **Is the USP test for the absence of *E. coli* suitable for screening for BCC?** Not less than 1 g is used to inoculate soybean–casein digest broth, which is mixed and incubated at 30–35 °C for 18–24 hours. MacConkey broth is inoculated with an aliquot from the soybean–casein digest broth and incubated at 40–42 °C for 24–48 hours. Subculture on a plate of MacConkey agar, and incubate at 30–35 °C for 18–72 hours. Growth on the plate indicates the possible presence of *E. coli*. This is confirmed by an identification test. Although this test seems more promising in that most *B. cepacia* strains grow on MacConkey agar, the selective enrichment in MacConkey broth at 42 °C may be too selective for many BCC strains.
- **Test for the absence of *S. aureus*, *C. albicans*, *Salmonella* species, and *Clostridium* species.** All these four tests for the absence of specified microorganisms are unsuitable for the detection of BCC members.
- **Test for the absence of bile-tolerant Gram-negative bacteria.** The test for the absence of bile-tolerant Gram-negative bacteria is capable of isolating members of the family *Enterobacteriaceae*, e.g. *E. coli* and non-fermenters, e.g. *P. aeruginosa*, so it may be capable of detecting BCC members. Not less than 1 g is used to inoculate soybean–casein digest broth, which is mixed and incubated at 20–25 °C for time sufficient to resuscitate the bacteria without encouraging the multiplication of the bacteria (usually two hours but not more than five hours). Enterobacteria enrichment broth-Mossel is inoculated with an aliquot from the soybean–casein digest broth and incubated at

30–35 °C for 24–48 hours followed by subculture on *violet red bile glucose agar* and incubated at 30–35 °C for 18–24 hours. The product complies with the test if there is no growth on the plate.

- **Test for the absence of BCC (Proposed).** The following test for the absence of BCC has been proposed by the USP in the September–October 2018 Pharmacopoeial Forum. Not less than 1 g is used to inoculate soybean–casein digest broth, which is mixed and incubated at 30–35 °C for 18–24 hours. *Burkholderia cepacia* Selective agar (BCSA), widely used in clinical microbiology laboratories supporting the care of cystitis fibrosis patients, is inoculated by streaking on from the broth and incubated at 30–35 °C for 48–72 hours. The product complies with the test if there is no growth on the plate. Colonies on the plate would be identified to determine if they are BCC members.

11.3.5 Method Suitability

If one or more of these tests for the absence of specified microorganism are used to screen for BCC, then the ability of the test to recover a representative BCC strain, e.g. *Burkholderia cenocepacia* ATCC BAA-245, should be demonstrated as outlined in the USP <62> method suitability testing.

11.4 Risk-Based Microbial Testing of Non-sterile Drug Products

Directing your microbial testing program toward non-sterile drug products with a higher risk of microbial contamination makes sense with respect to both economics and patient safety and is defendable from a regulatory point of view. Testing recommendations for low-, moderate, and high-risk products summarized from the PDA Technical Report No. 67 *Exclusion of Objectionable Microorganisms from Non-sterile Pharmaceutical and OTC Drug Products, Medical Devices and Cosmetics* are given in Sections 11.4.1–11.4.3.

11.4.1 Low-Risk Products

Low-risk pharmaceutical drug products may include solid oral dosage forms such as compressed tablets and power-filled and liquid-filled capsules, lip gels, and rectal suppositories.

- Products pass the microbial testing if the microorganisms identified in USP <1111> and in the relevant product monographs were *not* isolated and enumeration counts are below the specified limit.

11.4.2 Moderate Risk Products

Moderate risk pharmaceutical products may include vaginal suppositories, ointments and creams, topical lotion, and oral liquids.

- Products pass the microbial testing if the microorganisms identified in USP <1111> and in the relevant product monographs are *not* isolated, other colonies observed on selective medium for the absence of specified/objectionable microorganisms and/or soybean–casein digest agar after general enrichment are *not* objectionable (see risk decision tree, Figure 11.3), and enumeration counts are below the specification limit.

11.4.3 High-Risk Products

High-risk products include nasal sprays, inhalants, otic products, and topical products used on broken skin.

- Products pass the microbial testing if the microorganisms identified in USP <1111> and the relevant product monographs are *not* isolated, other colonies observed on selective medium for the absence of specified/objectionable microorganisms and one or more nonselective media such as soybean–casein digest agar or blood agar after general enrichment are *not* objectionable (see risk decision tree, Figure 11.3), and enumeration counts are below the alert limit.

11.5 Sources of Objectionable Microorganisms

Based on the analysis of drug product recalls, the author believes that the origin of objectionable microorganisms isolated from non-sterile drug products, in descending order, is pharmaceutical ingredients ≥ ingredient water > process equipment > manufacturing environment > manufacturing personnel (Cundell 2005; USP <1115>).

11.5.1 Pharmaceutical Ingredients

Pharmaceutical ingredients, especially excipients with an animal, plant, or even mineral origin, may be contaminated with bacteria from fecal matter including pathogenic strains of *E. coli*, *Salmonella*, and other member of the family *Enterobacteriaceae* as well as fungi and Gram-positive, spore-forming *Bacillus* species from the air and soil. Ingredients that are synthetically derived or from an animal, plant, or mineral origin that are further processed have a lower risk level (see Chapter 2).

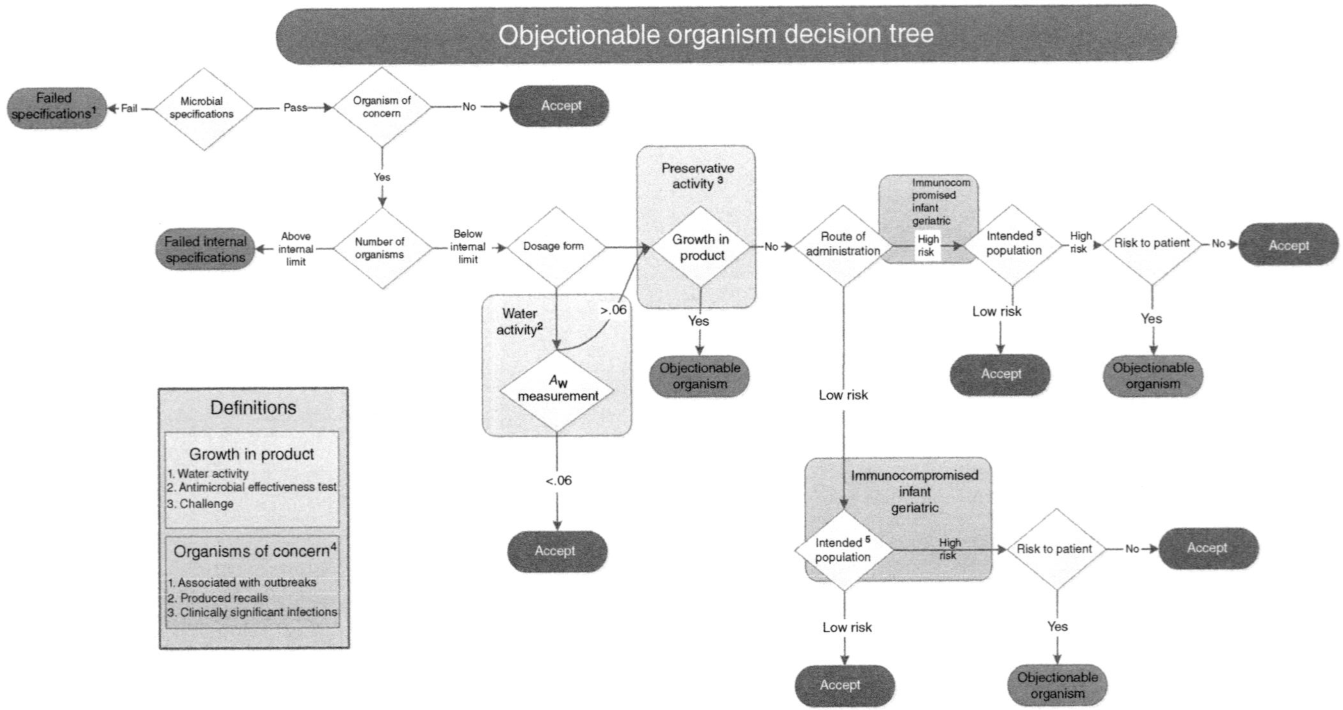

Figure 11.3 Objectionable microorganism risk decision tree. *Source:* Courtesy of the PDA.

11.5.2 Ingredient Water

The design and operation of pharmaceutical water systems have significantly improved over the past decades so that source of risk has receded but not disappeared. However, water systems may be the source of Gram-negative, non-fermentative bacteria known as opportunistic pathogens and for their ability to overcome preservative systems in multiple-dose non-sterile aqueous drug products and grow in the product. Aqueous products have been defined in USP <51> *Antimicrobial Effectiveness Tests* as product with water activities greater than 0.6. In practice, these Gram-negative bacteria will not grow in drug products with water activities below 0.9 (Cundell and Fontana 2009).

11.5.3 Manufacturing Personnel

Operators working in non-sterile drug product manufacturing facilities typically wear gowns, shoe covers, and hairnets. The wearing of makeup and jewelry, and food and drink and prohibited. They are excluded from the workplace, if they are sick and have skin infections. Many unit operational steps are conducted in closed tanks and equipment. When operators handle product directly or take samples, they should wear gloves and facemasks to avoid contaminating the product. In general, the recall record does not implicate skin-borne bacteria suggesting current hygiene practices are effective.

11.5.4 Process Equipment

The sanitary design, cleaning, and storage of processing equipment are important in terms of preventing both chemical and microbial contamination of non-sterile drug products. Microbial monitoring of product-contact surfaces during cleaning validation studies is highly recommended. Typically, the physical removal of product residuals will also remove microorganisms but the formation of biofilms especially in valves and piping must be considered.

11.5.5 Manufacturing Environment

Environmental sources of contamination may, for instance, come from the dust or soil particles brought into the room either by personnel or material flow or via the air ventilation system. Typical representatives are spore-forming, Gram-positive rods such as *Bacillus* spp. and molds. The risk of microbial proliferation in the manufacturing environment would more likely be triggered by residual water in poorly dried process equipment, and on floor or wall surfaces. This is typically the case in equipment washing rooms or waste water from sinks or cleaning in place systems that are insufficiently dried. Even if microbial contamination of non-sterile products from the environment to unacceptable levels is rare, such events may occur especially when water leakages occur as reported by Roesti (2012).

Intrinsic or Extrinsic Microbial Contamination?

One of the major challenges of an epidemiological investigation of a microbial contamination outbreak with a non-sterile drug product, cosmetic, or medical device is to determine if the product contamination is intrinsic or extrinsic.

Intrinsic contamination, as the term implies, usually occurs during manufacture and the contaminating microorganism may grow during the shelf life exceeding the product specifications and/or being identified as an objectionable microorganism. In this situation, unopened containers, as well as opened containers of the product, will be shown to be contaminated with the microorganism and the microorganism may have been found in pharmaceutical ingredients, pharmaceutical water systems, during environmental monitoring, or even in other products manufactured at that site.

With multiple-use products, extrinsic contamination may occur when a user or multiple users mishandles and inadvertently contaminates the product or the product will enhance the growth of the user's own microflora at the application site. Strain typing may confirm the single source of intrinsic contamination whereas with extrinsic contamination, although the contaminant may be limited to a single species, will be the result of geographically separated, multiple strains of the same species.

Classic examples of extrinsic contamination are toxic shock syndrome associated with exotoxin-producing *S. aureus* growing in high absorbent tampons during use in the 1980s (CDC 1990) and *Fusarium keratitis* caused by the fungal contamination of a reformulated contact lens solution in use in the mid-2000s (CDC 2006).

11.6 Risk Assessment to Determine if a Microorganism Is Objectionable in a Non-sterile Drug Product

This section summarizes the flowchart from PDA Technical Report No. 67 Exclusion of Objectionable Microorganisms from Non-sterile Pharmaceutical and OTC Drug Products, Medical Devices and Cosmetics (Figure 11.3). It should be noted that the authors of this chapter do not supply a list of objectionable microorganisms, as they will depend on the specific dosage form, product formulation, product attributes, intended use, and the target patient population. Only the manufacturer has sufficient information to make the decision on what microorganism is objectionable in which drug product.

The decision tree is as follows:

Step 1: Does the non-sterile drug product contain microorganisms of concern? If the answer is yes, and the product exceeds the microbial enumeration test limit for the dosage form, recommend to the Quality Control Unit that the batch be rejected.

Step 2: If the batch does not exceed this limit, does the water activity exceed 0.6. If no, recommend that the batch be accepted for release as the product will not support microbial growth. An exception to this recommendation would be the presence of *Salmonella* in the product that can survive at low water activities and cause enteric infection with low counts.

Step 3: If yes, it does exceed 0.6, is the water activity sufficient for the growth of contaminating microorganism or would the growth be prevented by the antimicrobial preservative system in the formulation of the product? If no, recommend to the Quality Control Unit the rejection of the product due to the presence of an objectionable microorganism. An example would be members of the BCC that may have the ability to grow and overcome the antimicrobial preservative system.

Step 4: Is the route of administration, a low or high risk to the patient? Based on the indications is the target of the drug product a high-risk patient population? For dosage form directed toward infants, the immune-compromised and the elderly, recommend to the Quality Control Unit the rejection of the product. An example would be oral syrups containing honey as an ingredient that may contain low levels of the spore-forming *Clostridium botulinum* that may cause infant botulism.

Sections 11.6.1–11.6.3 provide additional insights on the points to consider in a risk assessment of objectionable microorganisms

11.6.1 Microorganism of Concern

To identify if the isolated microorganisms as microorganisms of concern, i.e. potential for classification as objectionable in some dosage forms, the sources of information described in Section 11.2 may be used.

This research should be carried out in order to characterize the patient risk (e.g. infection and allergy) of the isolate for the intended route of administration.

11.6.2 Potential Microbial Proliferation in the Drug Product

Independent of the pathogenicity of the isolated microorganism, the capacity of microorganisms to pursue metabolic activity or proliferate in the product is a serious concern.

Actually, if the microorganisms survive or grow in product, they may affect the chemical, physical, functional, and therapeutic attributes of a pharmaceutical drug product. The single physical attribute that is most likely to control microbial growth is water activity.

Microbial degradation of the active ingredient within a drug product may reduce the therapeutic activity and create toxic degradative products, while degradation of nonionic surfactants and thickening agents will affect their functionality and preservatives, and their resistance to microbial contamination (Bloomfield 1990).

The nature of the product influences the survival or growth capability of the microorganisms. The most relevant factor is the water activity, not the water content. Water activity (a_w) is the relative availability of water in the drug product. It is defined as the ratio of vapor pressure of water in product to vapor pressure of pure water at the same temperature. The USP informational chapter <1112> provides a very good overview of the water activity levels in drug products and microbial growth.

Below the a_w level of 0.60 microorganisms cannot grow. Additionally, the water activity of the product is very important when determining the storage conditions, i.e. when there is a high water activity the product risks spoilage during its storage and shelf life. In general, bacteria outcompete fungi at high water activity, no bacterium encountered in a pharmaceutical setting will grow below 0.85 while most fungi grow above 0.75.

The drug products with the lowest risk related to a_w are nonaqueous liquids or dry solid oral dosage forms because they do not support spore germination or microbial growth. The most critical ones are aqueous drug products such as nasal sprays, if not preserved.

Some microorganisms are known to be resistant against some preservative systems, such as *Pseudomonas* species more resistant to quaternary ammonium compounds. When a microorganism is isolated in a preserved drug product, a challenge test to verify the efficacy of the product's antimicrobial components against the isolate is good practice.

11.6.3 Risk Level in Targeted Patient Populations

Although the microbiological quality requirements in USP <1111>, especially the microbial enumeration limits, may be considered suitable for all patient populations, the exclusion of objectionable microorganisms will have a disproportionate impact on impaired patient populations. These populations include infants, immunologically suppressed patients, transplant recipients receiving immune-suppression drugs, patients recovering from invasive surgical procedures, and the elderly. In the cases of infants and elderly, oral liquids and syrup may be prescribed instead of compressed tablets because of the inability to swallow a tablet, exposing the patient to a more risk of potential infection. Perhaps orally disintegrating tablets are a better solution for these target populations than multiple-use oral liquids or syrups that are more susceptible to microbial contamination.

As discussed in Chapter 2, patient populations may be broadly classified as healthy, moderately impaired, and immunocompromised. The medical status of the recipients of a drug product will influence the risks of infection.

Surprisingly, the number of immunocompromised patients receiving drug products is largely unknown. This lack of information is a detriment when it comes to microbial contamination risk analysis. A recent letter to the JAMA editors suggested that 4% of US adults self-reported that they have been told at one time by a health professional that they are immune-suppressed. Of these adults surveyed, 2.8% reported current immunosuppression (Harpaz *et al.* 2016). To this microbiologist, the numbers were revealing and must be considered during objectionable microorganism risk assessments because microbial specifications do not make any distinction as to medical status for the recipient of a drug product. It is important to know the indications of a drug product and the administration and dosage instruction as found in the package insert. The default microbiological requirements in USP <1111> have stricter microbial enumeration requirement for more invasive drug products and the absence of specified microorganism requirement is broken down by dosage form.

Pharmaceutical products specifically directed toward higher risk patient population should have stricter objectionable microorganism exclusion requirements than product dispensed to healthy individuals.

11.7 Case Histories

11.7.1 Allopurinol Tablets for Cancer Patients

In March 2009, the Hong Kong Board of Health reported that four batches of allopurinol tablets manufactured by a local pharmaceutical company were found to be grossly contaminated with the fungus *Rhizopus microsporus* ($>10^3$ CFU/g). At least five patients at Queen Mary's Hospital receiving aggressive cancer therapy and treated with anti-gout drug allopurinol for the common side effect hyperuricemia contracted intestinal mucomycosis and died (Cheng *et al.* 2009). How could compressed tablets, which are considered having a low risk for microbial contamination, be the cause of this fungal outbreak and patient deaths?

The tablets were manufactured at the local facility using a wet granulation that was dried in a tray dryer oven at 50 °C for 4 hours to a water content of 3%. The granulation was then held at 20 °C for up to 14 days prior to tablet compression. A typical formulation is allopurinol, 100 or 300 mg, corn starch, FD&C Yellow No. 6 Lake (yellow tablets only), lactose, magnesium stearate, and povidone.

A probable source of the mold *R. microsporus* was the filler corn starch used in the tablet manufacture as it was found to contain 2 CFU of *Rhizopus*/g. Although the ascospores of *R. microsporus* are thermotolerant and would survive 4 hours at 50 °C during tray drying, it appears less likely that a granulation dried to 3% water content and stored at 20 °C for 5–14 days prior to tablet compression would become highly contaminated with *R. microsporus*. Strain typing the fungus found in the drug product and those infecting the cancer patients to demonstrate a match would have been helpful.

11.7.2 Laxative Recommended for Infants

On 16 July 2016, the FDA announced voluntary nationwide recall of all non-expired lots of oral liquid docusate sodium manufactured by PharmaTech LLC, Davie, Florida, in one pint (473 ml) bottles of the laxative and distributed by Rugby Laboratories for contamination with *B. cepacia* and linked to a five state outbreak. Laboratory evidence later linked the *B. cepacia* to the company's purified water system. In a 10 August 2016 update, the CDC confirmed 60 cases from 8 state outbreaks using molecular typing and recommended that clinicians and their patients not use any brand of liquid docusate sodium as a stool softener or other medical reason.

11.7.3 Alcohol-Free Mouthwash in Hospital Settings

There are multiple reports in the clinical literature of nosocomial infections in hospitals from use of bacterially contaminated mouthwash. All of these outbreaks were associated with the use of alcohol-free mouthwash in intensive care units. It is widely acknowledged that the preservative systems used in alcohol-free mouthwash are not as effective. The same mouthwash was often used in general medical wards in the same hospital without infection highlighting the higher risks associated with aggressively managed hospital patients in intensive care units. From August 1996 through June 1998, 74 patients at 2 Arizona hospitals that had been on ventilators contracted *B. cepacia* respiratory infections from an intrinsically contaminated alcohol-free mouthwash (CDC 1998). Seven years later, from April through August 2005, 116 patients from 22 hospitals in 9 southern states were infected with *B. cenocepacia* from an intrinsically contaminated alcohol-free mouthwash (Kutty *et al.* 2007). This phenomenon has been reported outside the United States. This included an April 2012 report of a cluster of infection of patients undergoing mechanical ventilation in an Ecuadorian hospital implicating an alcohol-free mouthwash and a 2012 report of six patients in a German hospital developing ventilator-associated pneumonia due to BCC from an intrinsically contaminated alcohol-free mouthwash (Winterfield *et al.* 2012).

11.7.4 Non-sterile Alcohol Wipes in an Intensive Care Unit

On 5 January 2011, the Triad, the manufacturer of alcohol prep pads, swabs, and wipes, recalled multiple products for the Gram-positive, spore-forming bacterium *B. cereus* contamination. The previous fall, a Colorado children's hospital noticed an unusual cluster of bloodstream infections associated with the installation of central line catheters in pediatric cancer patients receiving chemotherapy. Non-sterile alcohol wipes were contaminated with the objectionable bacterium *B. cereus* presumably derived from either the alcohol or the textile used to manufacture the wipes (Cundell 2015).

11.7.5 Metformin Hydrochloride Oral Solution

On 24 November 2017, Sun Pharmaceuticals recalled Riomet Oral Solution, manufactured by a contract manufacturer, due to contamination with the soil-borne fungus *Scopulariopsis brevicaulis*. The recall notification from the FDA website stated:

> Sun Pharmaceutical Industries is recalling two lots of Riomet (Metformin Hydrochloride Oral Solution), which were found to be contaminated with *Scopulariopsis brevicaulis*. Use of the affected Riomet potentially could result in a risk of infection, especially in the immune-compromised patient. The most plausible portal of entry of *Scopulariopsis brevicaulis* is the respiratory tract, where it may cause pneumonia, sinusitis and disseminated infections. The contamination was discovered during sample preparation for the Antimicrobial Preservative Effectiveness Testing being performed as part of the 12-month stability study interval.
>
> Riomet (Metformin Hydrochloride Oral Solution) is indicated to treat type 2 diabetes mellitus in adult and children aged 10 and above. Riomet is packaged in 118 mL (4 fl. oz.) and 473 mL (16 fl. oz.) bottles.

Scopulariopsis brevicaulis is best known for non-dermatophyte nail infections and the cause of more limited invasive infection of immune-compromised individuals. The publication by Iwen *et al.* (2012) reported 32 cases of proven invasive *Scopulariopsis* infections in the clinical literature from 1974 to 2012. They recommended ITS region base sequencing for the fungal identification. The recall did not mention if the TCYMC exceeded the 100 CFU/ml limit for an oral liquid or if the stability sample failed the AET. Presumably the FDA felt that the presence of the objectionable microorganism justified a Class II recall. If you had an objectionable microorganism list in your company, you may include fungi *Aspergillus, Fusarium*, and perhaps *Scedosporium* species, but would you include *Scopulariopsis*?

11.7.6 Comforts for Baby Water with Fluoride

On 4 December 2017, the Kroger Company recalled multiple lots of Comforts for Baby Purified Water with Fluoride Added in 1-gal clear plastic containers after receiving complaints about mold in the product. The recalled lots have sell-by dates of 26 April 2018–10 October 2018 so the recall may represent six months of production. According to the FDA announcement, testing by Kroger identified the mold *Talaromyces (Penicillium) marneffei*. Larone (2011) states that the fungus causes deep-seated infections that can be focal or disseminated mainly in immune-compromised (HIV-positive) patients who live in Southeast Asia where *P. marneffei* is endemic. The fungus is thermally dimorphic forming flat, powdery, tan colonies on sabouraud dextrose agar (SDA) at 25–30 °C, later becoming reddish yellow with a red soluble pigment defusing into the medium. At 35–37 °C on SDA, the colonies are soft, white to tan, dry, and yeast-like. Recent reports identify the fungus as an emerging pathogen in non-HIV-infected children and adults (Chan *et al.* 2016). The bamboo rat and the soil from their burrows are considered important enzootic and environmental reservoirs of *T. marneffei*.

The question must be asked how did this fungus contaminate purified water sold expressly for infants? The labeling states that the product distributed by Kroger Co., Cincinnati, Ohio, is purified water minerals added for flavor (potassium bicarbonate, calcium chloride, and magnesium chloride) plus fluoride processed by steam distillation. The manufacturer's name does not appear on the label. Are they located in China or the South East? Possible contamination routes are the storage of the purified water, the packaging components, minerals, and the filling operation.

11.8 Conclusions

Considering that patient populations are expanding to include many more immunosuppressed individuals, aggressively treated surgical patients, infants, and the elderly, the importance of excluding objectionable microorganisms from non-sterile drug products has become more critical. This chapter has defined the term objectionable microorganism, provided examples from the technical literature, and recommended screening strategies and microbial contamination risk assessment tools. If pharmaceutical companies really want to mitigate the risk of recipient infection, they must employ broadly trained, experienced microbiologists who can communicate with management, understand the limitation of microbial testing, product formulation, manufacturing processes, and how products are administered. The objective of excluding objectionable microorganisms from non-sterile drug product, not only is a critical GMP requirement and ensures recipient safety but makes business sense in terms of maintaining the reputation and economic viability of your company.

Bibliography

Bloomfield, SF, 1990, 'Microbial contamination: spoilage and hazards', in Denyer, S, and Baird, R (eds), , Ellis Harwood, New York, pp. 29–52.

Brown, VI, Lowburg, EJL, 1965, 'Use of an improved Cetrimide agar medium and other culture methods for *Pseudomonas aeruginosa*', *J. Clin. Path.*, vol. 18, pp. 752–756

CDC, 1990, 'Historical perspectives reduced incidences of menstrual toxic shock syndrome – United States, 1980-1990', *MMWR*, vol. 39(25), pp. 421–423.

CDC, 1998, 'Nosocomial *Burkholderia cepacia* infection and colonization associated with intrinsically contaminated mouthwash', *MMWR*, vol. 47(43), pp. 926–928.

CDC, 2006, 'Update: *Fusarium keratitis* – United States, 2005–2006', *MMWR*, vol. 55(20), pp. 563–564.

Chan, JFW, Lau, SKP, Yuen, KY, Woo, PCY, 2016, '*Talaromyces (Penicillium) marneffei* infection in non-HIV-infected patients', *Emerg. Microbes Infect.*, vol. 5, pp. e19, https://doi.org/10.1038/emi.2016.18.

Cheng, VCC, *et al.*, 2009, 'Outbreak of intestinal infection due to *Rhizopus microspores*', *J. Clin. Microbiol.*, vol. 47(9), pp. 2834–2843.

Cundell, AM, 2005, 'Risk-based approach to pharmaceutical microbiology', in Miller, MJ (ed), , DHI/PDA, River Grove, IL/Bethesda, MD, pp. 29–62.

Cundell, T, 2013, 'Mould contamination in pharmaceutical drug products and medical devices', *Eur. Pharm. Rev.*, vol. 18(6), pp. 1–18.

Cundell, T, 2015, 'Bacterial contamination of alcohol swabs', in Madsen, RE, and Jornith, MW (eds), , DHI/PDA, River Grove, IL/Bethesda, MD, pp. 117–120.

Cundell, AM, Fontana, A (eds), 2009, Water Activity Application in the Pharmaceutical Industry, DHI/PDA, River Grove, IL/Bethesda, MD.

Harpaz, R, Dahl, RM, Dooling, KL, 2016, 'Prevalence of immunosuppression among U.S. adults, 2013', *JAMA*, vol. 316(23), pp. 2547–2548.

Holt, JG, Krieg, NR, Sneath, PHA, Staley, JT, Williams, ST, 2000, , 9th ed., Lippincott Williams and Williams, Philadelphia, PA, pp. 787.

Iwen, PC, Schutte, SD, Florescuri, DE, Noel-Hurst, RK, Sigler, L, 2012, 'Invasive *Scopulariopsis brevicaulis* infection in an immuno-compromised patient and review of prior cases caused by Scopularsis and Micrococcus species', *Med. Mycol.*, vol. 50, pp. 561–569.

Kutty, PJ, *et al.*, 2007, 'Multistate outbreak of *Burkholderia cenocepacia* colonization and infection associated with the use of intrinsically contaminated alcohol-free mouthwash', *Chest*, vol. 132, pp. 1825–1831.

Lambe, DW, Stewart, P, 1971, 'Evaluation of Pseudosel agar as an aid in the identification of *Pseudomonas aeruginosa*', *Appl. Microbiol.*, vol. 23(2), pp. 377–381.

Larone, DH, 2011, , 5th ed., ASM Press, Washington, DC.

Lowbury, EJL, 1951, 'Improved culture methods for the detection of *Pseudomonas pyoccyanea*', *J. Clin. Path.*, vol. 4, pp. 66–72.
Martin, M, *et al.*, 2012, 'Outbreak of *Burkholderia cepacia* complex caused by contaminated alcohol-free mouthwash', *Anesthetist*, vol. 61(1), pp. 25–29.
Moldenhauer, J, 2017, *Objectionable Organisms*, IVT Network. http://www.ivtnetwork.com/article/objectionable-organisms
Molina-Cabrillana, J, *et al.*, 2006, 'Intrinsically contaminated alcohol-free mouthwash implicated in a nosocomial outbreak of *Burkholderia cepacia* colonization and infection', *Infect. Control Hosp. Epidemiol.*, vol. 27, pp. 1281–1282.
Neza, E, Centini, M, 2016, 'Microbiologically contaminated and over-preserved cosmetic products according Rapex 2008–2014', *Cosmetics*, vol. 3, pp. 3.
Pfeiler, E, 2017, 'Organizational Overview and Recommendations for Non-sterile Products', CDER SBIA Webinar Series Division of Microbiology Assessment 15 March 2017.
Roesti, D, 2012, 'Objectionable microorganisms in non-sterile pharmaceutical drug products: risk assessment and origins of contamination', in Moldenhauer, J (ed), , vol. 6, DHI/PDA, River Grove, IL/Bethesda, MD, pp. 137–158.
Sutton, S, Jimenez, L, 2012, 'A review of reported recalls involving microbiological control 2004-2011', *Am. Pharm. Rev.*, vol. 15(1), pp. 42–57.
Torbeck, L, Raccasi, D, Guilfoyle, DE, Friedman, RL, Hussong, D, 2011, '*Burkholderia cepacia*: this decision is overdue', *PDA J. Pharm. Sci. Technol.*, vol. 65, pp. 535–543.

12

Data Integrity and Microbiological Excursion Handling

David Roesti[1] and Marcel Goverde[2]

[1]*Novartis Pharma Stein AG, Stein, Switzerland*
[2]*MGP Consulting GmbH, Binningen, Switzerland*

CONTENTS

12.1 Data Integrity

Decision-making under cGMP rules is based on information that is truthful, accurate, and complete. In the past years, there was a high increase in health authority observations and enforcement actions concerning failed compliance to data integrity principles (Figure 12.1 data from Platco and Cundell (2017)).

As reported in the PDA TR-80 (2018), the most common trends in regulatory observations on data integrity breaches include:

- Failure to perform required testing. Sampling or testing not performed but records have been generated as if these have been carried out.
- Falsification of data or cGMP records, e.g. reporting failing results as passing, writing test results in CoAs that were not executed.
- Deletion or overwriting of data.

Pharmaceutical Microbiological Quality Assurance and Control: Practical Guide for Non-Sterile Manufacturing, First Edition. Edited by David Roesti and Marcel Goverde.

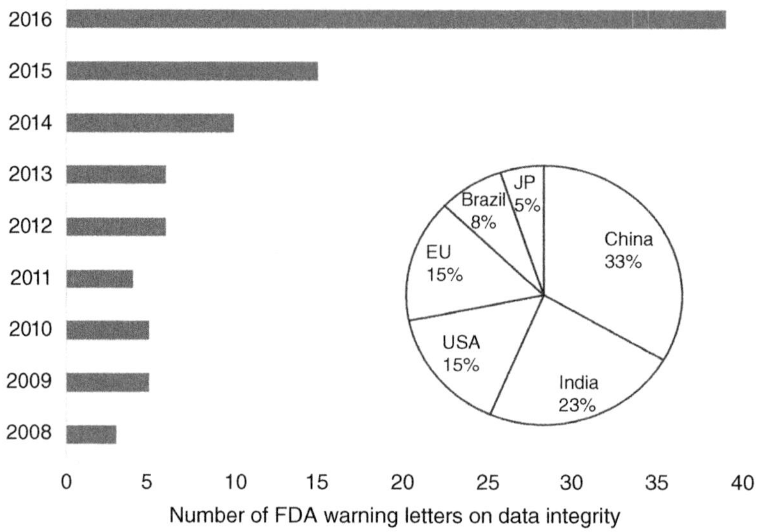

Figure 12.1 Graphical representation of the number of FDA warning letters addressing data integrity from 2008 to 2016 as well as proportion of countries receiving these warning letters in 2016. Source: Data from Platco and Cundell (2017).

- Deficiency in addressing data manipulation/falsification in excursion investigations.
- Failure to ensure that laboratory data include complete records such as, e.g. unexplained deletion of data records.
- Failure to configure computerized systems to meet the requirements for the security and control of data.
- Failure to document laboratory records contemporaneously and/or deliberate falsification of manual records.
- Performing unreported sample test injections.
- Failure to validate analytical methods.

The increased concern on data integrity issues has resulted in the publication of detailed guidelines in the past three years (PIC/S Draft Guidance 2016; WHO 2016; FDA 2018; ISPE 2018; MHRA 2018). Based on the actual health authority guidelines, data generated electronically or on paper under cGMP environments must follow the ALCOA or ALOCA+ principles (Figure 12.2).

As reported in health authority observations, the review paper of Platco and Cundell (2017) and the PDA TR-80, the following breaches in the ALOCA+ principles may happen for data generated by microbiological testing:

- Attributable:
  - Other analysts than those that performed the experiments sign off the raw data worksheet.
  - Different analysts share the same user ID and password.

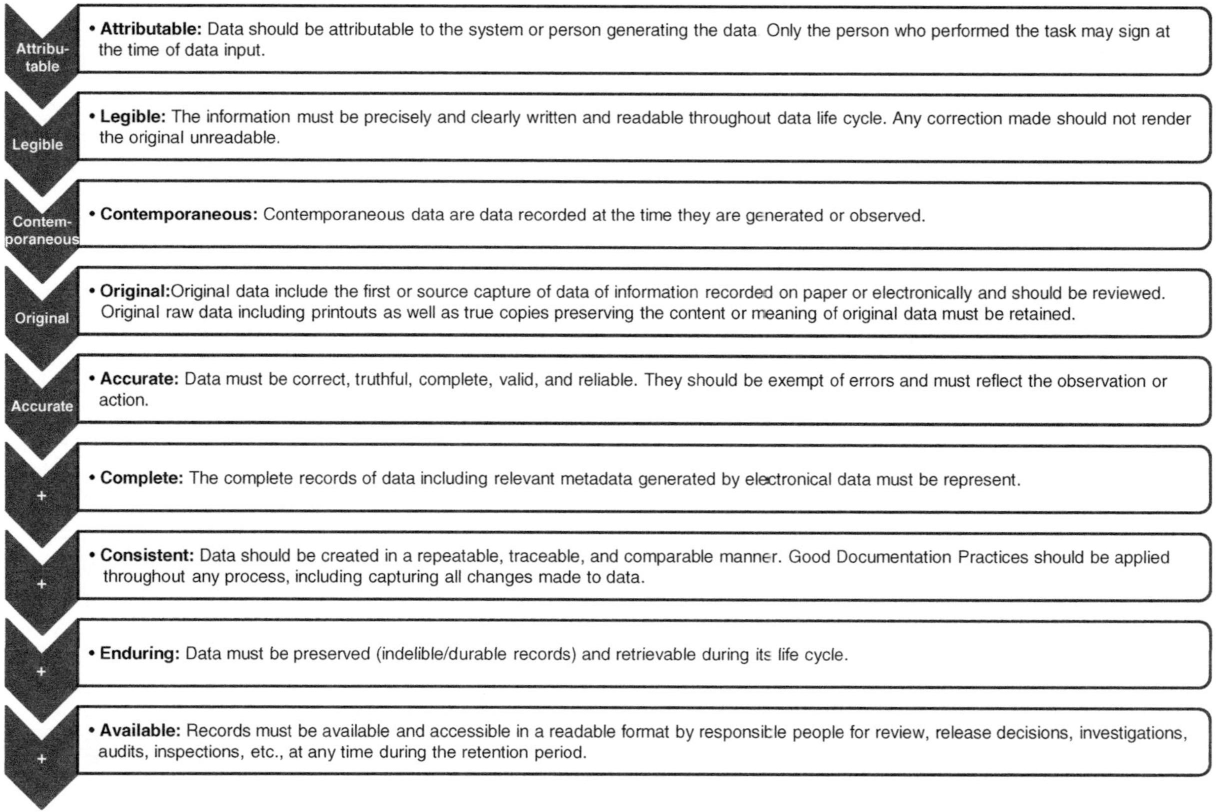

Figure 12.2 Data integrity principles based on the ALCOA+ concept.

- Missing date and signature for an action performed.
- Another device or equipment is documented than the one used.
- Readout of plates was performed by a different person than the one documented.

- Legible :
  - Use of correction fluids to correct data.
  - Masking original data by a handwritten correction.
  - Original electronic data not readable following the software update.
- Contemporaneous:
  - Backdating or predating data sheets.
  - Documentation of raw data at a later time point.
- Original:
  - Original data hand-copied in new raw data sheets (to improve the readability) with subsequent discarding of the original raw data sheets.
  - No unique numbering of the raw data sheets.
  - Data are reported on a stand-alone and uncontrolled computer system.
- Accurate:
  - The calculation for the number of microorganisms in 1 g of product did not take into account the dilution factor.
  - Wrong calculation of CFU means or log-reductions.
  - Observations or corrections are not documented in a way that they can be understood at a later time point.
  - Modifications to the data are not explained.
- Complete:
  - Electronic data deleted without backup to free up hard drive space.
  - Some part of the raw data was thrown away.
- Consistent:
  - Raw data template was inaccurately modified by the analyst independently of a controlled change process.
  - Raw data order was changed.
  - Attached documents are not in the correct order.
- Enduring:
  - Paper raw data records were stored in a room with high humidity level.
  - No indelible ink was used.
  - No copy of printouts on thermal paper.
- Available:
  - The requested raw data could not be presented during the audit as it was not found.

The general definitions of original data of the ALCOA+ principles consider the reporting of data on paper or electronically and that they remain consistent, enduring, etc. The Petri plates do not fully fill the principles of an original data record as per ALCOA principles. Nonetheless, the Petri plates may be considered as original data on the day that the method requires that the plates are to be read and recorded. After reading, if these same plates are subsequently stored, it is not possible to confirm the original results because the microbial counts may increase during their storage since even under refrigeration microbial colonies continue to grow but at a slower rate.

The interpretation of test results or the number of colonies tested is prone to a certain subjectivity and variability since microbiological methods are classically performed manually and based on the visual evaluation of an analyst performing the test.

Periodic training and qualification of analysts including periodic vision check and/or hiring of experienced personnel should limit to an acceptable level the variability of microbiological test results caused by analyst handling and reading as well as ensure that the defined good documentation practices are understood and followed. Re-qualification of analysts enumerating colonies on plates may include enumeration of exemplary agar plates containing different levels of microbial counts and comparing the results obtained with either an automated colony counter or the counts of a senior analyst or for laboratories with a high amount of personnel the mean counts of all personnel. Acceptance criteria should take into account an acceptable difference tolerance (e.g. 5 or 10%, mean ±2-times standard deviation). Analysts who fail re-qualification would require retraining on microbial enumeration with focus topics such as, for instance, on how to differentiate merging colonies.

To further improve data integrity and reduce subjectivity, alternative methods for reading of plates such as the use of automated plate readers and/or high-resolution photographs of the plate may be used. These systems have, however, inherent challenges such as difficulties to count colonies embedded in the agar gel from pour-plated dishes, satellite colonies may be counted, difficulty to differentiate overlapping colonies, difficulty to differentiate particles from colonies, and interpretation of the photo may vary from one individual to another. Automated enumeration methods that actually stack images to capture the colonies growing in time may overcome some of these challenges (see Chapter 13).

It should be noted that contemporaneous recording of actions during execution of the microbiological testing cannot be followed in all cases in order to avoid jeopardizing the aseptic status required during testing. It is acceptable that recording of actions takes place immediately after the working session upon exiting the laminar flow hood instead of constantly having to go back and forth from aseptic to non-aseptic areas to write down test actions executed.

It is common practice that the raw data sheet that contains the original data as well as the data entry from the raw data sheet to the laboratory information management system (LIMS) are reviewed for accuracy and completion by a qualified analyst or that has not executed or evaluated the test or the supervisor.

The greatest data integrity threat with microbiological testing resides with falsification of data and intentional omission of testing. To control this risk, the following actions may be taken:

- Company culture
  - Defining and applying strict ethical corporate standards and complete endorsement by the company leadership has the most significant contribution to eliminating fraud. When these are defied by short-term overaggressive business goals or cultural bad practices, fraud may become the rule and not the exception.
  - Companies must have a zero tolerance for intentional fraud as well as careless work habits leading to unintentional data integrity breaches. It should be noted that not just intentional but also unintentional data integrity violations can lead to jurisdictional consequences.
  - People should be encouraged to openly talk about their own errors made during testing as well as on how to improve data integrity issues. Honesty and goodwill must be treated fairly by the management and analysts should not fear of doing something wrong.

- Applying a rigorous quality management system
  - Clear standard operating procedures should describe what is expected in terms of good documentation practice and data integrity. These procedures should then be formally trained to all associates. Very important is a regular retraining on data integrity especially if there is a frequent change within the personnel.
  - Systematic review of the raw data recording sheets or entries in the LIMS by specialized personnel.
  - Trending of microbiological data should also consider data integrity risk. For instance, unexplained significant lower counts in the trend analysis of one analyst as compared to others or from one time period to another might indicate under reporting of counts in order to avoid exceeding action levels (thus, falsification).
  - A QA oversight of laboratory activities may be in place to evaluate if the rules on data integrity are adequately followed. This may consist of a regular (e.g. weekly or monthly) and surprise verification of all tests performed by the analysts and could include verification that the colonies or absence of colonies on the plates are conform to what is reported on the raw data worksheets or LIMS and that the sample and material/equipment inventory is correctly reported.

– Regular internal audits as well as appropriately certifying and auditing third-party contract laboratories (refer to Chapter 17) or manufacturing sites are also key.

For the compendial sterility test that combines criticality of the test and higher risk of misinterpretation of results, it is now a standard practice to perform a contemporaneous evaluation of the sample for microbial growth by a second analyst. Nonetheless, applying uncritically a contemporaneous reading by a second analyst (four-eye principle) for all samples and for all microbiological tests is to the author's opinion unnecessary and will not improve data integrity. Indeed, both analysts can agree to falsify results, a second analyst may oversee data integrity mistakes, enumeration of many colonies may vary even among qualified personnel, large amounts of resources are spent by hiring of additional personnel for the manual verifications where in fact these resources could be more wisely used to introduce automated reading methods with fully DI-compliant computerized systems.

Does It Make Sense to Carry Out a Systematic and Contemporaneous Enumeration of Agar Plates by a Second Analyst?

A strict application of the ALCOA+ principles may be interpreted as the necessity to execute a contemporaneous enumeration of agar plates by a second analyst and having both analysts reporting their results on the raw data sheet (four-eye principle). Whereas contemporaneous evaluation of sterility test growth in turbid media by two analysts is becoming the standard, four-eye principle for enumeration of agar plates is a real challenge:

- Precision in counts may vary from one analyst to another (even if they are trained and qualified) as colonies may overlap, swarm over media, etc., allowing a certain room for interpretation.
- By tolerating no differences in counts, a high amount of noncritical discrepancies will be generated consuming resources unreasonably.
- Microbiology is a "logarithmic science," sample size is statistically weak, and testing procedures have an inherent variability; so a difference of 20 or 21 count has absolutely no relevance.

There are no guidelines which refer to the acceptable difference between counts simply because no one can scientifically define this precisely. The different pharmacopoeia actually allows recoveries to vary by a factor 2. Therefore, the authors do not believe that a systematic and contemporaneous enumeration of agar plates by a second analyst significantly improves the level of data integrity in microbiological tests performed for non-sterile products.

As an alternative to a contemporaneous *enumeration*, a contemporaneous *verification* by a second person that the testing activity is performed correctly may be executed for high-risk tests with at least the following control points:

- The reading of results is correctly executed according to the SOP.
- The result on the agar plate is correctly transcribed on the GMP recording sheet, i.e. if growth is observed, this is captured in the GMP sheet.
- The description of the sample corresponds to the description on the GMP recording sheet.

In order to estimate the data integrity risk of microbiological tests and then adjusting the level of mitigation actions based on the risk intensity, risk matrix tools may be used (PDA TR-80; Platco and Cundell 2017; Tidswell and Sandle 2018, or see Chapter 2).

The following risk matrix tool may be used to justify the need to introduce contemporaneous verification for microbiological tests (see Table 12.1).

- **Low risk**: no need for contemporaneous verification by a second analyst.
- **Medium risk**: to be defined case by case: contemporaneous verification by a second analyst or via QA oversight on a periodic basis or no verification required.
- **High risk**: contemporaneous verification of all samples by a second analyst.

In Table 12.2 an example of a risk analysis for tests performed in non-sterile manufacturing is summarized.

In conclusion, data integrity will remain a hot topic for microbiology testing laboratories in the next years with a strong focus of potential falsification of test data. Many of these microbial test-specific data integrity challenges can be overcome by introducing automated plate readers or alternative microbiological test

Table 12.1 Risk matrix example for microbiological tests.

		Risk of result misinterpretation and non-detection of failure		
		Low	**Medium**	**High**
Risk on product quality and patient safety	High	Medium risk	High risk	High risk
	Medium	Medium risk	Medium risk	High risk
	Low	Low risk	Low risk	Medium risk

Table 12.2 Examples on risk analysis of a selection of tests performed in non-sterile manufacturing.

Microbiological test	Product quality/ patient safety risk	Misinterpretation risk	Resulting data integrity risk	Rationale
Tablet manufacturing: Environmental monitoring air grade D manufacturing area	L	M	L	Low water activity of tablets and no direct product contact
Tablet manufacturing: Monitoring product contacting equipment surfaces	M	M	M	Direct product contact risk. Low water activity of tablets and testing of final product would reduce the risk
Purified water used as excipient for tablets	M	M	M	Direct component of product. Low water activity of tablets and testing of final product would reduce the risk
Microbiological examination of final drug product (tablet)	M	M	M	Final release test, low water activity would reduce the number of vegetative cells
Microbiological examination of final drug product (nasal spray)	H	M	H	Critical drug product based on the route of administration and high water activity
Microbiological examination using automated methods of final drug product (nasal spray)	H	L	M	No risk of misinterpretation by use of automated method using electronic record and audit trails
Growth promotion tests	L	L	L	Low risk of misinterpretation as the aspect of the colonies of the test microorganisms is known and similar counts expected from inoculum. No direct product risk

methods that use computerized systems that meet the current data integrity and computerized systems quality standards. Chapter 13 provides a deeper insight of such methods. In the meanwhile, the data integrity risks related to classical manual reading and reporting of microbiological tests can be controlled through high ethical company culture and a robust cGMP system.

12.2 General Concept for Microbiological Excursion

In microbiological testing, excursions occur when measurements differ from the acceptance criteria or normal/expected values. In cGMP environments, an excursion requires a thorough investigation, a root cause assessment and actions to resolve the excursion, and prevent reoccurrence in the future (FDA 2006a; EudraLex 2013).

Five types of excursions may occur in microbiological testing:

- **Out of specification (OOS).** Result that does not comply with the determined specifications (e.g. microbiological examination of non-sterile product test result of 230 CFU/g where the specification is not more than 100 CFU/g). Specification criteria are correlated to product quality attributes. If the OOS is confirmed, the product batch cannot be released.
- **Exceeding the action level.** Result that does not comply with the determined action level (e.g. 173 CFU/m^3 with an action level of 100 CFU/m^3). Action levels are generally defined for monitoring of the environment, utilities, or intermediate product solutions. Exceedance of action levels signals an apparent drift from normal operating conditions and requires immediate actions by previously defined documented measures. If the action level is confirmed, batch release is to be decided case by case as the excursion may not directly impact the product's final quality.
- **Exceeding the alert or out of expectation (OOE) level.** Result that does not comply with the determined alert/OOE level (e.g. 78 CFU/m^3 with a level of 50 CFU/m^3). Alert or OOE levels are generally defined for final drug product where specifications are given, monitoring of the environment, utilities, or intermediate product solutions. Exceedance of alert or OOE levels signals a possible drift from normal operating conditions and requires actions by previously defined documented measures. Generally, there is no risk for the product and no risk assessment for the batch release is needed.
- **Out of trend (OOT).** In general, several results (or data points) that deviate from the expected trend (e.g. increasing number of microorganisms in a product that had demonstrated so far excellent results). OOTs may be defined for all types of testing and is an early warning of a potential degradation or loss of control within the environment, the utility, raw material, or product tested. If the OOT is confirmed, batch release is to be decided case by case as the excursion may not directly impact the product's final quality.

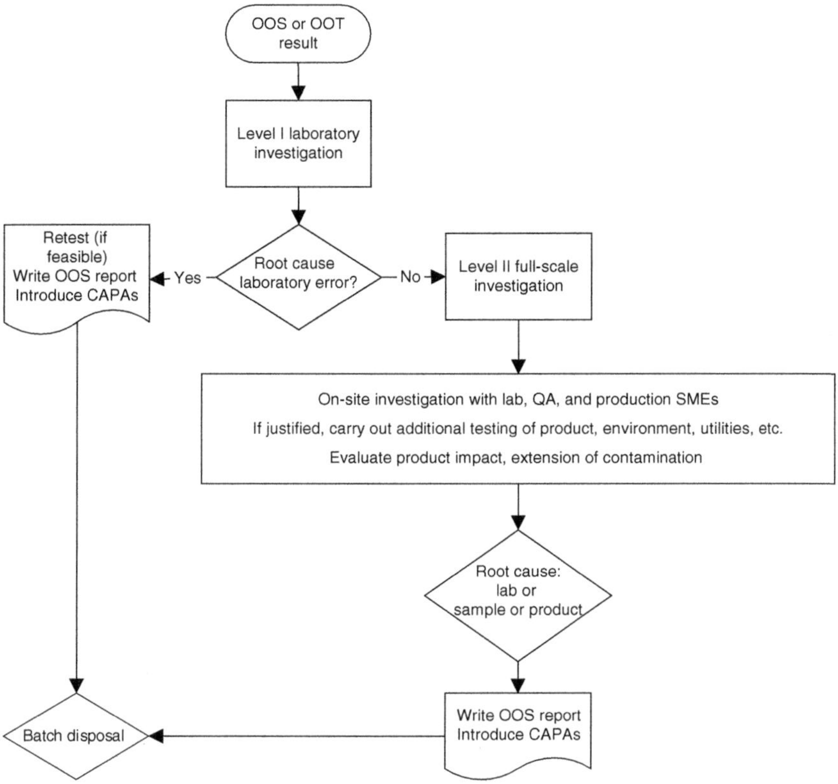

Figure 12.3 Typical process flow for OOS, OOT, or OOE investigations in the microbiology laboratory based on the FDA guidance for industry (FDA 2006b) and MHRA (2017). CAPA, corrective and preventive action; OOS, out of specification; OOT, out of trend; QA, quality assurance; SME, subject matter expert.

Typically, when an excursion occurs, the investigation to determine the root cause takes place in two steps (Figure 12.3).

12.2.1 Level 1 Investigation

This step is composed of a laboratory investigation to determine if an analytical error or contamination of the sample occurred during laboratory testing. It is common practice that this step is composed of an identification of the contaminants detected as well as an investigation in the lab generally performed by a reduced team composed of the laboratory analysts and laboratory head. Level 1 investigations should be completed as soon as possible (within one week is considered standard).

Checklists may be used for level 1 laboratory investigations. In Table 12.3 an example of a generic checklist for microbiological OOS is given.

Table 12.3 Example of microbiology laboratory checklist.

Was the test performed according to the suitable method?
Did the negative and/or positive controls comply?
If several dilutions have been tested, does the dilution ratio make sense?
What is the result of the microbial identification?
Did a similar type of contamination result occur during the working session on other independent samples?
Was the sample container integer and correctly identifiable?
Was the sample stored under the appropriate conditions and remained unspoiled prior to testing?
Was the right sample taken for the analysis?
Was the sample preparation and weighing correctly executed?
Was the correct growth medium used?
Did the growth medium pass the incoming controls (pH, absence of contamination, and growth promotion test)
Was the growth medium stored adequately and remained unspoiled?
Is the shelf life of the growth medium exceeded?
Was the growth medium container integer prior to testing?
Was the laminar flow working correctly during the test session?
Has the laminar flow been cleaned and disinfected according to procedures?
Do the environmental monitoring results of the laminar flow as well as laboratory meet the requirements?
Are the water baths cleaned and disinfected according to the procedure?
Are incubators cleaned and disinfected according to the procedure?
Where the correct incubation temperature and time parameters used?
Were laboratory glassware, consumables, and equipment stored, cleaned, and used according to the procedure?
Is the analyst trained and qualified for the test?
Has an adverse trend in counts been associated with the analyst that performed the test?
Did the analyst wear the appropriate gowning/gloves?
Has the analyst followed aseptic procedures during the test?
Is there a calculation error?
If duplicate samples were tested, is there a significant difference in counts between the two plates?
Was there an error in transcription of raw data in the LIMS system?
Is there a data integrity issue?

If a genuine laboratory failure is found with supportive evidence, then retests may be carried following QA approval, the original deviating result invalidated, and the deviation report written with root cause and corresponding actions. If no laboratory error can be identified, then the second step of investigation begins. In more critical cases (e.g. objectionable microorganism found in final product testing) level 2 investigation is started immediately in parallel to the level 1 investigation.

12.2.2 Level 2 Investigation

Immediate actions may be required if the product's quality is potentially impacted by the excursion. This may consist, for instance, of putting the batch in quarantine and assessing potential impact for the batches produced in the same campaign or already released on the market. Level 2 investigations should be completed within an acceptable timeframe (within 30 days is standard practice).

When carrying out a root cause investigation at this stage, the first step is to ensure that the investigation team is composed of the appropriate people. A complete investigation team would at minimum be composed of the investigation lead, subject matter experts or engineers relevant to the process, the operators or analysts that executed the process that failed, and quality assurance representatives.

An investigator should go to the area where the excursion occurred and interview directly the operators that were involved in the process. It will then be easier to understand the process, to ask for further clarification and verify if there are incoherencies between the responses of the operators and the process. In some cases, it may be more efficient to carry out interviews without the presence of direct management so that operators talk more openly. In all cases operators should not be considered as "potential culprits" and root cause investigation should not be a prosecution practice. In most cases operators themselves actually want to help to find the root cause and know the small details in the processes far better than management or even some technical process experts. A "finger-pointing" company culture reduces the chances of finding the true root cause and effectively eliminating the issue.

The investigation should consider all elements that would help determine the root cause including, e.g. monitoring results, additional testing if supportive of hypothesis, batch and training records, SOPs, excursions or unplanned events that occurred during the deviating result timeframe.

For various reasons (e.g. dominant personality in team, lack of experience, time pressure, and complexity of the case) there may be a tendency to make assumptions that are contradictory to the facts which would lead to failing to find the true assignable cause. Therefore, by complex and sensitive investigations it may be helpful that an independent facilitator coordinates and structures the investigation.

Several root cause investigation tools may be helpful to structure the investigation. The following steps may be followed sequentially:

1) **Problem statement.** This first step is to name the object and excursion. This may sound trivial but if the problem is wrongly stated, the investigation will be positioned in a wrong direction from the start. The statement should be specific and deal with one problem at a time. At this level no discussion on the potential causes should arise.

   Example: *Exceeded action level in purified water from sampling point X1.*
2) **Gather the facts.** This step serves to specify the problem and tighten the problem specification. Only factual information should be used and there should not be any speculation of causes or opinions at this moment. The "Is/Is Not" tool may be used for this step (see Table 12.4).

Example:

Table 12.4 Example of an "Is/Is Not" tool.

	Is	Is Not
What?	>300 CFU/ml of *Sphingomonas maltophilia* (action level 100 CFU/ml)	No *Staphylococcus* species, no molds, no *Bacillus*, no *Micrococcus*
Where?	Sampling point near compounding box of line 2	Lines 1 and 3 of the same loop were also tested and no contamination occurred
Number of occurrences?	Two exceeded action levels, the first on 6 March 2018 and the second on the 23 March 2018 with 150 CFU/ml of *S. maltophilia*	Prior to 6 March 2018 and after 24 March 2018 no contamination
Are trends ok?	Until 6 March, trends are ok	Increase in microbial contamination levels prior to 6 March

3) **Define the timelines.** This step serves to define the relevant timelines prior to, at the time, and after the case. This can be done by verifying logbooks, batch records, interviewing personnel, etc. It is important at this step to capture changes that occurred during the relevant time period.

   Example:

   - *1 March 2018: Sanitization of the piping system and compounding vessel*
   - *2 March 2018: Water sampling of the point of use; result complies*
   - *6 March 2018: Water sampling of the point of use; exceedance of action level*

- *20 March 2018: Investigation, root cause inconclusive as action sanitization of the piping system and compounding vessel*
- *23 March 2018: Water sampling of the point of use; exceedance of action level*

4) **Brainstorming and categorizing root cause assumptions.** This step serves to identify potential factors causing the excursion to occur. For the brainstorming session it is important that all relevant team members are present. At this step, there should not be any discussion on the relevance of the assumption in order to freely list out the highest amount of hypothesis. Brainstorming moderators should ensure that all members have the possibility to freely make hypothesis without being directly judged. Possible causes may be categorized using a Fishbone (Ishikawa) diagram, invented by Kaoru Ishikawa (e.g. Ishikawa 1989; Van Vliet 2013), as shown in Figure 12.4. In order to explore the cause/effect relationships underlying a particular excursion or to drill down general assumptions to single root causes, the "Whys" approach may be used as shown in the example below. The "Why" should be asked until there is no possibility to further drill down.

Example:

Example of the "Whys" approach:

Problem:	*Exceeded action level in purified water from sampling point X1.*
Question 1:	*Why was there an exceedance of this sampling point?*
Answer 1:	*Because the sanitization of the sampling point X1 was not long enough.*
Question 2:	*Why was the sanitization not long enough?*
Answer 2:	*Because the operator has performed sanitization for a few seconds only*
Question 3:	*Why did the operator perform the sanitization shorter?*
Answer 3:	*Because he was not trained on the SOP that describes that sanitization should be performed for at least 2 minutes.*
Question 4:	*Why was he not trained on the SOP?*
Answer 4:	*Because this SOP was missing in his overall training program.*
Question 5:	*Why was this SOP missing in his overall training program?*
Answer 5:	*Because there is no clearly defined list of SOPs for operators performing the sampling.*
CAPA	*A SOP-list for each working field will be created.*

5) **Compare possible causes with the facts.** At this step, possible causes are systematically verified against facts (e.g. from the "Is/Is Not" table, interviews, or batch records). This step enables to eliminate unreasonable causes:

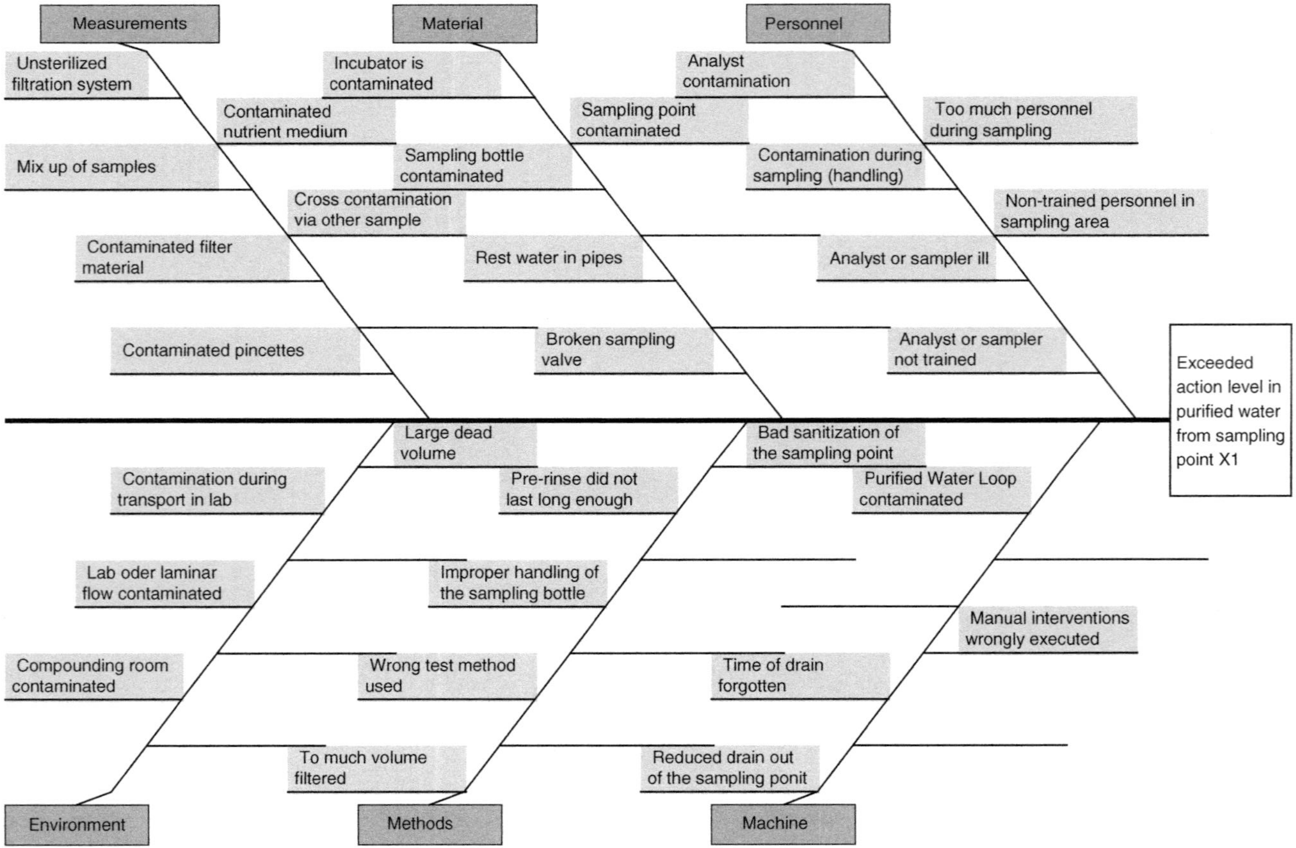

Figure 12.4 Example of an Ishikawa chart.

- If the facts contradict the root cause hypothesis, the cause is not the real root cause and therefore can be excluded.
- If the facts do not contradict the root cause, the latter may be kept.
- If the root cause requires additional supportive data, then these should be retrieved to provide additional facts.

For an example see Table 12.5

Table 12.5 Example of comparing possible causes with the facts.

Root cause hypothesis	Facts	Do the facts contradict the root cause?
Whole purified water loop is contaminated	Other sampling locations of the same water loop complied and no *Sphingomonas maltophilia* was detected	Yes, root cause not to be considered
Contaminated sample bottle	Bottle was integer and washing/sterilization performed according to validated procedure	Yes, root cause not to be considered
On the sampling prior to the deviating sample, draining was forgotten (leads to rest water in the pipes of the sampling point)	Only this sampling point is affected by the excursion. No active blowing in the pipes with compressed air, only passive draining. Sampler interview confirmed that sampler has not performed manual draining of sampling points after sampling	No
Sanitization of the sampling point with hot water not long enough	Only this sampling point is affected by the excursion. *S. maltophilia* is heat sensitive. Sampler interview confirmed that sanitization was not performed according to SOP (it should be more than two minutes and was done only a few seconds)	No

6) **Confirm root cause and define CAPAs.** This final step should define the root cause based on the elements determined during the investigation and introduce corrective and preventive actions (CAPAs). These will be documented in a QA-approved deviation report that will close the excursion.

   The **root cause** is defined as the fundamental reason or cause for the excursion. Contributing factors may be associated to the root cause. **Contributing factors** consist of conditions that would affect the likelihood and the intensity or severity of the excursion. Eliminating a contributing factor would not eliminate the excursion.

Depending on the evidence gathered, the root causes may be classified as follows:

- **True root cause.** Direct evidence demonstrates that the excursion is truly triggered by the root cause.
- **Most probable root cause.** Indirect evidence that the excursion may be triggered by the root cause. Often the case for microbiological excursions due to the limitations of the microbiological testing and monitoring (e.g. single contamination event, microbial amounts and populations changed at the time of test results, limited sample size and location, monitoring is only a snapshot at one moment, the whole volume of sample has been totally used in the original test, microbial counts evolve in time within the sample).
- **Root cause unknown.** There is no evidence that supports any hypothesis related to the excursion. Not having a clue on what might have caused an excursion would increase the risk of recurring contamination events even if, for instance, holistic sanitization actions have been accomplished as a preventive action.

Different types of actions following investigations may be introduced:

- **Corrective action.** Action to eliminate the cause of the excursion and its reoccurrence.
- **Preventive action.** Action to eliminate the cause of potential excursions in order to prevent reoccurrence. Preventive actions reduce the probability that a potential problem occurs.

To help define CAPAs that are efficient, easy to implement and cost effective, and to evaluate the mitigation effect, risk assessment tools such as FMEA may be used. Reoccurrence of excursions may demonstrate that systems or procedures may not be under control, may affect significantly product quality or company resources, and might cause loss of confidence with health authorities. The effectiveness of CAPAs may also be demonstrated with an improvement in the microbiological trend. Therefore, in a cGMP environment, it is key to evaluate the effectiveness of the CAPAs by, e.g. assessing which excursions are recurring, if process variation is under control, or that trends are improving.

Example:

Root cause assessment:	The most probable root cause of the microbial contamination is the non-execution of the manual draining of the sampling point on the previous sampling which possibly led to residual water in the sampling point pipes in which the water microorganism *Sphingomonas maltophilia* could proliferate. A contributing factor was that the hot water sanitization of the sampling point pipes was not executed with the minimum duration of 2 minutes. Missing SOP in the training program of the operator responsible for sampling made him uncertain of the exact procedure to follow.

Product assessment:	Batches B5, B6, and B7 of product XY123 were manufactured during the time the deviation occurred. All three product batches were microbiologically tested and complied with the specifications. The water from the deviated point of use was not utilized as a product excipient and all other sampling points of the water loop met the purified water requirements. This point of use was utilized as process water for the cleaning/disinfection of the equipment external surfaces (non-product contacting) and the floor/walls on the cleanroom. All surfaces are disinfected following cleaning. The environmental monitoring results of the cleaned/disinfected surfaces met the requirements and no microorganism originating from water was identified. Therefore, batches B5, B6, and B7 can be released from a microbiological point of view.
Short-term corrective actions:	• Extended sanitization of the contaminated water sample point of use. • Include the sampling SOP on the training plan of the operator responsible for the excursion. • Training manager to verify that all role profiles have received the adequate training. • Training of operator responsible for the excursion on sampling with focus on sanitization and draining after sampling.
Short-term preventive action:	• Retraining of all other operators performing water samples on the sampling SOP with focus on sanitization and draining after sampling. • Temporary increase in microbiological monitoring of sample point of use.
Long-term preventive action:	• Install an automatized draining of the sampling point pipes. • Create generic SOP-list for each work field.

7) **QA decision:** Finally, the deviation report is reviewed and approved by QA, following which the batches under quarantine can be released. During the QA review the following points should be considered:

- The investigation report was prepared correctly and is complete.
- The report is clearly and comprehensibly formulated and documented.
- All decision-relevant actions have been completed.
- The investigation is complete and scientifically correct. The cause has been found and is scientifically sound. If no root cause was found, all the points investigated have been properly documented.
- The corrective actions were identified, assigned, and executed or scheduled.
- Preventive actions have been identified, assigned, and a schedule for their execution has been created.
- The excursion was correctly classified (in terms of criticality, cause, etc.).
- If there was a negative trend, appropriate actions (e.g. QA oversight and temporary increased monitoring) have been initiated.

- If appropriate, effectiveness checks were started to prove that the implemented measures led to the desired result or effectiveness check actions were set for mid-term evaluation at a defined date.
- If necessary, was a risk assessment written for the release of the products?
- If necessary, was the excursion extended to other product batches?
- Closure of the excursion by QA.
- Decision on the release of the product batches and the room/equipment.

12.3 Considerations for Excursions

12.3.1 Excursion with Environmental Monitoring

Environmental microbiological monitoring controls the quality of the cleanroom over time. In general, monitoring refers to spot checks. Even if settling plates are used for four hours, the air volume actually measured is somewhere in the per mil of the air quantity in the room (for further details, see Chapter 8). Within this context, it is always challenging to establish if an environmental monitoring excursion is just a single incident or a general, recurring problem.

If action levels are exceeded, investigations must be carried out immediately to evaluate if the environment remains under the state of control defined.

For excursion with environmental monitoring it is advisable to distinguish between critical and noncritical excursions. Noncritical excursions are sampling points without direct influence on the product quality (e.g. floor, wall, and surfaces not in contact with the product and gowning areas). The critical excursions are assumed to have a possible impact on product quality (e.g. surfaces in contact with the product and air at the filling line with product directly exposed to the environment). In both cases, the appropriate deviation process, which is described in a SOP, must be initiated. As a rule, a more intensive investigation is carried out for action level excursions compared to exceeding the alert level; excursions with exceedance of the alert level may consist of verifying if the excursion is a single incident or if there is indication of an adverse trend (see Chapters 8 and 10).

After opening the excursion and informing QA and production, the investigation should take into consideration the following aspects:

- Identification of the contaminants: depending on the internal strategy or the criticality of the room, all or the most frequent isolates are identified. The level of identification may go as discriminating as the species (critical sampling points such as filling) or just to the genus or morphological group.
- Check for a sampling error (e.g. qualification of sampler, interview, and correct gowning and handling). A sampling error should be rare.
- Transport error: check for secondary contaminations during transport or incubation. Be aware that Petri dishes are not microbiologically tight, i.e.

there is always the possibility that microbes might enter the Petri dish during transportation or incubation.

- Check for laboratory error. A checklist such as the one shown in Table 12.3 can be used to check for errors during lab handling. Specific checks are:
  - Are the plates correctly closed or sealed?
  - Check the microbiological quality of transport boxes.
  - A visual check of the agar plate can also give an indication of a sampling error, e.g. if a clear fingerprint or drop is present.
  - Although usually no errors occur during incubation in the laboratory, it should be checked (e.g. check the incubator for contamination; was there an excessive film of moisture on the plate?).
- The original plate should be stored (e.g. at 2–8 °C) until closure of the excursion and/or taking a picture for documentation is advisable.
- Trending of historical data: in order to clarify whether this is an individual incident or a systematic problem, the data from the last samplings should be considered. It is recommended to define clear rules for trending (see Chapters 8 and 10 or Pfohl *et al.* 2005a, 2005b; Goverde and Roesti 2018).
- Resampling: Most excursions require resampling (it is advisable to resample the entire room/equipment to see the whole picture). This information is needed to check if there is general problem or a single incident. However, compliant results should never be used to invalidate the first measurement since the room or equipment are not in the same condition (e.g. cleaned again) as at first measurement.
- Production root cause investigation: It should start as soon as possible by a team composed of production, QC, QA, and further persons if needed. As a base, a checklist can be used to investigate the most common points (Table 12.6). If the root cause or contributing factors are found, then the relevant CAPAs should be implemented.
- Affected products: In the case of a critical excursion, a risk assessment of the products on the market should take place within a defined time period (e.g. three days). It is important to define which batches are affected. This must be decided on a case-by-case basis. For example, all batches since the last compliant sampling plus the batches that have been produced since the excursion can be challenged. In this risk assessment, various points can be considered, such as product history, pH, antimicrobial properties, preservative, water activity, application area, patient population, etc. In certain cases, product testing of critical batches is advisable especially if the product is not tested batchwise. All batches that are still under the control of the manufacturer should be quarantined until the excursion is resolved.
- The deviation report is reviewed and approved by QA.

For further reading, see the detailed case in Chapter 8.

12.3.2 Excursion with Water Testing

Many excursions in water microbiological monitoring come from contamination during sampling. They generally result from contaminated hoses at the sampling point or non-aseptic manipulation during the sampling step. The

Table 12.6 Example of a checklist for the root cause investigation by the production.

General information on the room or equipment, such as	• Which process is running in the room/on the equipment? • For which product is it used? • What is the criticality of the product (e.g. sterile, inhalation, and oral) • Are there other comparable rooms or equipment of this type? If Yes, were there also excursion observed?
General investigation by production	• Interview with operator: Were there some anomalies/special activities on this day? • Checking the logbook: Were there some anomalies on that day or the days before which could explain the excursion? • Were there some other excursions during this time period? Might there be a correlation? • Was the material introduced into the controlled room according to procedure? • Was there an uncommon material flow? • Where there any risk materials (e.g. cardboard or wooden pallets) in the room?
Sampling	• Was the sampler correctly qualified? • Were there some issues evident during sampling? • Was the sampling performed during special activities (e.g. repair, service, or cleaning)?
Cleaning and disinfection	• Was the cleaning/disinfection correctly performed according to the SOP? E.g. cleaning frequency, correct product and dilution, within expiry date, correct equipment, and correct storage. • Could a recontamination of the cleaned equipment happen during transport? • Which quality of water was used for the cleaning? Is the water controlled for microbial quality? Where there any exceeding microbial levels of the water testing? • Is there any antimicrobial treatment of the equipment (e.g. drying at high temperatures)? Or was the sterilization of the equipment correctly performed? • Was the contact time of the disinfection followed? • Was the room optically clean and dry? • Are there some obvious leaks (e.g. hose and silicon joints)?

Table 12.6 (Continued)

Personnel (operators, cleaning staff, and visitors)	• Are all concerned people correctly qualified and trained? • Were the people correctly gowned (overall, shoes, hood, mouth protection, gloves)? • Were the hands correctly disinfected? • Was the personnel flow executed according to the procedure? • Did the operator clean the equipment/room beforehand? Is she/he experienced? • Were there more people in the room than usual? • Were there technicians or visitors present? • Was there any misbehavior of the people observed?
Disturbance, maintenance, technique, etc.	• Were there some maintenance activities performed? • Were there some disturbances (e.g. temperature, humidity, pressures, HVAC, facility alarms HVAC, and air lock)? • Did the HVAC run correctly? • Did the room (air locks, emergency door, and windows) remain airtight?

Source: Adopted from Goverde and Roesti (2018).

greatest concern, however, would be in cases where a biofilm has matured and is shedding microorganisms in the whole water loop or at a critical point of use (e.g. water used as product excipient).

The investigation of exceed levels for water should take into consideration the following aspects:

- Identification of the contaminants.
- Using a checklist (see Table 12.3) a sampling and/or lab error must be evaluated.
- Training and qualification status of operator that has taken the sample.
- The point of use and sampling points concerned should be checked for cleanliness, leak-proofness and proper use of the sampling equipment (e.g. hoses), draining of sampling point, etc. It is also possible to take a swab test of the concerned sampling point to check for biofilm formation, for example, in the hose used.
- If relevant, reanalyze the point of use and sampling points by taking a fresh sample as soon as possible and carry out tests for objectionable microorganisms (e.g. test for *Pseudomonas aeruginosa* or *Burkholderia cepacia*) in addition to the total aerobic counts. Alternatively, all bacteria found when analyzing the additional samples may be identified. If possible, the additional samples should be collected before taking any further measures at the sampling point to confirm the excursion.

- Trending of sampling point or entire loop to check for recurring deviations or negative trends.
- Check where the water is applied and what is the resulting risk of contamination of equipment or product.
- The potential risk for the products prepared with this water has to be assessed and products must be kept under the plant's control until investigation completion. Further investigations (e.g. determination of the bioburden in preparations or products containing the quality of water concerned) may be made to prove that the microbiological quality of the preparation or product had not been compromised. If there is any risk to the product or the patient, respectively, the quality of water concerned may no longer be used without additional sanitization measures (e.g. ozone sanitization, filtration and heating).
- If an objectionable microorganism or an excessive count of microorganisms is confirmed in the water system, it must be assessed which additional sanitization measures (e.g. ozone, heating, filtration, mechanical elimination of biofilm) are required prior to use of water.

12.3.3 Excursion in the Growth Promotion Test

Growth media are composed of complex raw materials of varying quality as they are from natural origin. It is therefore not uncommon that growth promotion tests may fail. Generally, the failure concerns 1–2 microorganisms only and only gross errors made in the growth medium preparation could explain failed growth promotion tests of all test microorganisms. This is unlikely in a cGMP environment.

The investigation of failed growth promotion tests should take into consideration the following aspects:

Lab-related investigations:

- The microbial culture used to inoculate the growth medium should be checked. If it is a self-prepared cryogenized culture, the investigator should verify the shelf life. If it is a purchased lyophilized culture (e.g. Bioballs™), the suppliers' certificate of conformance should be checked. Also check if there were any abnormalities during transport or storage of the culture.
- Ask the analyst if there was an issue during resuspending and if dilution volumes were correct. Especially, the rehydration of lyophilized or cryogenized cultures can be tricky for certain species (e.g. Gram-negative rods).
- If possible, the microbial suspension should be retested (with a higher number of replicates) on another medium lot to verify reoccurrence of failure due to the test strain.
- Check if the analyst has been appropriately trained and if recurring errors or failed growth promotion tests occurred with the same analyst.
- Did the analyst clearly understand the procedure and was this procedure well described and understandable.

- Check if the pH of the growth medium was correctly measured or if the pH meter was adequately calibrated.
- Also verify if there has been a potential analytical error such as more than 50% difference in counts from one duplicate plate as compared to the other.
- Check if there were no cross contaminations (e.g. other microbial colonies on the plate).
- Verify if the incubators functioned correctly and were calibrated at the time of testing.
- In case of anaerobic incubation verify that the anaerobic jars are integer and that anaerobic conditions are fulfilled.
- Also, the pipettes used for the dilutions should be verified (calibration status and correct volume is pipetted).

Growth medium preparation-related investigation:

- For self-prepared media was the water quality meeting the purified water grade or are there any other issues with the water quality used.
- For self-prepared media was the correct amount of powder used.
- Did the decontamination or sterilization cycle function correctly.
- Was the glassware used to prepare the medium free of potential growth-inhibiting substances?
- If dehydrated medium was used, ask the supplier if there were any changes on the formulation or if they changed supplier of the raw material.
- Visually verify if no evident degradation of the product is visible (e.g. change in typical color, riddles on the surface or bubbles in the agar, condensation or higher humidity than usual).
- Verify if the growth medium pH was in conformance.
- Check if previous lots of the same growth medium or other medium types from the same supplier failed growth promotion tests, i.e. perform trending.
- Even if they are above 50%, do other test microorganisms show slightly lower recovery in the range 50–70%?
- Check the transport and/or storage of the medium (for both, dehydrated medium as well as ready-to-use medium).
- Does a repeat growth promotion test again demonstrate low microbial recovery?

If the investigation results in a laboratory error, then the initial growth promotion test may be invalidated and the repeat results considered as the final valid one. If it is confirmed that there is a growth medium failure or that the root cause is unknown, the growth medium lot is considered out of compliance and should not be used and discarded. The growth medium supplier should be in this case notified and further clarification initiated. Auditing of the growth medium manufacturer may also be part of remediation actions. If for a particular growth medium supplier recurring quality issues occur or if inappropriate procedures were highlighted during the audit, further utilization of the growth media from the supplier in question should be discontinued.

12.3.4 Excursion with Microbiological Examination of Non-sterile Products

These excursions are the most complex to investigate as many factors may impact the final microbiological quality of non-sterile drug products. Another aspect is that microorganisms might not be homogeneously distributed in product items and when low-level amounts of objectionable microorganisms are found in a final product formulation, finding a root cause is like looking for a needle in a haystack.

One of the key elements for these complex investigations is the setting up of the right investigation team members and use of structuring tools (refer to Section 12.2). Basically, the investigation should verify if one of the microbial controls failed (refer to Chapter 1 for an extensive list of such controls).

To continue the investigation of excursions with microbiological examination of non-sterile products, one should take into consideration the following aspects:

- Was the sampling performed using aseptic techniques, this includes operator handling, gowning, and sampling utensils?
- Using a checklist (see Table 12.3) a sampling and/or lab error must be evaluated.
- Identification of the contaminants. Especially, for drug products where objectionable microorganisms are relevant, the identification should be most accurate, although biochemical or proteomic identification is suitable, the microbiological expert should carefully analyse the results, and, in some cases, genomic identification should be applied.
- Gowning and behavior of operators that were involved in the manufacturing. Furthermore, an interview with the operators can help to identify special situations or unexpected events.
- Were the cleaning and disinfection steps of the cleanroom and equipment respected? Are cleaning utensils (e.g. buckets) conform?
- Trending of microbiological data of the product in question or product manufacturing line.
- Were the microbial product hold times respected?
- In addition to putting the deviating batch under quarantine for the time of investigation, potential contamination extent to other batches produced during the campaign or before needs to be assessed. It also needs to be assessed if product batches already released on the market are affected.
- Another key aspect is to generate additional data that would support root cause hypotheses. The list below provides some examples of tests that may be carried out.

- Retest of original sample if material remains or retained sample (not to invalidate original result but to provide supportive information). In certain cases, a sampling of several or even each bag of the batch in question might be advisable.
- Testing of product intermediates such as granulate, coating solutions, bulk product prior to packaging, etc. This might include bioburden testing and testing of the water activity of these intermediates.
- Testing of all individual drums if the bulk product has been held prior to further processing.
- Testing other sample containers from the same container lot.
- Microbiological quality of drug substances, excipients, and primary packaging used. If frequency testing was applied, lots that were untested but used in the deviating product must be tested for microbial contamination.
- Trend results of the water loop and, if supportive, retest of all water sampling points that are relevant for the manufacturing of the deviating product.
- Environmental monitoring trend results and, if supportive, additional testing in the manufacturing area/or on equipment product contacting surfaces.
- For products or product intermediates with a water activity that would enable microbial growth, a microbial challenge test with the isolated contaminant may be helpful to evaluate of it can grow in the solution of concern and how fast it can grow.
- Testing of cleaning solutions and utensils (e.g. mops), especially if they stay for a long time in a bucket.
- Additional testing of solutions, material, growth media, and equipment used in the lab (refer to Section 12.2 for lab investigation).
- Verification of the sample container physical integrity (e.g. cracks and deformed lid). Recovery of microorganisms from the inner surface of the containers (e.g. by aseptically adding buffer in the emptied container and testing the buffer with membrane filtration).
- Further information might be important for the decision of the batch such as water activity of the product or the specific batch, the patient population for the drug product and its route of administration.

Depending on the outcome of the investigation and if the contamination is a genuine product contamination and if the root cause is not limited to the batch in question, then a market quality excursion notification to the respective health authorities for a recall is to be initiated (FDA 2006b).

Bibliography

EudraLex, 2013, *The Rules Governing Medicinal Products in the European Union. Volume 4. EU Guidelines for Good Manufacturing Practice for Medicinal*

Products for Human and Veterinary Use. Part I – Basic Requirements for Medicinal Products, Chapter 1 Pharmaceutical Quality System, European Commission.

FDA, 2006a, Guidance for Industry – Quality Systems Approach to Pharmaceutical CGMP Regulations, Food and Drug Administration.

FDA, 2006b, Guidance for Industry – Investigating Out-of-Specification (OOS) Test Results for Pharmaceutical Production, Food and Drug Administration.

FDA, 2018, Guidance for Industry – Data Integrity and Compliance with Drug cGMP – Questions and Answers, Food and Drug Administration.

Goverde, M, Roesti, D, 2018, *Guidance for Deviation Handling of Microbiological Environmental Monitoring Excursions in Non-sterile Pharmaceutical Manufacturing*, ECA Pharmaceutical Microbiological Working Group.

Ishikawa, K, 1989, Introduction to Quality Control, originally printed in Japan as *Dai-3-pan Hinshitsu Kanri Nyumon*, 3rd ed., JUSE Press.

ISPE, 2018, ISPE GAMP® RDI Good Practice Guide: Data Integrity – Key Concepts, International Society for Pharmaceutical Engineering.

MHRA, 2017, Out of Specification and Out of Trend Investigations, Medicines & Healthcare Products Regulatory Agency (MHRA).

MHRA, 2018, 'GXP' Data Integrity Guidance and Definitions, Revision 1: March 2018, Medicines & Healthcare Products Regulatory Agency (MHRA).

PDA, 2018, *Data Integrity Management System for Pharmaceutical Laboratories*, Technical Report No. 80, Parenteral Drug Association.

Pfohl, M, Fricker, P, Kohler, D, Berchtold, M, 2005a, 'Trending von Umgebungskontroll-Daten für den Herstellbereich sterile Formen, Teil 1', Pharm. Ind., vol. 67(5), pp. 598–602.

Pfohl, M, Fricker, P, Kohler, D, Berchtold, M, 2005b, 'Trending von Umgebungskontroll-Daten für den Herstellbereich sterile Formen, Teil 2', Pharm. Ind., vol. 67(6), pp. 705–708.

PIC/S Draft Guidance, 2016, *Good Practice for Data Management and Integrity in Regulated GMP/GDP Environments*, Pharmaceutical Inspection Convention, Pharmaceutical Inspection Co-Operation Scheme, PI 041-1 (Draft 2), 10 August 2016.

Platco, C, Cundell, T, 2017, 'Data Integrity Issues in Microbial Testing', *American Pharmaceutical Review*, October 2017.

Tidswell, EC, Sandle, T, 2018, 'Microbiological test data-assuring data integrity', PDA J. Pharm. Sci. Technol., vol. 72, pp. 2–14.

Van Vliet, V, 2013, *Kaoru Ishikawa*. https://www.toolshero.com/toolsheroes/kaoru-ishikawa (Retrieved 24 January 2019).

WHO, 2016, *Annex 5 – Guidance on Good Data and Record Management Practices*, WHO Technical Report Series No. 996, World Health Organization.

13

Rapid Microbiological Methods

Michael J. Miller

Microbiology Consultants, LLC, Lutz, USA

CONTENTS

13.1 Introduction

Non-sterile pharmaceutical preparations, by definition, may contain microorganisms. As such, manufacturers must fully understand the potential impact to the product and the recipient of the product to ensure there is a relatively low or nonexistent risk for an undesirable outcome.

For example, certain microorganisms may affect the packaging and/or therapeutic activity of the product, rendering it inactive, unstable, unsafe, or unable to deliver its intended dose and treatment. More importantly, the presence of an opportunistic pathogen or objectionable microorganism could result in a significant clinical event, such as an infection or death, depending on the

Pharmaceutical Microbiological Quality Assurance and Control: Practical Guide for Non-Sterile Manufacturing, First Edition. Edited by David Roesti and Marcel Goverde.

ability of the organism to survive or proliferate in the product composition, the product's route of administration, method of application, and the age and/or physiological state of the target recipient (e.g. neonates, infants, the elderly, or immunocompromised patients).

For these reasons, non-sterile drug manufacturers are expected to demonstrate their products are free of microorganisms that could cause harm either to the product or to the patient or consumer. Current practices for conducting release testing as per compendial requirements (such as USP 61 for a quantitative assessment of non-sterile drug products and USP 62 for the absence of organisms that might be considered objectionable) are usually performed to demonstrate products are microbiologically safe for distribution.

Regulatory requirements clearly stipulate these expectations. For example, the current Good Manufacturing Practice for finished pharmaceuticals in U.S. Code of Federal Regulations section 21 CFR 211.113 requires manufacturers of non-sterile dosage forms to establish and follow written procedures designed to prevent objectionable microorganisms in drug products not required to be sterile (21 CFR 211.113).

Additionally, 21 CFR 211.84 requires each lot of a component, drug product container, or closure with potential for microbiological contamination that is objectionable in view of its intended use to be subjected to microbiological tests before use (21 CFR 211.84).

Furthermore, 21 CFR 211.165 specifies there shall be appropriate laboratory testing, as necessary, of each batch of drug product required to be free of objectionable microorganisms prior to the release of said batches for distribution (21 CFR 211.165).

Additional guidance on controlling the bioburden types and levels in non-sterile pharmaceutical preparations may be found in USP <1111>, *Microbiological examination of non-sterile products: acceptance criteria for pharmaceutical preparations and substances for pharmaceutical use*; USP <1115>, *Bioburden control of non-sterile drug substances and products*; and PDA Technical Report #67, *Exclusion of objectionable microorganisms from non-sterile pharmaceuticals, medical devices, and cosmetics.* These same guidance documents also provide recommendations for acceptable levels and types of microorganisms in non-sterile products, depending on their composition, route of administration, intended use, and patient/consumer population. For example, the risk of microbiological contamination is greatest for inhalants and nasal sprays while the lowest risk may be considered for oral tablets and powder-filled capsules. Further guidance may be found in Chapters 2 and 11.

Therefore, it is understandable that in addition to testing a non-sterile finished product for release, applicable samples upstream of the finished product should be considered for microbiological analysis, including, but not limited to, incoming materials and components (e.g. active pharmaceutical ingredients, excipients, and water), in-process materials, personnel and the manufacturing environment, equipment, and processes. Accordingly, these assessments should effectively determine the numbers of viable microorganisms in each test sample and reveal the presence (or absence) of objectionable microorganisms, as appropriate for the sample under evaluation.

13.2 The Current State of Microbiology Testing

For most companies, conventional or traditional microbiology testing is performed using growth-based media. For the analysis of non-sterile finished product, testing according to the pharmacopoeias is usually followed. For example, USP <61>, *Microbiological examination of non-sterile products: microbial enumeration tests*; and USP <62>, *Microbiological examination of non-sterile products: tests for specified microorganisms*, or similar chapters in the European or Japanese Pharmacopoeias, are utilized.

USP <61> provides a quantitative enumeration of mesophilic bacteria and fungi that may grow under aerobic conditions. The tests described in the chapter are designed primarily to determine whether a non-sterile substance or preparation complies with an established specification for microbiological quality. Established specifications may be derived from the recommendations in USP <1111>, other applicable guidance documents, and/or risk assessments performed by a company in reference to microbial quality for a particular test sample. USP <62> allows for the determination of the absence of, or limited occurrence of, specified microorganisms or organisms that might be considered objectionable.

The methods described in the compendia are predicated on the need for viable microorganisms in the test sample to proliferate in a suitable microbiological medium and under specific incubation parameters, such as time and temperature. For example, the determination of total aerobic microbial count (TAMC), USP <61> requires incubating a portion of the test sample at 30–35 °C on soybean–casein digest agar for up to three days (for bacteria) or five days (for fungi). Similarly, the determination of total combined yeasts and molds count (TYMC) requires the use of sabouraud dextrose agar incubated at 20–25 °C for up to 5 days.

As can be seen for the incubation parameters described above, the use of traditional or conventional microbiological methods presents challenges in terms of the time to result. Days are required to allow viable microorganisms to sufficiently grow on agar plates while providing visual colony forming units (CFU) large enough to be seen by the naked eye.

Additionally, the ability to detect stressed or dormant microorganisms that may be present in the test sample might be hampered by the medium and incubation conditions used during these tests. Specifically, microorganisms that are stressed due to nutrient deprivation or following exposure to sublethal concentrations of antimicrobial agents, such as preservatives, may not replicate when cultured on artificial media because the medium and incubation parameters are not optimal for the resuscitation and subsequent proliferation of organisms that may be present.

For these reasons, the modern microbiological laboratory performing testing on non-sterile products and formulation components should look toward developing innovative approaches to the detection, quantification, and identification of microorganisms. Actually, this is recommended in the compendia as both USP <61> and USP <62> state that alternative microbiological procedures, including automated methods, may be used, provided that their equivalence to the Pharmacopoeial method has been demonstrated.

To support this position, the USP, European Pharmacopoeia (Ph. Eur.), and Japanese Pharmacopoeia have developed chapters that provide guidance on the selection and validation of alternative or rapid microbiological methods (RMMs) that may be used in place of the compendial tests for pharmaceutical products, including those that are intended to be non-sterile.

For example, Ph. Eur. chapter 5.1.6, *Alternative methods for control of microbiological quality*, explains that alternative methods for the control of microbiological quality have shown potential for real-time or near real-time results with the possibility of earlier corrective action. These new methods, if validated and adapted for routine use, can also offer significant improvements in the quality of testing. The chapter also indicates that alternative methods may be used for in-process samples of pharmaceutical products, particularly for the application of process analytical technology (PAT), for environmental monitoring (EM) and for industrial utilities, such as the production and distribution of water, thereby contributing to the microbiological quality control of these products.

13.3 Rapid Microbiological Methods

Many RMM technologies provide a greater level of sensitivity, accuracy, precision, and robustness when compared with conventional, growth-based methods. Furthermore, RMMs may be fully automated, offer increased sample throughput, operate in a continuous data-collecting mode, provide significantly reduced time-to-result (e.g. from days or weeks to hours or minutes), and for some systems, realize real-time data acquisition and trending.

Most rapid method technologies can detect the presence of a wide variety of microorganisms (e.g. bacteria, yeast, and mold) or a specific microbial species (e.g. *Pseudomonas aeruginosa* vs. *Staphylococcus aureus*), enumerate the number of microorganisms present in a sample, and can identify microbial cultures to the genus, species, and subspecies or strain levels. The manner in which microorganisms are detected, quantified, or identified will be dependent on the specific technology, analytical method, instrumentation, and software used.

For example, growth-based RMM technologies rely on the measurement of biochemical or physiological parameters that reflect the growth of microorganisms. These types of systems require the organisms in a sample to proliferate, either on a solid or liquid medium, in order to be detected and/or quantified. Viability-based systems use fluorescent stains and/or cellular markers for the detection and quantification of microorganisms without the need for cellular growth. Cellular component-based technologies rely on the analysis of cellular targets (e.g. ATP or endotoxin) or the use of probes that are specific for microbial sites of interest. Nucleic acid-based technologies may utilize PCR-DNA amplification, RNA-based transcription-mediated amplification (TMA), 16S rRNA typing, gene sequencing techniques, or other novel applications. Spectroscopic methods make use of light scattering and other optical techniques to detect, enumerate, and identify microorganisms. Finally, micro-electro-mechanical systems (MEMS), such as microarrays, biosensors, and lab-an-a-chip or microfluidic systems offer significantly smaller-sized technology platforms as compared with benchtop instrumentation.

In some cases, non-growth-based detection systems may require an initial enrichment phase in order to generate a sufficient level of signal to be detected. For example, most ATP bioluminescence methods do not have a level of sensitivity to detect 1 attomole of ATP, which is the level normally expected to be found in a single, viable bacterial cell. As such, the test sample may need to be "enriched" in a growth medium to generate adequate bacterial growth, which would translate to a minimum amount of ATP that can be detected by the system at the required level of sensitivity.

13.4 Applications for Non-sterile Pharmaceutical Drug Products

It is fair to state that virtually any microbiological test that is currently used to support the development, manufacture, and release of non-sterile pharmaceutical drug products can be replaced by one or more of the available rapid methods currently available to the industry.

For example, applications may include, but are not limited to, raw material and component testing, in-process bioburden evaluations, pharmaceutical-grade water testing, EM (e.g. viable surface and air testing, as applicable to the non-sterile manufacturing process and facility requirements), finished product release testing (e.g. bioburden and presence/absence tests for specified or objectionable organisms), microbial identification, and contamination control and remediation.

In particular to in-process testing, rapid methods may be used to support forward processing manufacturing decisions in real time or close to real time.

For example, in-process bioburden testing, which is difficult to utilize using conventional methods due to the long time to result, may be enhanced by using a rapid method in which contaminated processes or unit operations may be quickly identified, thereby facilitating a go/no go decision on whether to continue manufacturing a particular batch of product. This information may also provide a greater level of confidence that the final batch will meet the required microbial specifications for release. In the event there is a contamination issue, the same or a similar rapid method could be used during the investigative process to identify the source of contamination as well as to confirm any remediation steps have been successful before resuming manufacturing. This scenario may reduce the overall downtime a facility might experience when dealing with such an incident.

Because some rapid methods are considered to be quantitative in nature (i.e. these will provide an actual cell count, either as a CFU or some other signal specific to the technology or method) while others are qualitative (i.e. the result is either a positive or negative outcome, such as a positive response for the presence of viable microorganisms or a target organism the end user is looking for), the end user must decide what output is most relevant to his or her application and testing requirements. Specifically, a quantitative method may be required when determining whether a test sample meets the compendial requirements for an acceptable number of microorganisms in a non-sterile product at the time of release. However, it may be possible to utilize a qualitative method for this same purpose, even though an actual cell count may not be attainable.

Let us use the compendial requirement for a bioburden assay as an example. If a compendial specification is not more than 1000 CFU, it may be possible to correlate a qualitative result to match this required level of microbiological quality by diluting a test sample to this same specification level (i.e. diluting 1 : 1000). If the qualitative result is negative (no growth or microbial recovery), the user has demonstrated that the number of microorganisms in the original test sample was less than 1000, and the compendial specification has been met. I have previously designated this strategy as a "dilute-to-spec" method (Miller 2012).

However, if the number of microorganisms in the original sample is very close to the specification level, this strategy may not be appropriate to use because the qualitative result can provide either a positive or a negative result, depending on the actual distribution of microorganisms or other factors. Therefore, an understanding of the historical bioburden level in the test sample may help to resolve this potential issue; otherwise, confirmatory testing may be required. This is exactly the approach GSK previously utilized with a qualitative ATP bioluminescence technology to release a non-sterile nasal spray that had a quantitative microbial limits specification. As long as the qualitative result showed no recovery of microorganisms, the lot was accepted and released. However, if a positive qualitative result was obtained, the required specification was confirmed using the quantitative, compendial assay. Because almost every lot of product had no recoverable microorganisms, this rapid method strategy allowed the company to release the nasal spray in one-half the time required for a final result using the compendial assay. It should also be noted that the FDA's first PAT approval was for this rapid method and application. GSK also used the same strategy for testing biological indicators (Dalmaso 2006).

13.5 Technology Review

To help the reader in deciding what rapid methods may be compatible with his or her test samples and required applications, the following technology overview is provided.

Although it is not possible to describe all of the rapid methods that are commercially available or are currently in development, specific technologies have been chosen based on their currently accepted use in the pharmaceutical industry as they relate to testing non-sterile products and formulation components. It should be noted, however, that this author is not recommending any technology over another, as the most appropriate rapid method to utilize will be based on a number of factors, including, but not limited to, the type of analysis that will be performed (e.g. qualitative, quantitative, or identification), the scope of the application, required level of sensitivity, specificity or range of microorganisms to be detected, the number and type of samples to be evaluated, the required time to result or detection, sample size requirements, and sample composition and compatibility with test preparation and handling. These requirements are usually defined by the intended user and are captured in a formal user requirements specification (URS). The URS would describe all of the functions the rapid method must meet which will form the basis for ultimately selecting a method for validation and routine use. In fact, the URS supports the strategies for how the method will be validated via test protocols and meeting acceptance criteria.

Separately, a risk assessment should be performed that identifies any potential issues once the new method is employed for routine use. It would be appropriate to perform the risk assessment after developing the URS, as user specifications may impact the manner in which the risk assessment is performed. For example, if the URS requires a limit of quantification at the single cell or CFU level, the risk assessment should address any impact on product quality, GMPs, and/or finished product release decisions if the system is not capable of attaining this level of sensitivity. Other examples of risk assessment considerations may include the potential for false-positive or false-negative results, demonstrating equivalence to existing methods or the supplier's ability to support the system during its life cycle.

A risk assessment for a new RMM would be performed no different than risk assessments for other new applications. First, identify potential hazards (e.g.

the RMM instrumentation does not function or gives incorrect data, computer system errors, or the availability of consumables or necessary reagents). Next, determine the likelihood of occurrence, the severity of harm, and the ability to detect each of the hazards identified. Each risk is then analyzed against predefined quantitative (i.e. a risk score) or qualitative (low/medium/high) criteria. The output is a quantitative risk score or a qualitative risk ranking. Any identified risks that are unacceptable (i.e. above a certain risk score or risk ranking) would need to be reduced or controlled to an acceptable level. Once appropriate controls are implemented and during the life cycle of the RMM, it will be necessary to ensure that no new risks have been introduced and the controls put in place are effective.

To supplement the information provided below, the reader is encouraged to review the case studies presented in PDA's *Encyclopedia of Rapid Microbiological Methods* (Miller 2005–2012) and the comprehensive index of rapid method products at Rapid Micro Methods (2018). The information provided in the subsequent sections of this chapter has been adapted directly from the rapidmicromethods.com website, with the owner's permission. The reader may also reference the website's RMM Product Matrix (http://rapidmicromethods.com/files/matrix.php), which provides comprehensive comparison tables with the following information: the name and company of the RMM, scientific method, applications, time to result, sample throughput, sample size or type compatible with each system, level of sensitivity, organism libraries (for microbial identification systems), and testing workflow.

13.5.1 Growth-Based RMMs

Rapid methods that rely on the growth of microorganisms in liquid or solid conventional media can be used for a variety of applications including bioburden testing, presence/absence testing, EM, water analysis, and microbial identification. Most systems will quickly detect growing organisms or enumerate microcolonies using specialized scientific principles, such as measuring changes in electrical conductivity or impedance, detecting byproducts of respiration (e.g. oxygen consumption or carbon dioxide production), utilization of carbohydrates and other substrates, or sensing autofluorescence of microcolonies following excitation from specific wavelengths of light.

Impedance microbiology is based on the premise that microbial growth results in the breakdown of larger, relatively uncharged molecules into smaller, highly charged molecules (e.g. proteins into amino acids, fats into fatty acids, and polysaccharides/sugars into lactic acid).

Growth may be detected by monitoring the movement of ions between electrodes (conductance), or the storage of charge at an electrode surface (capacitance). Impedance systems can detect changes in measurable electrical threshold in liquid media (during microbial growth) when microorganisms proliferate in containers that include electrodes. Growth may, therefore, be detected prior to the liquid media showing any signs of turbidity. Impedance systems were utilized as one of the original rapid methods for screening antimicrobials and formulations containing preservatives in the early 1990s. The Sy-Lab BacTrac system is an example of this technology that is available today. Although a quantitative result (i.e. in terms of a CFU or another signal) is not possible, the data could be used to estimate the concentration of viable cells in the original sample, or simply to determine the presence or absence of microorganisms.

Microorganisms, when grown in liquid culture, produce carbon dioxide. In a closed container, microbial growth may be detected by monitoring changes in the amount of carbon dioxide present. BioMérieux's BacT/ALERT allows carbon dioxide that is generated during microbial growth to diffuse into a liquid emulsion sensor and the resulting color change alerts the user that the presence of microorganisms has been detected. Another example of this type of RMM is the Becton Dickinson (BD) Diagnostic Systems BACTEC FX. Because both methods are qualitative in nature, the user should determine if these methods are appropriate for detecting specific levels of microorganisms in non-sterile test samples.

A number of identification technologies measure the ability of microorganisms to utilize biochemical and carbohydrate substrates dehydrated in a microtiter plate format. These types of RMM systems monitor changes in kinetic reactions or turbidity, the latter indicating microbial growth. One technology, the Biolog Omnilog system, includes a tetrazolium violet dye in the same wells that contain a dehydrated carbon source and if a microorganism utilizes that carbon source, the well will turn purple. The bioMérieux VITEK 2 uses an optical system that monitors changes in each well using different wavelengths in the visible spectrum, and detects turbidity (microbial growth) or colored products as a result of substrate metabolism. In both systems, the resulting data (normally in the form of positive and negative responses) are compared with an internal database or reference library for that specific platform and a microbial identification is provided. Because it may be necessary to determine the microbial identification of isolates recovered from non-sterile products, these methods may be appropriate to use for this purpose. It should be noted, however, that isolated colonies (or a pure culture from liquid media) would be required in order to use these methods because mixed cultures can provide an invalid response or incorrect microbial identification.

Rapid Micro Biosystems' Growth Direct provides a sensitive digital imaging method for detecting microbial growth on agar surfaces. The technology detects the yellow-green fluorescent signal emitted by growing microcolonies when illuminated with blue light. Cellular autofluorescence in this spectral region is a property of all microbial cells due to the presence of ubiquitous fluorescent biomolecules including flavins, riboflavins, and flavoproteins. The system will provide a microcolony count in about half the time it would take for the same organisms to grow into a CFU, the size of which is countable by a laboratory technician. Because this method is a quantitative assay, the technology can be used to provide an actual cell count during bioburden studies. Moreover, some in the industry regard this type of technology as an automated compendial method, because the system essentially provides a nondestructive, countable CFU signal, which is the same as what is recovered in the compendial assays. The method requires the test sample to be filterable; therefore, viscous formulations, lotions, and ointments may need to be evaluated with more sample-compatible technologies or a strategy for converting the sample into a filterable matrix may need to be developed.

bioMérieux's EviSight Compact TOTAL VISION system relies on the growth of microorganisms on standard Petri plates. Microcolonies are detected and counted via high magnification imaging. The image processing software analyzes information on size and shape and provides a CFU count every 30 minutes. Similarly as the Growth Direct system, the EviSight Compact may be regarded as a nondestructive and automated alternative assay for bioburden testing.

The Biolumix and Soleris Neogen systems employ a variety of broth media that will encourage the growth of target microorganisms. The vials contain unique dyes in which microbial growth is detected by changes in color or fluorescence. An optical sensor detects these changes, which are expressed as light intensity units. The vials are also constructed with two independent zones: an upper incubation zone and a lower reading zone. The two zones eliminate masking of the optical pathway by the test sample and/or by microbial turbidity. This technology may be used as an alternative to the compendial test for specified microorganisms, as long as the technology provides consumables and media that are specific for the intended target organisms. The current system provides assays for the detection of *Escherichia coli, Salmonella, Pseudomonas, Staphylococcus,* and *Bacillus cereus,* as well as a variety of other organisms.

The Bactest Speedy Breedy is a portable respirometer that monitors pressure changes relating to gaseous exchanges within a closed culture vessel as a result of microbial respiration. The system provides real-time analysis of positive and negative pressure changes in the vessel headspace, facilitating the detection of viable microorganisms. Because this is a qualitative method, the same limitations for obtaining a cell count or estimation of viable cells is the same.

13.5.2 Cellular Component-Based Technologies

Cellular component-based RMMs detect specific cellular components or the use of probes that are specific for microbial target sites of interest. For example, adenosine triphosphate (ATP), fatty acids, surface macromolecules, bacterial endotoxin, proteins, and nucleic acids have been used as targets for RMMs for the detection, quantification, and identification of microorganisms. Similarly as for growth-based technologies, a wide range of microbiology applications may be realized, including providing a quantitative bioburden assessment and the detection or identification of microorganism species.

Bioluminescence is the generation of light by a biological process and is most commonly observed in the tails of the American firefly *Photinus pyralis*. In the presence of D-luciferin and luciferase, bioluminescence occurs when ATP is catalyzed into adenosine monophosphate (AMP). One of the reaction byproducts is light (in the form of photons), which can be detected and measured using a luminometer. Because all living cells store energy in the form of ATP, cellular ATP has been used as a measure of microorganism viability in a number of RMM systems. Current bioluminescence technologies use ATP-releasing agents to extract cellular ATP from microorganisms that may be present in a sample, and following the addition of bioluminescent reagents, can detect the amount of light emitted from the sample. Depending on the technology, the amount of light emitted can indicate the presence of microorganisms, or may be correlated with viable cell counts. However, if a sample is expected to contain very low levels of microorganisms, it may be necessary to allow these microorganisms to replicate during an enrichment phase (e.g. either in liquid culture or on agar medium) in order to increase the amount of ATP that can be detected (refer to the prior discussion on ATP levels earlier in this chapter). Quantitative and qualitative ATP bioluminescence systems are currently available; for example, the Millipore Milliflex Rapid system is a quantitative technology that detects ATP arising from microcolonies that have developed on an agar surface while qualitative technologies will provide a measurement of relative light units indicating the presence of microorganisms in the original sample. One qualitative system, the Celsis Advance II system, utilizes an additional substrate, adenosine diphosphate (ADP), which can be converted to ATP in the presence of an enzyme, adenylate kinase; the latter that is extracted from cells in addition to cellular ATP. The additional substrate and enzyme reaction helps to increase the total amount of ATP in the reaction by 1000 times, thereby increasing the sensitivity of the assay. The qualitative ATP method used by GSK for the rapid release of its non-sterile nasal spray was Pall Corporation's Pallchek System. Unfortunately, this handheld ATP system was removed from the marketplace a few years ago and is no longer available for purchase.

The analysis of fatty acids recovered from microorganisms is currently being used as a rapid microbial identification method. The cellular membrane contains lipid biopolymers, and the fatty acid profiles are unique for different types of organisms including bacteria and fungi. In the MIDI Sherlock MIS system, fatty acids that are extracted from a microbial culture are analyzed using gas chromatography and the resulting peaks compared with an internal library or database. Similarly as for the previous growth-based microbial identification systems discussed, cells from an isolated colony or pure culture are required to avoid incorrect or invalid results.

Matrix-assisted laser desorption ionization, time of flight (MALDI-TOF) mass spectrometry is a recent addition to the RMM toolbox for microbial identification. When microorganisms are exposed to an energy source, such as a laser, they are disintegrated and generate a variety of charged ions. The ions are separated in an electrical field based on their mass-to-charge ratio. As an analyzer detects the separated ions, molecular weight patterns or spectra are generated. These patterns are based on the macromolecules normally expressed on the surface of a particular microbial species. The workflow is relatively easy to perform: intact cells from a primary culture (e.g. an isolated colony from an agar plate) are spotted onto a stainless steel target plate and allowed to co-crystallize with a UV-absorbing matrix. After drying, the plate is placed into a mass spectrometer and exposed to a laser. The matrix absorbs energy from the laser, and cellular macromolecules are desorbed, ionized, and analyzed. The resulting mass spectra are compared with an internal database to provide a microbial identification. Multiple instruments are currently available to perform this type of analysis to identify bacteria, yeast, mold, and mycobacteria. These include the Bruker MALDI Biotyper (the first instrument that was introduced for microbial identification using MALDI) and the bioMérieux Vitek MS.

13.5.3 Viability-Based Technologies

Viability-based technologies differentiate viable cells from dead cells and can target specific microorganisms using nucleic acid, enzymatic, or monoclonal antibody probes. In many cases, direct labeling of single cells is possible with no cell growth requirement, facilitating time to result in hours or even minutes. Because these methods do not require growth, the enumeration of stressed, fastidious, dormant, or viable, but non-culturable organisms may be higher than that obtained using conventional, growth-based methods. Applications are fairly broad-based, including quantitative raw material, in-process sample and finished product bioburden testing, water analysis, and EM. The same technologies have been validated as alternative sterility tests, although this is outside the scope of analyses required for non-sterile pharmaceuticals.

An example of a viability-based rapid method employs flow cytometry. Cells contained within the test sample are labeled with a viability marker and then passed through a flow chamber, essentially one cell at a time. Fluorescence and light scatter signals are detected and individual cells are counted as they pass through a laser beam. The process of labeling and detecting viable cells in these systems can be accomplished in as little as a few minutes. Although the sample size compatible for flow cytometry systems is generally small (e.g. <1 ml), both aqueous and many types of nonaqueous samples are well suited with this method. Existing systems provide a quantitative assessment of the test sample;

however, although single cell detection is possible, obtaining a single cell quantitative result usually cannot be obtained with good accuracy and/or precision. For this reason, the commonly accepted limit of quantification for flow cytometry is between 10 and 50 viable cells. Examples of available technologies include the BD Diagnostics FACSMicroCount and the bioMérieux Chemunex D-Count and BactiFlow.

Solid-phase cytometry is similar to flow cytometry except that cells are captured on a solid surface such as a 0.45 μm filter. The filter is subsequently stained with a viability substrate followed by laser excitation, resulting in a quantitative result. Single cell enumeration with excellent accuracy and precision has been demonstrated by a number of companies and for a variety of test samples. One limitation of this technology is that the test sample must be filterable and the captured material on the membrane (i.e. microbial cells and components of the test sample) will not produce background noise or interference that may result in a false-positive or false-negative response. However, the potential for these types of responses would be determined when validating the method for its intended use, particularly when demonstrating method suitability.

A common workflow involves passing the test sample through a membrane filter and then labeling the filter with a nonfluorescent viability substrate. Within the cytoplasm of metabolically active cells, the substrate is enzymatically cleaved to liberate a free fluorochrome. Only viable cells with intact membranes have the ability to perform this cleavage and retain the fluorescent label. The entire membrane surface is subsequently laser-scanned and labeled microorganisms are quantified. The process of labeling and detecting viable cells in this system can be completed in 10–90 minutes, depending on the technology. Examples of solid-phase cytometry systems include the bioMérieux Chemunex ScanRDI and the LumiByte BV MuScan. Because this technology can use fluorescent dyes and microbe-specific probes, such as antibodies, rRNA, or peptide nucleic acid (PNA), simultaneous enumeration and detection of target microorganisms is possible. For example, the LumiByte BV MuScan utilizes DNA probes (FISH), PNA-probes, antibodies, and target specific dyes to detect and quantify specific microorganisms within a few hours, in addition to providing an accurate cell viable count. Examples of test kits for specific organisms include *Salmonella* spp., *Legionella pneumophila, Campylobacter jejuni, Listeria monocytogenes, E. coli, Enterococcus* spp., *S. aureus*, yeast, and fungal spores.

13.5.4 Spectroscopy-Based Technologies

Optical spectroscopy is an analytical tool that measures the interactions between light and the material being studied. Light scattering is a phenomenon in which the propagation of light is disturbed by its interaction with particles. For example, "Mie scattering" is one form of light scattering in which scattered light is proportional to particle size. Therefore, many particle counters employ Mie scattering to detect, count, and size particles in an environment, such as those used in cleanrooms and other microbiologically controlled areas.

Mie scattering is also the basis for a number of rapid methods used to detect, count, and size viable cells in the air and water. When particles from an air or water sample are processed through one of these systems, particles within a 0.5–20 μm range are sized and counted. At the same time, a UV laser that intersects the particle beam will cause biological material, such as microorganisms, to intrinsically auto-fluoresce, due to the presence of NADH, riboflavin, or dipicolinic acid. These types of technologies are highly sensitive, and provide viable and nonviable detection and counting results continuously and in real time. Additionally, there is no need for reagents, staining, labeling, or cellular growth. Examples of technologies that are specifically designed for air sampling include the BioVigilant IMD-A, Particle Measuring Systems BioLaz, and TSI's BioTrak system. Some of these systems can sample and evaluate up to 1 m^3 of air within 35 minutes. For non-sterile manufacturers who have implemented EM programs that include active air monitoring, these technologies may offer real time and continuous monitoring capabilities while reducing the need for resources to perform these time-consuming and resource-intensive evaluations. Additional uses of these technologies may include the investigation of EM excursions and to aid in the qualification of controlled or classified areas if non-sterile manufacturing processes require such levels of control.

Optical spectroscopic methods designed for testing pharmaceutical-grade water include the BioVigilant IMD-W and the Mettler Toledo 7000RMS Microbial Detection Analyzer. These operate in a similar fashion as the air monitoring systems described above; however, water is introduced into the instruments, either as a stand-alone sample or as an online connection to a water distribution system. The potential for these technologies to rapidly and continuously alert the user to a potential contamination issue is significant, given that the industry routinely uses water to manufacture non-sterile pharmaceuticals at risk (i.e. although the water is tested for viable organisms, this raw material is generally utilized in manufacturing long before the microbiology test results are available). Detecting an out-of-specification trend in water systems may also indicate suboptimal sanitization practices and/or the development of biofilms in distribution loops, storage tanks, or water generation unit operations (e.g. activated carbon and ion exchange beds, filtration units).

The use of an intrinsic fluorescence RMM offers multiple benefits including some of the fastest times to result, single cell detection, and the elimination of consumables, reagents, and media. However, there is a trade-off between time to result and sensitivity level versus the ability to physically recover microorganisms, should they be present. The current technologies, with the exception of only one, do not have a practical way in which detected organisms can be captured for subsequent analysis, such as microbial identification. Even for the one method that can purportedly capture airborne microorganisms on a gelatin filter, there is no guarantee that captured organisms will proliferate on a growth medium, given the potential for airborne organisms to be stressed, injured, or be in a state of physiological dormancy. In these cases, when excursions or out-of-trend findings are realized in real time, more conventional air monitoring systems may be utilized to try to capture and recover contaminants in the same area, assuming the organisms will be able to grow under the stated incubation conditions. Research associated with these types of technologies in pharmaceutical environments has already been published (Miller *et al.* 2009a, 2009b).

This brings up another consideration. For the reasons stated above, some alternative RMM technologies, such as those relying on intrinsic fluorescence and light scattering, may detect viable microorganisms when conventional methods may not. In this case, prior historical trending data, which were based on a CFU, may not be appropriate to use with the new RMM. As such, new trending levels independent of the prior CFU counts may be set based on a statistical evaluation of the RMM data and where appropriate, new action or alert levels scientifically developed. FDA microbiologists have previously addressed and supported this strategy in a 2006 publication (Hussong and Mello 2006). The authors stated the following: *Often, new methods rely on a completely different body of information, some may be direct measurements, some indirect. In either event, previous acceptance criteria may not be applicable. Therefore, implementation of newly developed, or more rapid, microbiology methods may also require the establishment of new acceptance criteria.*

Another type of light scattering is Raman, where information about molecular vibrations and rotations of molecules may be determined. Rapid methods based on Raman have been developed for the detection of microorganisms in a variety of sample matrices and in air, and because organism has its own unique Raman spectrum, this can be used as a fingerprint for identification. However, most Raman systems will detect both viable and nonviable cells, which would present a challenge for industries (such as non-sterile pharmaceutical manufacturing) in which viability must be confirmed. This scenario is different from what was previously discussed above in relation to the detection of stressed, injured, or dormant microorganisms. In this case, dead cells may be identified as viable, creating a false-positive situation in which safe drugs could be misconstrued as being contaminated. One technology vendor has overcome this issue by employing a viability-staining step prior to conducting Raman spectroscopy. The mibiC GRAM^{RAY}instrument allows for a Raman signature to be generated only from particles that have been determined to be viable cells, enabling the simultaneous detection, enumeration, and identification of organisms in a sample within minutes. The workflow for this technology requires the sample to be spotted on a metal carrier. Viability staining and automated image analysis using dark field illumination detects viable particle quantity, shape, and size for particles ranging from 0.5 μm and larger. Raman spectroscopy is then performed on each viable particle, and a spectral signature is provided. The spectral signatures are then statistically correlated to a library of known microorganisms. The supplier also purports the method to be nondestructive, allowing for further analysis of the recovered microorganisms. Raman coupled with viability staining may offer a solution for the quantitation and identification of contaminants in filterable samples within a matter of minutes.

13.5.5 Nucleic Acid Amplification-Based Technologies

Nucleic acid amplification-based rapid technologies utilize a number of gene amplification and detection platforms, including polymerase chain reaction (PCR), TMA, 16S rRNA typing, and gene sequencing. Most of these methods will detect the presence of a target microorganism or generate data that can be used to determine the identification of an isolate, from the genus level down to the subspecies and/or strain level.

To maintain correct RNA structure and ribosome function, the 16S sequence of rRNA is highly conserved at the genus and species level. However, the non-conserved fragments within the rRNA operon (the spacer and flanking regions of the 16S sequence) can be used to differentiate strains within a particular species. One rapid technology makes use of the DNA sequences that encode for the rRNA operon for microbial identification and strain differentiation. In the Hygiena Riboprinter, restriction enzymes, such as EcoRI or PvuII, are used to cut recovered and purified DNA into fragments. The fragments are then separated according to size by gel electrophoresis and immobilized on a nylon membrane (this is commonly referred to as a Southern Blot technique). The double-stranded DNA is denatured to single-stranded DNA, and the membrane is subsequently hybridized with a DNA probe (derived from an *E. coli* rRNA operon). Finally, an antibody–enzyme conjugate is bound to the probe and a chemiluminescent agent is added. Light emitted by the fragments is captured, and the image pattern is compared with patterns stored in the system database. If the pattern is recognized, a bacterial identification is provided. The pattern can also be used to determine if the same strain has been previously observed. This may be helpful when investigating the source of an environmental isolate or a contaminated non-sterile product.

Most nucleic acid amplification technologies that detect a specific or target microorganism employ PCR as their fundamental scientific principle. Small fragments of DNA (primers) are used to find a specific sequence (target) on a sample of DNA. A heat-stable enzyme (e.g. Taq DNA polymerase) makes millions of copies of the target sequence, which can then be easily detected and used to determine if a specific microorganism was present in the original sample. A typical workflow is as follows: following the extraction of DNA from microorganisms in a sample, all the reagents necessary for PCR (the specific DNA primer, DNA polymerase, and nucleotide bases [A, T, C, G]) plus a fluorescent dye (e.g. SYBR Green, which binds to double-stranded DNA) are hydrated with the DNA sample and processed in a thermocycler. The thermocycler goes through a series of heating and cooling steps to facilitate binding of the primer onto the target DNA sequence (if it is present), followed by elongation of the double-stranded DNA and then amplification of the DNA fragments. Each amplification cycle results in two copies of DNA. This is repeated and the number of copies of DNA per amplification cycle increases exponentially (i.e. 2–4 to 8–16, and so on). Real-time quantitative PCR can track the increase in fluorescence from one complete PCR cycle to the next, indicating that the target DNA sequence was present in the test sample because it is being replicated and amplified over time. This would indicate the target microorganism was also present in the original test sample. Additional detection phases may also be employed, in which the temperature of the amplified DNA sample is raised to the point where the DNA strands separate (denature), releasing the fluorescent dye and lowering the signal. This change in fluorescence can be plotted against temperature to generate a melting curve. Similar detection techniques using different types of probes and dyes (e.g. Taqman or FRET probes) are also available, depending on the supplier's system. Because each target organism generates its own specific melting curve related to its endpoint fluorescence detection, the system can determine if a specific target organism was present in the original sample. A number of systems use PCR as the scientific principle for microbial detection; these include the Hygiena BAX system and Pall Corporation's GeneDisc. The BAX system can detect *Salmonella, L. monocytogenes, C. jejuni/coli, E. coli* O157:H7, *Enterobacter sakazakii, S. aureus, Vibrio* spp., and yeast and mold. The GeneDisc has a USP Specified Microorganism assay which can detect the presence of *E. coli, Salmonella, P. aeruginosa, S. aureus, C. albicans,* and *Aspergillus brasiliensis.* Additional GeneDisc assays include STEC, *E. coli* 0157, *Listeria, Legionella, Enterococcus,* and *Cyanobacteria.*

Most PCR systems can only detect the presence of a single-target DNA sequence at a time; however, some newer technologies can perform multiplex detection assays, meaning they can detect multiple DNA targets simultaneously and in some cases, within the same reaction vessel or tube. This is made possible by using different primers and probes/dyes at the same time.

Unfortunately, PCR is not without its limitations. Although the time to result can be as early as a few hours to detect the presence of a particular organism, the potential for false positives is present when the starting material is DNA. DNA is ubiquitous in the environment and residual DNA or DNA from nonviable cells could result in a positive assay when that target microorganism was never present in the original test sample. Therefore, care must be taken to avoid DNA contamination when performing the test. Additionally, the test sample may need to be enriched in media before performing PCR to ensure any DNA detected and amplified came from the target organism(s) in the original sample under investigation.

To overcome the potential for false positives, a different gene amplification technique called TMA has been developed. TMA targets single-stranded RNA, which is less stable in the environment as compared with DNA. TMA also uses the enzyme RNA polymerase to produce many more of RNA copies per amplification cycle, as compared with only two copies of DNA from a PCR cycle, representing up to a 10 billion fold increase of copies in as early as 15–30 minutes. There are a number of additional advantages when using TMA as compared with PCR. For example, PCR requires a series of temperature changes to facilitate amplification where TMA is performed at one temperature (e.g. 41 °C), eliminating the need to use a thermocycler. There is also improved amplification reliability and sensitivity because there are usually thousands of rRNA copies per cell instead of a single DNA target. A few companies offer TMA and related amplification test kits and instrumentation, including the BioFire FilmArray system, which utilizes multiplex PCR, multiple primers and a closed-system pouch that isolates, purifies, amplifies and detects either RNA or DNA.

The use of a DNA-intercalating dye may also be utilized to reduce the potential for amplifying genetic material from dead microorganisms during PCR. Nocker *et al.* (2007) describes the use of propidium monoazide (PMA; a membrane impermeable dye) to penetrate dead microbial cells whose membranes have been compromised. Once the PMA gains access to the cell, it will intercalate into double-stranded DNA. This will prevent the DNA from being amplified during PCR. The authors concluded that performing PCR after PMA treatment would only allow the amplification of DNA from viable cells (i.e. those with intact cell membranes). A similar study by Kramer *et al.* (2009) used PMA in combination with real-time PCR or Thermo Fisher's LIVE/DEAD BacLight Bacterial Viability Kit in combination with flow cytometry for the determination of viable probiotic bacteria in a lyophilized product containing *Lactobacillus acidophilus* and *Bifidobacterium animalis* ssp. *Lactis*. The authors demonstrated PMA real-time PCR and flow cytometry determination of viability could complement a standard plate count method; the latter which considers only the culturable part of the population.

PCR and TMA have been used for the detection of a specific target microorganism. When microbial identification is required, the industry commonly turns to gene sequencing. The use of 16S rDNA is the current standard for taxonomic classification. Therefore, if we can extract DNA from an isolated microbial colony, amplify the DNA using PCR and sequence the first 500 base pairs of the 16S rRNA gene, we can compare the sequence to an internal database to assist with a microbial identification. Current systems accomplish this task using a genetic analyzer and automated workflows. The most commonly used instrument in microbiology labs for this purpose is the Applied Biosystems MicroSEQ. The starting material for gene sequencing is usually an isolated colony from an agar plate from which a test sample has already been analyzed.

It has also been proposed that gene amplification techniques can be used to estimate the number of microorganisms in the test sample based on the number of cycles it takes to detect the presence of a target sequence. However, there may be too many technical hurdles to overcome to get this to work reliably and repeatedly. For example, in order to detect a wide range or microorganisms, multiple nucleic acid primers would be required. Next, steps would need to be taken to minimize false positives, such as using RNA as the starting material. Third, there would need to be some correlation between the number of cycles required to detect the amplified copies to a viable cell count. Until these challenges are adequately addressed, nucleic acid amplification technologies may be limited to the detection of specific microorganisms and for microbial identification purposes.

13.6 Validating Rapid Microbiological Methods

Validation is the act of demonstrating and documenting that a procedure operates effectively and repeatedly. Because rapid methods are considered alternatives to existing or compendial microbiology methods, the end user is responsible for demonstrating the rapid method is at least equivalent to, or is non-inferior to, the method intended to be replaced. As such, validating a rapid method should encompass the entire process that starts with the decision to change the non-sterile product microbiological testing program and continues through ongoing routine use of the validated method. Since most rapid method technologies consist of equipment, instrumentation, software, communication interfaces, and, of course, the actual analytical method that detects microorganisms, the entire system must be appropriately qualified. Therefore, the validation plan should provide a roadmap for all of the activities that will be required to demonstrate that the rapid method system is suitable for its intended use.

There are a number of validation documents that provide guidance on method development, method suitability testing, equivalency studies, the use of statistics, feasibility or proof-of-concept studies, training, technology transfer, and the establishment of validation plans. The three most commonly used guidance documents are as follows:

- PDA Technical Report #33, Evaluation, Validation and Implementation of Alternative and Rapid Microbiological Methods (PDA 2013)
- USP Chapter <1223>, Validation of Alternative Microbiological Methods
- Ph. Eur. chapter 5.1.6, Alternative Methods for Control of Microbiological Quality

Each of the documents provides a review of validation strategies, although the 2013 revision to PDA TR33 is considered to be the most comprehensive in terms of content. USP <1223> was significantly revised in 2015 and is significantly different from its prior version. In some instances, the new USP chapter appears to conflict with some of the teachings in PDA TR33 and the very recent update to the 2017 revision to Ph. Eur. chapter 5.1.6. For these reasons, the reader is encouraged to review recent publications (Miller 2015a, 2015b, 2017a, 2017b) comparing and contrasting each of the three documents.

Prior to making a commitment in purchasing a specific RMM system, most companies will perform feasibility or proof-of-concept studies to demonstrate product or test samples will be compatible with the technology of interest. This also assumes the end user has already developed user requirements and has matched the appropriate RMM for his or her application(s). Feasibility or proof-of-concept studies may be performed in-house (after renting an instrument or obtaining an instrument on loan) or at the RMM supplier's facility. The purpose of this testing is to determine if the test sample matrix (e.g. the finished product, in-process sample, raw material, or other substance) is incompatible with the rapid method or interferes with the outcome of the analysis. For example, a test sample that does not contain any viable microorganisms should provide a similar result when introduced into the RMM system (i.e. there are no false positives). Similarly, a microorganism challenge introduced into the test sample and subsequently analyzed by the RMM should detect, and, when applicable, enumerate the microorganisms at a statistically significant level as what was originally inoculated. If it does not, the sample matrix may mask the presence of viable microorganisms, resulting in a false-negative response. Any sample matrix that presents false-positive or false-negative results should be investigated and resolved before a commitment is made to purchase and validate the RMM system.

Concurrent with these activities, the end user should audit or assess the supplier no differently than if they were qualifying a new supplier of raw materials or other equipment or services.

Additionally, a risk assessment should be conducted to identify any new risks or hazards that may be associated with the implementation of the rapid method. A discussion of risk assessments is provided earlier in this chapter.

The next step is to develop the validation master plan (VMP) for all of the activities that will be required to demonstrate that the rapid technology is validated and suitable for its intended use. The validation plan should identify all project deliverables, responsible parties for each phase of test execution, review and approval processes, and the documentation required to satisfy the expectations of the validation strategy.

A functional design specification (FDS) or similar document should be developed that will describe all of the functions and requirements for the RMM system and what will be tested to ensure that the system performs as specified in the end user's original URS. In essence, the FDS is the validation roadmap, telling the end user how and where they will assess whether the RMM system meets the user requirements (as specified in the URS).

Examples of what the FDS may address and require testing for include, but are not limited to, the following list. Note that the end user's FDS will be based on the testing required to demonstrate that the RMM is able to meet the requirements as outlined in the related URS.

1) Purpose, scope, and description of the RMM
2) Documentation
   a) User manuals
   b) Guidelines
   c) Standards
   d) SOPs
3) Physical specifications
   a) Size
   b) Electrical power
   c) Voltage frequency
   d) Operating temperature
   e) Environmental requirements
   f) Utility requirements
4) Computer system specifications
   a) Processor, hard drive, RAM, and video graphics
   b) Network address and connections
   c) Operating software
   d) Printer ports
   e) Software and algorithms
   f) Databases
5) Security specifications
   a) User ID and password
   b) Access to data
   c) Record retention
   d) Audit trail
   e) Administrative control
   f) Data view and print reports
   g) Data transfer to a dedicated server
   h) 21 CFR Part 11
6) Functional specifications
   a) Accuracy
   b) Precision (repeatability)
   c) Specificity

   d) Limit of detection
   e) Limit of quantification
   f) Linearity
   g) Range
   h) Ruggedness (including intermediate precision and reproducibility)
   i) Robustness
   j) Equivalence

7) Functions that will not be used (or tested)
8) System customization
9) Alarm configuration and error handling

The FDS should also point to specific test scripts where each requirement will be evaluated and verified against pre-established acceptance criteria (e.g. what will be performed during the IQ, OQ, and PQ phases of the validation plan).

The installation qualification (IQ) establishes the equipment is received as designed and specified, that it is properly and safely installed with the correct utilities in the selected environment, and that the environment is suitable for the operation and use of the equipment. The IQ can be carried out by the RMM supplier (during initial commissioning) and/or by the end user, especially if the end user's company requires a more extensive IQ program.

The operational qualification (OQ) provides documented verification that the equipment performs effectively and reproducibly as intended throughout the anticipated or representative operational ranges, defined limits, and tolerances. The OQ is the focal point for the majority of the computer system validation (CSV), including hardware, software, and security testing.

The performance qualification (PQ) provides documented evidence that the instrumentation, as installed, consistently performs in accordance with predetermined criteria and thereby yields correct and appropriate results. The PQ will include validating the method, method suitability for the product or test samples to be routinely evaluated, and a demonstration of equivalence to the existing method.

Method validation requires the use of a standardized panel of test microorganisms and/or other reference materials that are used to demonstrate an appropriate level of accuracy, precision or repeatability, specificity, limit of detection or quantification, linearity, range, ruggedness, and robustness. Specific testing requirements will depend on whether the rapid method will provide a quantitative, qualitative, or identification result. The selection of microorganisms to use for this purpose should be relevant to the user requirements (e.g. what organisms require to be detected based on the test sample and its intended use and patient population) and what might provide the greatest challenge to the RMM system (e.g. slow-growing organisms should be selected for growth-based methods).

The USP, Ph. Eur., and PDA recommend the following studies to be performed (Tables 13.1–13.3).

Method suitability studies must also be performed and these are very similar to what current compendial requirements are for microbiological testing: that the presence of a particular product, material, or sample matrix does not

Table 13.1 Validation criteria for qualitative methods.

Validation criteria	USP <1223>	Ph. Eur. 5.1.6	PDA TR33
Accuracy		[a]	
Specificity	✓	✓	✓
Limit of detection	✓	✓	✓
Ruggedness	✓		✓
Robustness	✓	✓	✓
Repeatability	✓[b]		

[a] Accuracy can be used instead of the limit of detection test.
[b] USP does not provide guidance on this criterion.

Table 13.2 Validation criteria for quantitative methods.

Validation criteria	USP <1223>	Ph. Eur. 5.1.6	PDA TR33
Accuracy	✓	✓	✓
Precision	✓	✓	✓
Specificity	✓	✓	✓
Limit of detection	✓	[a]	✓
Limit of quantification	✓	✓	✓
Linearity	✓	✓	✓
Operation (dynamic) range	✓	✓	✓
Ruggedness	✓	(Addressed as intermediate precision)	✓
Robustness	✓	✓	✓
Repeatability	✓[b]	(Addressed under precision)	(Addressed under precision)

[a] May be needed in some cases.
[b] This may be covered under Precision, although USP does not provide additional guidance.

Table 13.3 Validation criteria for microbial identification methods.

Validation criteria	USP <1223>	Ph. Eur. 5.1.6	PDA TR33
Accuracy	✓	✓	✓
Precision	✓		✓
Specificity		✓	
Robustness		✓	

significantly impact the performance of the rapid method, which may include background noise, interference, false positives, and/or false negatives. Each of the guidance documents provides recommendations on how to perform these types of studies.

Finally, equivalence testing will determine how similar the new method test results are when compared with the existing method. The new method is run in parallel with the existing method for a specified period of time or number of product batches or test samples and then the results are statistically compared with each other. Practical examples of how to utilize equivalence and non-inferiority statistical models when validating RMMs may be found in one publication by this author (Miller *et al.* 2016).

The end user should determine the most appropriate strategy for the duration and extent of equivalence studies, which may be influenced by the critical nature of the test method, the material being analyzed, the statistical methods used when interpreting the resulting data, regulatory expectations, and/or other quality requirements. However, the three guidance documents are not necessarily aligned in terms of whether actual product or test samples should be included during equivalency testing. For example, PDA TR33 recommends including product while USP <1223> generally infers that product should not be used because it is the analytical method of detecting microorganisms that must be shown to be equivalent (although one section in the USP chapter assumes at least one product should be included during equivalence studies for a particular technology, and then only method suitability can be performed for subsequent products). Ph. Eur. chapter 5.1.6 is somewhere in-between, in that equivalency can be demonstrated either with or without product. The Ph. Eur. actually states *alternatively, and in some cases additionally, equivalence testing can be carried out by the parallel testing of a predefined number of samples or for a predefined period of time.* However, Ph. Eur. does not provide any additional clarity on what this risk assessment should address. The reader is encouraged to review the equivalence testing recommendations in each guidance document, as well as in a series of comparative reviews this author has previously published (Miller 2015a, 2015b, 2017a, 2017b). Furthermore, discussions with regulators who will review your validation protocols and/or data may provide additional guidance on whether product or test samples should be included during equivalence studies.

Companies have utilized the recommendations from one or more of the existing validation guidance documents or have developed their own validation strategies. In any case, firms should discuss their validation plans with the relevant regulatory authorities up front to ensure there is agreement on the intended testing strategy. Additionally, formal policies exist in which proposed validation plans may be submitted to the regulators for review and approval. Specifically, comparability protocols may be submitted to the FDA and a post-approval change management protocol may be submitted to European Medicines Agency (EMA) national authorities as a Type 2 variation. In each, the validation testing plan, methods, and acceptance criteria are submitted for review. When approved, the end user would conduct the studies (as outlined in the submitted and approved proposal) and demonstrate the acceptance criteria have been met. In most cases, data does not have to be submitted to FDA (except in cases where the method impacts a biologics product or Biologics License Application (BLA)). However, data are always submitted to EMA as either a Type 1A or 1B variation. Whether the data are submitted as a Type 1A or 1B will be determined on the method, affected product, and/or other factors. For these reasons, it is always suggested to consult with the relevant regulatory authority on expectations for these types of activities.

Some rapid technologies may be considered as automated traditional or compendial microbiological test methods, especially when the test results are presented as a CFU. As such, these technologies may be qualified for their intended use without the need for demonstrating certain method validation requirements. For example, at least accuracy and precision assessments should be performed, in addition to method suitability and equivalence/comparability studies. Examples of technologies that may fall into this category include methods that provide a CFU count on agar plates or membranes but which do so in a more rapid timeframe as compared with the visual detection of microbial colonies. Some of these growth-based methods were discussed earlier in this chapter. As such, a verification that the plate counts from the rapid method and the conventional method are statistically the same (or non-inferior) may be sufficient instead of performing a full validation as recommended by USP, Ph. Eur., and PDA. In any case, a risk assessment should be performed to support a reduced validation strategy compared with other RMMs that provide a test result that has a signal different from the CFU.

There are a number of publications providing practical examples of how RMMs can be used as alternatives to conventional or compendial bioburden assays. For example, a 2017 publication by Gordon *et al.* (2017) demonstrates the successful validation of an alternative bioburden testing method based on viability staining and the detection of cellular fluorescence from growing microorganisms using the Millipore Milliflex Quantum system. All three validation guidance documents (USP, Ph. Eur., and PDA) were considered when developing a validation plan comprising Robustness, Ruggedness, Repeatability, Specificity, Limit of Detection and Quantification, Accuracy, Precision, Linearity, Range, and Equivalence. A panel of 15 microorganisms including ATCC strains and facility isolates were utilized in either an unstressed or stressed physiological state, the latter representing a worse-case scenario for a growth-based RMM (i.e. the longest time to grow and be adequately detected/enumerated). Three product solutions were used during equivalence studies and non-inferiority statistical analyses demonstrated the rapid method was no worse than the conventional method.

13.7 Developing a Business Case for Rapid Methods

The initial costs associated with feasibility studies, validation testing, and installation activities for a new, rapid method can be significant. However, it is unfair to only focus on the up-front costs when evaluating a new technology, as there can be substantial long-term cost savings or cost avoidances that may be realized. Therefore, the end user should develop a comprehensive economic analysis to support the decision to purchase, validate, and ultimately implement a rapid method for routine use.

Financial models, such as return on investment (ROI), payback period, and net present value may be used to support the decision to purchase a new rapid method. Using one of these models, it is relatively simple to compare the overall costs associated with the conventional method and the proposed RMM, while taking into account the potential cost savings by implementing the RMM. The information obtained should be used to complement the technical and quality justifications for qualifying the RMM. Practical examples of how to conduct these financial exercises with case study-calculated ROIs are presented in publications on this topic (Miller *et al.* 2009a, 2009b).

Briefly, when performing a financial assessment of this nature, the overall costs associated with the conventional method and the proposed RMM are compared. These can include direct and indirect costs (e.g. consumables, cost of labor, depreciation of equipment, and maintenance programs), potential cost savings, and/or cost avoidances associated with the RMM as well as the RMM capital, training, and validation costs.

When all of the elements associated with the costs and savings for both the conventional method and the RMM have been collated, this information can then be used in financial models to calculate whether there is an economic advantage for implementing the RMM. Some RMMs may provide a positive ROI while others may provide a negative ROI. However, the technical and quality attributes afforded by implementing a RMM may outweigh a negative ROI, depending on the need for a rapid technology and its impact on patient outcomes.

13.8 Summary

This chapter provides guidance for how to identify, validate, and implement RMMs for a variety of applications associated with the manufacture and microbiological testing of non-sterile pharmaceutical products. Because non-sterile products, by definition, may contain microorganisms, testing non-sterile dosage forms is of significant importance to ensure that objectionable organisms will not be passed on to the recipients of said products; namely, the patients and consumers who trust our industry in providing safe and effective medicines. The continued use of nineteenth-century microbiological methods is counterproductive in our goal of continuously improving our products, processes, and understanding of how microorganisms impact pharmaceuticals, especially from the perspective of end-user safety. For these reasons, the industry should consider moving microbiology into the twenty-first century by implementing next-generation alternative and rapid method technologies.

What would the modern microbiology lab look like? Because each end user will have specific needs and testing requirements based on the composition of the test sample(s), the required time to result, the compatibility of the test sample(s) with the desired technology, and other factors, it is not possible to draw conclusions that can be applicable to all laboratories. However, an example laboratory would employ the following:

1) Quantitative finished product release testing (e.g. USP 61) will utilize solid-phase cytometry, flow cytometry, ATP bioluminescence methods, or rapid growth-based technologies, depending on the desired time to result. Real-time or close to real-time results will require nongrowth-based systems. Similar systems would be used for raw material testing, realizing just in time manufacturing capabilities, as the results would be available during the same shift.
2) Qualitative finished product release testing (e.g. USP 62) will employ nucleic acid amplification technologies, such as PCR or TMA, for the rapid detection of specified or objectionable organisms. Methods shall be developed that eliminate the potential for false positives from dead cells when PCR is used. If a longer time to result is acceptable, the lab will utilize rapid growth-based methods that can detect the presence of specific microorganisms.
3) Intrinsic fluorescence and light scattering techniques will be used for real-time EM and in-process water testing. If near real-time results are acceptable, solid-phase cytometry and flow cytometry can be used.
4) Gene sequencing and MALDI-TOF mass spectrometry will be the standard for microbial identifications.
5) If commercially available, Raman spectroscopy coupled with viability staining will be used for microorganism detection, enumeration, and identification, as long as the sample matrix is compatible with the system workflow.

6) Automated systems will be used for labs that are required to analyze large numbers of samples; however, their use will be dictated by the required time to result as many automated systems are growth-based.
7) As MEMS technologies (e.g. lab-on-a-chip, biosensors, and microarrays) evolve for the pharmaceutical QC laboratory, these will be employed for multiple applications. However, automation and sample throughput may take a back seat when these miniaturized systems are implemented.

Bibliography

CFR, 21 CFR 211.113. Code of Federal Regulations. Title 21, Volume 4. Current Good Manufacturing Practice for Pharmaceuticals. Production and Process Controls. Control of microbiological contamination. Revised 2017.

CFR, 21 CFR 211.165. Code of Federal Regulations. Title 21, Volume 4. Current Good Manufacturing Practice for Pharmaceuticals. Laboratory Controls. Testing and release for distribution. Revised 2017.

CFR, 21 CFR 211.84. Code of Federal Regulations. Title 21, Volume 4. Current Good Manufacturing Practice for Pharmaceuticals. Control of Components and Drug Product Containers and Closures. Testing and approval or rejection of components, drug product containers, and closures. Revised 2017.

Dalmaso, G, 2006, 'Rapid steam sterilization biovalidation using biological indicators and the Pallchek Luminometer', in Miller, JM (ed.), Encyclopedia of Rapid Microbiological Methods Volume II, PDA/DHI Books, Bethesda, MD/River Grove, IL, pp. 251–272.

European Directorate for the Quality of Medicines & Health Care. Chapter 5.1.6, Alternative Methods for Control of Microbiological Quality. 2017; Supplement 9.2: 4339–4348.

Gordon, O, Goverde, M, Staerk, A, Roesti, D, 2017, 'Validation of Milliflex® quantum for bioburden testing of pharmaceutical products', PDA J. Pharm. Sci. Technol., vol. 71, pp. 206–224.

Hussong, D, Mello, R, 2006, 'Alternative microbiology methods and pharmaceutical quality control', Am. Pharm. Rev., *vol.* 9(1), pp. 62–69.

Kramer, M, Obermajer, N, Bogovic MB, Rogelj, I, Kmetec, V, 2009, 'Quantification of live and dead probiotic bacteria in lyophilised product by real-time PCR and by flow cytometry', Appl. Microbiol. Biotechnol., vol. 84(6), pp. 1137–1147.

Miller, MJ (ed.), 2005–2012, Encyclopedia of Rapid Microbiological Methods, Vols. 1–4, PDA and Davis Healthcare International.

Miller, MJ, 2009a, 'Breaking the rapid microbiological method financial barrier: a case study in RMM return on investment and economic justification', BioPharm Int., *vol.* 22(9), pp. 44–53.

Miller, MJ, 2009b, 'Rapid microbiological methods and demonstrating a return on investment: it's easier than you think!', Am. Pharm. Rev., *vol.* 12(5), pp. 42–47.

Miller, MJ, 2012, 'Case study of a new growth-based rapid microbiological method (RMM) that detects the presence of specific organisms and provides an estimation of viable cell count', Am. Pharm. Rev., vol. 15(2), pp. 18–25.

Miller, MJ, 2015a, 'A fresh look at USP <1223> validation of alternative microbiological methods and how the revised chapter compares with PDA TR33 and the proposed revision to Ph. Eur. 5.1.6', Am. Pharm. Rev., vol. 18(5), pp. 22–35.

Miller, MJ, 2015b, 'Rapid methods update: revisions to a United States Pharmacopeia chapter', Eur. Pharm. Rev., *vol.* 20(4), pp. 38–43.

Miller, MJ, 2017a, 'A comprehensive review of the revised European Pharmacopoeia chapter 5.1.6 (part one)', Eur. Pharm. Rev., vol. 22(5), pp. 20–24.

Miller, MJ, 2017b, 'A comprehensive review of the revised European Pharmacopoeia chapter 5.1.6 (part two)', Eur. Pharm. Rev., *vol.* 22(6), pp. 27–30.

Miller, MJ, Lindsay, H, Valverde-Ventura, R, O'Connor, MJ, 2009a, 'Evaluation of the BioVigilant IMD-A, a novel optical spectroscopy technology for the continuous and real-time environmental monitoring of viable and nonviable particles. Part I: Review of the technology and comparative studies with conventional methods', PDA J. Pharm. Sci. Technol., *vol.* 63(3), pp. 244–257.

Miller, MJ, Walsh, MR, Shrake, JL, Dukes, RE, Hill, DB, 2009b, 'Evaluation of the BioVigilant IMD-A, a novel optical spectroscopy technology for the continuous and real-time environmental monitoring of viable and nonviable particles. Part II: Case studies in environmental monitoring during aseptic filling, intervention assessments and glove integrity testing in manufacturing isolators', PDA J. Pharm. Sci. Technol., *vol.* 63(3), pp. 258–282.

Miller, MJ, *et al.* 2013, Evaluation, Validation and Implementation of Alternative and Rapid Microbiological Methods, PDA Technical Report No. 33, PDA, Bethesda, MD.

Miller, MJ, Van den Heuvel, ER, Roesti, D, 2016, 'The role of statistical analysis in validating rapid microbiological methods', Eur. Pharm. Rev., *vol.* 21(6), pp. 46–53.

Nocker, A, Sossa-Fernandez, P, Burr, MD, Camper, AK, 2007, 'Use of Propidium Monoazide for live/dead distinction in microbial ecology', Appl. Environ. Microbiol., *vol.* 73(16), pp. 5111–5117.

Rapid Micro Methods, 2018, rapidmicromethods.com

Sawant, A, *et al.* 2014, Exclusion of Objectionable Microorganisms from Nonsterile Pharmaceuticals, Mmedical Devices, and Cosmetics, PDA Technical Report No. 67, PDA, Bethesda, MD.

United States Pharmacopeial Convention, 2017a, '<1223>, Validation of alternative. Microbiological methods', USP 40/NF35, pp. 1756–1770.

United States Pharmacopeial Convention, 2017b, 'USP <1111>, Microbiological examination of nonsterile products: acceptance criteria for pharmaceutical preparations and substances for pharmaceutical use', USP 40/NF35, pp. 1415–1416.

United States Pharmacopeial Convention, 2017c, 'USP <1115>, Bioburden control of nonsterile drug substances and products', USP 40/NF35, pp. 1423–1430.
United States Pharmacopeial Convention, 2017d, 'USP <61>, Microbiological examination of nonsterile products: microbial enumeration tests', USP 40/NF35, pp. 117–123.
United States Pharmacopeial Convention, 2017e, 'USP <62>, Microbiological examination of nonsterile products: tests for specified microorganisms', USP 40/NF35, pp. 123–130.

14

Validation of a Rapid Microbiological Method for the Microbiological Examination of Non-sterile and Nonfilterable Drug Products, APIs, and Excipients

Oliver Gordon[1] and David Roesti[2]

[1] *The Francis Crick Institute, London, United Kingdom*
[2] *Novartis Pharma Stein AG, Stein, Switzerland*

CONTENTS

Pharmaceutical Microbiological Quality Assurance and Control: Practical Guide for Non-Sterile Manufacturing, First Edition. Edited by David Roesti and Marcel Goverde.

14.1 Introduction

An alternative rapid microbiological test method based on the Celsis Advance system according to USP <1223>, Ph. Eur. 5.1.6, and PDA Technical Report No. 33 was validated for the microbiological examination of non-sterile and nonfilterable drug products, excipients, and APIs. The alternative test method, which for reasons of simplicity will be referred to as Rapid MET throughout this book chapter, replaces both quantitative and qualitative microbiological testing of non-sterile products by combining both these requirements in a single test.

The current test methods for the microbiological examination of non-sterile products have been harmonized and are described in Ph. Eur. 2.6.12, 2.6.13, USP <61>, <62>, and JP <4.05>/I, <4.05>/II. The incubation conditions for the microbial enumeration tests (METs) are 30–35 °C for 3–5 days for total aerobic microbial count (TAMC), and 20–25 °C for 5–7 days for total yeasts and molds count (TYMC). In addition, tests for specified microorganisms composed of enrichment and followed by one or two selection steps have to be performed in order to demonstrate the absence of specified microorganisms in 1 or 10 g of product. In general, these tests last two to three days and in some cases up to six days. Therefore, microbiological examination of non-sterile products may take up to a week before a final result is available. Shortening the incubation time, however, would enable to align microbiological testing with manufacturing concepts such as LEAN manufacturing. Furthermore, the necessity to perform two quantitative (TAMC/TYMC) and several qualitative tests requires a multitude of different nutrient media and substantial hands-on time through the subculturing steps.

One of the major difficulties concerning rapid microbiological method application to non-sterile products is that these in general are nonfilterable. Most RMM systems on the market enabling precise enumeration are, however, based on membrane filtration (Gordon *et al.* 2011; Miller 2012b). Semiquantitative methods or indirect enumeration methods of nonfilterable products via the detection of by-products from microbial metabolism such as CO_2 have recently been developed (Miller 2012a). However, these methods only provide a rough estimation of the microbial count and pharmaceutical products' quality concepts such as trending or expectation or alert levels may not be applicable.

Another approach would be to perform a direct inoculation of the product and instead of providing a microbial count, a presence/absence result for microbial growth may be obtained. Because it is a presence/absence test, the rapid method should only be implemented for products which have an excellent microbiological quality reflected by a record of usually being negative for microbial growth in routine testing with the compendial method. For such products, both enumeration and specified microorganisms tests could be combined in one single test; if the absence of microbial growth is demonstrated for

1 g product, all specifications are met.[1] Thereby, significant reduction of hands-on time could be achieved.

The Celsis Advance System was selected for this purpose, which detects microbial growth based on ATP bioluminescence and may also be used for nonfilterable products. The reagent kit with which the Celsis Advance is operated in this case (called AKuScreen) amplifies the ATP bioluminescence reaction by the addition of ADP, which is converted to ATP by the cellular enzyme adenylate kinase. Adenylate kinase is present in all known microorganisms, and catalyzes conversion of ADP to ATP. Thus, if the enzyme is present in the sample, addition of excess ADP leads to formation of more ATP and hence amplification of the ATP bioluminescence signal. Detection by ATP bioluminescence requires a much smaller amount of microorganisms than visual detection, therefore significantly reducing the necessary incubation times. Intensity of ATP bioluminescence is expressed as relative light units (RLU).

14.1.1 Workflow Rapid MET

Different growth conditions were compared during feasibility studies and the following test setup for routine was defined based on the obtained results:

- 10 g of the product tested is dissolved in 90 ml of dilution buffer. Dilution technique and buffer composition as used for the compendial test methods may be applied.
- 90 ml liquid nutrient medium (TSB supplemented with 4% polysorbate 80 + 0.5% soy lecithin), ~10 g sterilized glass beads of approximately 1 mm diameter, and a sterile magnetic stir bar are added into each of two bottles. 10 ml of the product dilution (corresponding to 1 g drug product/excipient) are transferred into each bottle. These two bottles are the test samples for the Rapid MET with the Celsis Advance. Supplemented TSB was selected because it has been proven superior in recovering spiked microorganisms in a large variety of product-specific method suitability tests (D. Roesti personal observation).
- One bottle is incubated at 20–25 °C, the other one at 30–35 °C. The incubation time is at least 72 hours. Studies including a wide variety of different stressed in-house isolates indicated that this incubation time ensures detection of all relevant strains tested.

1 In case that a product required the absence of *Salmonella*, either 10 g product had to be used for the Rapid MET or the compendial test for the absence of *Salmonella* had to be performed in parallel to the rapid method. In general, the absence of *Salmonella* is mostly required for products from natural origin. It is of note, however, that most of such products are expected to have a high bioload and therefore be unsuitable for the Celsis Advance System Rapid MET application.

- After the incubation period, the bottles are transferred onto magnetic stirrers and stirred for 20 minutes in order to disperse aggregating microorganisms.[2]
- 50 µl of the samples are measured in duplicate with the Celsis Advance System.

14.1.2 Evaluation of Test Results

Microbiological acceptance criteria are based on Ph. Eur. 5.1.4, USP <1111>, and JP G.4. The compendial method complies, if the microbial limits of the drug product/excipient of interest are not exceeded in the TAMC and TYMC, and if the specified microorganisms are not detected. In the Rapid MET, 1 g of product is evaluated for microbial growth with the Celsis Advance and if no microbial growth is detected, product may be released for MET and absence of specified microorganism in 1 g product (except for *Salmonella*). In case that microbial growth is detected in the Rapid MET with the Celsis Advance, the contaminant(s) will be identified in order to exclude specified or objectionable microorganisms. Furthermore, compendial MET (TAMC/TYMC) will be performed in order to assess whether the level of contamination is above acceptance criteria.

14.2 Method Validation

14.2.1 General Validation Strategy

The aim of the validation was to compare two methodologies (Rapid MET and compendial method) and demonstrate equivalence based on defined validation parameters. The validation parameters consisted of robustness, ruggedness, repeatability, specificity, limit of detection (LOD), accuracy and precision (according to Ph. Eur. 5.1.6 and USP <1123>), and equivalence in routine operations. Demonstration of equivalence between two methodologies was deliberately performed without focusing on a particular type of product. Thus, most experiments were performed without using any product (an exception was the validation parameter "Equivalence in routine operation"). Suitability of a particular product for the Rapid MET is assessed in an additional, product-specific

2 During the feasibility studies, the in-house isolated mold *Penicillium* sp. was not reliably detectable without such pre-treatment; the mold formed such dense aggregates that several cases occurred in which no cell material was by chance pipetted in the assay cuvettes of the Celsis Advance instrument. The glass bead treatment enabled sufficient dispersal of the *Penicillium* sp. and detection with the Celsis Advance System within 72 hours. The combination of small beads and magnetic stirrer had proven most effective when compared to horizontal shaking, overhead shaking, and chemical dispersal through Triton X-100.

study. By demonstrating that the Rapid MET performs at least equivalent to the compendial method and demonstrating that a product in focus for testing with the Rapid MET is suitable for application of that method, it can be ensured that the ability to assess microbiological quality of that product is not compromised by using an alternative to the compendial test method.

Because our application of the method consists of demonstrating the absence of culturable microorganisms in 1 g product, low-level inocula of approximately 1–5 CFU were applied in most validation experiments. Not only pharmacopoeial test microorganisms were used but also in-house isolates. In-house isolates were used in stressed state. Not only does this represent "worst-case" scenarios for detection but it also mimics the actual situation: most contaminants in sterile and non-sterile products are expected to be stressed, e.g. following a treatment with disinfectants, exposition to heat, or dehydration. Therefore, the aim of stress-protocols (e.g. by application of heat or by nutrient depletion) is not to kill all the microorganisms, but rather force surviving microorganisms into a stressed but still viable state with prolonged lag-phase. Upon start of incubation, these microorganisms can, however, recover from their stressed state and return to their normal growth behavior. The used stress protocols were published methods (Gray *et al.* 2010).

14.2.2 Statistical Data Evaluation

14.2.2.1 Fisher's Exact Test and Chi-Square Test

Data were evaluated with several statistical methods. Since the Rapid MET provides qualitative data (presence/absence), Fisher's exact test was used for most evaluations. Fisher's exact test determines whether categorical variables are independent. In our case, the question to be answered was whether successful detection of microorganisms was independent of the test method used (rapid method or compendial method). The null hypothesis was that no relation between the experimental outcome and the used analysis method existed. The null hypothesis was rejected when the p-value was below 0.05 corresponding to a confidence level of 95%. Because Fisher's test belongs to the class of exact tests, it can also be applied for relatively small sample sizes (Fisher 1922). However, because of its methodology, the test is basically limited to comparison of only two data sets.

If more than two sets of data had to be compared (e.g. for the validation parameter "Repeatability," which required evaluation of results from different days and daytimes), the Chi-Square test was used. Similar to Fisher's exact test, the Chi-Square test determines whether categorical variables are independent. The Chi-Square statistic quantifies how much the observed distribution of positive/negative results varies from the theoretical distribution one would expect if no relation between the experimental outcome and the test method exists. This procedure is approximate; it only gets accurate with a certain sample size (Pearson 1900). Therefore, it was only used if Fisher's exact test could not be applied and also with a higher sample size.

14.2.2.2 Sample Size and Test Power

One of the most important aspects when designing experiments is the choice of an adequate sample size. If sample size is chosen too low, the experimental outcome may be biased through randomness, which in our context mainly means that the ability to detect statistically significant differences is compromised. The ability to detect a statistical difference is often expressed through test power. In reference to USP <1010>, a test power of ≥0.8 can in general be considered acceptable. In our validation, the largest acceptable difference between the rapid method and the compendial method was defined as 30% in reference to USP <1227>.[3] For all experiments, the confidence level was defined as 95%, because 5% possibility of type I error seemed acceptable (Note: Type I errors represented our "risk" of failing a statistical acceptance criterion due to random data fluctuations, although in reality no difference between both methods under examination existed). In general, test power is dependent on the sample size, the largest acceptable difference between the methods under evaluation and the confidence level. Table 14.1 summarizes the impact of these parameters.

Moreover, as can also be inferred from Table 14.1, the sample size dictates which observed effect size leads to a significant difference in the statistical test. With big sample sizes, even minimal differences in recovery lead to statistical

Table 14.1 Variables affecting test power.

Variable factor	Constant factors	Effect on test power
Higher sample size	Largest acceptable difference, confidence level	Higher test power
Lower sample size	Largest acceptable difference, confidence level	Lower test power
Higher largest acceptable difference	Sample size, confidence level	Higher test power
Lower largest acceptable difference	Sample size, confidence level	Lower test power
Higher confidence level	Sample size, largest acceptable difference	Lower test power
Lower confidence level	Sample size, largest acceptable difference	Higher test power

3 Two different methodologies must by definition show differences in the obtained results; particularly if these are (micro-) biological assays. This is also the case for every traditional test method; even such small variables as different operators or nutrient media lots inevitably lead to minor differences. Therefore, a boundary in relation to the reference method should be defined, above which the alternative method operates in a verified and consistent manner.

significance; whereas with small sample sizes also big differences between the data sets of interest may not become statistically significant.

For quantitative data, an additional factor impacting test power is the standard deviation (the higher the standard deviation, the lower the test power). For qualitative data, a factor with a similar impact is the overall success rate. Regarding a microbiological validation, the success rate would represent the overall frequency of microbial detection and is thus dependent on the inoculum used for the validation experiments. The higher the overall success rate, the higher the test power. The reason is that with a higher mean inoculum (e.g. 10 CFU), a negative test result is most probably not due to randomness; whereas with very low mean inocula (e.g. 1 CFU), some samples will by chance not be inoculated. Thus, with low mean inocula and accordingly low overall success rates, it is more difficult to estimate whether a negative test result is due to a weakness of the methodology or due to random absence of microorganisms in the inoculation suspension. Consequently, lower overall success rates result in a lower test power.

In the present study, we applied low-level inocula of 1–5 CFU for 16 different microbial strains. The use of such low inocula inevitably leads to different overall success rates for the different microorganisms represented in the validation. Because the overall success rate for a specific strain remains unknown until the actual validation experiments are performed, it is not possible to give more than an estimate for the required sample size prior to generation of the validation data. This initial sample size subsequently may have to be increased if indicated by the obtained experimental results.

In general, qualitative statistical tests demand for a high number of replicates to reach adequate test power. Such high numbers of replicates can well be accomplished when allowed to pool results obtained for several microbial strains. We have made the experience, however, that evaluations for single microbial strains are often required and sole reliance on pooled data not accepted. Because we envisioned using 16 different microbial strains for demonstration of specificity alone, the overall sample size of the study would have been overwhelming and not justified for our intended application.[4] Therefore, we developed a modified test power procedure, which acted as a tool to decide whether or not sample size should be increased for an individual microbial strain in a particular experiment. To that end, we chose 14 replicates as starting sample size for strain-specific evaluations, which was the lowest reasonable value derived from simulations

4 We believe that the use of a broad microbial spectrum is one of the most important aspects when validating an alternative microbiological method. Concentration on only a few strains may permit to work with higher sample sizes per strain, but we believe that this diminishes the overall weight of the study. Because of the high product volume tested with the Rapid MET as compared to the current method, the risk of diminished ability to evaluate product quality was in general considered low.

and applying our modified test power procedure, which is summarized in the following paragraphs. If indicated by the obtained experimental data, the sample size was increased based on the following definitions:

If the numerical recovery of the rapid method was superior or equivalent to the numerical recovery of the compendial reference method (e.g. both methods detected microbial growth in 12 out of 14 replicates), sample size was not increased.

If the numerical recovery of the rapid method was lower than the numerical recovery of the reference method (e.g. the rapid method detected microbial growth in 12 out of 14 replicates, but the reference method detected growth in 13 out of 14 replicates), our modified test power calculation was performed. If indicated (calculated test power < 0.8), sample size was doubled.

We saw justification to use the lowest value (14 replicates) because of the closeness between the rapid and the compendial method and the fact that the 1 g product required by the Rapid MET often exceeds the product amount tested with the compendial MET. The modified test power calculation took into account the largest acceptable difference of 30% as well as the generated data. To that end, the success rate of the reference method was used as first proportion and 70% of that success rate as second proportion. Furthermore, the confidence level was adjusted according to the generated *p*-value. This, of course, does not represent a formal test power calculation in a strict statistical sense, but rather was used as a tool to decide whether additional data should be generated for providing evidence that the requirement of at least 70% recovery of the rapid method compared to the reference method was fulfilled, while keeping overall sample size at a manageable level.[5] In applications regarded as more critical, the initial sample size could be increased in order to further lower the probability of random sampling error (in case of a sterility test, the authors would, for instance, recommend to use at least 30 replicates per microbial strain as starting sample size).

Table 14.2 shows examples of modified test power calculations with different hypothetical data. In these examples, Method A represents the compendial

5 Our approach based on probability calculations. Briefly, the assumption was made that the compendial reference method has 100% recovery and negative test results are only due to random spiking with sterile inoculation suspension. Probabilistic evaluations for different experimental outcomes were performed, assuming that the alternative method had a recovery of 70% of the compendial reference method. Probabilities for all experimental outcomes which would not trigger an increase in sample size with our definitions outlined above were summed up, resulting in a probability which can also be interpreted as an experimental power. Therefore, our evaluation was rather based on the numerical values. When assuming Poisson-distributed microorganisms, our simulations indicated that the probability to not detect a recovery of less than 70% for an individual strain with a mean microbial count of 1.5 CFU was ~20%, which can be interpreted as a test power of 0.8 as suggested by USP <1010> (for higher microbial numbers, our detection probability increased and was more than 0.9 for mean microbial inocula higher than 2.5 CFU). We considered this approach appropriate for our purpose.

Table 14.2 Examples of our modified test power calculation based on hypothetical data.

Method A +/− (success rate)	Method B +/− (success rate)	*p*-Value Fisher's exact test	Minimal success rate for Method B to fulfill 70% recovery of Method A	Test power	Evaluation
13/1 (0.93)	9/5 (0.64)	0.16	0.65 (0.93 × 0.7)	0.45	Increase sample size
26/2 (0.93)	18/10 (0.64)	0.02	0.65 (0.93 × 0.7)	Not necessary since significant difference	Significant difference
13/1 (0.93)	10/4 (0.71)	0.33	0.65 (0.93 × 0.7)	0.71	Increase sample size
26/2 (0.93)	20/8 (0.71)	0.08	0.65 (0.93 × 0.7)	0.72	Increase sample size
39/2 (0.93)	30/12 (0.71)	0.02	0.65 (0.93 × 0.7)	Not necessary since significant difference	Significant difference
13/1 (0.93)	11/3 (0.79)	0.6	0.65 (0.93 × 0.7)	0.78	Increase sample size
26/2 (0.93)	22/6 (0.79)	0.25	0.65 (0.93 × 0.7)	0.89	Sufficient test power
13/1 (0.93)	12/2 (0.86)	1	0.65 (0.93 × 0.7)	0.88	Sufficient test power
7/7 (0.5)	6/8 (0.43)	1	0.35 (0.5 × 0.7)	0.67	Increase sample size
14/14 (0.5)	12/16 (0.43)	0.79	0.35 (0.5 × 0.7)	0.83	Sufficient test power
5/9 (0.36)	3/11 (0.21)	0.68	0.25 (0.36 × 0.7)	0.54	Increase sample size
10/18 (0.36)	6/22 (0.21)	0.38	0.25 (0.36 × 0.7)	0.39	Increase sample size
20/36 (0.36)	12/44 (0.21)	0.14	0.25 (0.36 × 0.7)	0.34	Increase sample size
35/63 (0.36)	21/77 (0.21)	0.04	0.25 (0.36 × 0.7)	Not necessary since significant difference	Significant difference

For all calculations which took into consideration data of more than only one microbial species, classical *post-hoc* test power calculation was applied.

reference method, whereas Method B represents an alternative method. The largest acceptable difference was defined as 30%. If the test power was not sufficient, sample size was doubled and the ratio of positive and negative results was kept constant (e.g. 26 instead of 13 positive test results) in order to simplify interpretation of the examples.

14.2.2.3 Equivalence Tests

For some validation parameters demonstration equivalence instead of absence of a statistically significant difference had to be provided. While statistical tests like the Chi-Square test evaluate whether a significant difference between different data sets exists, statistical equivalence tests demonstrate the ability of the method of interest to operate within a predefined equivalence boundary. In our case, this equivalence boundary was one-sided (we did not mind if the Rapid MET performed superior to the compendial method), and it was defined that the Rapid MET had to reach at least 70% recovery of the compendial reference method (USP <1227>). We used Fisher's exact test modified for demonstrating one-sided equivalence of success rates (as published by Rasch *et al.* (1998)) for such purposes. The null hypothesis was that the methods did not perform equivalent. The null hypothesis was rejected if the *p*-value of Fisher's exact test modified for one-sided equivalence of success rates was lower than 0.05 (since the test was performed at a confidence level of 95%). Rejection of the null hypothesis meant acceptance of the alternative hypothesis and thus that the methods performed equivalent regarding the boundary of 70%.

14.2.3 Experimental Conditions for Validation Experiments

Unless justified in the text, the general test conditions for the validation were as detailed in Sections 14.2.3.1 and 14.2.3.2.

14.2.3.1 Rapid MET

A low inoculum of the microorganism of interest contained in 10 ml buffer was spiked into glass bottles containing 90 ml liquid nutrient medium (TSB supplemented with 4% polysorbate 80 + 0.5% soy lecithin), ~10 g sterilized glass beads of ~1 mm diameter, and a sterile magnetic stir bar. The bottles were incubated at 30–35 °C (if the microorganism of interest was a bacterium, *Candida albicans* or *Aspergillus brasiliensis*) or at 20–25 °C (if it was a yeast or mold). The incubation time was not more than 72 hours. Following incubation, samples were treated for 20 minutes on a magnetic stirrer. Subsequently, samples were tested for the presence/absence of microorganisms with the Celsis Advance (using the AKuScreen kit).

14.2.3.2 Compendial Method

A low inoculum of the microorganisms of interest contained in 1 ml 0.9% NaCl solution was transferred into a Petri dish and covered with ~20 ml TSA (if the microorganism of interest was a bacterium, *C. albicans* or *A. brasiliensis*) or SDA (if it was a yeast or mold). SDA plates were incubated at 20–25 °C for at least 7 days, TSA plates at 30–35 °C for at least 5 days. Microbial growth was visually evaluated by a qualified analyst.

14.2.4 Method Validation Results

14.2.4.1 Robustness

Robustness describes the reliability of the method in routine use. The application of "small but deliberate variations in method parameters" must not lead to significantly different results.

Both Ph. Eur. 5.1.6 and USP <1223> state that Robustness/Ruggedness determination is best suited to be demonstrated by the supplier of the method. Robustness was shown by the supplier of the rapid microbiological method and submitted in a Drug Master File which was accepted by the FDA in May 2010. A selection of robustness parameters covered by the supplier was different reagent reconstitution volumes, reagent reconstitution times, reagent temperatures, sample volumes, reagent injection volumes, and instrument temperatures. Results by the supplier on robustness parameters were reviewed and two additional robustness parameters were identified and therefore included in the validation. These two additional robustness parameters were the incubation time and the length of the glass bead treatment.

Penicillium sp. and *Escherichia coli* were chosen as microbial representatives for this validation aspect. Stressed *Penicillium* sp. represented the worst-case microorganism regarding detection with the Celsis Advance due to its slow growth rate and tendency to form dense aggregates; *E. coli* was chosen due to its exceptionally fast growth rate. One to five CFU of the test strains were inoculated. Results were evaluated using Fisher's exact test at a confidence level of 95%. The results obtained after 72 hours incubation (representing the reference incubation time) were compared to results obtained after 66 and 120 hours incubation, respectively. No significant differences were observed and the test power criterion was passed (Figure 14.1). By demonstrating that also 66 and 120 hours of incubation did not lead to significantly different results, appropriateness of the target incubation time of 72 hours was further supported.

The glass bead treatment which assists in dispersion of aggregating microorganisms (e.g. *Penicillium* sp.) was developed in preliminary studies to the method validation, which indicated that a treatment time of 15 minutes is effective. This duration may, however, also be subject to small variations. The results obtained after 15 minutes glass bead treatment (representing the reference glass bead treatment time) were compared to the results obtained after 10 and 20 minutes glass bead treatment, respectively (Figure 14.1c). Because *E. coli* does not form aggregates, only *Penicillium* sp. was used for that validation aspect. For the comparison of 15 minutes treatment versus 20 minutes treatment, no significant differences were detectable and our test power

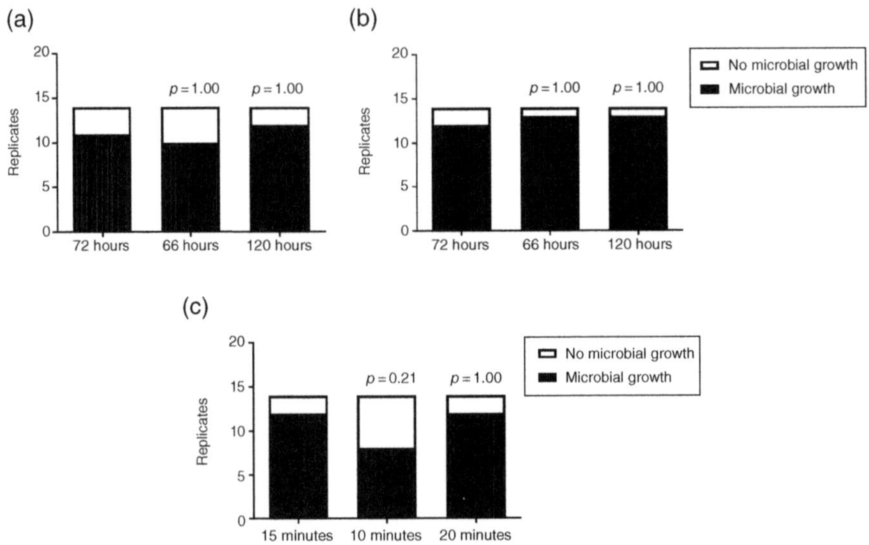

Figure 14.1 Robustness toward different incubation and glass bead treatment times. Robustness toward different incubation times is shown for *Penicillium* sp. (a) and *Escherichia coli* (b). (c) Robustness toward different glass bead treatment times for *Penicillium* sp.

criterion was passed. However, regarding the comparison of 15 minutes treatment versus 10 minutes treatment, the difference was not yet significant, but our test power criterion was not passed, indicating that additional data should be generated. In the light of the obtained data, we refrained from increasing sample size, but concluded that 10 minutes treatment may not be sufficient. Therefore, we decided that duration of the glass bead treatment should be increased from 15 to 20 minutes for routine use as well as for the remaining validation experiments. By performing the glass bead treatment for 20 minutes, it is ensured that minor variations from the target treatment time do not have an impact, since even 15 minutes treatment was shown to be sufficient. Longer treatment times do not have an adverse effect – the treatment is by far not harsh enough to lead to destruction of single microbial cells.

Robustness toward different incubation times was therefore successfully demonstrated for fast-growing and slow-growing aggregate-forming microorganisms. For routine testing, an incubation time of minimum 72 hours will be applied for the Rapid MET on basis of the Celsis Advance. Robustness toward a variation of minus 6 hours and plus 48 hours was demonstrated.

14.2.4.2 Ruggedness

Ruggedness describes the reproducibility of test results through analysis of samples under different routine circumstances (alteration of analysis parameters, which represent unavoidable changes). Ruggedness is normally expressed as the lack of influence of operational and environmental variables of the microbiological method on the test results. A selection of ruggedness parameters covered by the supplier was different instruments, operators, reagent lots, and cuvette sizes. These parameters were considered adequate and no additional experiments were performed.

14.2.4.3 Repeatability

Repeatability describes the reproducibility of test results through analysis of samples under routine circumstances at different daytimes and on different days (using the same analyst with the same equipment). Due to its slow growth rate and the tendency to form dense aggregates, *Penicillium* sp. represented the worst-case microorganism regarding repeatability. Thus, by demonstrating repeatability for detection of *Penicillium* sp., repeatable detection of fast-growing and/or non-aggregate-forming microorganisms was warranted.

1–5 CFU inocula of *Penicillium* sp. were repeatedly applied on different days and daytimes and several such data sets evaluated for significant differences with Chi-Square test (Figure 14.2). In total, four test runs (two in the morning and two in the afternoon of four different days) each consisting of 20 replicates were performed. Repeatability was acceptable despite a challenging test setup (low numbers of stressed, slow-growing aggregate former). Furthermore, the ability of the assay to deliver acceptable results when repeatedly performed under varying circumstances is shown by each experiment performed during validation.

14.2.4.4 Specificity

The specificity of a method was defined as the potential to detect a broad range of microorganisms, which for a growth-based RMM mainly depends on the

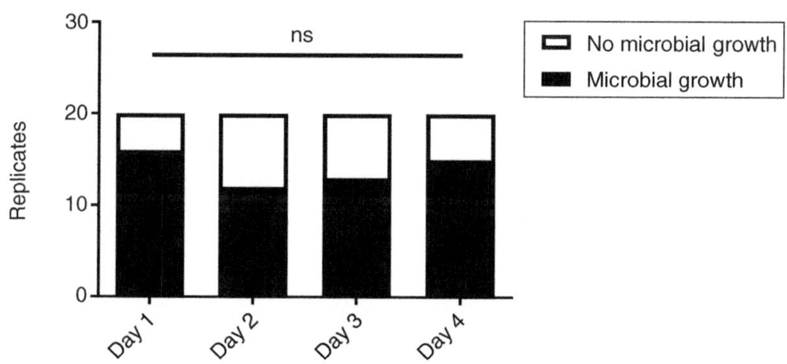

Figure 14.2 Repeatability on different days and daytimes. Results for repeatability on different days and daytimes with *Penicillium* sp. No significant differences were detectable with Chi-Square test (ns, no significant difference).

Table 14.3 Strain selection used for validation of specificity.

Yeast/Mold	Sporulating bacteria	Gram-positive bacteria	Enterobacteria (Gram-negative)	Waterborne Gram-negative bacteria
Candida albicans	*Bacillus subtilis*	*Staphylococcus aureus*	*Escherichia coli*	*Pseudomonas aeruginosa*
Aspergillus brasiliensis	*Bacillus clausii*	*Staphylococcus epidermidis*	*Salmonella abony*	*Burkholderia cepacia*
Penicillium sp.	*Bacillus licheniformis*	*Staphylococcus warneri*		*Pseudomonas stutzeri*
		Kocuria rhizophila		*Stenotrophomonas maltophilia*

fertility of the nutrient medium used. In our validation, specificity was shown by detection of 16 different microorganisms, including Gram-negative rods, Gram-positive sporulating bacteria, Gram-positive cocci, yeasts, and molds. All specified microorganisms mentioned in the harmonized method for Microbiological Examination of Non-sterile Drug Products as described in Ph. Eur. 2.6.13 and USP <62> were included, except from *Clostridium sporogenes* (anaerobic microorganism that would not grow under the defined test conditions). Furthermore, stressed in-house isolates were used. All test microorganisms were spiked with a low inoculum (1–5 CFU) and are summarized in Table 14.3. The 14 replicates per strain were generated in two independent test runs each.

Data for each individual strain was evaluated using Fisher's exact test at a confidence level of 95%. Pooled data from all strains were evaluated for a statistical difference using Chi-Square test at a confidence level of 95%. Furthermore, pooled data for all strains were evaluated for statistical equivalence using Fisher's exact test modified for demonstration of equivalence of one-tailed success rates. If the test power was below 0.8 for an individual strain, additional data were retrieved (Tables 14.4 and 14.5).

14.2.4.5 Limit of Detection

The LOD is defined as the lowest number of microorganisms that can be detected under the stated experimental conditions. Since for the Specificity validation a low inoculum of 1–5 CFU was used, a suitable detection limit as compared to the compendial MET of the Ph. Eur. 2.6.12 and USP <61> was already demonstrated with that series of experiments (see Tables 14.4 and 14.5). Moreover, in the compendial MET, for typical non-sterile and nonfilterable dosage forms such as nonaqueous preparations for oral use, often a diluted product amount may be tested (e.g. 100 mg leading to a maximum detection level of <10 CFU/g).

Table 14.4 Strain-specific specificity results.

Microorganism	Result Rapid MET Growth/No growth	Result Compendial MET Growth/No growth	*p*-Value Fisher's exact test	Test power	Interpretation of Fisher's exact test
Pseudomonas aeruginosa ATCC 9027	12/2	11/3	1	N/A[a]	No significant difference
Staphylococcus aureus ATCC 6538	9/5	12/2	0.38	0.58	No significant difference. Generation of additional data indicated
Additional data for *S. aureus*	19/9	23/5	0.36	0.80	No significant difference
Bacillus subtilis ATCC 6633	13/1	10/4	0.33	N/A[a]	No significant difference
Escherichia coli ATCC 8739	14/0	14/0	No value[b]	N/A[a]	No *p*-value can be calculated. Data suggest sufficient recovery
Burkholderia cepacia ATCC 25416	13/1	12/2	1	N/A[a]	No significant difference
Aspergillus brasiliensis ATCC 16404	14/0	14/0	No value[b]	N/A[a]	No *p*-value can be calculated. Data suggest sufficient recovery
Candida albicans ATCC 10231	14/0	11/3	0.22	N/A[a]	No significant difference
Salmonella abony NCTC 6017	10/4	14/0	0.098	0.43	No significant difference. Generation of additional data indicated
Additional data for *S. abony*	23/5	25/3	0.70	0.95	No significant difference
Staphylococcus epidermidis stressed in-house isolate	10/4	11/3	1	0.81	No significant difference

(Continued)

Table 14.4 (Continued)

Microorganism	Result Rapid MET Growth/No growth	Result Compendial MET Growth/No growth	*p*-Value Fisher's exact test	Test power	Interpretation of Fisher's exact test
Kocuria rhizophila stressed in-house isolate	12/2	14/0	0.48	0.84	No significant difference
Bacillus clausii stressed in-house isolate	9/5	11/3	0.68	0.68	No significant difference. Generation of additional data indicated
Additional data for *B. clausii*	19/9	21/7	0.77	0.86	No significant difference
Penicillium sp. stressed in-house isolate	8/6	8/6	1	N/A[a]	No significant difference
Stenotrophomonas maltophilia stressed in-house isolate	11/3	5/9	0.054	N/A[a]	No significant difference
Staphylococcus warneri stressed in-house isolate	14/0	14/0	No value[b]	N/A[a]	No *p*-value can be calculated. Data suggest sufficient recovery
Bacillus licheniformis stressed in-house isolate	11/3	12/2	1	0.85	No significant difference
Pseudomonas stutzeri stressed in-house isolate	5/9	3/11	0.68	N/A[a]	No significant difference

[a] Rapid method numerically equivalent or superior, therefore no test power calculation.
[b] *p*-Value cannot be calculated if both methods recovered microbial growth in all replicates (no 2×2 contingency table can be formed).

Table 14.5 Specificity results pooled data from all strains.

Result Rapid MET Growth/No growth	Result Traditional MET Growth/No growth	*p*-Value Chi-Square test	Test power Chi-Square test	*p*-Value Fisher's exact test modified for demonstration of equivalence of one-tailed success rates	Interpretation of results
212/54	212/54	1	1	3.11×10^{-10}	No significant difference. Statistical indication of equivalence.

In conclusion, for each individual strain included in the validation a recovery of >70% was successfully demonstrated. Taking into account the data generated for all strains, no statistically significant difference regarding recovery was detectable with Chi-Square test. Furthermore, through Fisher's exact test modified for demonstration of equivalence of one-tailed success rates, statistical proof of equivalence against a 70% boundary was provided.

Based on the acceptance criteria defined in Ph. Eur. 5.1.4 and USP <1111>, complete absence of specified microorganisms may be required in 1 or 10 g product depending on the product's route of administration (which also is the product amount used for Rapid MET). Therefore, for the LOD study all specified microorganisms mentioned for the growth promotion tests in the Microbiological Examination of Non-sterile Drug Products of the Ph. Eur. 2.6.13, Ph. Eur. 5.1.4, USP <62> and USP <1111> were included except from *C. sporogenes* (anaerobic microorganism that would not grow under the defined test conditions) and *Salmonella* (absence required in 10 g product and therefore most likely out of scope for our Rapid MET application). *Penicillium* sp. was also included in the LOD validation since it represents a worst-case microorganism in terms of detection (low growth rate and strong aggregation). In conclusion, the test strains for LOD validation were *Pseudomonas aeruginosa, Staphylococcus aureus, E. coli, C. albicans,* and stressed *Penicillium* sp. These test strains were serially diluted to extinction (~50 CFU, 5 CFU, 0.5 CFU, and 0.05 CFU per replicate). The Rapid MET was used according to the previously described method. For the compendial method, the test was performed according to the Ph. Eur. 2.6.13 and USP <62> and consisted of enrichment in liquid media followed by selection through selective media. For *Penicillium* sp., 1 ml of test suspension was transferred into a Petri dish and ~20 ml SDA was poured. The SDA plate was then incubated for at least seven days.

Two independent test runs with 10 replicates per dilution and microorganism were performed. From the pattern of replicates, positive or negative for microbial growth, the 95% confidence interval of the MPN was calculated for each test run with each microorganism using a MPN-table. The 95% confidence interval of the MPN of both methods was evaluated for overlapping. This approach to determine the LOD is referenced in USP <1223>. The ten-replicate MPN-table was obtained from FDA Bacteriological Analytical Manual Appendix 2 (Blodgett 2010).

No significant differences were detectable regarding the LOD (Figure 14.3). For additional evidence, the MPN/g results were plotted and statistically compared using paired *t*-test. No statistically significant difference between the

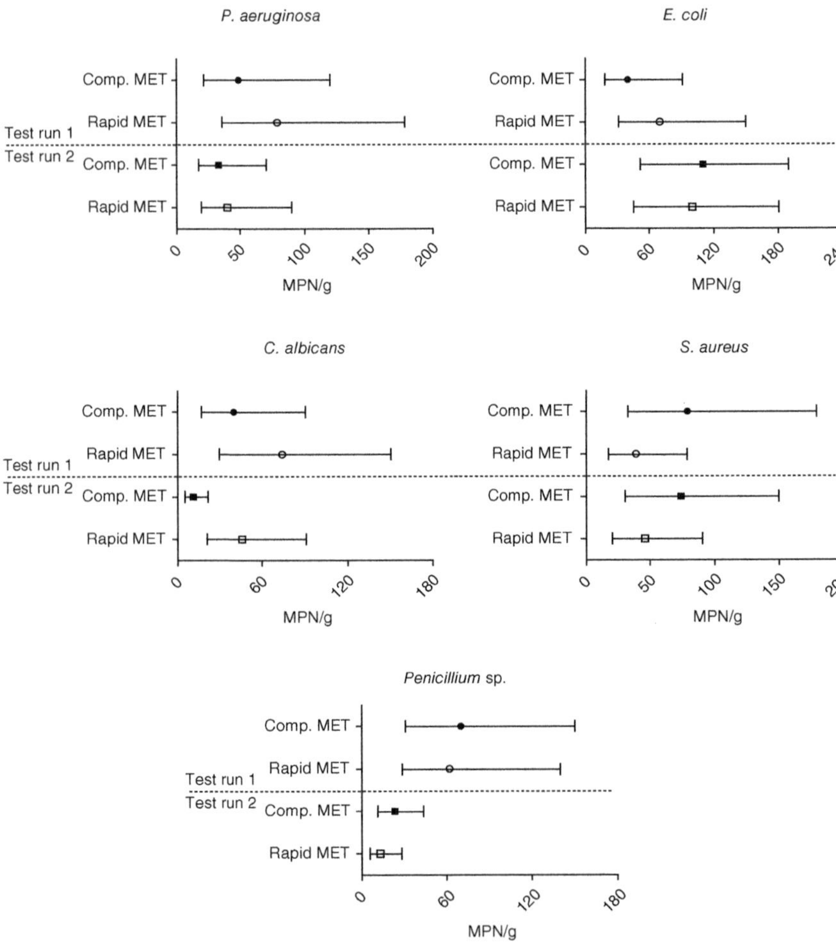

Figure 14.3 Limit of detection. Results of limit of detection compendial method versus Rapid MET. MPN/g as well as the 95% confidence interval of the MPN/g value is indicated. When the 95% confidence intervals overlap there are no significantly different detection limits.

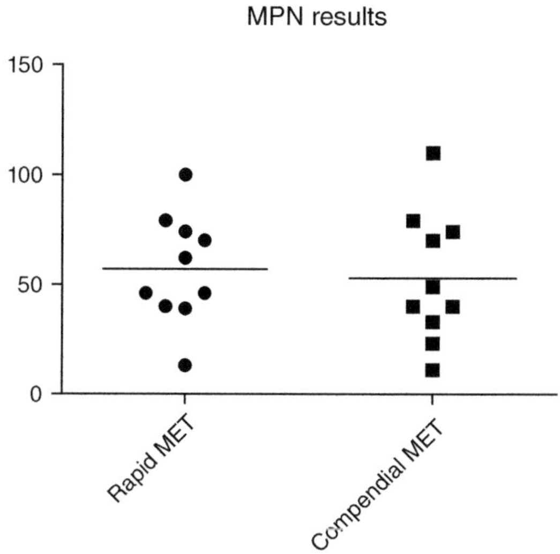

Figure 14.4 Comparison of the obtained MPN/g values. Plotting of the MPN/g results obtained through the detection limit validation. Individual MPN/g values as well as the means are shown.

Rapid MET and the compendial method was detectable and the mean of the obtained MPN/g values was nearly identical for both methods (Figure 14.4).

14.2.4.6 Accuracy and Precision (According to Ph. Eur. 5.1.6)

We decided to insert this validation parameter since the Rapid MET is also planned for registration through European authorities, where according to Ph. Eur. 5.1.6 Accuracy and Precision is also required for qualitative methods. Accuracy and Precision is expressed as the relative rate of false-positive and false-negative results between the rapid method and the compendial method using a standardized, low-level inoculum. The Rapid MET was further challenged for false-negative results by using the microorganisms *Penicillium* sp. and *Bacillus clausii*. *Penicillium* sp. represents the worst-case microorganism regarding detection with the Celsis Advance due to its slow growth rate and tendency to form dense aggregates. *Bacillus clausii* was used as endospore. Thus, in order to multiply and become detectable, *B. clausii* first had to enter into an active state, which strongly depends on the used nutrient medium and incubation conditions. In order to assess the rate of false-positive test results, negative controls were used in which no microorganisms were inoculated.

14.2.4.6.1 Definition of False Negative A false-negative result arises when a test result is negative although the sample has been inoculated. It is of note, however, that some samples would have most probably by chance remained sterile due to the low inoculum (no microorganisms present in the microbial suspension used for inoculation). In order to challenge for false-negative results, samples were inoculated with 1–5 CFU of the test organisms.

14.2.4.6.2 ***Definition of False Positive*** A false-positive result arises when microbial growth is detected in a negative control. This may either happen because of a microbial contamination or because of an artifact (e.g. ATP-contamination). In order to challenge for false-positive results, samples were inoculated with sterile buffer.

No statistically significant differences between the compendial method and Rapid MET for false-positive and false-negative results were detectable using Fisher's exact test and our test power criterion was passed; thus, comparable accuracy and precision was successfully demonstrated and the rates of false-positive and false-negative results were similar. Exemplary data for *Penicillium* sp. and sterile buffer are shown in Table 14.6. No false-positive results due to ATP contamination were observed.

14.2.4.7 Equivalence in Routine Operation

The objective of this test was to demonstrate the equivalence in routine use of the Rapid MET with the Celsis Advance system as compared to the compendial method. Since the Rapid MET with Celsis Advance system is a qualitative method, only the capacity to detect microbial growth (=presence/absence of microorganisms) was evaluated. Thus, the compendial MET was counted as positive for microbial growth if either a MET (TAMC and TYMC) or a test for specified microorganisms was positive for microbial growth. The experimental setup included samples for which rather few positive test results were expected as well as samples for which several positive test results were expected. Five routine-relevant products were selected, consisting of hard-gelatin capsules, excipients, and film-coated tablets. Thirty random samples from different batches per product were analyzed in parallel according to the compendial

Table 14.6 Example statistical significant tests comparing the rate of false-positive and false-negative results of the compendial versus the rapid MET using *Penicillium* sp. as test strain.

	Rapid MET Growth/No growth	Compendial MET Growth/No growth	*p*-Value Fisher's exact test
False-negative rate	6/14	7/13	1
False-positive rate	0/20	0/20	NA (both methods have no positive results)

Ratios were statistically compared using the Fisher's exact test.

Table 14.7 Equivalence in routine operation.

	Rapid MET Growth/No growth	Compendial MET Growth/No growth
Product 1 (hard-gelatin capsule)	18/12	1/29
Product 2 (excipient)	0/30	1/29
Product 3 (excipient)	30/0	30/0
Product 4 (film-coated tablet)	0/30	2/28
Product 5 (film-coated tablet)	20/10	1/29
Total	68/82	35/125
p-Value Fisher's exact test modified for demonstrating one-tailed equivalence of success rates	$<10^{-15}$	
Interpretation Fisher's exact test modified for demonstrating one-tailed equivalence of success rates	Statistical indication of equivalence	

method described in Ph. Eur. 2.6.12, 2.6.13, USP <61>, <62>, and according to the Rapid MET routine test setup described Section 14.1.

The Rapid MET detected microbial contamination more often than the compendial MET (Table 14.7). This finding is not surprising considering that the rapid method may test a higher amount of product compared to compendial MET since dilution of the product to <1 g is not valid. The *p*-value of Fisher's exact test modified for demonstrating one-tailed equivalence of success rates was clearly below 0.05, indicating statistical equivalence regarding the boundary of 70% recovery.

14.3 Suitability Test (Product-Specific Method Validation)

Additional testing has to be performed for each drug product/API/excipient which is in scope of the Rapid MET in order to demonstrate that the method is also suitable for that product. The suitability study includes the sample effects study and the suitability of the test method.

14.3.1 Sample Effects Study

The sample effects study determines if the drug product/API/excipient itself interferes with the Celsis AKuScreen assay, e.g. by adding turbidity to the sample which impedes the detection of the bioluminescence reaction or by being bioluminescent itself. The procedure for determining sample effects was based on the supplier's recommendation. Briefly, RLU values generated by the product are measured to assess bioluminescent background, and a standardized amount of ATP is measured with or without product to evaluate whether the presence of product significantly diminishes the signal. We regarded a reduction of the ATP signal through the presence of product by more than 30% as significant.

Furthermore, we used the RLU background of the product to determine a product-specific positive/negative discrimination threshold. The results obtained through measurement with the Celsis Advance are RLU. The RLU values *per se* do not have significance for the intended application, since the system is solely used as a presence/absence method. However, RLU values have to be defined above which a sample is to be considered positive for microbial growth. The approach used by the system supplier is to consider a sample positive for microbial growth if the RLU values are several times higher than the negative control. We applied this approach for validation experiments which did not involve the product – in these cases, a sample was considered positive for microbial growth if RLU values were 10 times higher than the negative control.

However, we did not consider this approach optimal when testing products. First, different products have very different background RLU levels (e.g. capsules in general create stronger background than tablets). Second, the RLU levels of the negative control are prone to a certain extent of variation. In our case, the negative control would be nutrient medium, and throughout the validation we saw that RLU values of nutrient medium were subject to batch-to-batch variations. With maximum 800 RLU, this variation was rather low in our case. However, if a multiple of the RLU value obtained for the negative control is used for discriminating whether a test sample is positive or negative for microbial growth, variability of this discrimination threshold is much higher. Thus, there would be a risk that product background may be above the discrimination threshold if RLU values of the negative control would be low; or below the discrimination threshold if RLU values of the negative control would be high.

For the reasons presented above, we decided to introduce product-specific discrimination thresholds. These are determined in the product-specific method suitability through the sample effects study. With these experiments, luminescent background generation or masking of ATP bioluminescence is assessed. The product-specific discrimination threshold is defined based on the RLU background of the product of interest according to the following workflow.

Three product batches are tested to determine the product RLU background. The highest obtained result is used for the workflow shown in Figure 14.5 in order to mitigate the risk of false-positive results. If further product dilution is indicated, this is achieved through increase of the nutrient medium volume

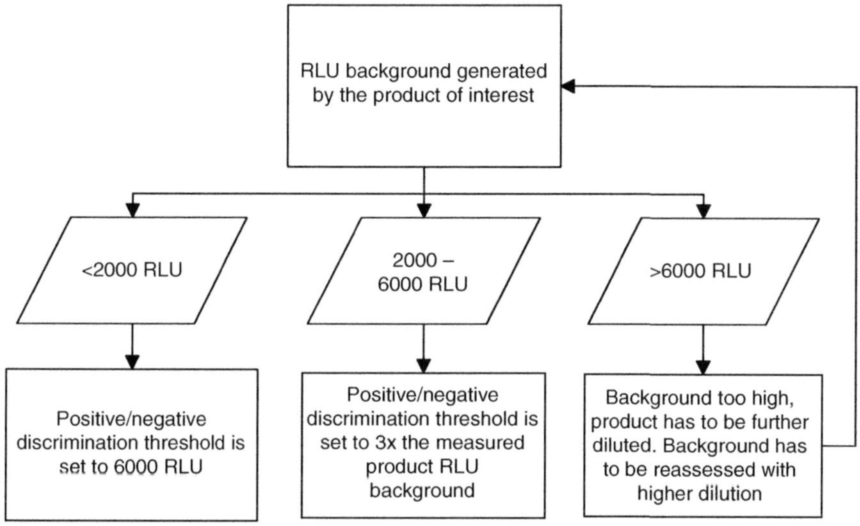

Figure 14.5 Definition of the product-specific positive/negative discrimination threshold.

used for incubation, since the Rapid MET does not allow testing of a smaller product amount than 1 g. The RLU values of even low-level contaminations of slow-growing microorganisms are clearly above the maximum discrimination threshold of 18,000 RLU (3 × 6,000). For example, the average RLU value of positive samples of the LOD 0.5 CFU inoculum of *Penicillium* sp. was 363,031 RLU. Thus, there is no risk that low-level contaminations would lead to false-negative results, due to a too high product-specific discrimination threshold.

The above-presented sample effects concept was already applied for a film-coated tablet final dosage form which is intended as pilot drug product for the registration of the Rapid MET. In that study, we successfully demonstrated that the film-coated tablet did not mask the bioluminescence signal and the average bioluminescent background was 218 RLU. Therefore, samples would be regarded as being positive for microbial growth if more than 6,000 RLU are measured with the Celsis Advance, and the following workflow would then be applied (Figure 14.6).

14.3.2 Suitability of the Test Method

Suitability is shown through detection of a range of microorganisms in the presence of the product of interest, therefore demonstrating acceptable microbial recovery. Ten grams of the product is dissolved in 90 ml of the dilution buffer and stirred. Test microorganisms as the ones requested for the compendial methods (*E. coli* ATCC 8739, *P. aeruginosa* ATCC 9027, *S. aureus* ATCC 6538, *Bacillus subtilis* ATCC 6633, *A. brasiliensis* ATCC 16404, and *C. albicans* ATCC

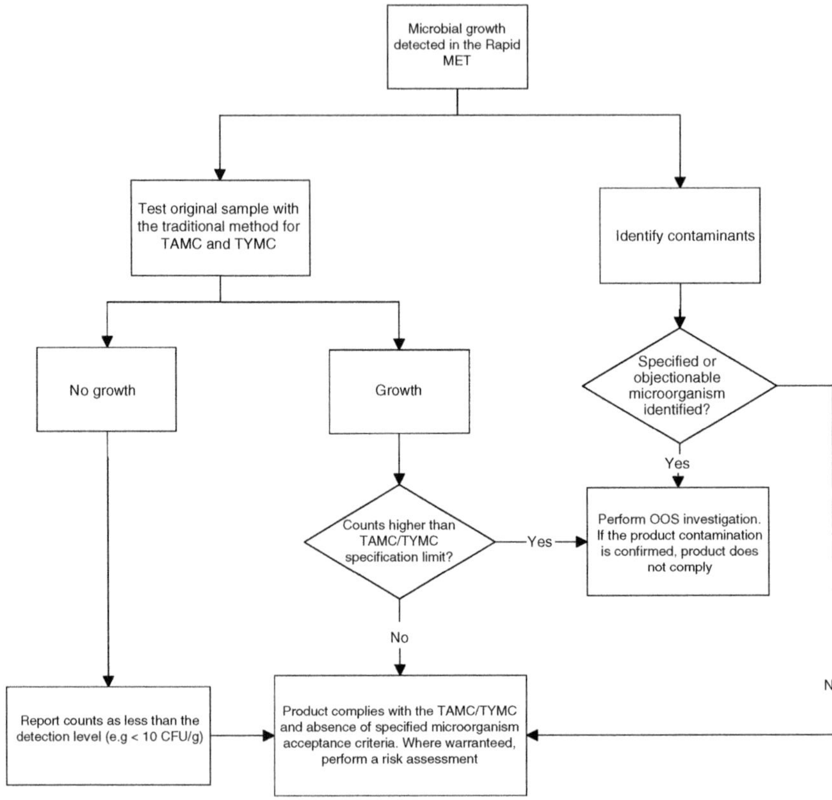

Figure 14.6 Schematic overview of result evaluation with the Rapid MET if growth occurs (OOS, out of specification).

10231) are inoculated individually in the product dilution. Then, 10 ml of the product dilution (corresponding to 1 g product) harboring <100 CFU of the test microorganism are added to glass bottles containing 90 ml TSB + 4% polysorbate 80 + 0.5% soy lecithin, glass beads, and a magnetic stir bar. The bottles are incubated for not more than 72 hours either at 20–25 °C or at 30–35 °C, depending on the type of microorganism. Following incubation, samples are treated for 20 minutes on a magnetic stirrer. Subsequently, samples are tested for the presence/absence of microorganisms with the Celsis Advance (using the AKuScreen kit). Three independent test runs are performed and samples are measured in duplicates. Acceptance criteria are that all test microorganisms are positively detected with the Celsis Advance system and the test microorganism is confirmed with an identification of the recovered microorganism.

14.4 Discussion

The Rapid MET was validated according to USP <1223>, Ph. Eur. 5.1.6, and PDA Technical Report No. 33. Validation was comprised of the validation parameters robustness, ruggedness, repeatability, specificity, LOD, accuracy and precision, and equivalence in routine operation. For the validation, a combination of pharmacopoeial ATCC strains as well as a broad selection of in-house isolates was used. In-house isolates were used in stressed state. Results of the rapid method were statistically compared to the compendial method regarding the USP <1227> acceptance criterion of ≥70% recovery. We developed a modified test power calculation as a tool to confirm the appropriateness of the used sample size to detect such a difference. Furthermore, equivalence of the rapid method as compared to the compendial method was demonstrated in a statistically verified manner. The Rapid MET on basis of the Celsis Advance system (using the AkuScreen reagent kit) was therefore successfully validated as an alternative method to the compendial test for microbiological examination of non-sterile products.

Whereas one can expect that the overall microbial spectrum isolated may slightly differ between two test methods due to factors such as, for instance, the use of different growth media or incubation conditions, the extent of this difference may be estimated through a thorough validation comparing the overall equivalence of the methods based on predefined parameters. Adequate recovery of a range of relevant microorganisms provides solid evidence that the method of interest represents a suitable alternative to the compendial method. Furthermore, it should be verified that the most critical microorganisms (e.g. specified or objectionable microorganisms) that can be isolated with the compendial method should also be recovered in the rapid microbiological method for product release testing. Statistical approaches can be applied for definition of sample sizes and testing of hypotheses with a high probability of correctness. While test power calculations are rather straightforward for quantitative methods, treatment of qualitative data is more difficult. Furthermore, sample sizes for strain-specific qualitative evaluations should still remain at a practical level. We applied a modified test power approach to decide whether sample size should be increased based on our obtained validation data.

Calculation of test power often requires specialized software, which may not be easily accessible to every firm. Furthermore, advanced statistical knowledge may be required. PDA Technical Report No. 33 provides a table which helps microbiologists not familiar to test power calculations by suggesting sample sizes for quantitative experiments; to the knowledge of the authors similar guidance is not available for qualitative experiments. We are of the opinion that our approach to start with a relevant sample size, which may be increased based on the obtained results, represents a reasonable compromise between a still practical sample size and low possibility of experimental bias through random results.

The Rapid MET allowed reduction of the incubation time from 3 to 7 days (TAMC) and 5 to 7 days (TYMC) to 72 hours enabling a faster throughput time of product release testing. The incubation time of 72 hours may seem rather long; however, it confirms our experience that growth-based rapid methods allow to reduce the incubation time approximately by a factor 2–3 compared to the compendial reference method, if a challenging validation approach with stressed isolates is used. Furthermore, our 72 hours also include a generous safety margin. In addition to the reduced incubation time, the main advantage of our use of the Celsis Advance System is that it covers MET and absence of specified microorganisms in one single test, allowing for a significant reduction of the hands-on time and growth media storage. Furthermore, the readout is performed by a validated system and therefore there is reduced challenge on data integrity as for compendial method when the readout is performed by only one person, and there is a possibility for automation and direct data integration into a LIMS system. Internal benchmarking has shown that hands-on time may be reduced by up to 20–30%, depending on the number of tests for specified microorganisms required by the product. With the Rapid MET, 1 g of drug product is always being used by default for testing. Therefore, in case that no microbial growth is detected with the Rapid MET, an additional test for specified microorganisms is not necessary since the absence of microbial growth automatically excludes the presence of any specified microorganisms in 1 g of drug product. The specified microorganisms should, however, be included in the product-specific method suitability study. To provide optimal growth conditions for both bacteria and yeasts/molds, two different incubation temperatures are tested (20–25 and 30–35 °C), although the same medium is used for both tests.

USP <1111> and Ph. Eur. 5.1.4 recommend the absence of bile-tolerant gram-negative bacteria in some products (e.g. inhalants). Bile-tolerant gram-negative bacteria are currently not a well-defined homogenous group but rather defined only on the capacity to grow in bile-salt containing media. If growth occurs in the Rapid MET test, a microbial identification is not sufficient to determine if the contaminant is a representative of that group of specified microorganisms. For this reason, the compendial test for bile-tolerant Gram-negative bacteria has to be performed in parallel. As an alternative, rather than performing the test in parallel by default, a retest with the compendial method for the absence of bile-tolerant Gram-negative bacteria could be performed if growth is observed in the Rapid MET. It is of note that in order to also cover the test for the absence of *Salmonella*, the product-specific suitability test and release testing had to be performed with 10 g product.

Although the Rapid MET with the Celsis Advance will mainly be used for nonfilterable products, testing of filterable products may also be achieved through direct inoculation or filtration of the product on a 0.22 or 0.45 μm filter membrane followed by transfer and incubation of the membrane in the Rapid MET growth medium.

One of the reasons why the Rapid MET with the Celsis Advance was preferred to other existing systems is that the product to be tested can be incubated in a large volume of nutrient medium. Indeed, using a higher volume of nutrient medium can mitigate antimicrobial activity as well as the background of products. The validation work for the Rapid MET was conducted using 90 ml liquid nutrient medium, which now is used as a standard for rapid microbiological examination of non-sterile products. Nevertheless, it was demonstrated that also a nutrient medium volume of 190 ml instead of 90 ml could be used without negative impact on detection limit of the assay (data not shown). If required, even higher volumes may be used, but they would first need to be validated.

One of the main factors affecting microbial recovery and therefore validation of a growth-based alternative method is the nutrient medium. In the Rapid MET with the Celsis Advance, TSB supplemented with 4% polysorbate 80 and 0.5% soy lecithin should be used for routine testing. The use of liquid growth medium is mandatory when performing tests with the Celsis Advance. TSB, as well as its solid equivalent TSA, is a rich nutrient medium offering good growth promotion for a wide variety of microorganisms (Smith *et al.* 1974; MacFaddin 1985). We decided not to use sabouraud dextrose broth for the 20–25 °C incubation temperature, since our method requires the use of products with low bioload; therefore, selection for yeasts/molds from a diverse microbial spectrum is not necessary. Furthermore, TSB can undoubtedly recover a broader microbial spectrum than SDB; thus, we saw it as a more suitable nutrient medium for a presence/absence test. As a matter of fact, TSB incubated at 20–25 °C is the pharmacopoeial incubation condition for detecting yeasts and molds in the sterility test, further indicating that growth-promoting properties for these types of microorganisms should be appropriate. During method validation no evidence for inferior recovery of a slow-growing and stressed mold (*Penicillium* sp.) was observed, and also pharmacopoeial *C. albicans* as well as *A. brasiliensis* strains were adequately recovered. Some mold species, however, have the tendency to form dense aggregates, which may hamper detectability with the Celsis Advance because only a small aliquot of the sample is actually tested for ATP bioluminescence. We have overcome this problem of aggregation by adding the glass bead treatment step, which better homogenizes the microbial cells and therefore ensures that an adequate number is present in the aliquot used for ATP bioluminescence detection.

In case that no microbial growth is detected in the Rapid MET with the Celsis Advance, product may be released for TAMC/TYMC and the absence of specified microorganism in 1 g product. In case that microbial growth is detected in the Rapid MET with the Celsis Advance, the contaminant(s) will be identified. With this approach, it can be assured that also a low-level contamination of specified or objectionable microorganisms would be detected. Furthermore, the original sample of the product would be retested with the compendial MET in order to assess whether the level of contamination is above acceptance criteria (Figure 14.6). This retest is performed only to provide a count estimate to the detected microbial contamination. If the compendial MET would not recover any microorganisms, the initial microbial finding from the Rapid MET would not be invalidated, but reported as being below the detection limit of the compendial MET (e.g. <10 CFU/g). Furthermore, where warranted, adequate risk assessment for a low-level contamination of that microorganism would have to be performed.

Nevertheless, additionally performing the compendial test for enumeration is time-consuming if microbial growth is detected frequently; thus, products which often exhibit microbial growth may be out of scope for this application. For such products, other RMMs allowing for precise enumeration of microorganisms present in nonfilterable products may be more suitable. For instance, the use of an automated, rapid MPN methodology, which allows for testing of a representative amount of product or enumeration of microcolonies within pour-plated nutrient agar, could represent possible solutions for higher bioload products. Alternatively, product solution could be diluted to the specification limit and a presence/absence test be applied. Consequently, if growth is observed, the limit would be considered exceeded. However, the drawbacks of this approach are the need to perform enumeration tests and absence of specified microorganisms tests in parallel, instead of combining both of these tests in one. Likewise, alert or expectation levels are difficult to define and each finding is a potential out of specification result. Finally, trend analysis of microbial bioload would be challenging and the presence of objectionable microorganisms below the specification level could not be evaluated with such an approach.

14.5 Conclusion

The Rapid MET with the Celsis Advance system was successfully demonstrated to be a possible alternative to the compendial method described in Ph. Eur. 2.6.12, 2.6.13, USP <61>, <62>, and JP 4.05/I and 4.05/II for the microbiological examination of non-sterile products. Our statistical evaluation concept allowed for a robust and scientifically sound validation approach. We defined a process for the product-specific method suitability as well as determination of a product-specific threshold, which if exceeded points toward microbial growth. The Rapid MET using the Celsis Advance can be applied to nonfilterable products of good microbiological quality, allowing for a reduced incubation time to 72 hours and a substantial reduction of hands-on time and improvement in data integrity.

Acknowledgments

The authors would like to thank Markus Mordstein and Jennifer Gray for valuable discussions and Melanie Nast, Sabrina Scheper, Laura Benz, and Catharina

Nord for great technical support. We also thank Alexandra Staerk, Marcel Goverde, and Anne Smiejan-Roesti for critical review of the manuscript.

Bibliography

Blodgett, R, 2010, 'Most Probable Number Determination from Serial Dilutions (Internet)', http://www.fda.gov/Food/ScienceResearch/LaboratoryMethods/BacteriologicalAnalyticalManualBAM/ucm109656.htm

Fisher, RA, 1922, 'On the interpretation of Chi-Square from contingency tables, and the calculation of P', *Journal of the Royal Statistical Society*, vol. 85(1), pp. 87–94.

Gordon, O, Gray, JC, Anders, HJ, Staerk, A, Neuhaus, G, Schlaefli, O, 2011, 'Overview of rapid microbiological methods evaluated, validated and implemented for microbiological quality control', *European Pharmaceutical Review*, vol. 16(2), pp. 9–13.

Gray, JC, Staerk, A, Berchtold, M, Hecker, W, Neuhaus, G, Wirth, A, 2010, 'Growth promoting properties of different solid nutrient media evaluated with stressed and unstressed micro-organisms: Prestudy for the validation of a rapid sterility test', *PDA Journal of Pharmaceutical Science and Technology*, vol. 64, pp. 249–263.

MacFaddin, JF, 1985, , *Media for Isolation – Cultivation – Maintenance of Medical Bacteria*, vol. 1, Williams and Willkins, Baltimore, London.

Miller, MJ, 2012a, 'Case study of a new growth-based rapid microbiological method (RMM) that detects the presence of specific organisms and provides an estimation of viable cell count', *American Pharmaceutical Review*, vol. 15(2), pp. 18–25.

Miller, MJ, 2012b, 'The RMM Product Matrix (Internet)', http://rapidmicromethods.com/files/matrix.php

PDA, 2000, *The Evaluation, Validation and Implementation of New Microbiological Testing Methods*, PDA Technical Report No. 33, PDA, Bethesda, MD.

Pearson, K, 1900, 'On the criterion that a given system of deviations from the probable in the case of a correlated system of variables is such that it can be reasonably supposed to have arisen from random sampling', *Philosophical Magazine*, Series 5, vol. 50(302), pp. 157–175.

Ph. Eur. 7th Edition chapter 2.6.12, *Microbiological Examination of Non-sterile Products: Microbial Enumeration Test.*

Ph. Eur. 7th Edition chapter 2.6.13, *Microbiological Examination of Non-sterile Products: Test for Specified Microorganisms.*

Ph. Eur. 7th Edition chapter 5.1.4, *Microbiological Quality of Non-sterile Pharmaceutical Preparations and Substances for Pharmaceutical Use.*

Ph. Eur. 7th Edition chapter 5.1.6, *Alternative Methods for Control of Microbiological Quality.*

Rasch, D, Herrendörfer, G, Bock, J, Victor, N, Guiard, V, 1998, *Verfahrensbibliothek, Versuchsplanung und Auswertung*, Oldenbourg Verlag, Munich.

Smith, RF, Blasi, D, Dayton, SL, 1974, 'Evaluation of media for isolation of yeasts from oral, rectal and burn wound specimens', *Applied Microbiology*, vol. 28(1), pp. 112–116.

USP 34-NF29 chapter <1010>, *Analytical Data – Interpretation and Treatment.*

USP 34-NF29 chapter <1223>, *Validation of Alternative Microbiological Methods.*

USP 35-NF30 chapter <1111>, *Microbiological Examination of Nonsterile Products: Acceptance Criteria for Pharmaceutical Preparations and Substances for Pharmaceutical Use.*

USP 35-NF30 chapter <1227>, *Validation of Microbial Recovery from Pharmacopeial Articles.*

USP 35-NF30 chapter <61>, *Microbiological Examination of Non-sterile Products: Microbial Enumeration Test.*

USP 35-NF30 chapter <62>, *Microbiological Examination of Non-sterile products: Tests for Specified Microorganisms.*

15

An Ex-Regulator's View of the Microbiology QA/QC Functions in the US Pharmaceutical Industries

David Hussong

Eagle Analytical Services, Houston, TX, USA

CONTENTS

15.1 Foods, Drugs, and Quality

A monk in Florence, Italy, first developed standards for pharmaceuticals in Western cultures around 1498. This publication, *The Nouvo Receptario*, became the legally authorized standard for local apothecaries (Bender and Thom 1999). By comparison, the first *United States Pharmacopoeia* was published in 1820. By then, many other countries had already prepared compendia that established standards for the preparation of medicines. In the nineteenth century, drugs were prepared by local pharmacies. Local residents knew the pharmacists, and it was likely the pharmacist knew the customers. A trust existed in those relationships. Expectations existed locally for most foods and medicines, and many goods were prepared within short distances reached by horse-drawn carriage, railroad, or boat. Establishing standards for the preparation of pharmaceuticals around 1820 was very timely since

Pharmaceutical Microbiological Quality Assurance and Control: Practical Guide for Non-Sterile Manufacturing, First Edition. Edited by David Roesti and Marcel Goverde.

Figure 15.1 Illustration of the United States Pharmacopoeia I from 1820 (left). Founding Fathers of the USP (right). Artist: Robert Thom. Painting commissioned by USP, 1957. *Source:* Courtesy of USP.

transportation and industrial manufacturing were about to advance rapidly (Figure 15.1).

The development of industrialization and distribution infrastructure created needs for testing and standards unlike what existed before the twentieth century. Most drugs were oral or topical, and either dry or as elixirs, either of which were preserved by their reduced water activity. These physical properties made the products lower microbiological risk. In the 1800s, the role of microorganisms in medicine and spoilage were only just being recognized. The manufacturing efficiency that evolved with industrial advances and improved shipping systems allowed producers to distribute goods across great distances. Due to industrialization, foods and drugs were prepared and shipped greater distances over longer times on widespread transportation systems to meet consumer needs in other cities or states, and even countries. Too often, production practices were not adequate to assure the needs of the consumers. Toxic colorants, flavorings, and preservatives often were added to the consumer goods to add or maintain the appearance of quality but to the detriment of the consumer's health. Many manufacturers simply were unaware of the toxicity of the materials they added to consumables, and some did not care.

Upton Sinclair's publication of *The Jungle* in 1904 described unsafe practices for the processing of foods, and it raised public awareness of the hazards associated with food processors. Such awareness encouraged Harvey Wiley of the US Department of Agriculture to direct testing to establish the risks of impure foods and the hazards of toxic additives. These tests used human volunteers (the "poison squad") who consumed items containing toxic additives. Data from these experiments plus the public concerns from

"The Jungle" formed the basis for the political push to pass the Pure Food and Drugs Act of 1906. The Act provided for a Bureau of Chemistry in the US Department of Agriculture for testing foods. Eventually, the Bureau of Chemistry reorganized to form the Food and Drug Administration shortly before the passage of the Food, Drug, and Cosmetic Act of 1938 (Janssen 1981).

With the emphasis that has been placed on drug product quality in cGMP programs, concepts relating to assurance of product quality in the United States have developed since the first "modern GMP regulations" were published in 1963 (Janssen 1981). These concepts took basic manufacturing and laboratory practices, and applied them to buildings, equipment, personnel, components, master formula and batch records, production and control procedures, product containers, packaging and labeling, laboratory controls, distributions records, and stability and complaint files. The scope of manufacturing practices covered planning, formulating, component acceptance, process development, process flow, process monitoring, and product acceptance. Records and data management were included as elements of cGMP. These pretty much covered the basics of industrial pharmaceutical manufacturing practices integrated with laboratory testing. Testing practices in the 1960s were largely intuitive. Quality assessment was based on technology of the time and statistical treatments were applied as needed.

The improvement of chemistry testing technology since 1906 has been dramatic, particularly with the advent of chromatographic methods. Concerns about toxic and carcinogenic substances in foods, drugs, and the environment grew as the test methods became more sensitive and able to detect them. Public awareness and political sensitivity to it resulted in *The Food Additives Amendment* of 1958 (the "Delaney Clause") that was incorporated into the Food, Drug, and Cosmetics Act. This amendment prohibited the FDA from approving the use of any food additive found to cause cancer in animals or humans. There was no tolerance for detectable carcinogens in foods and that philosophy was extended to drugs. Some pesticides were found to cause cancer in test animals, and often these were used in agriculture and industrial settings. Essentially, most agricultural products had at least some trace of these compounds. The Delaney Clause became problematic as chemical assays became more sensitive and could detect pesticides at levels considered less than a health risk. This put the Delaney Clause, with its "zero-tolerance" standard, in conflict with the US EPA's Federal Insecticide, Fungicide, and Rodenticide Act (FIFRA), which permitted the presence of these chemicals based on a risk/benefit determination. In 1985, the National Academy of Sciences (NAS) undertook study of the conflicting regulations. Their report, issued in 1987, recommended consistent standards between agencies based on risk rather than absolutism (Janssen 1981; National Research Council 1987).

In the author's opinion, FDA's roots in chemistry drove most of FDA's regulatory perspectives. Chemistry standards developed from the "poison squad" studies to address risk to the consumer from toxins. Most of the microbiology regulatory methods for pharmaceutical quality evolved from hygienic testing of food, water, and wastewater. More recent microbiology developments arose from aerospace contamination control, biodefense, clinical, or food methods. Unfortunately, many of the classical microbiology methods in cGMP were perceived with an expectation that they would yield results with consistency and precision found in chemistry methods. However, there is a lack of precision inherent to growth-based microbiology tests that arise from how samples are collected, how microbial populations are dispersed, and how microorganisms respond to the various test parameters. These can be (and usually are) very different between microbes and chemicals. After all, microorganisms are living creatures that are subject to genetic and physiological variations.

15.2 Why Microbes Confound a Census

Microbes grow and survive in a niche that is often unique to an overall material or environment. For example, natural waters have highly variable populations near the surface, near the bottom, at interfaces, and within the water column. A great amount of this variability is evident in biofilms, which can show population dynamics as competing populations move between their planktonic state and the biofilm matrix (Boles and Horswill 2012; Houry *et al.* 2012). Natural and potable water environments tend to foster surface interactions that offer greater gas exchange with ambient air, and offer opportunities for growth of more respiratory populations. As indicated by Palmer *et al.* (1976), planktonic bacteria may reside in vertically stratified populations and vary by the sizes of particles that microorganisms bind onto. Pharmaceutical waters may be generalized as displaying greater shear forces due to flow, and reduced residence time for batch production systems. Pharmaceutical water does not have a significant amount of particles or organic materials when compared to natural waters. However, bacteria will employ the same survival strategies regardless of the water systems they populate. In many ways, pharmaceutical air systems offer similar environmental variation at vent covers, surfaces, and the air column (Wiens 1976). When counting microbial populations, these variables interact to influence the results (Postgate 1969).

Also of importance in population counts is the metabolic state of microorganisms relative to the counting method. The adaptive nature of many species cause them activate or inactivate systems based on their environment. Through the 1960s and 1970s, environmental microbiologists published observations that bacteria were not forming colonies when transferred from water to agar media. Centers for Disease Control (CDC) studied *Pseudomonas* species (now *Burkholderia*) associated with pulmonary infections of patients using nebulizers, and found that some strains were killed (reduced by as much as two or three logs) when transferred from water to trypticase soy broth medium. Survivors would grow and subsequently flourish upon further transfer. Additionally, the authors observed that these strains were more readily cultivable if transitioned through dilute media before transfer to counting media (Carson *et al.* 1973). This meant that commonly grown microorganisms might not be detected from their natural environments unless laboratory media incorporated transitioning steps before counting.

More recently, the portion of resident microorganisms detected by commonly used sampling and cultivation methods was shown to be a small subset of the total cleanroom population. Surveys of cleanrooms conducted using media with low nutrient content detected more microorganisms by one or two orders of magnitude (Nagarkar *et al.* 2001). Traditional environmental monitoring methods have not adapted to these cultivation methods for oligophilic bacteria, possibly in large part because cultivation of oligophilic bacteria can take weeks.

For the environmental monitoring program and the quality unit, this means that environmental counts from commonly used methods are only a representation or indicator of the microbial state of the cleanroom. Due to the time necessary to incubate samples using growth-dependent monitoring methods, any data will represent the environment several days or even a week before. These growth-based data are to be viewed as a "snapshot" of what may have been the microbiological state of control when the samples were collected. As a result, investigating environmental anomalies will be challenged by the likelihood that the area was cleaned since sampling.

Other sampling influences are shown in the sample collection methods. One example showed that contact plate samples detect only a portion of the resident surface flora, and up to five consecutive samplings of the same location yields about 80% of the cultivable population (Tidswell *et al.* 2005). Favero *et al.* (1968) studied methods to sample surface-borne microorganisms and concluded that no single method will reliably provide a full estimate of the microbial population: it is the nature of the surface (size, shape, and composition) that determines whether a method is suitable at all (Niskanen and Pohja 1977). Additionally, Lemmen *et al.* (2001) noted that surface sampling methods (swab vs. contact plates) yielded different recovery frequency rates for gram-negative versus gram-positive cells. Such innate variability confounds the precision of microbiological population measures using cultivation-based methods.

The quality unit needs to be aware of the difficulties of making judgments about the state of the cleanroom or process controls when the majority of the microbial population is not detectable. For this reason, just like finished product testing, failure to detect microorganisms does not mean they are not present – it is absence of evidence and not evidence of absence. The quality managers cannot rely on data showing low counts (or no counts) as a reason to become too relaxed in other areas of process vigilance.

15.3 Microbiological QA Decisions

As pharmaceutical therapies evolved and formulations became more complex, processing requirements increased as well. Injection products as well as ophthalmic and surgical products have manufacturing standards relating to microbiological characteristics (e.g. sterility and pyrogenicity). As discussed, methods to assess the chemical quality of parenteral drugs evolved since the FD&C Act. Airborne particle measurement technologies (sizing and counting) and particle control methods have developed out of the radioactive dust control needed for the Manhattan Project (which developed the atomic bomb) and NASA space flight research. Some of these environmental management concepts translated well to manufacturing controls for parenteral products and electronics manufacturing. However, testing and metrics for these controls offer certain philosophical dilemmas, especially for manufacturing parenteral drugs.

Microbiological control and cleanroom particle control are related – but not *directly* related. HEPA filters offer great levels of efficiency for their intended purpose, the removal of particles. Since particle measures are physical methods, they are not subject to biological variability. Only systems that exclude personnel and other environmental inputs can rely exclusively on HVAC controls for particulate and microbial quality assurance. For the product attribute "sterile," which is an absolute microbiological condition, there is no adequate measure: the compendial sterility test has weaknesses, as do all other tests. Product measures of sterility are indirect and statistically limiting. For example, the compendial sterility assay relies on visible turbidity to detect the growth of microbial contaminants. There is no guarantee that contaminating microorganisms will grow under the chosen conditions (medium, temperature, time, and physiological state of the cells). For these reasons, the compendial sterility tests are indirect and subjective measures. When non-sterile products are manufactured, microbiological controls and product tests are subject to the same constraints but using different methods.

A microbiological quality assurance risk exists at every step of the manufacturing process: source materials, facility controls, formulation, processing, finished product testing, and shelf life (stability) tests. These risks arise from all the factors described above. Since microbiological quality attributes are a small component of a drug's specification, it is unsurprising that professionals with a chemistry degree rather than a microbiology degree lead most QC and QA departments. As such, the chemistry perspective is foremost in decision-making that is important in QA activities related to microbiology. Biology is chemistry with a "twist." Chemistry is physics with a "twist." Physics is math with a "twist." These philosophies arise from math, which is precise and unforgivingly devoid of "twists" or art. There is very little "art" in chemistry, whereas about half of microbiology is "art." Therein is the dilemma for data interpretation in microbiology.

15.4 Who Has Responsibility for Quality?

It was common to use the terms "Quality Control" (QC) and "Quality Assurance" (QA) interchangeably, especially within the context of small companies several decades ago. However, QC and QA departments are part of an overall "quality management" program, and smaller organizations may find the overlap creates difficulty when defining authorities. The terminology grew from environmental studies spurred by Super Fund research sponsored by the US EPA (Cross-Smiecinski 2002). However, these responsibilities must not create conflicts in decision-making (which this author observed in the early years of the generic drug industry in the United States). The distinction between QC and QA is important, but in the drug cGMP regulations, 21 CFR 211.22 uses nonspecific terminology under the catch-all heading, "Responsibilities of the quality control unit."

A chemist from the National Bureau of Standards, John K. Taylor (not to be confused with John M. Taylor, a previous FDA Associate Commissioner for Regulatory Affairs) offered clarity in the distinctive roles of QA and QC. Dr. Taylor defined Quality Assurance as *A system of activities whose purpose is to provide to the producer or user of a product or service the assurance that it meets defined standards of quality with a stated level of confidence.* He defined Quality Control as *The overall system of activities whose purpose is to control the quality of a product or service so that it meets the needs of the users. The aim is to provide product of quality that is satisfactory, adequate, dependable, and economic.* In summary, Dr. Taylor asserted that Quality Assessment is composed of *The overall system of activities whose purpose is to provide assurance that the overall quality control job is being done effectively. It involves a continuing evaluation of the products produced and the performance of the production system* (Taylor 1987).

It is common to view "quality" as a system associated with release of a finished product: something was made and testing shows that it meets established criteria. However, the notion of quality since the era of Deming[1] in recent years has become a lifecycle activity based on the "Plan, Do, Study, Act" (PDSA) principles. These are the same as Shewhart's[2] "control circle/cycle/wheel" and many other terms to describe a continuous system of actions to improve processes until the product is no longer manufactured. This holistic philosophy engages the entirety of those involved in the processes associated with a product or service throughout its lifecycle. As a lifecycle function, no individual has ownership of a product's quality, but when a problem arises, the quality assurance director bears the weight of resolving the issue. When the question or complaint is microbiological in nature, the entire team must work together to help the QA Director, but that team will need the training and experience of a microbiologist to investigate the nature of the problem and recommend corrective actions.

Since the industrial revolution, industrial manufacturing has become a complex structure that rarely occurs between sunrise and sunset of a given day. For sterile manufacturing, the preparation of materials and equipment requires organizing all components of the process, testing, and approving them. Then, they must be stored and protected to remain suitable for use in a sterile process. Skilled and practiced operators are entrusted to assemble the needed items and operate the system while other personnel prepare records that confirm the proper performance of each activity. This requires planning and integration of the disciplines necessary for sterile product manufacturing, and for that matter, similar planning is needed for most other microbiologically sensitive products.

15.5 Product and Process Planning

Pharmaceutical manufacturing is a significant undertaking. There is significant risk and great potential benefit associated with a drug product. When a pharmacist prepares a dose, the patient receives something that is supposed to mitigate a medical issue. When a batch of drug is prepared, the product is intended help a large population. The risks of manufacturing error are magnified by the increased number of users (scale) and the greater amount of time to distribute and consume the product (shelf life). When the product

1 William Edwards Deming (14 October 1900 to 20 December 1993).
2 Walter Andrew Shewhart (pronounced like "shoe-heart," 18 March 1891 to 11 March 1967).

requires the attribute of sterility, the microbiological controls can be very demanding and appropriately so considering a failure could be catastrophic for the patient.

Sterility of a finished product may be necessary for several reasons. In the food industry, sterilization is performed as a preservation method, particularly in canning processes. For pharmaceutical manufacturing, sterility is generally necessary to protect the user from exposure to infectious agents. For non-sterile drug products, microbial control is generally accomplished by good manufacturing practice (GMP), low water activity of the formulation, and/or antimicrobial preservatives. Similar philosophies are employed during food production; however, refrigeration and pasteurization are more common microbial control strategies for foods (Pflug 2010).

To assure the safety and effectiveness of a therapeutic product, carefully designed studies are performed and evaluated. Chemists, physicians, toxicologists, and statisticians are the typical leads in early drug developmental studies. As the development of a drug product approaches the point of commercialization, the transition from experimental development to industrial production creates novel challenges and opportunities. The PDSA model begins with establishing the industrial objectives that include determination of the optimal formulation and product presentation. The US FDA expects sterility of drug products for injection and those applied to or near the eye. It is desirable that products applied to open wounds are sterile, but this is not a requirement (FDA 2006).

All products whether intended as sterile or non-sterile need to be prepared with adequate controls to avoid "filth" and "insanitary conditions." While microorganisms may be present in non-sterile products, there should not be too many microorganisms or the wrong species. FDA regulations for nonsterile drug products reflect these concerns in batch release testing (21 CFR 211.165), component (21 CFR 211.94), and facility and equipment requirements (21 CFR 211.113). Raw material and packaging requirements are intended to assure the use of suitable materials that are free of adulteration that could degrade the product or harm the patient.

While some microbiological latitude should be expected when manufacturing non-sterile products, some common sense in microbiology and engineering remains important. For example, fresh salad will have a microbial load of about 10^6 CFU/gram. The number of microorganisms in salad is an indicator of its origins and freshness. Clean and sterile are different and should not be confused. The greater health risk comes from the presence of the wrong microorganisms. Refrigeration and storage time will control general spoilage of food, but the source and handling are necessary controls to prevent the presence of pathogens. Similar concerns are reflected in the regulatory expectations that require adequate written procedures designed to prevent objectionable microorganisms in drug products. The compendia recommend criteria for many non-sterile drug products. In USP, these criteria are in USP Chapter <1111>, an informational chapter. There are numerical limits for the microbial load of non-sterile products (assuring the absence of adulteration indicated too many microorganisms), and test methods are found in USP Chapter <61> (Microbial Enumeration Tests). The criteria offered for finished products are small numbers, and small numbers are problematic for microbiological data. General Chapter <1111> discusses expected variability for counting populations when the raw data are small numbers to acknowledge that counts can actually reflect a range of results from the same population. A twofold variance from the criteria is a tolerable test result range for the single result. The analyst should consider trends and consistency of counts rather than data from singlicate tests.

There is no practical way to offer lists of species that would meet the expectations of the regulation relating to "objectionable microorganisms." However, the compendial tests for specified microorganisms are provided in USP Chapter <62> (Tests for Specified Microorganisms). This chapter offers test methods for organisms that indicate potential adulteration from types of contamination or generally insanitary conditions. USP Chapter <1111> recommends criteria that includes the absence of:

is an indicator of touch contamination and is capable of causing infections.

is an indicator of poor water quality contact and is a pathogen when delivered by certain pathways.

is an indicator of generally poor hygienic practices or adulterated materials.

is an indicator of adulteration that would be unacceptable for vaginal products and possibly others.

Salmonella species are enteric pathogens representing very poor hygienic conditions.

Bile-tolerant Gram-negative bacteria in certain inhalation products: Included among these are *Escherichia coli, Salmonella,* many of the pseudomonads and Enterobacteriaceae.

Again, while these are hardly an exhaustive list, they do provide a starting point to consider when materials or products might contain microbial flora from unacceptable sources or poor processing. Additionally, other microorganisms may present a risk for certain routes of administration or formulations. The manufacturer remains responsible for having appropriate procedures for preventing objectionable microorganisms, but good-quality materials and clean processes meet the expectations of the regulation. Chapter 11 provides a deep insight on objectionable microorganisms. The freedom from these specified organisms (and other outright pathogens mentioned in regulatory documents or monographs) represent an adequate starting point for microbiological quality attributes of non-sterile products.

Other microorganisms are problematic to certain product types. Products of biologic origin may include mycoplasmas. Test methods for mycoplasma are provided in USP Chapter <63>. Aqueous products often need antimicrobial preservatives to prevent the growth of microorganisms in non-sterile products. As noted in USP General Chapter <1231> (Water for Pharmaceutical Purposes), various species of "pseudomonads" can resist antimicrobial agents and are common to water systems. USP <62> offers test methods for *P. aeruginosa* (which is a potential pathogen that can flourish in water systems), but specific tests for other of the "pseudomonads" had not been in the compendia. Certain *Burkholderia* species (previously part of the *Pseudomonas* genus) are pathogenic for some patient groups, and some *Burkholderia* species can even metabolize benzalkonium chloride preservatives (Ahn *et al.* 2016). Recently, FDA advised drug manufacturers that *Burkholderia cepacia* complex (Bcc) poses a contamination risk in non-sterile, water-based drug products (Torbeck *et al.* 2011; FDA 2017). Even if regulators have requested that the absence of *B. cepacia* complex be part of the specification for some products, no compendial methods existed for detecting these bacteria. USP undertook to provide test methods for Bcc organisms and proposed they become General Chapter <60>. There are other of these "pseudomonads" that have been shown to overcome preservatives. For example, Borowicz *et al.* (1995) reported survival of *Stenotrophomonas maltophilia* in a buffer solution containing 1% sodium azide. However, *S. maltophilia* are less invasive and less commonly reported as the etiology of infections.

15.6 Quality and Documentation in Drug Manufacturing

Some production activities are very direct and uncomplicated. For pharmaceuticals, the range of activities is very diverse and planning becomes essential. FDA regulations continuously restate that these procedures must be established and followed. The individual steps in the procedures are checked and confirmed, and a "responsible person" verifies their performance with a signature. Similarly, investigations and corrective procedures are undertaken and then signed as the actions are completed. These must be executed and closed in a timely fashion with no deviations in the signature order and timing. Too often, an incomplete signature block or an investigation that has not been closed becomes the start of a cascade of inspectional observations, and microbiology deviations can remain open for a long time and become forgotten.

Open investigations can suggest a quality culture weakness that causes the investigator to look deeper into the quality systems and their documentation.

15.7 Summary

Quality is a system owned by no one group or person in a manufacturing setting. The entirety of the manufacturing organization bears responsibility for product quality. It is, however, directed and ultimately the responsibility of the quality assurance director.

Quality means that the product meets the needs of the consumer. The role of the regulator is to protect the consumer, while the producer provides for the consumer. For pharmaceuticals, many elements interplay in this dynamic: medicine, pharmacology, chemistry, engineering, microbiology, and then there are business aspects like storage and distribution.

The role of microbiology within this much larger context of quality illustrates some unique elements that contribute to pharmaceutical manufacturing. In all aspects of these disciplines there are risks associated with the use of medications. Manufacturing risks can be minimized by manufacturing design and vigilance in the practices of manufacturing. However, minimizing will not eliminate all risk. Especially in microbiology, a perspective of risk must be considered whether taking a solid oral tablet like a compressed tablet or eating salad for dinner.

Even the injection of a drug that is labeled "sterile" has risks imposed by drug preparation and administration. "Sterile" cannot be measured by testing, and while process validation measures only the responses within a defined range, it relies on projections farther into the unmeasurable regions of the absolute state called "sterile."

The quality unit can only address the measurable and move from there using risk-based perspectives from the PDSA principles or Shewhart's "control circle/cycle/wheel." This vigilant observation of process data and reaction to deviations are needed because microbiological inputs will continuously change. Environmental deviations should be expected and carefully sought. When deviations are detected, they should be assessed for risk and appropriate corrections made. With the realization that microbiological environmental data reflect events from several days before, and are a "snap shot" of only portion of the resident flora, assigning specific causes to data deviations are sometimes not possible even after weeks or months. Increased monitoring provides some evidence as part of increased vigilance, but future monitoring will not identify an event in the past. Only continuous quality activity can offer a reasonable expectation of microbiological product quality.

Bibliography

Ahn, Y, Kim, JM, Kweon, O, Kim, S-J, Jones, RC, Woodling, K, da Costa, GG, LiPuma, JJ, Hussong, D, Cerniglia, CE, 2016, 'Intrinsic resistance of *Burkholderia cepacia* complex to benzalkonium chloride', mBio, vol. 7(6), pp. 1–11. Available from: http://mbio.asm.org/content/7/6/e01716-16.full.pdf+html, last visited on 10 March 2019.

Bender, GA, Thom, RA, 1999, Pictorial Annals of Medicine and Pharmacy, Warner Lambert Company, Morris Plains, NJ.

Boles, BR, Horswill, AR, 2012, 'Swimming cells promote a dynamic environment within biofilms', PNAS, vol. 109(32), pp. 12848–12849.

Borowicz, JJ, Brishammar, S, Gerhardson, B, 1995, 'A *Xanthomonas maltophilia* isolate tolerating up to 1% sodium azide in Tris/HCl buffer', World J. Microbiol. Biotechnol., vol. 11, pp. 236–237.

Carson, LA, Favero, MS, Bond, WW, Petersen, NJ, 1973, 'Morphological, biochemical, and growth characteristics of *Pseudomonas cepacia* from distilled water', Appl. Microbiol., vol. 25(3), pp. 476–483.

Cross-Smiecinski, AJ, 2002, 'Quality assurance', in Hurst, CJ (ed.), Manual of Environmental Microbiology, 2nd edition, ASM Press, Washington, DC, pp. 158–165.

Favero, MS, McDade, JJ, Robertsen, JA, Hoffman, RK, Edwards, RW, 1968, 'Microbiological sampling of surfaces', J. Appl. Bacteriol., vol. 31, pp. 336–343.

FDA, 2006, *Guidance for Industry: Chronic Cutaneous Ulcer and Burn Wounds — Developing Products for Treatment.* Available from: https://www.fda.gov/downloads/Drugs/GuidanceComplianceRegulatoryInformation/Guidances/UCM071324.pdf, last visited on 10 March 2019.

FDA, 2017, 'FDA communication 5/22/2017'. Available from: https://www.fda.gov/drugs/drugsafety/ucm559508.htm, last visited on 10 March 2019.

Houry, A, Gohara, M, Deschampsa, J, Tischenko, E, Aymerich, S, Grussa, A, Briandet, R, 2012, 'Bacterial swimmers that infiltrate and take over the biofilm matrix', PNAS, vol. 109(32), pp. 13088–13093.

Janssen, WF, 1981, 'The story of the laws behind the labels', *FDA Consumer*, Available from: https://www.fda.gov/AboutFDA/WhatWeDo/History/Overviews/ucm056044.htm, last visited on 10 March 2019.

Lemmen, SW, Häfner, H, Zolldann, D, Amedick, G, Lütticken, R, 2001, 'Comparison of two sampling methods for the detection of Gram-positive and Gram-negative bacteria in the environment: moistened swabs versus Rodac plates', Int. J. Hyg. Environ. Health, vol. 203, pp. 245–248.

Nagarkar, PP, Ravetkar, SD, Watve, MG, 2001, 'Oligophilic bacteria as tools to monitor aseptic pharmaceutical production units', Appl. Environ. Microbiol., vol. 67(3), pp. 1371–1374.

National Research Council Committee on Scientific and Regulatory Issues Underlying Pesticide Use Patterns and Agricultural Innovation, 1987, Regulating Pesticides in Food, National Academy Press, Washington, DC, pp. 25–27. Available from: https://www.acsh.org/wp-content/uploads/2012/04/20040503_delaney.pdf, last visited on 10 March 2019.

Niskanen, A, Pohja, MS, 1977, 'Comparative studies on the sampling and investigation of microbial contamination of surfaces by the contact plate and swab methods', J. Appl. Bacteriol., vol. 42, pp. 53–63.

Palmer, FE, Methot, D, Jr., Staley, JT, 1976, 'Patchiness in the distribution of planktonic heterotrophic bacteria in lakes', Appl. Environ. Microbiol., vol. 31(6), pp. 1003–1005.

Pflug, IJ, 2010, Microbiology and Engineering of Sterilization Processes, 14th edition, Environmental Sterilization Laboratory, Otterbein, IN.

Postgate, JR, 1969, 'Viable counts and viability', in Norris, JR, Ribbons, DW (eds.), Methods in Microbiology, vol. 1, pp. 611–628, Academic Press, New York.

Taylor, JK, 1987, Quality Assurance of Chemical Measurements, Lewis Publishers, Chelsea, MI.

Tidswell, EC, Bellinger, M, McCullough, D, Alexander, A, 2005, 'Consecutive replicate contact plate sampling assists investigative characterisation of surface-borne bioburden', Eur. J. Paren. Pharma Sci., vol. 10(4), pp. 93–98.

Torbeck, L, Raccasi, D, Guilfoyle, DE, Friedman, RL, Hussong, D, 2011, '*Burkholderia cepacia*: this decision is overdue', PDA J. Pharm. Sci. Tech., vol. 65(5), pp. 535–543.

United States Pharmacopeia, 2018, 'In-Process Revision: <60> Microbiological Examination of Nonsterile Products – Tests for Burkholderia cepacia complex', *Pharmacopeial Forum*. Available from: http://www.usppf.com/pf/pub/data/v445/CHA_IPR_445_c60.xml, last visited on 10 March 2019.

US 21 CFR 211.113, 'Control of microbiological contamination', *Code of Federal Regulations*, Title 21, Food and Drug Administration.

US 21 CFR 211.165, 'Testing and release for distribution', *Code of Federal Regulations*, Title 21, Food and Drug Administration.

US 21 CFR 211.22, 'Responsibilities of quality control unit', *Code of Federal Regulations*, Title 21, Food and Drug Administration.

US 21 CFR 211.94, 'Drug product containers and closures', *Code of Federal Regulations*, Title 21, Food and Drug Administration.

USP 41-NF 36 chapter <1111>, 2018, 'Microbiological Examination of Nonsterile Products: Acceptance Criteria for Pharmaceutical Preparations and Substances for Pharmaceutical Use', in *USP 41–NF 36*. U.S. Pharmacopeial Convention, Rockville, MD.

USP 41-NF 36 chapter <1231>, 2018, 'Water for Pharmaceutical Purposes', in *USP 41–NF 36*. U.S. Pharmacopeial Convention, Rockville, MD.

USP 41-NF 36 chapter <61>, 2018, 'Microbiological Examination of Nonsterile Products: Microbial Enumeration Test', in *USP 41–NF 36*. U.S. Pharmacopeial Convention, Rockville, MD.

USP 41-NF 36 chapter <62>, 2018, 'Microbiological Examination of Nonsterile Products: Tests for Specified Microorganisms', in *USP 41–NF 36*. U.S. Pharmacopeial Convention, Rockville, MD.

USP 41-NF 36 chapter <63>, 2018, 'Mycoplasma Tests', in *USP 41–NF 36*. U.S. Pharmacopeial Convention, Rockville, MD.

Wiens, J, 1976, 'Population responses to patchy environments', Ann. Rev. Ecol. Syst., vol. 7, pp. 81–120.

16

Practical Guide for Microbiological QA/QC of Non-sterile Pharmaceuticals Manufacturing for EU

Ina Bach

Dr. Bach AG, Global Compliance Assistance, St. Gallen, Switzerland

CONTENTS

16.1 Introduction

Testing for microbiological purity is an essential part of quality control of drugs as well as medical devices. The highest requirements apply to sterile products, as they must be free from microorganisms, which must be demonstrated by the sterility test. However, even non-sterile drugs are subject to stringent requirements, as the presence of certain microorganisms in a drug may attenuate or inactivate the therapeutic activity of the drug while also endangering the patient's health. Therefore, the lowest possible microbial burden must be ensured in the production, storage, and placing on the market of pharmaceutical preparations

Pharmaceutical Microbiological Quality Assurance and Control: Practical Guide for Non-Sterile Manufacturing, First Edition. Edited by David Roesti and Marcel Goverde.

by applying the GMP guidelines (EudraLex 2015, Part I and its Annexes; ICH 2005, 2008; PIC/S 2018; WHO 2011, 2014; ZLG 2010).

However, this is only possible if appropriate quality requirements are already taken into account for the starting materials (active substances and excipients), the process equipment as well as the hygiene concept (hygiene regulations for environment and staff).

16.2 General Requirements

The microbiological purity requirements are laid down in the pharmacopoeias for non-sterile pharmaceutical preparations and pharmaceutical substances in Ph. Eur. chapters 5.1.4 and 5.1.8.

Testing methods to be used to control the microbiological quality are also described in the pharmacopoeia in Ph. Eur. chapters 2.6.12, 2.6.13, and 2.6.31.

Both the requirements and the test methods were the subject of international harmonization. The aim of the harmonization is to assert the same testing methods and requirements in the USP, the Japanese, and the European pharmacopoeia.

In addition, relevant documents for medical devices are published and can only be attained with a sufficient degree of assurance if:

a) The product-specific risk for a microbiological contamination is already systematically investigated during the development phase of a drug.
b) The specifications, test methods, the sampling and the test frequency for drug products, devices, active pharmaceutical ingredients (API), excipients, and primary packaging materials are established according to the product-specific and process-related risk as well as the nature and the mode of application of the relevant material.
c) A concept exists to monitor the personal hygiene, the premises, the equipment, the manufacturing process, and process media which supplements the product-related quality assurance measures.

As far as testing is concerned, a variety of product-related and facility, personal, and process-related microbiological tests have to be carried out to ensure the microbiological quality of pharmaceutical drug products or medical devices (Table 16.1).

Since a microbiological control area is an essential element of a comprehensive and well-established Quality Control Unit, all activities are more responsible for reviewing, approving elements, and collecting raw data in compliance with GMP requirements.

Therefore, it should be considered a principle and a matter of course to carry out activities of a microbiological control laboratory as an integral part of audits and self-inspections in a company also for the production of non-sterile drug substances (API's), excipients, and finished products.

Table 16.1 Product-related activities for non-sterile products.

Test activities	Requirements
Holding and standing time	If non-sterile aqueous intermediate solutions and suspensions (e.g. coating solutions) have to be stored for a prolonged period of time, the maximum permissible holding period of these has to be determined by means of a microbiological challenge test.
Antimicrobial effectiveness test (AET)	For all liquid and semisolid drug products (DP) the question arises whether the whole contents of a container is used in a single dose or whether the drug product is presented in a container for multiple use. The AET is performed in the development phase and also during the stability testing studies in order to show that the antimicrobial efficacy is granted up to the end of shelf life. The AET is not performed during follow-up stability and is not a routine release test.
Manufacturing/release testing (routine testing activities)	Whether the microbiological release testing has to be carried out and depends on the applications, antimicrobial or growth promotion properties, and the historical data of the drug substance (API), excipients, or DP.
Microbiological process controls (source of contamination)	If a non-sterile drug substance (API), excipient or DP is microbiologically contaminated (OOS result or OOE result), the root cause has to be disclosed and eliminated. For this purpose, microbiological step-by-step checks during the manufacturing process may be necessary.
Microbiological process controls (step-by-step controls)	All manufacturing steps in a non-sterile liquid drug substance (API) or DP and production steps in solid DP dealing with aqueous solution or suspension must be carefully evaluated and kept under control.
Test for microbial counts	Non-sterile drug substance (API), excipients, and DP have to fulfill microbiological purity requirements laid down by the harmonized text of the Pharmacopoeias. Furthermore, internal microbiological limits for primary packaging need to be established.
Cleanroom concept (room classification)	Description for production of non-sterile drug substance (API) and DP. It defines the requirements for each cleanliness zone regarding design, construction, qualification and operational requirements as well as requirements for personal and sanitization.
Microbiological monitoring in terms of cleanroom concept (air, surfaces, and personnel)	Based on the definition of cleanliness zone for non-sterile products and based on a risk assessment

(Continued)

Table 16.1 (Continued)

Test activities	Requirements
Microbiological cleaning validation	The efficacy of cleaning and disinfection of product-contacting surfaces needs to be validated in terms of the hygienic cleanliness. The microbiological cleaning validation deals with the efficacy of the applied disinfectants, the storage, and the respective environmental conditions for non-sterile products.
Evaluation and monitoring of disinfectants and detergents	Detergents and disinfectants, ready to use, have to be monitored for possible microbiological contaminations.
Monitoring of gases	Microbiological examination of compressed gases (e.g. air, carbon dioxide, and nitrogen) is done for initial qualification of gas distribution systems installed in manufacturing areas and a routine monitoring program should be implemented.
Water for pharmaceutical use	Water in various qualities is used in a variety of pharmaceutical processes; water is susceptible to microbial contamination. Therefore, the proper design of a treatment and distribution system for water for pharmaceutical use is of utmost importance, and the chemical, biological, and microbiological quality has to be constantly supervised. Minimum purity requirements adapted to the water quality have to be applied, including testing frequencies, sample strategies and procedures, interpretation of results, and handling of deviations/out-of-specification results. A detailed sampling plan should be part of the system-specific SOP.

16.3 Audit Assessment Tools of a Microbiological Laboratory

16.3.1 General Overview

This chapter provides guidance on the use of process and system assessment tools in preparing for and performing audits and self-inspections in microbiological laboratories. An assessment tool is designed to provide guidance and regulatory references to enable the auditor to identify common elements that are applicable to the specific audit and will focus the auditor on the critical elements of the process.

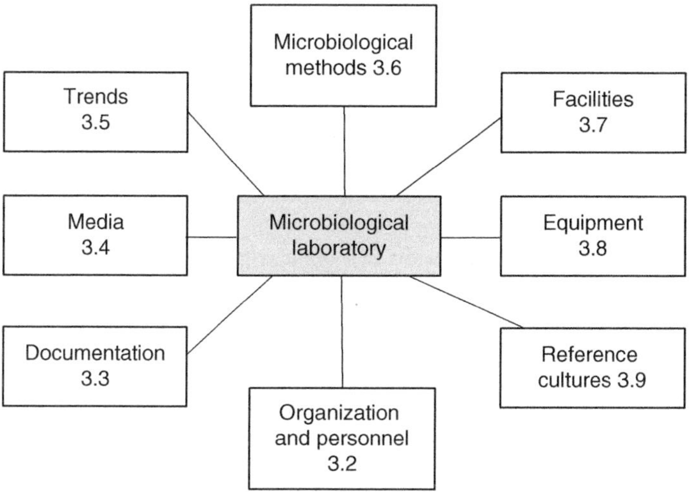

Figure 16.1 Overview of critical success factors for the inspection of microbiological laboratories.

They provide the auditor with an instructional guide and enhance consistency by assuring that critical systems, subsystems, processes and elements are addressed consistently and thoroughly during the audit/assessment (Figure 16.1).

16.3.2 Organization and Personnel

- Request and review organization charts of the microbiological department.
- Review the functionality and responsibilities for key positions – job description.
- Determine if facility/business unit operates under a corporate quality policy.
- Verify that the responsibilities and procedures applicable to the quality control unit are defined in a written procedure.
- Is there a new employee orientation and training program that also includes cGMP training sessions?
- Is there a continuous training program implemented?
- Written training records available and traceable documented?
- Trainers are qualified to provide training sessions.
- If consultants are hired to advise any activities in the microbiological laboratories, verify evidence of their education, training, and experience.
- Medical checks should be on a regular basis.

References: EudraLex 2015, chapters 2.8, 2.9, 2.10, 2.11, 2.16, 2.20, 2.23.

16.3.3 Documentation

Normal principles of good documentation practice apply and are implemented, because testing can be reported on log sheets, data sheets, logbooks, etc.

Attention: Most microbiological tests are extremely manual with little automation.

Take in consideration that most documentation in microbiology is simple, voluminous, repetitive, and laborious for review.

Parts to consider regarding documentation evidence are:

- SOPs for each test/activity existing.
- Test documentation available.
- Documentation for media preparation present.
- Charts, temperature printouts, and equipment log books available.
- Documents and raw data present for equipment qualification, maintenance, and calibration procedures.
- Availability of testing monographs.
- Method suitability of protocols and reports.
- SOPs for handling reference microorganisms, temperature monitoring systems for incubators present.
- A SOP covering safety and environmental issues (i.e. handling of cytotoxic or highly active substances).
- Environmental monitoring program for the microbiological laboratory is present and raw data are available.
- Continuously water testing program for water used for media preparation present.
- SOPs for sampling procedures for microbiological environment and personal monitoring and water testing program are implemented.
- SOPs for evaluation of cleaning and disinfection agents are available.
- SOPs for cleaning and disinfection are available.

References: EudraLex 2015, chapters 4.1–4.9, 4.18, 4.28, 4.29, 4.30, 4.31, 6.16, 6.17.

16.3.4 Culture Media

- The kind of culture media in use is defined in the pharmacopoeia (e.g. Ph. Eur. 2.6).
- What types of culture media are used? E.g. ready-to-use or as dehydrated powder for in-house manufacturing?
- All culture media should be subjected to appropriate controls to ensure fertility and growth promotion and, as appropriate, inhibitory and indicative properties.
- All culture media should be traceable with respect to when and where they are used.

- If ready-to-use media are used, the points to consider should be:
  - Each incoming batch tested for sterility and growth promotion (reference organisms and in-house flora).
  - Correct storage, following "First Expiry and First Out" (FEFO) principle and expiry date.
- If dehydrated powder for in-house manufactured culture media is used, the points to consider should be:
  - Each incoming batch to be tested for growth promotion (extended release test).
  - Each batch manufactured according to defined procedures and specified by the manufacturer with documentation.
  - Sterilization process validated and controlled.
  - Each manufactured batch with defined batch number is tested for growth promotion (reduced release test) and as appropriate inhibitory and indicative properties.
  - Full batch number traceability system is present.

References:

- EudraLex 2015, chapters 4.14, 4.22, 4.25, 4.26, 6.19
- Ph. Eur. 2.6, chapter 2.6.1
- PIC/S 2018, chapters 5.2–5.5, PI 012-3 chapter 11.3.2, 11.3.3
- Rieth 2012

16.3.5 Trends

In general, and in this context, microbiology is not an exact science because every test is destructive; it is not possible to repeat a test exactly on the same sample. Microbiological data derived from a single monitoring event represent not more than a snapshot of the environmental bioburden, which is subject to spatial–temporal variation. Therefore, reviewing microbiological trends is often more useful than individual results.

Points to consider should be in this case:

- In principle, requirements for trend analysis in the context of quality assurance is available and implemented.
- Regarding action, alert limits, and out-of-trending (OOT), there are clear definitions of procedures when limits are exceeded present.
- Are clear specifications available in which period of time trending is to be documented? (Weekly, monthly, quarterly, etc.)
- Existing trending program also include the evaluation of the in-house flora?

References: EudraLex 2015, chapters 6.9, 6.29, 6.32.

16.3.6 Microbiological Methods

In general, most microbiological methods used are described in the Pharmacopoeia, e.g. bioburden testing of raw materials, water, detection of objectionable microorganisms, e.g. *Escherichia coli, Pseudomonas aeruginosa, Salmonella* spp., endotoxin for purified and highly purified water, and preservative efficacy testing.

The focus should therefore be on the review of applied microbiological test methods as:

- The reported test method is detailed in the form as reported in the Pharmacopoeia (i.e. which method is used).
- Specifications of active substances, excipients, and drug products are correct and detailed in the statement of microbiological purity.
- Microbiological tests are carried out in accordance with written requirements (SOPs) and/or according to the guidelines of Pharmacopoeias.
- Verification of raw data consistency with data integrity and full traceability.
- Review and evaluate possible suitability test documents for microbiological test methods. **Note:** A strict approach regarding to linearity, specificity, precision, etc., are not appropriate for method suitability test. Hence, a strict case-by-case approach is appropriate.
- Bioburden testing under microbial-controlled environmental requirements.
- Validation of sample preparation expected (i.e. recovery) according to bioburden testing.
- In general, repeated testing should follow predetermined, modified OOS-procedures.
- Points to consider for enumeration test by filtration, plate count, or most probably number (MPN) method according to total viable aerobic count:
  - Clean bench is state of the art
  - Flame must be near exposed surface if no laminar flow cabin is used
  - No open doors during test
- Suitability of microbial enumeration test is adequate in that the recovery rate must be more than 50% of the inoculum.

References:

- Ph. Eur. 2.6, chapter 2.6.16
- EudraLex 2015, chapters 1.2ii, 1.3viii, 1.3x, 1.4iii, 1.4iv
- PIC/S 2018, PI 041-1

16.3.7 Facilities

Microbiological laboratories should be designed to avoid:

- Change contamination of tests; false-positive test results
- Cross contamination between tests
- Escape of hazardous organisms

- Infection of operators

A clear clean zone concept should be implemented and with regard to the classification of rooms, they should be suitable for use. An adequate environmental monitoring program with clear requirements for personal hygiene, cleaning and disinfection should be present.

Material and personal flow should minimize the risk of cross contamination in that:

- Removal (and decontamination) of waste containers and sharps bins.
- Removal (and decontamination) of bottles with contaminated solutions.
- Clear requirements of entering clean areas and leaving dirty areas.
- Appropriate staff live culture work and qualified for the work with living microorganisms.
- Presence of non-sterile items in clean area; sanitization of working areas after use.
- Incubation of agar in clean area.

References: EudraLex 2015, chapters 3.1, 3.26, 3.27, 3.29, 5.20, 6.5, 6.6.

16.3.8 Equipment

Generally, most microbiological equipment is essentially simple, so qualification requirements need not to be implemented. However, with the use of automated analysis equipment, and/or the use of rapid microbiological methods (RMMs), a requirement for qualification and validation that should not be underestimated must be taken into account. It is advisable in this connection to create a detailed validation master plan (VMP) for the microbiological laboratory.

Typical equipment in microbiological laboratory is:

- Autoclaves and incubators
- Laminar flow cabinets
- High-tech equipment like isolators, systems for identification of microorganisms like DNA sequencer, and automated systems
- Equipment for environmental sampling
- Temperature-controlled areas like refrigerators and cold storage rooms

Points to consider should be in this case:

- Basic principle of equipment qualification (DQ → IQ → OQ → PQ) applied.
- Equipment and relevant IT-systems in the microbiological laboratory that should be qualified are part of the VMP.
- An adequate and detailed calibration and maintenance plan of each necessary equipment is present (e.g. water bath, air sampler).

Review and evaluate the qualification documentation of an autoclave in that:

- Heat distribution and heat penetration is adequate.
- All loading possibilities are qualified regard to regulatory requirements incl. biological indicators.
- Temperature and pressure registration for routine sterilization is present.
- Clear requirements for revalidation is present (number of runs per load; frequency for revalidation).
- Clear requirements for recalibration of temperature probes and pressure transmitter (recommendation twice per year).
- Performance of steam quality tests and installed HEPA filters.
- User management is defined and if necessary, an audit trial is implemented.

References: EudraLex 2015, chapter 6.5, Annex 15.

16.3.9 Reference Cultures

Most biological tests require the use of reference cultures for positive controls and growth promotion testing. What should usually be considered within a microbiological laboratory area?

- Reference cultures should normally be traceable to the appropriate culture collection, i.e. ATCC, NCTC; not more than five passages are allowed after being cultivated from a lyophilized or cryo culture (i.e. if Reference cultures from the first generation is received).
- In general, the Pharmacopoeia provides details of which strains to be used for which tests. Is this implemented and respected?
- In principle, propagation of reference cultures must be under controlled conditions to ensure they remain fit for purpose.
- Reference organisms will usually be used in low numbers to ensure test sensitivity [typically 10–100 CFU (colony forming units) per test]. Correctly implemented and documented?
- Determinations and characterization of the in-house flora derived from contaminants are isolated from environmental and product samples. Requirements implemented and documented in SOPs?
- Carefully controlled storage areas and maintenance procedure for each reference culture.

References: EudraLex 2015, chapters 6.21–6.24, 6.19.

16.4 Regulatory Audits in the Microbiology Department

To ensure the quality and effectiveness for each pharmaceutical product, depending on its dosage form, microbiology is an important and essential part

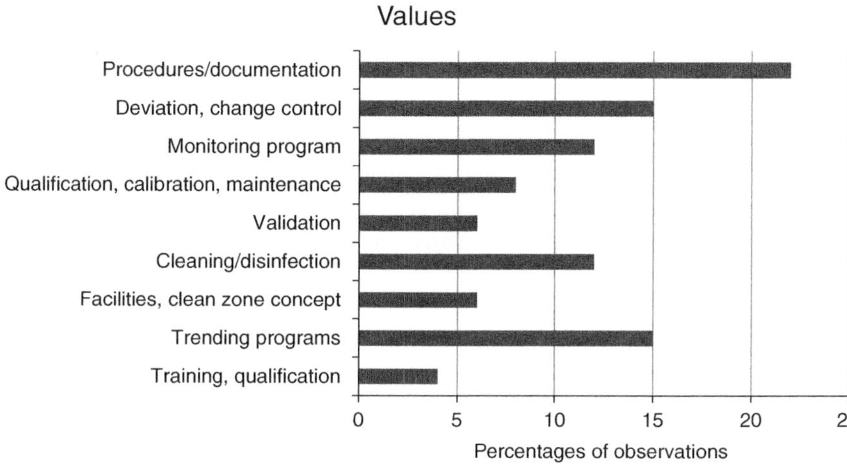

Figure 16.2 Overview of observations in specific areas within the microbiological laboratory in the last five years based on regulatory inspections.

of an established quality control. Microbiology is a science that deals with immense number of subjects. As a result, a microbiological control laboratory is an integral part of official inspections by agencies and should therefore be routinely checked for compliance with requirements, even in self-inspections.

Typical findings during the evaluation of a microbiological laboratory (Figure 16.2):

1) There is no link between the sterilization of media batches and the subsequent growth promotion testing. There is no systematic release of media batches.
2) The pH of agar plates was not measured due to the lack of appropriate equipment.
3) Not all loading plans in use were part of the initial performance qualification of the autoclave.
4) No clear personal and material flow in classified working areas was defined in the microbiological laboratory. Serious cross contamination issues.
5) Water contamination issues.
6) Environmental monitoring issues like:
   - No detailed corrective actions with OOS results
   - Absence of monitoring for critical work
   - No inactivating agent in contact plates and swabs used for environmental monitoring
7) The autoclave sterilization cycle for media is not charted (temperature and pressure of the cycle).
8) The pH of microbiological media after sterilization is not measured.

9) In the microbiological laboratory there is no incubator for 20–25 °C. Room temperature incubation is performed in the noncontrolled laboratory environment.
10) No trending is performed for environmental data.
11) The points of use designated for sampling are not referenced by a unique identification code (mapped on the distribution loops diagram).
12) The process of deviation handling is not adequately described in a SOP regarding classification of deviations and no time limit for the closure of each CAPA.
13) It cannot be assured that the justification for a change is sufficiently documented or that change control activities are fully considered and documented.
14) There are no work zones and defined, classified conditions for the work on the open product or on test approaches.
15) For the autoclave, which is used for sterilization, no qualification documents could be provided.
16) During the preparation of the culture media, a check of the growth characteristics by means of control microorganisms is not carried out systematically.
17) The work instruction for the enumeration for total aerobic microbial counts and specific microorganisms does not consistently meet the requirements of the Pharmacopoeia.
    - Changes in methods are not described and their suitability or equivalence to the Pharmacopoeia is not proven.
18) Poor traceability of test records. Attended documents do not contain all necessary data to guarantee a complete traceability, i.e. missing time information for the evaluation, number of approaches, passages, missing information about the filter type.
19) Lack of traceability of the reference microorganisms to an official standard. Purchased reference cultures (ATCC strains) and the stock cultures produced from them (max. four passages) were not traceable to a certified standard by available documentation. It was also unclear whether the cultures obtained were from a corresponding accredited institution (supplier qualification program).

16.5 Conclusions

In addition to effectiveness, safety and quality are the most important requirements for a drug. Testing for microbiological purity is an essential part of quality control of medicinal products. The highest requirements apply to sterile preparations, but also non-sterile drugs are subject to strict requirements. For non-sterile pharmaceutical preparations, the European Pharmacopoeia prescribes different requirements for the total number of microorganisms and the absence of certain types of microorganisms, depending on the type and location of application. This also includes starting materials and active ingredients.

But how to recognize microbial and other impurities in the product at a reasonable cost? Statistical principles in sampling are just a few of the important issues discussed in analytical quality control and pharmaceutical microbiology. Examples of risk assessments for laboratory and manufacturing microbiological activities should be implemented as a central element of microbiological quality assurance. Practical hygiene plans, information on nutrient media production as well as specifications and evidence for the validation of microbiological test procedures, and the qualification of essential equipment complete this consideration.

The current experience with official GMP inspections still shows an increasing focus on the field of quality control, where apart from physicochemical and analytic analytics, more and more pharmaceutical microbiology is being examined.

Bibliography

EudraLex, 2015, *The Rules Governing Medicinal Products in the European Union. Volume 4. EU Guidelines for Good Manufacturing Practice. Medicinal Products for Human and Veterinary Use. Part I – Basic Requirements for Medicinal Products (including its Annexes)*, European Commission.

ICH, 2005, *ICH Topic Q9. Quality Risk Management*, International Council for Harmonisation of Technical Requirements for Pharmaceuticals for Human Use.

ICH, 2008, *ICH Topic Q10. Pharmaceutical Quality System*, International Council for Harmonisation of Technical Requirements for Pharmaceuticals for Human Use.

Ph. Eur. 9th Edition chapter 2.6, 2018, 'Biological tests'.

Ph. Eur. 9th Edition chapter 2.6.12, 2018, 'Microbiological examination of non-sterile products: microbial enumeration tests'.

Ph. Eur. 9th Edition chapter 2.6.13, 2018, 'Microbiological examination of non-sterile products: test for specified microorganisms'.

Ph. Eur. 9th Edition chapter 2.6.31, 2018, 'Microbiological examination of herbal medicinal products for oral use and extracts used in their preparation'.

Ph. Eur. 9th Edition chapter 5.1.4, 2018, 'Microbiological quality of non-sterile pharmaceutical preparations and substances for pharmaceutical use'.

Ph. Eur. 9th Edition chapter 5.1.8, 2018, 'Microbiological quality of herbal medicinal products for oral use and extracts used in their preparation'.

Pharmeuropa 29.2 chapter 2.6.36, 2018, 'Microbiological examination of live biotherapeutic products: test for microbial enumeration of microbial contaminants'.

Pharmeuropa 29.2 chapter 2.6.38, 2018, 'Microbiological examination of live biotherapeutic products: test for specified micro-organisms'.

PIC/S, 2018, *Guidance to Good Manufacturing Practice for Medical Products*, PE 009.

Rieth, M, 2012, Pharmazeutische Mikrobiologie – Qualitätssicherung, Monitoring, Betriebshygiene, Chapter Referenzstämme, Wiley VCH-Verlag.

WHO, 2011, *Annex 5, Supplementary Guidelines on Good Manufacturing Practices for Heating, Ventilation and Air Condition Systems for Non-sterile Pharmaceutical Dosage Forms*, WHO Technical Report Series, No. 961, 2011.

WHO, 2014, *Annex 2, WHO Good Manufacturing Practices for Pharmaceutical Products: Main Principles*, WHO Expert Committee on Specifications for Pharmaceutical Preparations.

ZLG, 2010, *Aide Memoire 0712105, Inspektion von Qualifizierung und Validierung in pharmazeutische Herstellung und Qualitätskontrolle*, Zentralstelle der Länder für Gesundheitsschutz bei Arzneimitteln und Medizinprodukten.

17

Which Microbiological Tests Can Better Be Performed In-house and What Can Be Easily Outsourced

Thomas Meindl and Ingo Grimm

Labor LS, Bad Bocklet, Germany

CONTENTS

17.1 Introduction

Nowadays, the production and release of medications is so complex and time-consuming to monitor that hardly any manufacturer can produce efficiently, cost-effectively, and successfully without the support of service providers. This

Pharmaceutical Microbiological Quality Assurance and Control: Practical Guide for Non-Sterile Manufacturing, First Edition. Edited by David Roesti and Marcel Goverde.

trend even goes so far that "cloud companies" exist that are solely responsible for identifying new leads, handling regulatory processes, and supervising supply chain management. Such companies operate completely without their own production and testing equipment. This of course gives these forms of organization a great deal of flexibility in terms of resource management, but on the other hand they are 100% dependent on the delivery and reliability of their partners.

In the chapters of this book, various aspects of microbiological testing are presented. At first glance, the microbiological tests described in the various pharmacopoeias appear reasonably clear (at least when compared to the incredible amount of chemical and physical analytics and the resulting monographs). However, the microbiological tests are no less challenging.

The challenge in microbiology is that the described procedures are only described for a single matrix. Anyone who knows the complexity of the products on the pharmaceutical market quickly concludes that it is a great challenge to establish and finally validate suitable microbiological processes that do justice to the respective matrix of the product or preparation. In this context, the question naturally arises as to whether this know-how must be built up and expanded, trained and maintained in all pharmaceutical companies, or whether an organization can purchase this knowledge.

As in other disciplines along the value chain, considerations are made as to which test procedures remain in the company and which are placed with a service provider. In the range of possible services, the outsourcing of release-relevant testing involves the highest risk in economic terms, but also the highest savings potential. Once the tests have been validated, testing the end products in routine testing always follows the same test plan and thus offers a high potential of rationalization.

It is of course extremely important to find the right partner for outsourcing testing services, because incorrect tests can have an impact on patient safety. But also the reputation of the company toward health authorities and patient groups can be on stake. This risk can be divided into two groups. For example, the risk of testing non-sterile samples is usually significantly lower than for testing sterile samples, although it may economically make sense to outsource sterile samples because the quality control processes take place in significantly higher, very expensive cleanroom classes or maintenance-intensive isolators.

Which microbiological tests should therefore be given preference in this controversial area? The question is not easy to answer, because further considerations are necessary to make the right decision.

Thus, in addition to the type of test, there is always the question of the time frame within which certain results must be available. Various process steps require that results must be available very promptly, e.g. due to the lack of stability of the samples or the influence of room conditions on the growth of microorganisms. Should such factors play a role, suitable validation measures such as transport validation must be carried out. Such studies can then be used to determine the time frames within which the test must be conducted.

It does not make sense to outsource the collection of measurement data directly in the process. This means when so-called in-line or end-line measuring methods must be the method of choice. Or, when it comes to extremely time-critical in-process controls, the results of which have to be delivered ad hoc in order to be able to enter the next production step. However, in-line or end-line measurement methods have not yet become widely accepted in pharmaceutical operations. An exception is here the production of biopharmaceuticals, where in-line, atline, and end-line measurement methods, respectively, are becoming more and more popular. There are various technologies that can be used to support these ideas. However, the qualification of such procedures, the assurance of the safety of the test results, and the data integrity of the generated results require a high level of know-how and effort (see Chapter 13).

Another aspect for deciding whether a particular test should be transferred to a contract laboratory is the question of how often a particular test will be carried out. If there are only very few samples for a certain test in production operation, it is usually not very useful to establish the necessary test method. Because setting up the necessary laboratory, the required equipment, the costs for qualifying the equipment and premises, the personnel costs, the expenses for ongoing training operations, and the necessary costs for a backup solution in the event of illness or vacation do not make the establishment of an own laboratory reasonable.

This maybe illustrated by a practical example, such as an antimicrobial effectiveness test (AET, USP chapter <51>, Ph. Eur. chapter 5.1.3). This test is based on a study concept by which the antimicrobial potency of a preservative system or of the sample matrix in general is investigated. In general, a selection of microorganisms is inoculated in the sample and at defined intervals the samples are investigated for microbial growth. If a preservative system is to be challenged, the goal of the study is to design a system with the minimal concentration of the preservative system sufficient to control the growth of the selected panel of microorganisms either by their destruction or inhibition of microbial proliferation. In order to achieve meaningful results appropriate for the requirements of the respective site, it is highly recommended to implement in addition to the standard microbial species microbes which were collected during microbial monitoring. Especially, these microbes are the critical ones which can be expected to occur at the production site. To be able to perform an efficient AET study, a well-equipped and experienced microbiological laboratory is required. This laboratory needs to be able to cultivate a broad variety of microorganisms as well as to provide the expertise to maintain a broad stem collection of microbes. In addition, it is often required to have the expertise and equipment to be able to prepare more or less "exotic" media to cultivate some organisms, since these media are often not available at least not in small amounts at the usual suppliers.

In summary, these investments may be justifiable if they will be used for routine laboratory services. But studies such as an AET are nothing less than routine experiments. This also addresses the human resource factor. For these studies, experts are required, whose skills need to be kept at the required level by continuous training although the tests will only be performed on an irregular basis.

There are many other examples for these irregular "once a year" laboratory services. One other example is the validation of disinfectant measures by testing a disinfectant against environmental microbial isolates collected during hygiene monitoring on the present surfaces using the protocols (SOPs) installed.

In addition, we live in times in which demography is affecting all branches of industry, including the pharmaceutical industry. Well-trained technicians with sound good manufacturing practice (GMP) knowledge are hardly available on the laboratory market. Each company must decide how it can make the most efficient use of its personnel capacities.

17.2 Advantages and Disadvantages of Outsourcing

17.2.1 Specialization and Know-How in Analytical Testing

The question of whether there are technically simple tests that can be outsourced with comparatively little effort depends largely on the matrix of the sample. A very good example of this is endotoxin testing. The validation of this test was carried out in a very simple procedure for many years. For example, it was required that the positive endotoxin spike should be added to the matrix only after it was extracted from the sample. This positive product control had to be recovered. Recently, however, the regulatory requirements, especially in the European Pharmacopoeia, have changed with regard to these issues. It is now requested from health authorities that the stability of endotoxins in product matrices and the sample containers during storage must be evaluated. Sample hold time studies are executed to provide experimental data and consist of spiking product matrices with endotoxins and assessing if the endotoxins remain detectable after a certain storage period.

Of course, endotoxin challenges are not really relevant for non-sterile matrices. Here, in recent years the topic of so-called objectionable microorganisms became relevant. The "objectionables syndrome" is well addressed by the FDA in 21 CFR 211.113: *Appropriate written procedures, designed to prevent objectionable microorganisms in drug products not required to be sterile, shall be established and followed.* Following this requirement, a high demand of microbiological expertise is required especially since exact definitions of the term objectionables are still missing. Common features of objectionables are that they grow in the product and can adversely affect the properties of the product.

In addition, objectionables are expected to have the potential to cause infections due to the number and their pathogenicity (for further details see Chapter 11).

The objectionables topic finds recently more attraction because it is in a strong focus of the FDA due to several findings. Although objectionables may cause great financial and health damage, the topic is often addressed only after a caused accident and not in advance through a risk assessment. This may be due to the high demand of microbiological skills required for cultivation of some typical objectionables such as the most prominent objectionable *Burkholderia cepacia* complex (BCC). Wild-type isolates of this microorganism should be preferred in studies due to adaptation to the specific product.

Regarding this topic, it would be advisable to carry out, e.g. an AET study including objectionables directly in the process of product development, especially for non-sterile highly water-containing products. In these study historical bioburden results should be included as well as a stability program in order to confirm the absence of BCC. For samples with complex matrices such as oily preparations, proteins, glycopeptides, etc., testing is much more difficult and complex to validate than for simple samples such as ultrapure water. Thus, a meaningful outsourcing decision will no longer depend solely on the type of test, but on the matrix of the product.

If different matrices are to be investigated in the company, the various test methods would have to be established, qualified, and ready for use. In contrast to classical microbiology, such rapid methods are always associated with high investment costs for equipment, qualification costs, service and maintenance contracts, and costs for the development of know-how. Such methods hardly pay off for a manufacturing company with a small- or medium-sized microbiological laboratory, as these costs multiply by the number of matrices produced. On the other hand, a laboratory service provider can more easily serve the range of different technologies with regard to different matrix requirements, because of the bundling of market requirements and own know-how results in positive scaling effects.

Experience in microbiological testing of a broad variety of sample matrices is key to be able to design successful test strategies also for non-sterile products; especially semisolids, gels, and emulsions can be a challenge since in these matrices a lot of active moieties are present which can interact with microbes leading to difficulties in quantitative detection of these.

17.2.2 Challenges for Outsourcing

Of course, there is also the question of what disadvantages result from the outsourcing of microbiological tests. In principle, a manufacturing company is dependent on a service provider and its testing capability. This trust can only be strengthened by continuous relationship management, regular audits, and the reputation of the respective service provider. Furthermore, the requesting party has only indirect influence on the processes in the service laboratory and a failure of the service provider has a direct influence on the production and release processes and the delivery capability of the requesting party.

In pharmaceutical manufacturing according to GMP requirements quality oversight covering all critical processes is mandatory. To set up efficient and sustainable systems in the own facilities is a big challenge. To set up or control such systems at an external site is much more complicated. In this respect, audits by the manufacturing company at the contract research organization (CRO) site must have at least the same depth as an internal audit. For example, deviation and change managements have to include the interaction of the manufacturing company and the CRO in a way, that the manufacturing company is always and in time informed about any items with a potency to have an impact on the ordered tests. Of course, external quality oversight can never reach the same level as the internal processes.

As a necessary basis of an efficient cooperation between a pharmaceutical manufacturing site and an outsourcing laboratory partner, the test capacities of the laboratory have to be described fair and correctly including buffers for nonroutine extra samples or tests planed on short notice at nonroutine days such as weekends or bank holidays. These buffer capacities are part of the investment in such an outsourcing constellation. Another question is, whether samples can be tested by the outsourcing partner in the required time frame? For example, microbiology quality tests on water samples have usually to be performed in 24 hours after sampling. In addition, any putative impacts due to the transport of the samples to the testing laboratory have to be taken into account. To cover this issue, the outsourcing partner should offer a service to validate the transport. Ideally, the outsourcing partner should be equipped with transport facilities able to control and document temperature and humidity as required for specific samples.

17.3 Business Case and Hidden Costs

17.3.1 The New Challenges in Pharmaceutical Environment

The environment in the pharmaceutical industry has changed significantly in the last 10 years. In addition to the tightening of national and international regulatory requirements, ever-increasing demands with regard to patient safety, consumer protection, and international liability risks, the supply chain in the pharmaceutical business is also becoming increasingly complex.

Only a negligible small proportion of pharmaceutical companies still cover the entire process chain from the development of new active ingredients, support of clinical approval phases, market authorization for various markets, production of the medications, their packaging/filling, and marketing all by themselves. On the contrary, the competence of the individual market participants is purchased in an established supplier network. On the one hand, the constantly increasing complexity of the processes in each work step requires special expertise, which lies in the core business of the respective supplier. On the other hand, significant synergy effects will ultimately be reflected in the efficient design of the entire value chain and keep production costs, including quality control costs, competitively low.

17.3.2 Outsourcing to Strengthen the Core Business

Numerous processes that do not serve to strengthen the core business of the pharmaceutical company are outsourced to specialized suppliers with adequate quality standards and excellent process know-how. The internal process competence of work that does not directly add value will probably be reduced in the future, either for cost reasons or due to a lack of qualified personnel. Conversely, however, more is invested in the competence to accompany, monitor, and control these externally assigned processes through a sustainable professionalization of supply chain and external quality management.

Due to the expiration of patents, the efforts of health insurance companies to take over the lower costs of generics, and the lack of innovative products in the development pipeline, the pressure to innovate, especially in the area of non-sterile products, is increasing. This pressure requires more financial resources in the research area than before, and these financial resources must be secured through the leanest possible internal processes.

Furthermore, low-bioburden products require other and significantly more complex production lines that are closely related to aseptic production processes. This change in production requires high investments in materials and personnel, which must be generated via the return on investment (ROI). The installation of these new production lines in clean room classes ≥ D including the necessary supply of ultrapure media requires high personnel qualification capacities. These plants require a much more closely coordinated concept of environmental monitoring and a significant expansion of in-process controls. It should be borne in mind here that according to the formulated will of the corresponding guidelines, e.g. in the WHO Good practices for pharmaceutical microbiology laboratories guide (WHO 2010) that non-sterile tests shall also be carried out in at least cleanroom class D in the long term; it will probably no longer be possible to conduct such tests in a merely "controlled test environment" in the long term.

Furthermore, demographic developments in Europe shall result in the limited availability of competent testing staff. The shortage of skilled workers will force the awarding of testing services in the future.

On the one hand, following the financial crisis and the subsequent zero interest rate phase, high investments in property, plant, and equipment were made in production facilities; on the other hand, these investments were usually approved by management with a defined headcount cap. Conversely, this means that internal capacities are not built up and extensive testing and analytical processes must be transferred to external service providers, since the operation of the plants and the indispensable QC processes cannot be covered by internal capacities.

17.3.3 Which Testing May Be Outsourced

Inevitably, customers of testing institutes will try to establish their internal processes with the supplier. In the future, this will mean that this demand can only be met satisfactorily by service providers who are able to establish customer-specific processes and which also have the capacity (headcount as well as lab space) to deliver in the desired GMP quality grade. The individualized and harmonized dedicated service solutions will be a major guarantee of success for a functioning supply chain in the future.

The challenges described so far will place higher demands on the selection of laboratory service providers in the medium term and, in addition to scientific expertise, other success-critical factors will also have to be examined when selecting suppliers.

In the field of microbiological quality control, many fields of testing can be outsourced to external service providers. The portfolio ranges from microbiological incoming tests on, e.g. raw materials, active ingredients and excipients, nutrient media, single-use devices to microbiological tests as part of environmental monitoring, microbiological testing of ultrapure media, operational and personnel hygiene, and microbiological tests as part of a batch release.

Priority must be given to whether the projects are temporary projects, e.g. qualification of rooms or facilities, a backup solution for a busy company laboratory, or a complete and long-term outsourcing to an external supplier. In addition, it should already be clear in the project conception phase of the outsourcing which fields are to be assigned to an external service provider, so that the scope of the project becomes transparent for both sides and it is ensured that the partner also has the appropriate core competencies, matrix experience, and testing capacities.

The most fundamental decision is certain: whether tests of "high throughput" samples usually tested by a small selection of routine protocols should be transferred that are rather easy to process from a scientific point of view, but require high personnel and laboratory capacities (e.g. room, air, ultrapure steam, water, and personnel monitoring) by means of the number of test samples alone, or whether specialized tests should be transferred that require a high level of know-how on the part of the testing personnel and the test environment at a low sample throughput and must be extensively validated before routine use.

Regardless of the scope of the project and testing, however, it is important to name the persons responsible for the project and to define the responsibilities for the individual aspects of the project. In the case of extensive projects or a significant relocation of routine tests, both the laboratory managers and strategic purchasing are required on the part of the outsourcing company; on the part of the contract laboratory, the scientifically active persons and sales are required to act. In this constellation, all individual requirements for testing services can be discussed, assembled, and converted into a cost-transparent overall solution. The appointment of special contact persons (single person of contact, SPOC) has proven to be useful. The highest level of efficient support in this respect is achieved, if the contact person is a proven specialist in the respective field.

In practice, especially in microbiology, throughput times play an important role. The throughput times are logically related to the available capacities at the service provider, considering the limited incubation times.

When integrating service laboratories into these tests, the requesting party and client must reach agreements on the utilization and availability of laboratory capacities and personnel resources in order to ensure compliance with the timelines required by the requesting party through sufficient transparency.

In the context of the qualification of new suppliers in the GMP environment and during regular audits of an existing supplier relationship, it makes sense to check during the audit to what extent the test reliability of the order laboratory is given. In addition to the classical audit topics and the examination of technical, infrastructural easy-to-visualize backup solutions and redundant systems, the requesting party should also explain the employee structure, the training program, and the concepts for recruiting and for securing know-how and staffing. A staffing level that is too thin jeopardizes the testing capability of an external provider, even if the infrastructure is excellent. To be able to cover up this audit approach, in an optimal setup the audit team consists of a compliance expert as well as a test specialist to assess both GMP and scientific aspects.

17.3.4 Sample Hold Time and Transport

In addition to the pure execution of the tests, the service orientation of the contract laboratory is very important, which is why the business hours, which are especially necessary for microbiological tests on weekends and holidays, the basic availability and reaction time of the testing institute, the contact data of direct contact persons, individual annual meetings, the ability to innovate, automated reporting channels and status reports, and bidirectional interfaces between the LIMS systems of both companies are inherently important points that should be considered and discussed when setting up a sustainable outsourcing concept.

Ideally, a holistic service concept also includes the service provider not only taking over the testing of the products but also establishing and organizing sample logistics. Special attention must be paid to temperature-sensitive and mechanically unstable test samples.

As described before, the tests of purified water are time and temperature sensitive and thus a critical test for an outsourcing approach. Clearly defined requirements cannot be derived in the present European Pharmacopoeia, whereas in chapter <1231> of the current USP it is recommended to perform the tests within 2 hours of collection or in a window of 24 hours if refrigerated. This is already a recommendation based on which such an outsourcing project can be organized. Therefore, it is scientifically more profound to have an approach by which the outsourced test is started only after a validation study of the transport has been successfully performed. Such a study should be designed based on a risk assessment covering the conditions relevant for the project and at least include the microorganisms known to be critical. Critical are organisms which are difficult to detect or cultivate, stressed organisms and organisms known by the customer as environmental isolates. As a basis for a transport study, Bomblies (2011) claimed that the kinetic behavior of a particular microorganism under cooled storage conditions should be investigated in order to be able to define the appropriate individual transport conditions. By collecting this information, the risk of outsourcing can be minimized.

The solution to this problem lies in adapted and controlled, tracked logistics, so that the sample is not affected by the dispatch to the test laboratory and is therefore no longer representative.

17.4 Hidden Costs

As with all other outsourcing projects, the outsourcing companies expect business benefits from the relocation of microbiological testing services. These advantages can only be presented transparently if the costs of the purchased tests are clear. When comparing offers, various hidden costs must be considered. The outsourcing company is very well advised to check the following aspects with regard to cost transparency:

- Are there any internal costs for any necessary reregistration with registration authorities?
- Are the costs for any necessary validation or the costs for a method transfer described sufficiently transparent?
- Are surcharges levied for diversified quality standards such as tests within the scope of an accreditation, in the GMP environment or in the cGMP area?
- Are the costs for additional work steps such as complex sample preparation, hazardous material surcharges, disposal surcharges, and surcharges for the use of special nutrient media included?

- Are the costs for the investigation of OOS results sufficiently transparent?
- Are there any additional costs for the initial or regular audit of the supplier?
- Are costs for archiving the raw data included in the offer and are the necessary consulting services also included?
- Are regular throughput times mentioned in the offer or do express surcharges arise?
- What are the surcharges for weekend and holiday work?
- Does the supplier have shut down phases (e.g. during bank holiday). This question is particularly important if routine checks in the sense of a backup solution are to be relocated?

17.5 Quality Agreement Between the Contract Laboratory and the Requesting Company

17.5.1 Regulatory Environment

Quality agreements, wage inspection and responsibility delimitation contracts are defined by statutory regulations. These include the AMWHV (chapter 2, § 9), as well as various ICH guidelines (e.g. Q10 chapter 2.7) and the EU-GMP guidelines (e.g. Part I, chapter 7). Here you will find the specifications where quality-relevant parameters are to be defined and how.

Such a contract specifies in writing how compliance with the requirements of good manufacturing practice GMP is to be observed. The contract laboratory and the manufacturer can also define additional aspects outside GMP that are necessary for the regulation of cooperation. This frequently includes questions such as standard lead times of test samples, how results are communicated, connection of communication interfaces, aspects of sample logistics, reaction times, etc.

A well-designed contract can support the cooperation between requesting party and client very efficiently, because the framework of the cooperation is proactively stipulated and the division of responsibilities is clearly defined.

17.5.2 Regulatory Aspects and Contents of a Modern Quality Agreement

Once the contracting parties have been clearly defined in the first step, a short preamble should describe in which context the contract is concluded, what the specific subject matter of the contract is, and who is the requesting party, and who is the client (Figure 17.1). Since it concerns the regulation of quality-relevant aspects, the quality assurance system of the ordering party and the quality standard of the client should be described; furthermore, it makes sense to mention the responsible surveillance authorities.

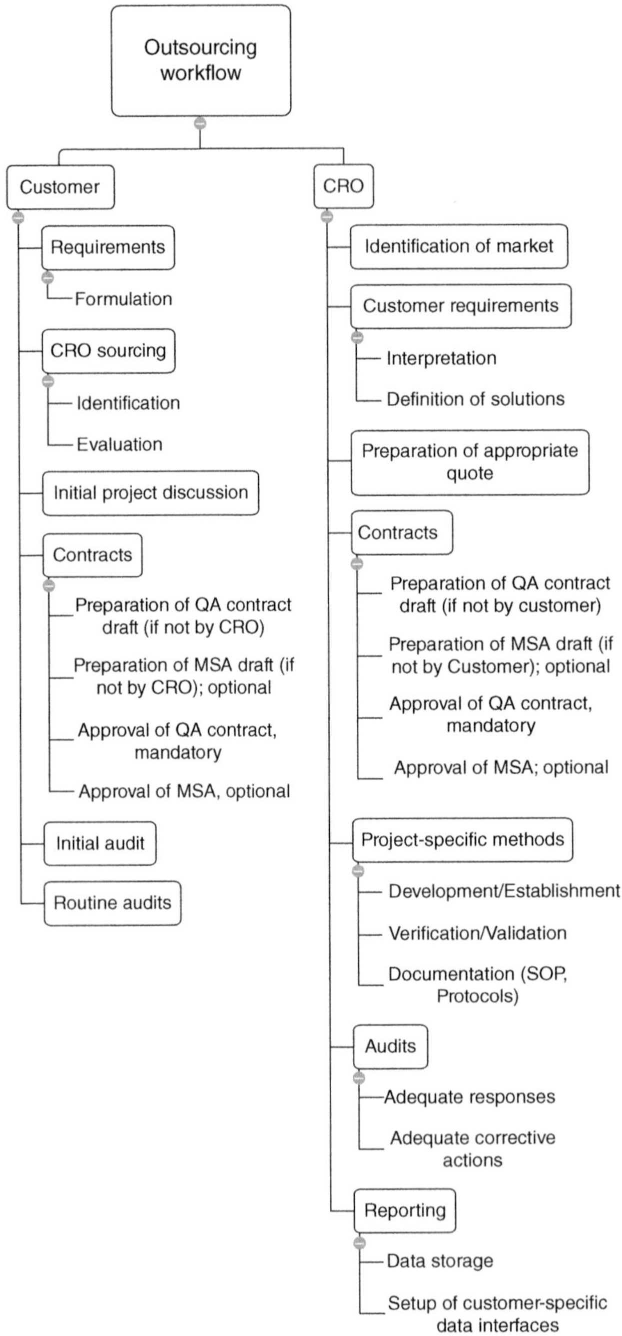

Figure 17.1 Overview of outsourcing workflow. The organization of an outsourcing project requires a set of formal processes and document exchanges between customer and CRO.

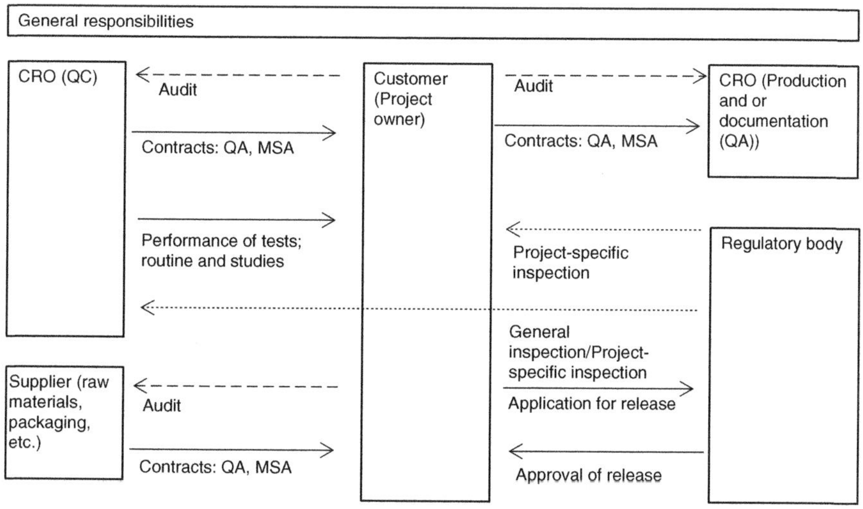

Figure 17.2 Overview of outsourcing responsibilities.

In order to regulate the respective responsibilities within the framework of the cooperation, both the duties of the client as well as those of the requesting party within the framework of the cooperation must be defined transparently (Figure 17.2). In the case of complex constructs, it makes sense to display them in table form.

In order that the requesting party can fulfill his audit obligation and that the audit planning is designed as efficiently as possible in advance, it makes sense to record the conditions applicable to audits, the duration, the number of participants, and the desired regularity.

Since sensitive data are always at stake in the quality context, a mutual obligation to secrecy should also be incorporated into the contract. Furthermore, the effective date of the contract, the written form requirements for changes, and the notice periods must be made transparent. As a rule, a severability clause and a reference to corresponding commercial agreements makes sense, as these regulate the aspects of rectification of defective service, warranty claims, liability claims and limitation of liability, place of jurisdiction, and payment terms.

In addition, a modern quality agreement includes various systems that specify the following points:

- Who are the responsible contact persons of the requesting party and the client?
- Which products of the requesting party and which tests of the client are part of the contract?
- Which products of the requesting party are subject to the supervision of the US-American health authority?

considered to be the trickiest ones. In the framework of setting up the system suitability test, e.g. rare culture media which are challenging to prepare and "exotic" growth conditions need to be used or developed. Once these parameters are established and shown to be used for valid testing, the method can be transferred to a routine laboratory. If the specifically required laboratory equipment is established there, the routine protocols should be usable for a generally trained technician. As mentioned before, if the routine assays need to be performed on a regular basis with a minimal frequency of at least once every month, the laboratory expertise can be kept on a level required to fulfill GMP standards.

17.7.3 Example 3: Setup Environmental Monitoring for Microbial Contamination

An efficient environmental monitoring setup is a critical requirement for hygienically safe pharmaceutical production. A basic requirement for this is a risk assessment, which can be set up, for example, by a failure mode and effects analysis (FMEA) study. In such a procedure the planned processes in the present environment are analyzed and evaluated regarding their risk potential. Important factors of this risk potential evaluation are possible impacts regarding microbial contaminations. Knowing the critical steps and the specific room and equipment conditions of the respective working area enables to define the ideal monitoring spots as well as to define the frequency of this process.

Besides this, it is also possible to implement new state-of-the-art measurement technologies such as continuous measurement of viable particles.

In general, sampling for monitoring has to be well performed in order to avoid contaminations due to false behavior during sampling. Thus, an excellent training for the sampling staff is required.

After the setup of the initial monitoring procedure, the routine monitoring process can be streamlined based on the results gained through the establishment of the project. To be able to conduct such a streamlining process, some experience in this kind of projects is required in order to keep the monitoring level at the required quality.

Finally, the routine monitoring may then be performed by the own trained technicians as long as the training is well performed.

Furthermore, the frequency of the monitoring process has to be considered. If the outcome of the risk assessment is that the monitoring has to be performed only in intervals of some weeks, a complete outsourcing of the monitoring should be considered since the sampling may never become a routine procedure. In addition, it should be considered, that, while the internal staff is performing the monitoring procedure, these technicians will not be here available to perform the key laboratory processes required to produce the own products. In this situation a cost-value calculation is highly advisable.

17.8 Conclusions

These days, pharmaceutical production is a complex, multilayer process challenged by constantly increasing regulatory requirements and contradicted by the perpetual demand of cost reduction or at least improvement of cost efficacy.

One of the most important cost factors in pharmaceutical production is the costs for human resources. This is becoming more and more precarious since less and less qualified personnel are available on the market. Thus, for every enterprise it is a compulsory task to use the available capacities in order to push the proprietary protocols and finally fulfill the strategic goals. As a consequence of this situation, the management is challenged to determine a priority list of tasks. In this list, it can be defined, which processes need to be kept proprietary and thus in-house, and which ones can be considered to be outsourced. It may be helpful for organization of the list to score the impact of each of the tasks. The impact is first of all triggered by the possible risks connected to the individual procedure. The higher the risk is, the greater the experience and skills of the putative laboratory need to be. In addition, the obligatory audit in case of outsourcing to check for compliance in highly skilled tasks requires auditors who understand the technical procedures and thus are able to discuss with the performing laboratory on the required expert level.

A second important factor is the frequency of the considered microbiological tests. Each test requires a certain amount of effort for setup and maintenance. This comprises the required laboratory space, the equipment, the material including consumables, and finally, most important, the people selected to perform the test. Every element of this non-exhaustive enumeration necessitates initial and ongoing qualification in line with appropriate documentation and quality oversight. These nonproductive but essential elements of GMP analytics become more cost efficient the more often a certain test needs to be performed. Thus, the prize for a certain analytical test including all of these listed nonproductive GMP essentials may even be lower in case of outsourcing. On that account, the producing company is always well advised to quantify the demand of a specific assay in order to be able to deliberate about whether it may be more preferable to set up a technology in-house or to outsource.

So, what is the final conclusion or advice? The good news is, nearly every microbiological test can be outsourced since none requires an extraordinary invest in innovative development for each single product matrix. Thus, microbiological tests are usually no unique and proprietary key competencies for pharmaceutical companies at least if these tests are compared to the challenge of the setup of a cell-based biochemical potency assay, which is usually very much individual for each active cell or protein-based product. Nevertheless, due to the fact that microbiological as well as biological assays usually tend to be more imponderable than chemical physical assays, a solid fundamental of experimental expertise covering a broad spectrum of matrices is a key prerequisite for a smooth and successful in time setup of an assay. This leads to the reasoning that the outsourcing arrangement with the putative partner has to be set up in a wise and hence efficient way. The investment is only worth, if the service can be offered in a sustainable manner. So, an essential aspect to build a sustainable partnership is the capacity of the performing laboratory to implement matrix-specific methods highly efficient and thus cost effective and to guarantee to offer the specifically established test designs for a sustainable period of time. If the latter cannot be assured, the risk of the ordering company is quite high to be forced to invest a second time in method transfer or a new validation. This has also an impact on regulatory issues. Any change in this respect has to be documented and when indicated, inspected and accepted by the responsive authorities.

Thus, well-managed outsourcing projects in microbiology can be an important part of a highly effective supply chain management.

Bibliography

AMWHV, 2006, *Verordnung über die Anwendung der Guten Herstellungspraxis bei der Herstellung von Arzneimitteln und Wirkstoffen und über die Anwendung der Guten fachlichen Praxis bei der Herstellung von Produkten menschlicher Herkunft (Arzneimittel- und Wirkstoffherstellungsverordnung), Abschnitt 2, § 9 Tätigkeiten im Auftrag,* Bundesministeriums der Justiz und für Verbraucherschutz.

Bomblies, L, 2011, 'Microbiological Water Samples – a Perishable Product. A Study for the Validation of Shelf Lives and Temperatures, PharmInd, vol. 73(1), pp. 159–162.

EudraLex, 2015, *The Rules Governing Medicinal Products in the European Union. Volume 4. EU Guidelines for Good Manufacturing Practice. Medicinal Products for Human and Veterinary Use. Part I – Basic Requirements for Medicinal Products – Chapter 7. Outsourcing activities.* European Commission.

ICH, 2008, *ICH Topic Q10. Pharmaceutical Quality System – 2.7 Management of Outsourced Activities and Purchased Materials,* International Council for Harmonisation of Technical Requirements for Pharmaceuticals for Human Use.

Ph. Eur. 9th Edition chapter 2.6.12, 2018, 'Microbiological examination of non-sterile products: microbial enumeration tests'.

Ph. Eur. 9th Edition chapter 2.6.13, 2018, 'Microbiological examination of non-sterile products: Test for specified microorganisms'.

Ph. Eur. 9th Edition chapter 5.1.3, 2018, 'Efficacy of antimicrobial preservation'.

Pharmeuropa 29.2 chapter 2.6.36, 2017, 'Microbiological examination of live biotherapeutic products: test for microbial enumeration of microbial contaminants'.

Pharmeuropa 29.2 chapter 2.6.38, 2017, 'Microbiological examination of live biotherapeutic products: test for specified micro-organisms'.

Pharmeuropa 29.2 monograph XXXX:3053, 2017, 'Live biotherapeutic products for human use'.

USP 41 NF 36 chapter <1231>, 2014, 'Water for pharmaceutical purposes', United States Pharmacopeia.

USP 41 NF 36 chapter <51>, 2018, 'Antimicrobial effectiveness testing', United States Pharmacopeia.

WHO, 2010, *Good practices for pharmaceutical microbiology laboratories guide*, Working document QAS/09.297/Rev. 2 July 2010, World Health Organization.

W

X

Z